Hibernia: Stations 1–9, 2000 (73 x 42 cm)
Oil on canvas, sediment on canvas, angle iron support

Created by Nova Scotian artist, Susan Fiendel, "Hibernia Stations 1–9, 2000", is one of a number of pieces of artwork inspired by her observations of marine habitats of the Scotian Shelf and the Grand Banks. Susan participated in a 25-day research mission on board the oceanographic research vessel, *CCGS Hudson*, in May 1999. Through her observations of the active process of science being conducted in the marine environment, including benthic video images, sidescan sonar output, and water, seafloor, and biological sample collection and processing, Susan has created artwork that depicts both scientific research activities and the beauty of marine habitat.

Offshore Oil and Gas Environmental Effects Monitoring

Approaches and Technologies

EDITED BY

Shelley L. Armsworthy

Peter J. Cranford

Kenneth Lee

Battelle Press

Columbus • Richland

Library of Congress Cataloging-in-Publication Data

Offshore Oil and Gas Environmental Effects Monitoring: Approaches and Technologies / edited by Shelley L. Armsworthy, Peter J. Cranford, Kenneth Lee.

p. cm.

Includes bibliographical references and index.

ISBN 1-57477-146-9 (alk. paper)

1. Offshore oil industry—Environmental aspects—Congresses 2. Offshore gas industry—Environmental aspects—Congresses. 3. Environmental monitoring—Congresses. I. Armsworthy, Shelley L., 1969- II. Cranford, Peter J., 1958- II. Lee, Kenneth, 1953- IV. Title

TD195.P4O38 2004
368.738′263–dc22

2004052187

Printed in the United States of America

Battelle Press
505 King Avenue
Columbus, Ohio 43201-2693, USA
614-424-6393 or 1-800-451-3543
Fax: 614-424-3819
E-mail: press@battelle.org
Website: www. battelle.org/bookstore

Contents

Part 3: Applications and Regional Experience

Acknowledgments

The editors wish to thanks the reviewers of the manuscripts, Dr. Peter G. Wells and Dianne Geddes for their advice on the format of this book, and Rosalie Allen Jarvis for logistical support toward production.

This book is the direct result of the *Offshore Oil and Gas Environmental Effects Monitoring Workshop* held at the Bedford Institute of Oceanography, Dartmouth, Nova Scotia, Canada on May 26–30, 2003. We wish to thank members of the workshop organising committee, including Dave Burley, Brian Giroux, Lesley Griffiths, Geoffrey Hurley, Jennifer Matthews, Jim McComiskey, Anne Muecke, Ted Potter, Eric Theriault, Brian Veitch, Debora Walsh, Peter Wells, and Urban Williams. We acknowledge the support of the following organisations for their initiative in sponsoring the workshop and the production of this book:

Canada-Newfoundland Offshore Petroleum Board
Canada-Nova Scotia Offshore Petroleum Board
Canadian Association of Petroleum Producers
Centre for Offshore Oil and Gas Environmental Research
Environment Canada
Environmental Studies Research Fund
Fisheries and Oceans Canada
National Energy Board
Petroleum Research Atlantic Canada
Program for Energy Research and Development

EnCana Corporation and Petro-Canada provided prizes for the best oral and poster presentations by students.

Foreword

Environmental effects monitoring (EEM) of land and water, associated with many industrial activities and environmental issues, has been a rapidly progressing field for the past three to four decades. Nowhere is this truer than in the marine environment with the offshore oil and gas industry. Monitoring (EEM) studies have been conducted in the Gulf of Mexico, the eastern Pacific off California and Alaska, the Beaufort Sea and other polar seas, the North Sea and northeast Atlantic off the UK and Norway, the Adriatic Sea, Caspian Sea, Arabian (Persian) Gulf, Australia offshore, and most recently in the northwest Atlantic off Canada. Effects monitoring has become part of the regulatory regime in several countries (e.g. USA, Canada, Norway), of assessing and controlling impacts of exploration, production, and decommissioning operations, and of assessing effects of accidental discharges and spills.

The design of EEM programs, choice of monitoring techniques, data analysis, selection of threshold effect levels, and application of EEM results by management have advanced greatly. This has gone hand in hand with recent advances in molecular diagnostic techniques, genomics, and community ecology. There is a concern and requirement to understand the extent and significance of short and long-term ecological effects of exploration and development, and to minimise impacts to living resources at source and in the far-field. Such monitoring activity, linked with marine research and funded by government and industry, has been reported at many workshops and conferences. There is a vast literature on EEM, much of it pertaining to the oil and gas industry.

The current volume presents papers and discussion of the *Offshore Oil and Gas Environmental Effects Monitoring Workshop* of May 2003, held at the Bedford Institute of Oceanography. The workshop attracted practitioners from a number of countries to discuss current approaches and technologies of EEM associated with offshore oil and gas, and the application of EEM results by environmental managers. This book represents an overview of current practice in the field of EEM in offshore waters, and illustrates many lessons about detecting change in marine ecosystems.

What is particularly notable is the progress in a number of areas of effects monitoring and marine ecotoxicology. EEM is now placed formally within the context of ecological risk assessment in both North American and European waters, a point made at the first Halifax offshore EEM workshop in 2000 (Gordon et al., 2000). There is more agreement with the core indicators for monitoring, protecting key resource species, ecosystem attributes, and other human uses of the coasts and seas. Specific species of fish and the benthos are often used as health effect indicators. Considerable attention has been paid to developing and applying powerful statistical techniques, along with power analyses, for sampling and detecting change attributed to offshore operations. This is being done with an increased understanding of the natural background and variability of indicators and indices. Modelling is increasingly used to link the components of EEM, and to predict the fate and impacts of produced water (PW) (containing metals, PAHs, nutrients) and components of drilling muds. There are innovative combinations of laboratory and field techniques; for example new biomarkers (e.g. genotoxicity, embryotoxicity, histopathology) have been developed for use in effects monitoring, toxicity identification evaluation approaches are being deployed on PW to identify fractions of highest toxicity potential, and in-situ techniques are used to assess effects on the water column such as using pelagic organisms, caged organisms (fish, mussels), and passive samplers (SPMDs).

There is further work to understand the fate and risks of specific contaminants (e.g. barite, bentontite, mercury and other metals, suspended particulates) released in drilling muds and cuttings, and to evaluate the impacts, real and perceived, of discharges on fisheries species and populations. Platforms both affect and produce habitat, so studies are considering both negative and positive impacts to habitat around rigs, as well as developing an understanding of the effects of the rigs/platforms themselves (i.e. reef effects on fish and seabirds, and interactions with chemical effects). Following from decades of research, a special focus is on the benthic environment with chemical, toxicity, and benthic community sampling, including understanding the recovery of benthic ecosystems impacted by pipelines and their construction. Importantly, as exploration moves further offshore in many oceans (e.g. west of Britain, and offshore Brazil), the potential impacts in the deep sea are being addressed using new approaches and techniques. Finally, as pipelines proliferate, studies are addressing their impacts on the behaviour of marine species that migrate (e.g. lobsters); behavioural studies on molluscs and mammals also address the effects of seismic energy and chemical wastes.

A conclusion of past studies is that the effects of discharges of offshore platforms, if detected, are generally localised. This finding is largely supported by the papers given at this workshop. But complacency does not protect the oceans, so the conclusion of local impact is still one to test, as unanticipated effects and events may occur, especially with regard to noise pollution and marine mammals. Recent National Research Council and United Nation studies on oil pollution show that global oil inputs from the offshore industry are measurable (approximately 20,000 metric tonnes/yr) but small relative to inputs from other sea-based activities (NRC, 2003; GESAMP, 1993; 2004). The hydrocarbon footprint of offshore activities is much less than that of the presence of rigs and their various chemical and waste discharges.

Finally, it is noteworthy that the studies and programs involving EEM associated with the offshore have greatly aided our overall understanding of the fate and effects of contaminants (chemical and physical) in marine ecosystems. Indeed, offshore EEM is helping to elucidate some of the primary principles of marine ecotoxicology, ecological risk assessment, and marine ecosystem health. The reader should consult and read the present volume with this in mind, extracting techniques and approaches useful for EEM not only related to the oil and gas industry, but also to other human activities and stressors in the ocean environment. The collective tools of EEM, combined with ecological risk assessment, are essential for maintaining productive and healthy coastal and offshore ecosystems in an increasingly industrialized world. Much is to be learned from the collective experience of EEM and the offshore.

The editors are to be congratulated for organising the workshop and producing this insightful and information-rich addition to the literature on marine environmental effects monitoring.

—PETER G. WELLS, PH.D.

REFERENCES

GESAMP. 1993. *Impact of Oil and Related Chemicals and Wastes on the Marine Environment*. GESAMP Reports and Studies No. 50, International Maritime Organization, London, UK. 180p.

GESAMP. 2004. (In press) *Estimates of Oil Entering the Marine Environment from Sea-Based Activities*. GESAMP Reports and Studies No. 75, International Maritime Organization, London, UK. 164p.

Gordon, D.C., Jr., L.D. Griffiths, G.V. Hurley, A.L. Muecke, D.L. Muschenheim, and P.G. Wells. (Eds.). 2000. *Understanding the Environmental Effects of Offshore Hydrocarbon Development*. Canadian Technical Report of Fisheries and Aquatic Sciences 2311. 82p plus appendices.

National Research Council. 2003. *Oil in the Sea III. Inputs, Fates, and Effects*. The National Academies Press, Wash., DC. 265p.

Preface

The scientific and technical papers that follow were contributed by participants of the *Offshore Oil and Gas Environmental Effects Monitoring (EEM) Workshop* held at the Bedford Institute of Oceanography in Dartmouth, Nova Scotia, Canada on May 26–30, 2003. The workshop brought together international stakeholders of the offshore oil and gas industry (scientists, environmental managers, and industry representatives) to share their knowledge and expertise. The enclosed peer-reviewed papers present recent information concerning environmental protection in the offshore oil and gas industry, and represent the diverse experience and knowledge of scientific researchers and managers. These papers encompass a broad spectrum of issues, but fall into three general categories: environmental management, EEM approaches and technologies, and regional experiences obtained through the application of EEM methodologies. Some of the topics addressed in this volume include:

- cost-effectiveness and utility of modelling pollution effects before they occur,
- plans for the management of environmental accidents,
- improving environmental management methodologies by linking EEM with ecological risk assessment,
- techniques used for detecting contaminants and predicting their fates,

- methodologies used for estimating biological effects,
- concerns regarding seismic activity effects on living resources,
- approaches to oil and gas development in sensitive and deep-sea environments,
- effective experimental designs for offshore EEM.

This collection of papers helps to advance our understanding of the environmental impacts of the offshore oil and gas industry, and future research needs identified by some contributors will continue to contribute to the protection of the marine environment and its resources. Controversial papers were not avoided, so as to bring topical issues into view and recognise the need for additional study. Two of the papers included in this volume (T. Ahlfeld, R. Green) represent invited keynote addresses presented at the workshop. It is our hope that the success of the workshop will be reflected in this book, and that these papers will stimulate future research into EEM.

The development of this book is a product of the assistance of numerous contributors and sponsors. A list of individuals and organisations to which we are indebted, are listed in the Acknowledgments section. This is an opportunity to thank Drs. Peter Cranford and Kenneth Lee for affording me opportunity to act as manuscript coordinator and editor. This has been a unique experience in communication, patience, and planning.

—SHELLEY L. ARMSWORTHY

A Perspective on Offshore Oil and Gas Environmental Effects Monitoring

Cranford, P.J. (cranfordp@dfo-mpo.gc.ca) and K. Lee (Fisheries and Oceans Canada, Dartmouth, NS Canada)

We must remember that in nature there are neither rewards nor punishments—there are consequences.
—Robert G. Ingersoll (1881)

Environmental effects monitoring (EEM) is a central component of environmental protection and management strategies designed to minimise the consequences of anthropogenic activities (GESAMP, 1991). Frameworks for environmental management generally consist of a linked series of activities that identify, critically evaluate, and address predictions of potential environmental effects (Fig. 1). Environmental impact predictions are inherently subject to some degree of uncertainty as a result of gaps in knowledge, the natural variability of ecosystems, and the influences of unforeseen factors. This uncertainty is addressed by monitoring programs that ensure actual effects do not exceed the predictions and are within predetermined acceptable limits. It is generally accepted that management responses to monitoring data should be timely and follow a predetermined course of action. Furthermore, the management framework includes feedback-loops between prediction, monitoring, and management responses that serve to strengthen our ability to protect and

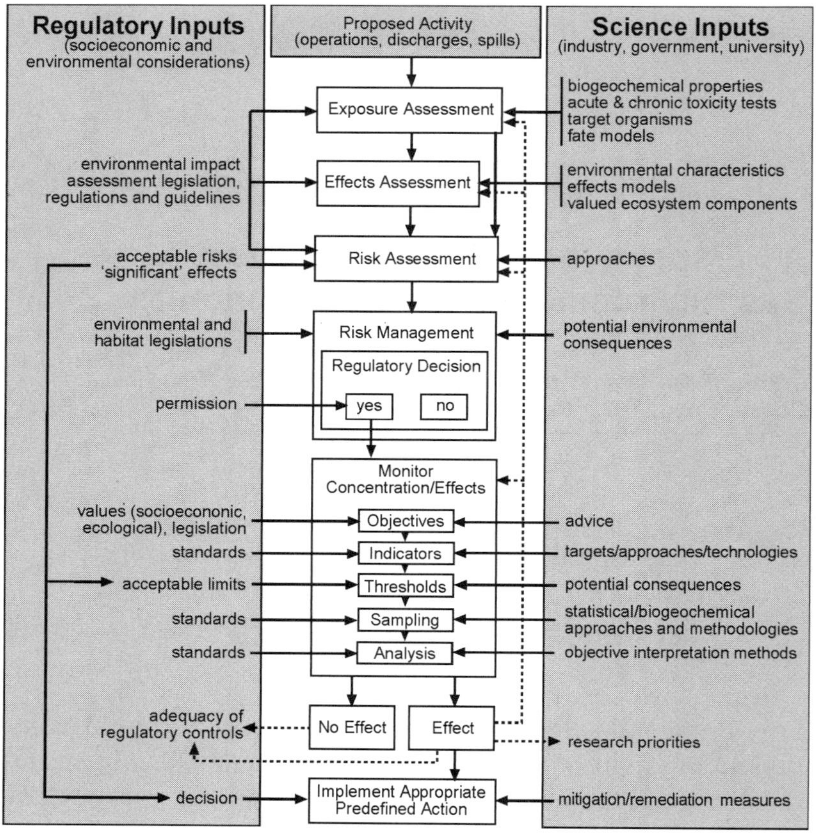

Figure 1. An integrated framework for marine environmental protection and management including inputs from regulatory management and science (feedback loops are illustrated by dashed lines).

manage the environment. Coordination between regulatory authorities and governmental and nongovernmental agencies influences operational regulations, the setting of scientific researches priorities, and the public's perception of industry environmental practice. There are numerous examples of EEM results driving regulatory controls on the offshore oil and gas industry, particularly in the direction of stricter guidelines for operational waste discharges. However, there are also recent examples of controls on some effluents being relaxed owing to contributions from both monitoring and applied research programs.

The papers published in this book (a subset of papers delivered at the *Offshore Oil and Gas Environmental Effects Monitoring Workshop* held at the Bedford Institute of Oceanography in Dartmouth, Nova Scotia on May 26–29, 2003) provide an overview of issues currently being addressed by offshore oil and gas EEM programs worldwide, identify some recent developments in our understanding of the industry's environmental performance, and highlight possible new directions, approaches, and technologies for advancing monitoring programs in the future.

The authors of the papers in this volume address a range of topics related to offshore oil and gas operations in relatively shallow to deep waters in the North Sea, the Atlantic Margin (west of Britain), eastern and northern Canada, the Gulf of Mexico, the Beaufort Sea, and the Caspian Sea. The papers are not limited strictly to discussions of EEM programs, but include other activities closely linked to EEM within environmental management frameworks. The contributions include the work of both scientists (government, university, and consultant) and environmental managers (industry and government) with emphasis on past, present, and future EEM programs. Including such a wide mix of participants provided ample opportunity for the expression of divergent points of view on subjects such as the best statistical design for EEM programs (Green, 2005) and potential impacts on seabirds (Burke et al., 2005). As this volume is intended for an audience with a background in science and in the environmental issues addressed, we felt it important that we not stray from presenting differing viewpoints. It is our hope that, by presenting these, and other studies in an open forum, further discussion and research in these important areas will be stimulated.

STATUS OF OFFSHORE OIL AND GAS EEM

Environmental monitoring has been defined as "the regular collection, generally under regulatory mandate, of biological, chemical, or physical data from predetermined locations such that the present status and any ecological changes attributable to [industry operations] can be quantified." (GESAMP, 1996).

EEM programs are designed around the testing of hypotheses that are based on concrete program objectives, including:

- to establish pre-development (background) environmental conditions to aid in the design of subsequent EEM,
- to determine natural trends in the environment and degree of variation during industry operations,
- to ensure conformity with regulations and specific management requirements (i.e. compliance monitoring),
- to ensure that contaminant concentrations and biological effects do not exceed predictions made in environmental impact assessments, (includes testing of impact assessment tools, such as contaminant fate models and other risk assessment methodologies),
- to identify spatial and temporal trends in contaminant distributions and biological effects, and ensure that contaminant concentrations and effects are not significantly greater than acceptable limits,
- to relate physical (e.g. current speed and direction), geological (e.g. sediment grain size) and chemical parameters (e.g. waste concentrations) to biological effects,
- to assess the effectiveness of the environmental management framework, including assessment of the effectiveness of any mitigation measures imposed,
- to monitor the rates of natural recovery and/or the effectiveness of remediation programs at impacted sites,
- to provide feedback for assessing the adequacy of regulatory standards and/or guidelines imposed on industry for environmental conservation and protection, and
- to help identify future research priorities.

EEM objectives are assessed on the basis of observations of the spatial extent and magnitude of site-specific environmental effects, while taking into account the presence of other potential sources of effects. An integrated monitoring approach is routinely undertaken during offshore oil and gas operations where a suite of chemical and biological analyses and observations are conducted in the receiving environment, including reference sites, and the environmental performance of the industry is based on the accumulated evidence. This "weight-of-evidence" approach is desirable over situations where decisions that could result in operational or regulatory actions are based on more limited chemical or biological investigations.

A variety of environmental concerns from offshore oil and gas operations can be addressed, at least partly, through monitoring. Petroleum reservoirs beneath the ocean are detected using seismic surveys that generate high sound levels that may affect invertebrates, fish, marine mammals, and sea turtles (Cott and Hanna, 2005; Miller et al., 2005). A long-standing concern over offshore oil and gas operations has been the potential biological effects associated with the exposure of organisms to low-level operational waste discharges. During oil and gas exploration and development operations, drilling waste discharges have been of primary concern. Hence, discharged drilling fluids (muds) have been a primary focus of environmental impact studies because their drift, dispersion, and dilution are much lower than those of dissolved discharges. In addition, during production operations, a high volume of produced water, which is recovered during extraction of the oil and/or gas, is discharged to the marine environment. This saline formation-water may contain metals, nutrients, radionuclides, hydrocarbons, and trace amounts of chemical agents (e.g. bactericides, corrosion and scale inhibitors) at elevated concentrations (Lee et al., 2005). Accidental spills of various size and type during exploration, extraction, and transport of oil and gas also present an environmental risk (Trudel, 2005), as does physical habitat alterations from the development and construction of submarine pipelines (Ellsworth et al., 2005).

In the United States, the Minerals Management Service (MMS) is charged with monitoring the outer continental shelf environment. The MMS has contributed significantly to research, monitoring, and assessments related to oil and gas operations over the past three decades, and the results have influenced regulatory controls on industry (Ahlfeld, 2005). As in other areas of the world, the focus of research in the U.S. has largely been on the long-term effects of development and production discharges on benthic communities. During the period when studies in the North Sea were directed toward understanding and monitoring the physical, chemical, and biological fates and effects of oil-based drilling muds, a large emphasis of many U.S. studies was related to water-based mud discharges. More recent studies at offshore sites discharging synthetic-based muds (SBM) have contributed to the U.S. Environmental Protection Agency publishing regulations in 2001 that permit controlled discharges of SBM associated with cuttings in most offshore areas of the U.S. As a consequence of this national coordinated effort, monitoring within the U.S. is only required to prove compliance with regulatory requirements.

EEM programs at Norwegian offshore installations are particularly notable for the long-term standardisation of EEM procedures, the regional (by latitudes) approach, and the open access to EEM data. Bakke and Nilssen (2005) discussed the advantages and disadvantages associated with this harmonised monitoring approach, which differs from that conducted off eastern Canada where independent EEM strategies are designed by the operator of each production installation for approval by regulators. This absence of a common EEM strategy for offshore installations in Canada makes spatial and temporal comparisons of biological effects at different installations largely unfeasible, and permits EEM programs to be influenced over the long-term by shorter-term perceived risks. For example, the Hibernia (located on the Grand Banks of Newfoundland) EEM program was designed with input from specialists during the period when benthic community impacts were believed to be more limited in geographic scale than were later detected (e.g. Olsgard and Gray, 1995). As a result, benthic community analysis – a major focus of offshore oil and gas EEM programs worldwide was not included in the Hibernia EEM program.

Variable site-specific monitoring needs identified during environmental impact assessments, combined with the variable nature of offshore operations (e.g. single to multiple drilling platforms, oil vs. gas wells, water depth), have precluded implementation of a common monitoring strategy in Canada.

A benefit of having a more flexible approach to EEM is the opportunity provided for testing and implementing new EEM methodologies and technologies designed to target site-specific valued ecosystem components. By integrating many site-specific and proven methodologies, EEM at the Terra Nova field on the Grand Banks of Newfoundland is reported to ". . . equal or surpass those being carried out in the North Sea, the Gulf of Mexico, and Australia" (DeBlois et al., 2005). A flexible EEM system may also result in significant operational cost savings. For example, Trudel (2005) described a three-tiered system spill-related environmental effects monitoring (EEM) program for marine birds and commercial fisheries, in which the level of monitoring is linked to the type and size of spill and the magnitude of environmental damage. In the case of the commercial fisheries EEM plan, allowance is given for the closure of the fisheries, if required, while minimising both the size of the area closed and the duration of the closure.

LINKING ECOLOGICAL RISK ASSESSMENT WITH EEM

A number of authors in this volume have emphasised the importance of coupling ecological risk assessment (ERA) approaches with EEM. Wells (2000) made the case for this at a previous Canadian workshop on the effects of offshore hydrocarbon development. The integrated management framework illustrated in Figure 1 shows links between these different activities, but the interaction is often unformulated, with ERA (when conducted) largely providing input to regulatory review/approval processes. Through the ability to prioritise potential sampling locations based on contaminant transport and ecological risk, contaminant fate models and ERA can play a greater role in the design of cost-effective EEM programs (see Berry and Wells, 2005; Sadiq et al., 2005; Warren and Mackay, 2005). These tools may also be useful for identifying optimal remedial actions.

Warren and Mackay (2005) make a case for incorporating modelling as a central component of monitoring programs. Environmental modelling exercises have often proved valuable for identifying gaps in knowledge that need to be addressed through additional data collection and research. By continuously comparing model predictions with environmental monitoring data, the authors believe that the two approaches will produce complimentary information that will demonstrate an adequate understanding of relevant abiotic and biotics processes, possibly improve our knowledge of cause-effect relationships, and eventually reduce the extent and expense of monitoring.

The accuracy of numerical models from the prediction of ecological risk requires knowledge of the contaminant levels as well as their transport, fate, and effects. Lee et al. (2005) has highlighted the complexity of this issue in terms of predicting the impacts of produced water discharges. In this case, preliminary studies have shown that the toxicity components within the produced water were significantly influenced by chemical kinetics following discharge. Furthermore, the full impact of long-term low level exposures to contaminants has not been resolved. While regulators have traditionally focused their attention on the application of acute biotests in EEM programs, the need for application of chronic toxicity data in risk assessment model is now recognised.

One objective of EEM is to provide information needed to test ERA predictions of potential environmental impacts. However, ERA and EEM concepts and approaches have largely been developed in isolation from each other, often resulting in outputs (predictions and measures, respectively) that cannot be directly compared or integrated. Smit et al. (2005) present a basis for more closely integrating ERA and EEM tools within management frameworks that would allow a more effective means of reducing uncertainty through the improvement of assessment models and the selection of appropriate monitoring parameters. The major challenge noted by the authors is the development of common outputs (test endpoints) from both tools. Such a development would permit ERA to evolve in response to monitoring results in much the same manner as regulatory controls on industry operations have been responsive to observed environmental impacts.

EEM APPROACHES AND TECHNOLOGIES

The ability to detect and quantify patterns in nature, including contaminant distributions and biological effects, is directly related to the statistical adequacy of the sampling program design (i.e. physical location and total number of sampling sites, number of replicates, timing of sampling). Decisions on the dispersion in space and time of samples and measurements are critical and need to be made based on the specific hypotheses to be tested by the EEM program. Within many fields of ecological study, certain sampling designs have become popular to the extent that they have become the standard methodology. The traditional design used at offshore oil and gas production facilities includes several transects extending out in different directions from the platform(s). Green (2005) examined the many choices that need to be made for the cost-effective implementation of this "bulls-eye" design concept with the intent of outlining the characteristics of a good design. His suggestion that the outer benthic sampling sites (reference) be limited to just 1000 m from the platform is a departure from current EEM approaches (see Bakke and Nilssen, 2005; DeBlois et al., 2005) and will be debated by those with divergence of views on the spatial scale of biological effects.

There is no opportunity to repeat a monitoring program that has employed a flawed statistical design. However, it is possible, and highly desirable, to repeat the statistical analysis of historical monitoring data using new or alternative approaches. A major shift in our perception of the spatial scale of benthic community effects around drilling platforms discharging cuttings containing oil-based drilling mud occurred in the early 1990s as a result of the employment of multivariate data analysis techniques. Environmental impacts associated with operational discharges of drilling wastes had been based primarily on the analysis of several single-index measurements of resident benthic community structure (e.g. diversity and evenness). This EEM approach resulted in a general consensus between operators, authorities, and the scientific community that there were no measurable biological effects beyond 500 m from a drilling platform discharging well cuttings containing oil-based drilling fluids (GESAMP, 1993). The introduction of multivariate statistical methods that simultaneously analyse data collected on numerous monitored parameters challenged this theory by demonstrating more widespread community effects at many North Sea oil fields, and by relating these effects to contaminants associated with the cuttings discharge (Gray et al., 1990; Olsgard and Gray, 1995). The cost to industry of this revelation was substantial, as it led to significant changes in operational waste discharge regulations in many countries, and the resulting need to adopt new drilling technologies and practices. As a result of stricter regulatory controls on drilling waste discharges and the increased transport to land or re-injection of operational wastes, the zone of impact around each oil and gas platform is expected to have decreased. There was no organised effort to obtain a consensus among participants at the *Offshore Oil and Gas EEM Workshop* on the current spatial scale of effects on benthic communities, however, the studies presented largely indicated that the benthic zone of impact was generally limited to within 500 m from the platform.

Multivariate statistical techniques that classify (cluster analysis) and compare (ordination) all benthic community and environmental variables are sensitive tools for assisting in detecting subtle community changes around drilling platforms. As a result, they have become a standard component of EEM data analysis strategies for many offshore oil and gas installations. A primary function of statistics is to increase objectivity in data interpretation. However, there is an ongoing debate among scientific experts regarding the potentially subjective interpretation of faunal changes detected using multivariate statistics (Bakke and Nilssen, 2005; Green, 2005). A basic problem for decision-makers is that these multivariate tests do not assign a relative scale of disturbance to the identified benthic community groups. The degree of impact needs to be established in order to assess the ecological, social, and economic significance of the observations prior to implementing corrective actions. Continued efforts toward the standardisation of taxonomic analysis (Bakke and Nilssen, 2005) and the development of training (i.e. calibration) data sets against which detected community changes can be compared (e.g. Savage et al., 2001) could enhance our ability to objectively interpret EEM results and to compare effects detected in different regions, and for different benthic communities and substrates.

Offshore oil and gas EEM programs have placed considerable emphasis on detecting changes in benthic communities attributed to lethal effects. Researchers have been striving to discern significant, but more subtle sublethal effects of waste discharges and to assess the associated risk to benthic and pelagic populations. Sublethal impacts can occur at contaminant concentrations that are several orders of magnitude lower than the lethal threshold, and the zone of impact may therefore be substantially larger than is indicated by changes in the structure of benthic communities. As noted by Green (2005) in his examination of EEM designs for offshore platforms, the locations of reference sites would have to be extended outward if the hypotheses to be tested included sublethal biological effects. The biological effects of drilling wastes have largely been attributed to chemical toxicity (hydrocarbons and metals), organic enrichment of sediment, and physical smothering. Studies of the sublethal impacts of discharged drilling muds to sea scallops show that an additional, and potentially more important, impact mechanism exists that could have consequences at the population level of some species. The physical effects of exposure to environmentally relevant concentrations (< 2 mg/l) of fine-grained suspended solids found in most drilling muds (bentonite and barite) caused significant impacts on scallop growth, reproduction, feeding rates, and food absorption efficiency (Armsworthy et al., 2005 and references cited). Similar results were reported by Barlow and Kingston (2001).

Water-based mud (WBM) discharges are not subject to the same strict regulations that control wastes containing hydrocarbons. WBM is released directly to the environment without treatment, and often as bulk dumps of spent drilling fluids. A model-based risk assessment for WBM discharges during exploration drilling predicted that the sublethal effects of WBM constituents may be detectable at substantial distances from the discharge site under certain oceanographic conditions (Cranford et al., 2003). As a logical next step after conducting laboratory-based ecotoxicology experiments, hazard assessments, and model-based risk assessments, in situ studies were conducted with caged bivalves to develop methodologies and technologies for monitoring meaningful sublethal effects at drilling sites (Cranford et al., 2005). Owing to the limited discharge of drilling wastes during this study, scallops were not significantly impacted, however, the methodologies developed proved successful for monitoring scallop growth and feeding behaviour and contaminant exposure conditions over short (days) to medium (months) time-scales.

The recent development of new guidelines for ERA by the Oslo and Paris Commission for the Protection of the Marine Environment of the North-East Atlantic (OSPAR) emphasises the importance of focusing on biological effects and has led to an increased need for biological parameters (biomarkers and bioindicators) in EEM that provide evidence of contaminant exposure and/or effects. One of the more useful attributes of some bioindicator responses is that they provide an early warning of the onset of more serious biological effects directly relevant to effects on the population. A greater reliance on biological effects monitoring within an environmental management framework will require the identification of threshold responses that signal a significant environmental and/or socioeconomic risk. Implementation of such an approach will require modification of current EEM strategies, which have generally consisted of measurements of chemical distributions, laboratory toxicity testing, and benthic community analysis (triad approach). Some resistance to the early implementation of biomarkers in experimental monitoring programs stems from the temptation to dismiss measurements made on standard test organisms imported to the study site and to the uncertainty of interpreting biomarker measurements at the organism or population levels. The use of caged mussels for detecting produced water exposure, contaminant dose and effects in pelagic organisms (Hylland et al., 2002) has been criticised because "mussels, of course, do not occur in the water column..." (Gray, 2002). The results of laboratory and in situ studies, when properly applied, form the basis for assessing the potential risks of contaminants to resident populations.

Ideally, risk assessments for pelagic organisms need to allow for the rapid dilution/dispersal rates of contaminants discharged to the ocean, the relevant transport pathways (e.g. advection, precipitation, buoyancy), and the behaviour of pelagic organisms (e.g. movements, feeding) (see Berry and Wells, 2005). Environmentally relevant exposure protocols need to be developed, but in the interim, empirical evidence of the potential biological effects of operational wastes on organisms of economic and ecological significance are needed to identify the requirements of EEM programs. Laboratory studies of the acute and chronic effects of constant produced water concentrations may be unrepresentative of actual exposure conditions, but are of value for assessing the relative sensitivity of different organisms and life-stages. Querbach et al. (2005) presented results of a laboratory study of the potential impacts of produced water on sensitive early life stages of economically important finfish and shellfish. The authors concluded that the toxicity of produced water is both species and life-stage specific, with effects detected on fertilization success, survival, and growth of the developmental stages of important resource species. Additional research is required on the actual fate and effects of produced water constituents before a conclusion on the risk to populations can be made, and to aid in identifying monitoring requirements.

A range of fish health bioindicators (condition indices, lesions, hisopathological alterations, enzyme induction, and blood cell counts) were investigated in American plaice collected before and during the development of the Terra Nova oil production site (Mathieu et al., 2005) with no adverse effects detected. The desire to use fish for bioindicator studies inevitably requires the investigator to choose between the problem of having no information on previous exposure conditions (sample resident population) and problems associated with modifying fish behaviour to control exposure (sample caged animals). In the absence of a more environmentally realistic monitoring protocol, a precautionary approach would suggest that the latter method is desirable because the animals are unable to move away from the study area (i.e. worst case scenario) and there is a greater chance of linking cause (exposure) to any measured effect. The objective interpretation of biomarker data that considers both the exposure conditions experienced by the test organism(s) and resident populations is fundamental to the inclusion of biomarkers within environmental monitoring and management plans.

The increasing focus on biological effects monitoring may eventually lead toward a greater tendency to search for causes only after an effect is detected. Any attempt to link effect to cause, or to test model predictions of waste exposure, requires precise and accurate methodologies and technologies for monitoring contaminant (or suitable tracer) concentrations. Cranford et al. (2005) report on the development of a highly sensitive barium extraction method that improves the quantitative recovery of barite (a major component of drilling muds) from sediment samples. Likewise, Milligan et al. (2005) present a method for quantifying bentonite, a constituent of most drilling muds not readily distinguished from naturally occurring clays by chemical analysis. The proposed method is based on the disaggregated inorganic grain size of suspended sediment samples. Observations of the distribution of bentonite around drilling platforms provide new insights into (1) important drilling waste transport and deposition processes, including flocculation, settling velocity, and resuspension, (2) the different transport pathways for different waste constituents (bentonite and barite), and (3) the accuracy of waste transport model predictions. Model predictions of suspended solid waste concentrations are very sensitive to waste settling velocity, resulting in a need for a comprehensive understanding of the processes controlling settling velocity in nature. Niu et al. (2005) present laboratory results describing the effects of a range of physical factors on the settling mechanism and speeds of cuttings samples, thereby improving the ability to establish realistic exposure scenarios of ecological risk analysis.

EMERGING ISSUES AND CHALLENGES

Increasing oil and gas exploration and production activities in the Arctic and deep-sea environments raise significant new challenges for EEM. Foremost to overcome are the technical challenges of working in these harsh and remote environments, and gaps in knowledge of the Arctic and deep-sea ecosystems. Skadshem et al. (2005) and Hartley (2005) discussed the uncertainties regarding the vulnerability of deep-sea life to toxicity responses and noise as compared with shallow water organisms. Information is needed to determine the relevance of extrapolating the lessons learned from decades of shallow-water oil and gas EEM to the deep-sea environment. There is also a greater need to characterise new sites slated for offshore development. For example, Boehm et al. (2005) recently reported significant levels of resolved petroleum-related hydrocarbons including alkylated and parent polycyclic aromatic hydrocarbons (Σ PAH) throughout the Absheron region. These background levels of PAHs were primarily attributed to petrogenic sources rather than pyrogenic (combustion related) PAH components. Probable sources include 1) natural seeps, 2) historic crude oil releases, and 3) terrestrial runoff of eroded petroleum sources rocks. For the support of EEM programs, prior to any development operations there is a need for study to enable one to distinguish the "environmental baseline" from potential future impacts.

Many countries are implementing legislation calling for an ecosystem-based management approach that is intended to resolve the currently fragmented management of ocean activities. This approach considers impacts on both the structure and function of the ecosystem and recognises the interdependence and interactions between all ecosystem components and all ocean industries (e.g. offshore oil and gas, fisheries, marine transportation, maritime defence, and ocean mining). The need to evaluate ecosystem level impacts within integrated oceans management requires enhanced monitoring programs that can effectively address the increased level of complexity inherent with moving EEM from organism, population and/or community approaches to the ecosystem level.

Sensitive effects indicators at all levels of biological organisation are needed to address broad conservation and protection management objectives, while providing information relevant to oil and gas impacts and the potential effects of other ocean industries. The general features of ecosystems that need to be protected (i.e. ecosystem conservation objectives) are only now being defined and these need to be set by society.

The participation of a wide range of stakeholders is needed during the development, implementation, and ongoing review and revision of EEM programs to ensure that marine environmental conservation and protection objectives are met using practical and effective methods. Science continues to play a vital and successful role in environmental protection. During the development of ecosystem-based management, the participation of the scientific community and other stakeholders will be vital for the provision of advice on the potential consequences of selecting particular ecosystem conservation objectives, performance indicators, and associated management targets. We face the challenge of synthesising relevant information from many scientific disciplines for the development of holistic and multiple endpoint monitoring approaches and methodologies that can describe marine ecosystem health and quality in relation to the collective human activities that need to be managed. Even more daunting is the need to do this in a manner that is user-friendly for managers and policy makers.

To continue to improve the balance between the appropriate level of environmental protection and cost-effective managerial controls on offshore wastes discharge, the future development of EEM programs needs to continue along three avenues:

1. *Interpretation.* The design of EEM programs and the interpretation of EEM data must use sound and objective science as the foundation to methodology. In some cases, less subjective approaches are needed for reaching overall conclusions on significant zones of impacts based on EEM data.

2. *Interaction.* Some important advances in our understanding of the industry's environmental performance and the effectiveness of regulatory controls have been made by comprehensive teams that interact closely with all stakeholders. These teams recognise the need for a multidisciplinary approach that simultaneously addresses many of the components of the environmental management framework (Fig. 1), including the need for improving links between modelling, ERA and EEM. The continued development of environmental protection and management strategies in some regions would benefit from the development of a more open and transparent process for sharing and communicating monitoring data.
3. *Integration.* Further steps toward integrated monitoring, in which a suite of chemical, biochemical, and biological analyses are conducted to assess effects at different trophic levels, are needed to address the needs of an ecosystem approach to ocean management. Integrated monitoring under this management framework will also have to account for the cumulative effects of all marine industries.

ACKNOWLEDGEMENTS

The Editors would like to thank the Organising Committee and all the agencies that sponsored the *Offshore Oil and Gas Environmental Effects Monitoring Workshop.* We are grateful to Anne Muecke, Leslie Griffiths, and Rosalie Allen Jarvis for their logistical assistance, and we would like to acknowledge all the workshop participants and contributing authors for their invaluable contributions to a memorable occasion. We thank Dr. Peter Wells for helpful comments on this manuscript.

REFERENCES

Ahlfeld, T. 2005. "Offshore Oil and Gas Environmental Effects Monitoring Investigations Conducted by the U.S. Minerals Management Service." In: S.L. Armsworthy, P.J. Cranford, and K. Lee (Eds.), *Offshore Oil and Gas Environmental Effects Monitoring: Approaches and Technologies.* Battelle Press (this volume).

Armsworthy, S.L., P.J. Cranford, K. Lee, and T. King. 2005. "Chronic Effects of Synthetic Drilling Mud on Sea Scallops (*Placopecten magellanicus*)." In: S.L. Armsworthy, P.J. Cranford, and K. Lee (Eds.), *Offshore Oil and Gas Environmental Effects Monitoring: Approaches and Technologies.* Battelle Press (this volume).

Bakke, T. and I. Nilssen. 2005. "Harmonised Monitoring of Offshore Drilling Waste Effects in Norway." In: S.L. Armsworthy, P.J. Cranford, and K. Lee (Eds.), *Offshore Oil and Gas Environmental Effects Monitoring: Approaches and Technologies.* Battelle Press (this volume).

Barlow, M.J. and P.F. Kingston. 2001. "Observations on the Effect of Barite on Gill Tissues on the Suspension Feeder *Cerastoderma edule* (Linné) and the Deposit Feeder *Macoma balthica* (Linné)." *Marine Pollution Bulletin.* 42:71-76.

Berry, J.A. and P.G. Wells. 2005. "Environmental Modelling of Produced Water Dispersion with Implications for Environmental Effects Monitoring Design." In: S.L. Armsworthy, P.J. Cranford, and K. Lee (Eds.), *Offshore Oil and Gas Environmental Effects Monitoring: Approaches and Technologies.* Battelle Press (this volume).

Boehm, P.D., C.L. Maxon, F.C. Newton, J.S. Brown and Y. Galperin. 2005. "Aspects of Polycyclic Aromatic Hydrocarbons in Offshore Sediments in the Azeri Sector of the Caspian Sea". In: S.L. Armsworthy, P.J. Cranford, and K. Lee (Eds.), *Offshore Oil and Gas Environmental Effects Monitoring: Approaches and Technologies.* Battelle Press (this volume).

Burke, C.M., G.K. Davoren, W.A. Montevecchi, and F.K. Wiese. 2005. "Seasonal and Spatial Trends of Marine Birds along Support Vessel Transects and at Oil Platforms on the Grand Banks." In: S.L. Armsworthy, P.J. Cranford, and K. Lee (Eds.), *Offshore Oil and Gas Environmental Effects Monitoring: Approaches and Technologies.* Battelle Press (this volume).

Cott, P. and B. Hanna. 2005. "Monitoring Explosive-based Winter Seismic Exploration in Waterbodies, NWT 2000-2002." In: S.L. Armsworthy, P.J. Cranford, and K. Lee (Eds.), *Offshore Oil and Gas Environmental Effects Monitoring: Approaches and Technologies.* Battelle Press (this volume).

Cranford, P.J., D.C. Gordon, Jr., C.G. Hannah, J.W. Loder, T.G. Milligan, D.K. Muschenheim, and Y. Shen. 2003. "Modelling Potential Effects of Petroleum Exploration Drilling on Northeastern Georges Bank Scallop Populations." *Ecological Modelling.* 166:19-39.

Cranford, P.J., S.L. Armsworthy, S. McGee, T. King, and K. Lee. 2005. "Scallops as Sentinel Organisms for Offshore Environmental Effects Monitoring." In: S.L. Armsworthy, P.J. Cranford, and K. Lee (Eds.), *Offshore Oil and Gas Environmental Effects Monitoring: Approaches and Technologies.* Battelle Press (this volume).

DeBlois, E.M., C. Leeder, K.C. Penney, M. Murdoch, M.D. Paine, F. Power, and U.P. Williams. 2005. "Terra Nova Environmental Effects Monitoring Program: From Environmental Impact Statement Onward." In: S.L. Armsworthy, P.J. Cranford, and K. Lee (Eds.), *Offshore Oil and Gas Environmental Effects Monitoring: Approaches and Technologies.* Battelle Press (this volume).

Ellsworth, S., J. Schmidt, D. Martin, and R. Shaul. 2005. "Live Bottom Impact Reduction, Mitigation, and Monitoring for the Construction of a Subsea Natural Gas Pipeline in the Gulf of Mexico." In: S.L. Armsworthy, P.J. Cranford, and K. Lee (Eds.), *Offshore Oil and Gas Environmental Effects Monitoring: Approaches and Technologies.* Battelle Press (this volume).

GESAMP (Joint Group of Experts on the Scientific Aspects of Marine Environmental Protection). 1991. *Global Strategies for Marine Environmental Protection.* Reports and Studies, No. 45, 34pp.

GESAMP (Joint Group of Experts on the Scientific Aspects of Marine Environmental Protection). 1993. *Impact of Oil and Related Chemicals on the Marine Environment.* Reports and Studies, No. 50, 180pp.

GESAMP (Joint Group of Experts on the Scientific Aspects of Marine Environmental Protection). 1996. *Monitoring the Ecological Effects of Coastal Aquaculture.* Reports and Studies, No. 57, 38pp.

Gray, J.S. 2002. "Perceived and Real Risks: Produced Water from Oil Extraction." *Marine Pollution Bulletin.* 44:1171-1172.

Green, R. 2005. "Marine Coastal Monitoring: Designing an Effective Offshore Oil and Gas Environmental Effects Monitoring Program." In: S.L. Armsworthy, P.J. Cranford, and K. Lee (Eds.), *Offshore Oil and Gas Environmental Effects Monitoring: Approaches and Technologies.* Battelle Press (this volume).

Hartley, J.P. 2005. "The Challenge of Deep Water Oilfield Environmental Monitoring West of Britain" In: S.L. Armsworthy, P.J. Cranford, and K. Lee (Eds.), *Offshore Oil and Gas Environmental Effects Monitoring: Approaches and Technologies*. Battelle Press (this volume).

Hylland, K, G. Becker, J. Klungsoeyr, T. Lang, A. McIntosh, B. Serigstad, J.E. Thain, K.V. Thomas, T.I.R. Utvik, D. Vethaak and W. Wosniok. 2002. *An ICES Workshop on Biological Effects in Pelagic Ecosystems (BECPELAG): Overview of the Programme.* ICES Council Meeting Documents. ICES CM 2002/X:02

Lee, K., K. Azetsu-Scott, S.E. Cobanli; J. Dalziel, S. Niven, G. Wohlgeschaffen and P. Yeats. 2005. "Overview of Potential Impacts from Produced Water Discharges in Atlantic Canada". In: S.L. Armsworthy, P.J. Cranford, and K. Lee (Eds.), *Offshore Oil and Gas Environmental Effects Monitoring: Approaches and Technologies.* Battelle Press (this volume).

Mathieu A., W. Melvin, B. French, M. Dawe, E.M. DeBlois, F. Power, and U. Williams. 2005. "Health Effect Indicators in American Plaice (*Hippoglossoides platessoides*) from the Terra Nova Development Site, Grand Banks, NL, Canada." In: S.L. Armsworthy, P.J. Cranford, and K. Lee (Eds.), *Offshore Oil and Gas Environmental Effects Monitoring: Approaches and Technologies.* Battelle Press (this volume).

Miller, G.W., V.D. Moulton, R.A. Davis, M. Holst, P. Millman, A. MacGillivray, and D. Hannay. 2005. "Monitoring Seismic Effects on Marine Mammals—Southeastern Beaufort Sea, 2001 01502002." In: S.L. Armsworthy, P.J. Cranford, and K. Lee (Eds.), *Offshore Oil and Gas Environmental Effects Monitoring: Approaches and Technologies.* Battelle Press (this volume).

Milligan, T.G., T. Tedford, D.K. Muschenheim, and C.G. Hannah. 2005. "Quantifying Fine-grained Drill Waste in Scotian Shelf Suspended Sediments." In: S.L. Armsworthy, P.J. Cranford, and K. Lee (Eds.), *Offshore Oil and Gas Environmental Effects Monitoring: Approaches and Technologies.* Battelle Press (this volume).

Niu, H., T. Husain, B. Veitch, N. Bose, and K. Hawboldt. 2005. "Transport Properties of Discharged Synthetic-based Drilling Cuttings." In: S.L. Armsworthy, P.J. Cranford, and K. Lee (Eds.), *Offshore Oil and Gas Environmental Effects Monitoring: Approaches and Technologies*. Battelle Press (this volume).

Olsgard, F. and Gray, J.S. 1995. "A Comprehensive Analysis of the Effects of Offshore Oil and Gas Exploration and Production on the Benthic Communities of the Norwegian Continental Shelf." *Marine Ecology Progress Series.* 122:277-306.

Querbach, K., G. Maillet, P.J. Cranford C. Taggart, K. Lee, and J. Grant. 2005. "Potential Effects of Produced Water Discharges on the Early Life Stages of Three Resource Species." In: S.L. Armsworthy, P.J. Cranford, and K. Lee (Eds.), *Offshore Oil and Gas Environmental Effects Monitoring: Approaches and Technologies.* Battelle Press (this volume).

Sadiq, R., B. Veitch, T. Husain, and N. Bose. 2005. "Prioritising Environmental Effects Monitoring (EEM) Programs: A Risk-Based Strategy." In: S.L. Armsworthy, P.J. Cranford, and K. Lee (Eds.), *Offshore Oil and Gas Environmental Effects Monitoring: Approaches and Technologies.* Battelle Press (this volume).

Savage, C., J.G. Field, and R.M. Warwick. 2001. "Comparative Meta-analysis of the Impact of Offshore Marine Mining on Macrobenthic Communities Versus Organic Pollution Studies." *Marine Ecology Progress Series.* 221:265-275.

Skadsheim, A., J.F. Børseth, A. Bjørnstad, L. Camus, E. Aas, and S. Sanni. 2005. "Hydrocarbons and Chemicals: Potential Effects and Monitoring in the Deep-sea." In: S.L. Armsworthy, P.J. Cranford, and K. Lee (Eds.), *Offshore Oil and Gas Environmental Effects Monitoring: Approaches and Technologies.* Battelle Press (this volume).

Smit, M.G.D., J.A. van Dalfsen, and C.C. Karman. 2005. "Linking the Different Worlds of Environmental Risk Assessment and Environmental Effect Monitoring." In: S.L. Armsworthy, P.J. Cranford, and K. Lee (Eds.), *Offshore Oil and Gas Environmental Effects Monitoring: Approaches and Technologies.* Battelle Press (this volume).

Trudel, K. 2005. "A Plan for Implementing Environmental Effects Monitoring During Oil Spills from Sable Offshore Energy Project Installations". In: S.L. Armsworthy, P.J. Cranford, and K. Lee (Eds.), *Offshore Oil and Gas Environmental Effects Monitoring: Approaches and Technologies.* Battelle Press (this volume).

Warren, C.S. and D. Mackay. 2005. "The Role of Mass Balance Models in Offshore Environmental Effects Monitoring." In: S.L. Armsworthy, P.J. Cranford, and K. Lee (Eds.), *Offshore Oil and Gas Environmental Effects Monitoring: Approaches and Technologies.* Battelle Press (this volume).

Wells, P.G., 2000. "Workshop Summary and Reflections." In: D.C. Gordon, L.D. Griffiths, G.V. Hurley, A.L. Muecke, D.K. Muschenheim, and P.G. Wells (Eds.). *Understanding the Environmental Effects of Offshore Hydrocarbon Development.* Canadian Technical Report of Fisheries and Aquatic Sciences. 2311:76-82.

PART 1

Environmental Management

Linking the Different Worlds of Environmental Risk Assessment and Environmental Effects Monitoring

Smit, M.G.D. (M.Smit@mep.tno.nl), J.A. van Dalfsen, and C.C. Karman (TNO-Environment, Energy and Process Innovation, Department for Ecological Risk Studies, The Netherlands)

ABSTRACT: In northwestern Europe, OSPAR regulations and guidelines form the internationally agreed framework for environmental regulation of activities at sea, including that of the offshore oil and gas industry. Both environmental risk assessment and chemical and biological effects monitoring are part of the evaluation of environmental impacts under OSPAR. The European Union Technical Guidance Document on risk assessment (EU-TGD; EC, 2003) is the OSPAR sanctioned approach for estimating and reducing environmental impacts from toxic substances (OSPAR agreement 20-2003). A key element of OSPAR's 1999, "Strategy on Environmental Goals and Management Mechanism for Offshore Activities" is the assessment of the effectiveness of measures adopted for the prevention and elimination of pollution of the maritime area. OSPAR has recognised that a coordinated chemical and biological effects monitoring programme is essential for identifying the nature and extent of environmental impacts. A guidance document for the design and conduct of monitoring programmes has been published (OSPAR, 2004). However, not all the tools and methods needed for the undertaking of such monitoring (e.g. water column impacts) have yet been fully developed or validated.

Although both tools have the same assessment endpoint (occurrence of biological effects in the environment), the results of environmental effect monitoring studies and environmental risk assessment models have not been rigorously compared in light of the different factors under consideration between the two methodologies. Nonetheless, validation is required to ensure that the assessments are reliable, ranging from field validation of (laboratory) test endpoints (field monitoring) to performance testing of assessment models.

Based on the two tier approach described, an overview of the development of the present regulations and requirements for NW Europe concerning risk assessment and monitoring is presented. A case is made of consideration to couple ecosystem management and risk assessment protocols to validate risk assessment models based on environmental effect monitoring studies.

INTRODUCTION

For the evaluation of environmental impacts of offshore oil and gas, industry-specific tools have been developed and are used by various countries all over the world. In general, these tools can be divided into two categories: Environmental Effects Monitoring (EEM) and Environmental Risk Assessment (ERA). ERA addresses (selected) stressors of the activity, while EEM addresses the possible (measured) effects in the environment. Figure 1 shows how both are related in the total framework of Environmental Impact Assessment (EIA).

Although both tools aim for the same goal: protection of the environment, until now, not much effort has been put into the integration of both tools. Both ERA and EEM have been developed independently. While ecologists are mainly involved in

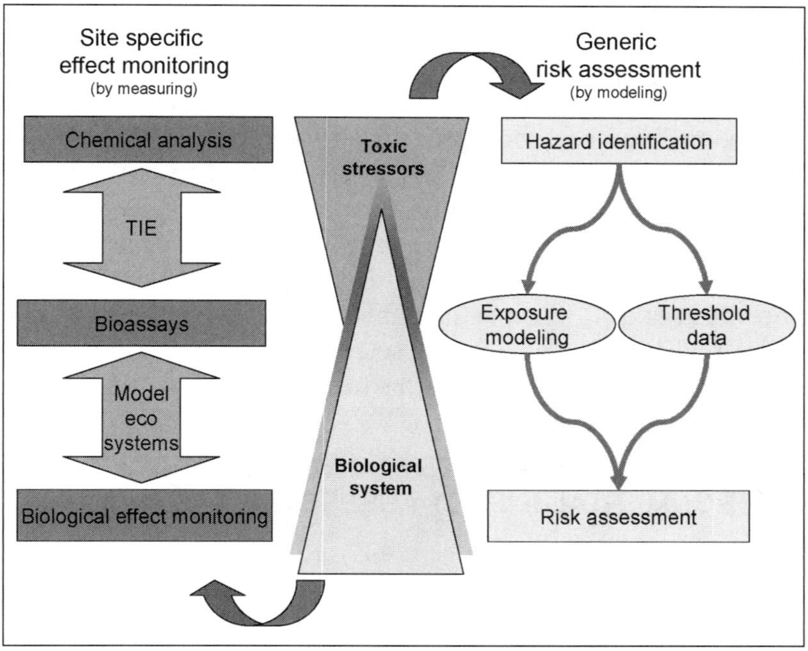

FIGURE 1. Evaluation of impacts using ERA and EEM.

environmental effects monitoring, environmental risk assessment is much more in the field of engineers and toxicologists. Regulators and operators consider both methodologies in offshore operations, but linking of the endpoints has so far not occurred. However, increased concern over the potential impact of oil and gas exploration in new offshore areas, which are often defined as sensitive due to vulnerable species and populations, has recently identified the need to critically evaluate the methodological links between environmental risk assessment, environmental effects monitoring, and ecologically relevant impacts.

This paper provides an overview of the state-of-the-art of assessment endpoints and test endpoints of ERA and EEM related to OSPAR regulations for the offshore oil and gas industry. The OSPAR Convention, introduced in March 1998, replaced the Oslo and Paris Conventions. However, Decisions, Recommendations and all other Agreements under the preceding Conventions continued to be applicable, unless terminated by ratifications under the 1992 OSPAR Convention.

Methodologies used under existing ERA and EEM protocols will be outlined and the strengths and weaknesses of each will be identified for evaluation. A framework will be presented to couple ERA and EEM by linking the test endpoints of both tools and by defining one common assessment endpoint. Developments needed for a better integration will be defined and discussed. It is expected that research over the next few years will be focused on the selection and development of tools for the development of this framework.

ASSESSMENT ENDPOINTS AND TEST ENDPOINTS IN ENVIRONMENTAL IMPACT ASSESSMENT (EIA)

In Cairns and Niederlehner (1995), two distinct uses of the term endpoint in ecological impact assessment are described: measurement endpoint and assessment endpoint. Assessment endpoints are defined as: "explicit definitions of the environmental values that are to be protected by an assessment. Depending on the specificity of the assessment, they may be very broadly and qualitatively defined (e.g. no loss of species) or very specifically and quantitatively defined (no more than one episode per 5 years in which ten or more fish are killed by an effluent release)." Test endpoints, (or measurement endpoints) are defined as: "the output of a toxicity test, series of field measurements, or other attempts to measure the responses of a biological system to a toxicant or other agent".

The best test endpoint to use is the test that supports the goals of the assessment. In general, this will be the test that, in combination with an extrapolation model, generates the best estimate of the assessment endpoint. This requires a clear statement of the environmental goals and development of extrapolation tools.

To invent the links between environmental effect monitoring and environmental risk assessment, identification and evaluation of the assessment endpoints and test endpoints used in both methodologies is described in the next two sections.

ENDPOINTS IN ENVIRONMENTAL RISK ASSESSMENT (ERA)

Characteristics of the Assessment Endpoint in ERA

Environmental Risk Assessment (ERA) has become a generally used tool in the evaluation of the potential environmental impact of chemical products or activities. According to the International Organization for Standardization (ISO) (ISO, 2002), the term risk can be interpreted as the combination of the probability of an event and the consequences of this event. In the case of environmental risk assessment, the probability of occurrence of the event is often equal to one (discharges are taking place). Therefore, the definition of risk in ERA mainly focuses on the consequences. The more widely accepted definition of risk related to the procedure of risk assessment was formulated in 1983 by the US National Research Council (USEPA): "the characterisation of the potential adverse health effects of human exposure to environmental hazards". The UK Department of Environment defines risk assessment as "the structured gathering of the information available about risks and the forming of a judgement about them". Risk assessment has since developed into a series of guidelines for conducting ecological risk assessment (USEPA, 1998). Within the European Union (EU), risk assessment is defined as: "A process of evaluation including the identification of the attendant uncertainties, of the likelihood and severity of (an) adverse effect(s)/event(s) occurring to man or the environment following exposure under defined conditions to (a) risk source(s)". A risk assessment comprises hazard identification, hazard characterisation, exposure assessment, and risk characterisation, and is an integrated part of the risk management procedure (Fig. 2). These four steps of the risk assessment process were first elaborated by the US national research council (USEPA, 1993) and were adopted by the EU (EC, 2003). The EU Environment and Climate Research and Development programme established the Concerted Action on Risk Assessment for Contaminated sites in 1996. A summery of European Commission (EC) guidance and practice was published in 1998 (EC, 1998).

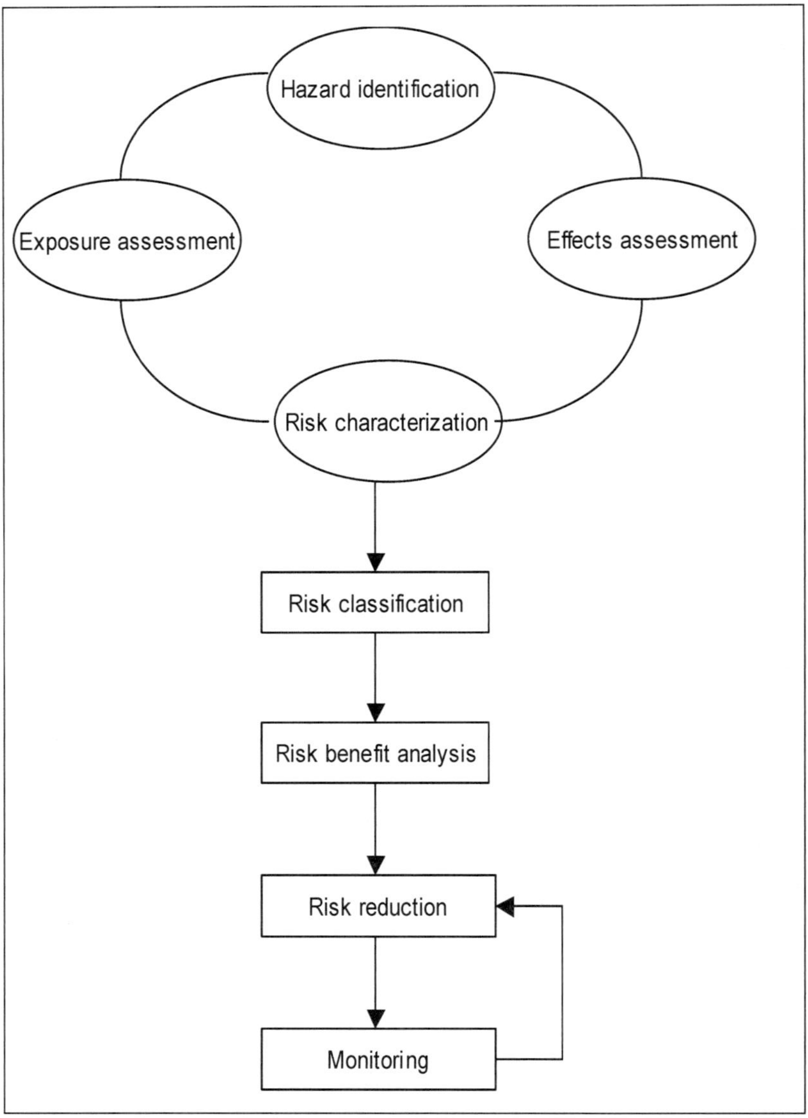

FIGURE 2. Steps in the risk management process.

The guidance on and requirements for environmental risk assessment for US Naval Operations has explicitly included the marine environment. In 2003, the EU published the updated Technical Guidance Document (EU-TGD) on risk assessment containing technical guidance specified to marine risk assessments (EC, 2003). OSPAR agreed to adopt the guidelines for risk assessment as described in the EU-TGD as the common EU/OSPAR approach on risk assessment methodology for the marine environment (OSPAR agreement 20-2003).

From the above mentioned definitions of risk, the following characteristics of the assessment endpoint of ERA can be derived. The assessment endpoint of risk assessment should include a quantification of the likelihood and severity of biological effects. Comparison of the results of the assessment with criteria set on the likelihood and severity of effects should take place to check whether the activity can be carried out within acceptable limits.

PEC:PNEC Ratio

If environmental risk assessment is applied to chemical exposure, the assessment will be based on the comparison of the exposure of (a part of) the ecosystem to a chemical with the sensitivity of (the same part of) the ecosystem for this chemical (through this specific exposure-route) (Suter, 1993) (Fig. 3). The exposure is represented by the Predicted Environmental Concentration (PEC), and can be obtained by actual field

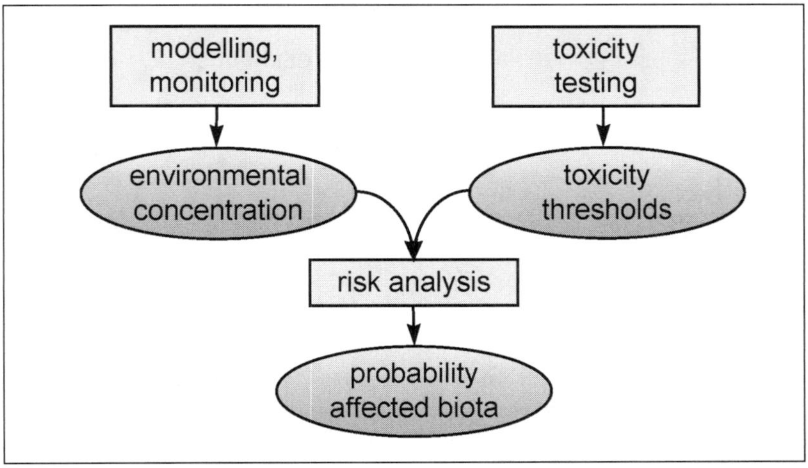

FIGURE 3. The general framework for Environmental Risk Assessment based on the comparison of an environmental concentration with the sensitivity of the environment.

measurements (monitoring data) or by estimations using environmental fate models. The toxicity threshold represents the sensitivity of the ecosystem, and is usually derived from standardised toxicity tests. These experimental test results are the test endpoints, which are most commonly used to derive the assessment endpoint in ERA. The first challenge is to have an adequate estimation of both the environmental concentration (PEC) and the toxicity threshold value. As illustrated in Figure 4, a basic set of data consists of a series of values being either exposure concentrations or measurements of toxicity, which can be summarised in a frequency diagram. To perform an assessment, choices have to be made on which test end-

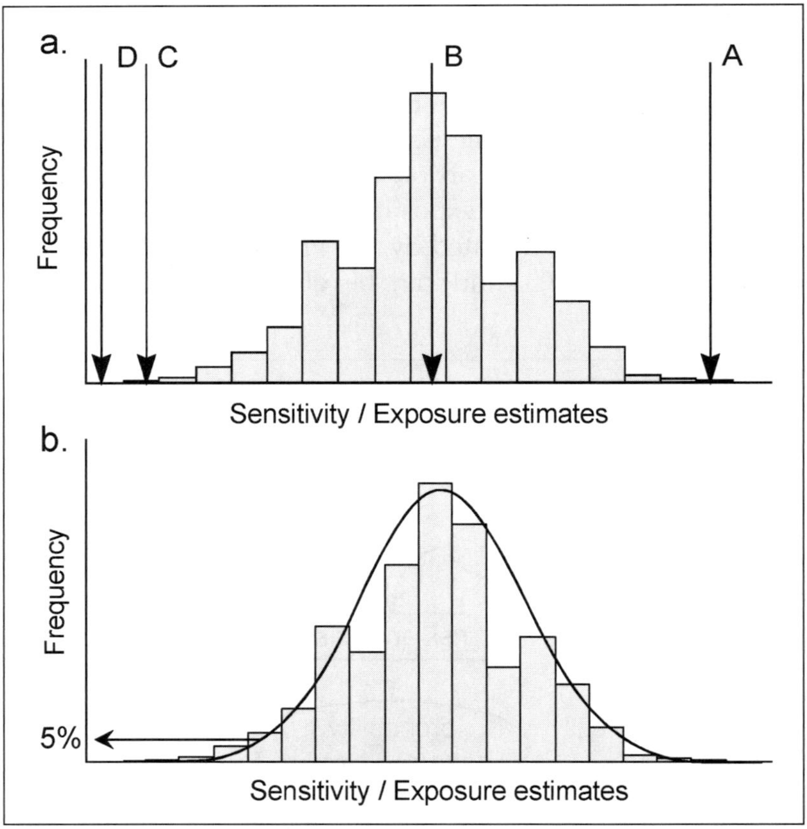

FIGURE 4. Derivation of PEC and PNEC values from available data. PEC and PNEC based on (a) point estimates and (b) probabilistic distribution of data points.

points to use and which procedure to follow to derive the value of the assessment endpoint.

The EU-TGD prescribes the use of a PEC:PNEC approach as a general tool for environmental risk assessment (EC, 2003). According to the EU-TGD, the definition of the Predicted No Effect Concentration (PNEC) is the concentration below which unacceptable effects on organisms will most likely not occur. The PEC:PNEC approach uses point estimates to represent both the exposure (PEC) and the sensitivity (PNEC). The PEC can be selected from all observed exposure concentrations in different ways. On the basis of precaution, the highest value from the dataset (or the 95th percentile) can be used to represent the environmental concentrations (Fig. 4a; option 'A'). To represent the most likely environmental concentration, it is also possible to use the median observed concentration to represent the PEC (Fig. 4a; option 'B'). To illustrate the methodology to derive a PNEC value, Figure 4a presents an overview of values for threshold concentrations (NOECs; No observed effect concentrations) for different species exposed to a specific toxicant. In the light of the precautionary principle, the use of the lowest figure from the set of available data would be preferred (Fig. 4a; option 'C'). Often, however, the lowest available toxicity threshold concentration is not regarded as being protective for the complete ecosystem. On the basis of uncertainties, such as variation among species or between the laboratory and field, one could therefore argue to apply an extrapolation factor to the lowest available toxicity value. This leads to a PNEC which is lower than the lowest available toxicity value. Sensitive species in the field and probably not present in the available dataset are also predicted by this value (Fig. 4a; option 'D').

With the PEC:PNEC ratio as an assessment endpoint of ERA, it is obvious that the definition of risk is related to the definition of the PNEC. A PEC:PNEC ratio higher than 1 indicates that unacceptable effects on organisms are not unlikely to occur; the higher the ratio, the more likely that unacceptable effects may occur. When we relate this endpoint to the definition of assessment endpoint in risk assessment as mentioned earlier ("quantification of likelihood and severity of effects"), it becomes clear that the PEC:PNEC ratio does not comply with this definition. The PEC:PNEC ratio is just an indication of the likelihood and not a quantification (see also Scholten et al., 2000). This is acceptable for identification of possible impacts and for prioritisation. However, it does not provide any characterisation of the expected impact and therefore, does not contain all the characteristics to be a proper assessment endpoint of environmental risk.

Typical risk assessment models, based on the PEC:PNEC approach, which are applied for the use of chemicals by the oil and gas industry, are EUSES (European Union System for the Evaluation of Substances) and CHARM (Chemical Hazard Assessment and Risk Management). Both models were specially developed for use in protective hazard assessments and prioritizing of chemicals, and are adopted by regulators in the EU. EUSES was developed in the EC for risk assessment of new and existing substances and pesticides (Vermeire et al., 1997) and was based on the Dutch USES model (Jager and Visser, 1994). Within OSPAR, the hazard assessment module of CHARM is the mandatory tool for selecting the most environmentally-friendly production, drilling, or work-over chemical (Thatcher et al., 1999).

Apart from EUSES and CHARM, several other models are being applied by different countries to manage the risks of discharges from the oil and gas industry. The Hydrocarbon Block Method (HBM), which has been forwarded as a method for environmental risk assessment for petroleum discharges (EC, 2003) and the Pollution Risk Offshore Technical EvalUation System (PROTEUS), which predicts the physical dispersion, chemical interactions, and environmental risk for discharges during exploration, production, and decommissioning, are also based on the comparison of PEC and PNEC.

Probabilistic Estimates

A recent development in ERA is the use of probability distributions instead of point estimates (e.g. Jager et al., 2001; Posthuma et al., 2002). To describe the PEC and/or the environmental threshold, a distribution (Fig. 4b) (Schobben and Scholten, 1993) is used that represents all the values from the frequency diagram in Figure 4a. In this case, all available information on the test endpoints is included in the assessment procedure. A frequency distribution of toxicity threshold values for different species to one specific component is often referred to as a species sensitivity distribution (SSD) (Posthuma et al., 2002).

Four different types of comparison of exposure and sensitivity are depicted in Figure 5. The traditional PEC:PNEC approach is presented in Figure 5a. The ratio of PEC and PNEC indicates whether unacceptable effects on organisms are likely to occur as a result of exposure to the specific chemical. It does, however, not provide a quantification of the environmental risk (severity and likelihood of effects). When a single value for the PNEC is compared to a distribution of PEC values (Fig. 5b), the term 'most likely' can be represented by the chance that the exposure concentration is higher than the PNEC. In case the SSD based on NOEC values is used to represent the sensitivity of the environment (Figs. 5c, d), the assessment endpoint risk will indicate change that a specific fraction of species is

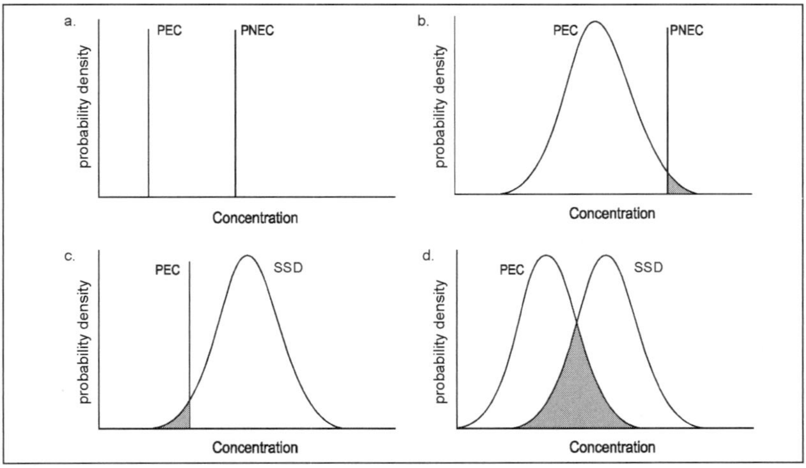

FIGURE 5. Four possible approaches for environmental risk assessment based on **(a)** point estimates, **(d)** probabilistic distributions, or **(b and c)** a mixture of both (SSD=Species Sensitivity Distribution).

exposed above their NOEC value. This assessment endpoint is in compliance with the definition of risk (severity and likelihood). Table 1 presents the different definitions of the assessment endpoint specified to the data used and methodology applied as presented in Figure 5.

TABLE 1. Definition of risk related to the specific ERA procedure referring to Figure 5.

CASE	RISK DEFINITION
A	Indication whether unacceptable effects on organisms are likely to occur
B	Probability that an ecosystem is exposed to a level where unacceptable effects on organisms will most likely occur
C	Probability that a (random selected) species is exposed to a level above its chronic threshold concentration
D	Probability that a species is likely to be exposed to a level above its chronic threshold concentration

Although the assessment endpoints provide a characterisation of the likelihood and severity of a defined effect, up to now probabilistic methods for risk assessment have not been adopted by the EU or OSPAR. It is to be expected, however, that these methods will become gradually incorporated in ERA procedures (Posthuma et al., 2002). For example an additional risk assessment module in CHARM already uses species sensitivity distributions and the principles as presented in Figure 4c. (Karman et al., 1996). Comparable to risk assessment in CHARM is the calculation of risks in the methodology for determination of the Environmental Impact Factor (EIF). This EIF for produced water is also mainly based on the main principles of risk and hazard assessment (PEC:PNEC) as described in the EU-TGD (Johnsen et al., 2000), but an additional calculation using SSDs is applied to translate the PEC:PNEC ratio into a quantitative risk estimate (Karman and Reerink, 1998). The EIF is calculated by DREAM (Dose-related Risk and Effect Assessment Model), a generic risk assessment discharge management tool prescribed by the Norwegian Pollution Authority (SFT) to be used by all oil companies producing off the coast of Norway (Smit et al., 2003).

Future Developments in RA

With use of SSDs in ERA, an assessment endpoint in line with the definition of risk is calculated based on several test endpoints (multiple toxicity values). However, the use of probabilistic methodologies still leaves subjects open for interpretation. The challenge that remains for ecologist and ecotoxicologists is the definition of (un)acceptable effects in relation to the most sensitive endpoints on the species level related to adverse effects to communities. ERA using SSDs completely focuses on the preservation of ecological structure. Important ecosystem dynamics like food web relations are not included. However, it is assumed that ecological functioning is also protected when the ecological structure is protected. Schindler (1987) and others concluded that functional measures were relatively insensitive and that changes in the composition of the community were better indicators of ecosystem stress. In stressed ecosystems, shifts in community composition often preceded changes in ecosystem function. However, predicting ecosystem function has a higher value because it integrates responses of component populations (Baird et al., 2001). Developments in risk assessment models should therefore focus on the translation from single species level to ecosystem level.

The concepts of such a risk assessment model are, next to the PEC:PNEC based EIF-module, incorporated in the DREAM model (Smit et al., 2003). For two representative species (a fish and a zooplankton species), body burdens for PAHs and alkylated phenols are calculated from exposure patterns (Karman, 2000). Effects that might occur from this increase in body burden are calculated as well. This part calculates effects on representative species directly from modelled exposure. Until now, the methodology to translate this effect on single species to effects on the community is still lacking in the model (Smit et al., 2003).

Tiered ERA from Screening to Environmental Impact Assessment (EIA)

ERA tools can range from very generic, to very specific. Often a stepwise approach is followed. Generic assessment usually involves a high level of uncertainty and a reduced amount of specific information. For specific ERA the opposite is valid. The first can be well applied for prioritising and comparing, the second is better suited for EIA. However, if this generic assessment approach indicates no unacceptable risks as a result of the activity, it can be concluded that it is safe to carry out the activity. When an unacceptable risk is indicated, a more specific risk assessment should be performed focusing on the stressor of concern. However, uncertainty will always be an important part of ERA. Sometimes the uncertainty associated with the risk value is too high to pass defined criteria for acceptance. In some of these cases, the use of the precautionary principle is preferred (Santillo et al., 1998). This principle promotes emission reduction in most cases.

From the above it becomes clear that different levels of risk assessment exist and that for these different levels, different suitable assessment endpoints can be defined. It has already been demonstrated that the PEC:PNEC ratio can be well used for screening and prioritising (as applied in CHARM and EUSES), and is very well suited for this because of the minimum amount of data needed. However, this ratio can not be used for impact assessment, because it does not provide a quantification of the likelihood or a characterisation of the severity of biological effects. Therefore, development of ERA models is needed, when assessment of environmental impacts is the final objective. A tiered approach can be followed to make the assessment more quantitative. The following steps can be identified in such an approach:

- Probabilistic risk assessment; this approach uses the same test endpoints as used in the PEC:PNEC approach but results in a more quantified risk estimate (probability that a species is exposed above its chronic NOEC).
- Mechanistic risk assessment; this approach uses new information on the test endpoints from the same toxicological experiments as used in the PEC:PNEC approach, but actual biological mechanisms are implemented.

- Ecological modelling in risk assessment; this approach uses the state of the art knowledge on biology, ecology, and toxicology and integrates principles to estimate effects on populations, food chains, and ecosystems.

It will be clear that the higher the tier of the risk assessment, the more specific the information needed on species, systems, and toxic properties. Hence, there must be a justification for such a tiered approach that can be found in the area or of the toxicant of concern. The higher the tier in risk assessment, the more comparable the results are to the results of monitoring; quantifying severity and extend of biological effects.

ENDPOINTS OF ENVIRONMENTAL EFFECT MONITORING (EEM)

Characteristics of the assessment endpoint of EEM

From the beginning of the oil and gas industry, concern grew for its impact on the environment. One of the first problems identified was the harmful effect of releases of contaminants in the marine environment on fisheries interests. The growing concern about the potential environmental impacts led to the development of Environmental Effect Monitoring (EEM) programs for all major components of development projects. For the offshore oil and gas industry, concerns were focused on the effects of produced water, drilling wastes and other discharges and, oil spills. The environmental impact of the offshore oil and gas industry has recently been summarised by Patin (1999). So far, effect monitoring has been the domain of marine ecologists, and has mostly been carried out locally and in relation to individual activities. For the most part, monitoring of effects in large areas or addressing cumulative impacts has not been done. For the latter, however, Norway is an exception. After revision of the Norwegian guidelines for offshore environmental monitoring in 1996, Norway adapted a regional approach in which both sediments and the water column were considered. As environmental awareness grew, new legislation was developed and regulations were quickly assessed.

Environmental monitoring aims to detect change in living systems over time. It is useful to describe, evaluate, and communicate the consequences of man's activity on nature by identifying whether effects occur at a specific distance from an area in which a certain activity takes place. In addition to the identification of spatial distribution, extent of substances released and biological effects, monitoring provides a means to assess the effectiveness of preventative measures and regulatory discharge limits. As a further protective measure, monitoring will also enable us to detect unforeseen impacts and new areas of concern. In summary, the assessment endpoint of environmental effect monitoring considers both the identification of spatial distributions (extent) and severity of (all) biological effects (OSPAR, 2004).

From Test Endpoint to Assessment Endpoint in EEM

At first, when the main interest of marine contamination was to ensure that food from the sea was safe for human consumption, it was possible to monitor these species for consumption without much knowledge of the ecosystem. However, when the goals of protection shifted toward ecosystem protection, resource managers were faced with either monitoring all species, which is obviously impossible, or selecting certain indicator species. Due to the complexity of marine ecosystems and gaps in knowledge, this is a complicated task (MacGarvin, 1994). Together with many others, Santillo et al. (1998) discussed the precautionary principle as a methodology, its use in the protection of ecosystems and the need to overcome this complexity.

Presently within EEM, the common approach is to measure the responses of the environmental system to an activity, or in case of pollutant discharge, to measure directly in the field, or use tests such as bioassays. Different sets of test endpoints are used in EEM using physical, chemical, and biological information (see Patin, 1999 for an extensive overview). Biological test endpoints widely used are abundance, biomass, diversity, and community structure, for which many indexes are developed. Physical and chemical test endpoints include sediment composition, environmental and body concentrations of metals and non metal (organic) substances and accumulation. These physical and chemical parameters are important as there is often strong correlation between the physical and chemical environment and the presence of biota.

Quantifying concentrations is essential to relate determined effects. However, environmental concentrations are not always good indicators for body concentrations or potential bioaccumulation. Environmental concentration of a substance is, therefore, often a poor indicator for potential toxic effects. This is even truer in knowing that in the field one has to deal with mixtures of toxicants of which the bioavailability may vary considerable. So preferably both are measured simultaneously. This makes the fate of a pollutant and the impact on the ecosystem very difficult to assess and subsequently the biological response difficult to measure in the field.

Until now, offshore oil and gas EEM has mainly focussed on drilling discharges, and therefore has concentrated on the benthic compartment and its related fauna. EEM aiming to identify effects of produced water releases is a relative new area of study. It is obvious that in offshore marine waters, it is not easy to measure effects on pelagic biota due to large dilution and dispersal of discharges. In addition, the tools and methods needed for such monitoring (e.g. water column impacts) have not been fully developed or validated. However, by setting limits to the concentrations in the discharges themselves, the risk of an environmental impact is reduced. Selection of components for which such limits are needed is based on hazard assessment. The value of such limits (e.g. for metals and hydrocarbons) is based on concentrations from field and laboratory experiments at which no toxic effect to fauna was observed or expected.

Within the EU and OSPAR region, a common approach is now applied for monitoring the environmental effect in the marine environment. The approach follows the OSPAR guidelines for monitoring the environmental impact of offshore oil and gas activities. These guidelines are developed to provide guidance on the design and conduct of monitoring programmes under the range of differences that exist between different producing areas. Consistency in the monitoring guidelines, with respect to both the selection of test endpoints and approach to sampling and analysis and to the application of quality assurance protocols, is emphasised. These guidelines were adapted by ASMO (OSPAR Assessment and Monitoring Commission) in 2001. The focus of these guidelines still lies on the monitoring of spatial and temporal changes in sea-bed contamination and benthic fauna community structure. A proposal to extend the guidelines to water column monitoring was recently discussed (OSPAR, 2004). The most important proposed and presently used guidelines are included in its Technical Annex 1. These drafted guidelines show that biomarkers (in caged and exposed fish and mussels) are proposed as test endpoints of water column EEM (ICES, 2001). These biomarkers can be indicators of chronic effects. However, the ecological relevance of biomarker responses remains to be fully defined. Defining the correlation between biomarker response and assessment endpoint monitoring (severity of biological effects) has been hampered by the absence of threshold levels and criteria to quantify ecological relevance of the impacts.

Future Developments in EEM

As discussed earlier in this paper the best test endpoint to use is the test endpoint that supports the goals of the assessment. In general, this will be the test that, in combination with an extrapolation model, generates the best estimate of the assessment endpoint (Cairns and Niederlehner, 1995). Measuring the assessment endpoint directly in the field would be best. Extrapolation would no longer be necessary and the uncertainty associated with it would be removed. Shifts in community structure, as used to indicate impacts on the seabed biota, can be such a test endpoint. Schindler (1987) and others concluded that changes in the composition of the community are good indicators of ecosystem stress. Shifts in community composition often preceded changes in ecosystem function in stressed ecosystems. However, it has been stressed by others that ecosystem function has a higher value because it integrates responses of component populations (Baird et al., 2001) and that test endpoints of ecosystem functioning should be developed. If biomarkers are going to be the prescribed test endpoint of EEM in the water column by OSPAR, extrapolation models are needed to translate these biomarker responses to ecologically relevant responses. If environmental concentrations are used as test endpoint, dose-effect relationships should be applied to estimate the value of the assessment endpoint. Next to the development of the right test endpoints in EEM programs, attention should be paid to the development of extrapolation models to put the results of the used test endpoints in the light of the defined assessment endpoint of monitoring. Whereas guidance is available for risk assessment procedures, no clear guidance is available for the interpretation of effect monitoring information.

COUPLE ERA AND EEM AND REDUCE UNCERTAINTY

Definition of a Common Assessment Endpoint of ERA and EEM

In the previous sections, the assessment endpoints of both RA and EEM were evaluated. Although RA predicts in advance, and EEM determines afterwards, it was indicated that both tools have the same aim, which is protection of the environment, and that subsequently the assessment endpoint are highly comparable. Until now, however, little effort has been invested in the integration of ERA and EEM. Figure 6 presents

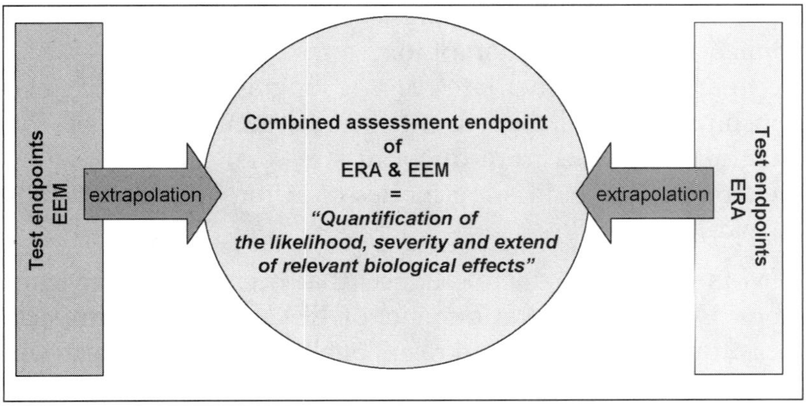

FIGURE 6. Definition of a combined assessment endpoint for EEM and ERA.

a basis for integration, which will facilitate cross-validation between the two tools.

When the combined assessment endpoint of ERA and EEM is defined as "a quantification of the likelihood, severity, and extend of relevant biological effects", both tools could be coupled if the test endpoints of both tools can be related to the same type of (relevant) biological effects. If, for example, biomarkers are used as a test endpoint in EEM, this test endpoint must have a relation to the test endpoints used in ERA and (after extrapolation) must comply with the assessment endpoint. These criteria must be kept in mind when developing test endpoints and extrapolation techniques for both ERA and EEM. Only then can ERA and EEM be used as complementary tools, which can also be used to support and validate each other and, with that, reduce their uncertainties.

Uncertainties in ERA and EEM

One of the main advantages of ERA is that it results in information on the possible effects of an activity before the activity has taken place. Critical processes in the activity plan can be indicated and evaluated, and if necessary, substituted by other processes with lower uncertainty or with less environmental harm. ERA can be a tool to implement cost effective mitigating measures beforehand and therefore it is incorporated in many risk managements frameworks (EC, 2003). However, ERA can only be successful when the selection of stressors to include is successful. ERA in the offshore oil and gas industry is mainly applied for the evaluation of toxic stress. However, it is possible to apply it to other stressors, as long as cause-effect relationships are available. Another uncertainty in RA is the integration of risks from different stressors, especially when the stressors have different modes-of-action (e.g. toxic versus physical effect).

EEM focuses on the actual effects that occur in the environment. One of the great benefits of EEM is that the impacts measured are an integrated result of all stressors present at the location. Therefore, the results of monitoring studies can be related to the integrated risk from the selected stressors in the risk assessment. Monitoring could reduce uncertainty in the ERA procedure associated with combining risks and selection of stressors. (e.g. Montagna and Harper (1996) indicated that effects of unexpected stressors like platform placement can overcome effects from chemical disturbance).

Monitoring can be carried out at different levels of the cause-effect chain. We distinguish monitoring of stressors (e.g. concentrations of chemicals) and monitoring of effects (e.g. biological community disturbances). The first provides information on the area of influence by the activity and, if the dose-effect relationships of this stressor are available, a possible environmental impact can be predicted. It does, however, not provide information on the actual effects occurring in the environment. Monitoring of stressors can be of value when emission reduction is one of the main goals.

The uncertainties in monitoring are associated with the appropriateness of test endpoints for effects, and the derivation of the assessment endpoint from them. Linking the changes in the test endpoints of monitoring to environmental impacts of human activities will fail if the natural variability of the parameters is not well understood and accounted for. Monitoring must be able to distinguish between activity induced and natural disturbances. Monitoring needs to take account of local and regional differences, operational practices, past and current field activities and discharges, and what is known about impacts from previous work. In effect monitoring it is therefore essential to have knowledge of background values, at the best of the undisturbed biological condition, as well as insight in the natural fluctuations and variability. Measuring effects on free swimming individuals happens to be very difficult and even more difficult to relate to the activity. Also, the effects measured in caged animals are only an indication of field effects, since the exposure pattern of these caged animals is far from realistic, and the natural behaviour of these animals is disturbed. Despite being one of the best-studied marine areas in the world, knowledge of the North Sea's variability is often insufficient to answer questions of whether observed changes are caused by human intervention or the result of natural variation. The extensive oil and gas production already has a long history in the North Sea, and together with the existing and growing intensity of other uses, such as fisheries, has resulted in the almost complete loss of non-impacted reference areas. The high variability in geomorphological, sediment, and biological characteristics such as species abundance and benthic community structure has hampered the development of general assumptions about effects directly related to an activity or discharge. For these reasons, and the need for various national authorities with different requirements to agree, no international standards for sediment quality have been established to date. Possible changes in test endpoints can be related to specified assessment endpoints in risk assessment, in order to reduce uncertainties. Therefore, endpoints in ERA and EEM for specific stressors should be developed in a way that they can be related.

Effective monitoring depends on the possibility of designing a program that is addressed to the scale of the effects in terms of spatial and temporal changes. Short-term and localized effects of disturbances, either by disposal of cuttings or releases of contaminants are relatively easy to detect. Bioaccumulation often resulting in long-term and chronic effects possibly showing up at large distances from the source of pollution, are more difficult to detect using mostly short term and local monitoring. As a result of long-term national and international monitoring programs around the North Sea, some examples of longer-term effects of human impact became evident and management measures were implemented. For example, the effects of TBT used in antifouling paint accumulating in gastropods (Hallers-Tjabbes and Boon, 1995) were the basis of an international ban on the use of these coatings. Also, as a result of monitoring drill cuttings and their effects, the disposal of oil-based muds was banned in the North Sea in the nineties. Risk assessment can be used to estimate long-term and chronic effects. However, development of risk assessment techniques is also necessary to reduce uncertainties, focus on integration of sources, and predict bioaccumulation and related effects.

From the above, it can be concluded that the uncertainties in ERA are more or less compensated by the strengths of EEM and vice versa. This is the most important motivation for integration of the two tools and defining a common assessment endpoint. The challenge that remains is the development of proper test endpoints.

The scheme presented in Figure 7 provides an overview of how risk assessment and monitoring can support each other, develop impact assessment to be more integrative, and reduce uncertainties. This will only be the case when the common assessment endpoint is formulated and proper test endpoints are defined.

Figure 7 visualises that prior to the activity, risk assessment can be used to estimate the possible effects that might occur from the planned activity. For this a sound evaluation of a process will be necessary indicating the most harmful stressors of the activity (hazard identification and assessment). A first level ERA can be carried out taking into account the protection of the most sensitive species in the ecosystem (e.g. evaluation using PEC:PNEC ratio). From this generic conservative approach, selection of stressors of no and possible concern at this specific location can be made. It must be kept in mind that the predictions using the PEC:PNEC approach are based on generic data and strongly conservative assumptions. They include much uncertainty and in no way are representative of what will be expected in the environment. Besides, if the toxicological endpoints for a conservative ERA are based on behaviour, there is no likelihood that EEM will show anything.

When this first step of conservative ERA results in a risk below acceptable levels, the activity can start and EEM should be carried out during or after the activity. When this EEM indicates effects, some unexpected stressors have been present or unexpected interactions occurred which were not included in the ERA procedure. This could result in an update of the ERA procedures or in the identification of other stressors (natural or anthropogenic). Another possibility is that the activity was not carried out in the right (proposed) way (e.g. higher discharged

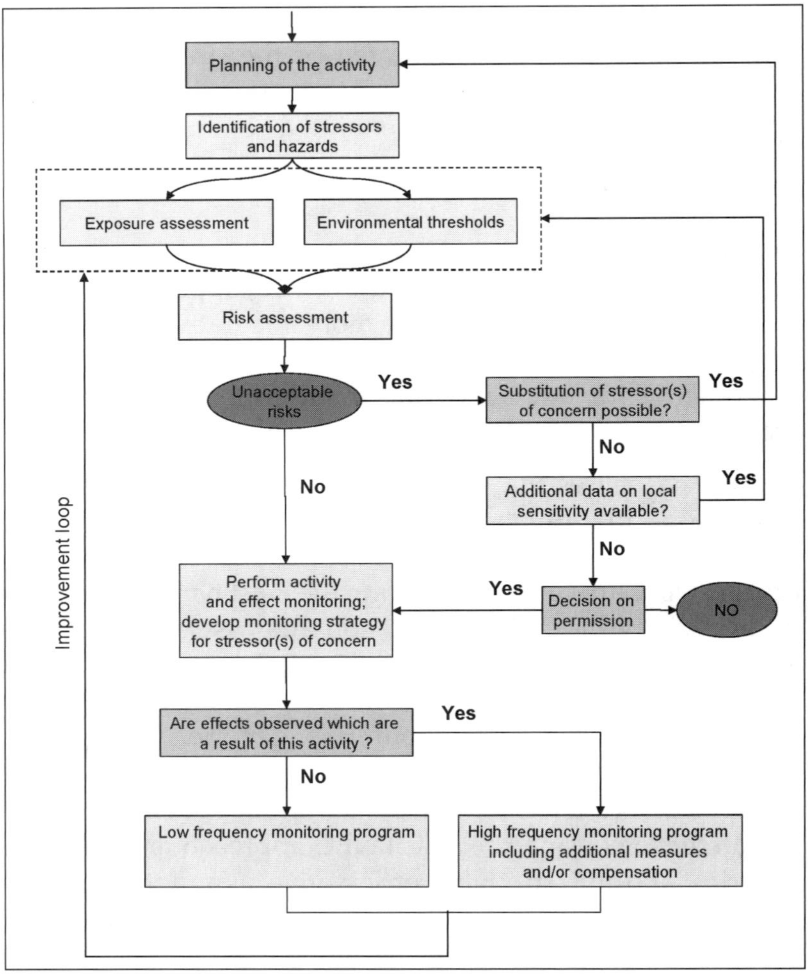

FIGURE 7. Overview of a flow scheme for coupling ERA and EEM.

volume). When the EEM indicates no effects, future monitoring can be carried out at low frequency.

When assessment indicates a risk beyond the accepted levels, stressor(s) causing this risk should be replaced by a less harmful alternative, or more specific information should be collected on the local environment. Baseline monitoring data can be used to gain this information. ERA can be updated by including this information. When updated results indicate no unacceptable risks, EEM can be carried out as mentioned above. When the risk is still too high, the stressor should be replaced. When this cannot be done, the regulating body could decide whether a permit can be granted or not. In this permit, additional mitigating measures or compensation activities can be defined based on expected impacts that are assessed in an extended risk and impact assessment program. EEM could be carried out more frequently and expected impacts should be compared to observed impacts in EEM.

Apart from the above scheme the following remarks should be taken into account:

- When biological effects occur but cannot be measured due to the absence of the right measurement tools, risk assessment protocols can be used to indicate the level of risk for effects.
- For an effective impact assessment, careful consideration must be given to the selection of assessment endpoints of risk assessment and environmental effect monitoring to facilitate comparisons.

Studies are now focused on the development of EEM techniques, including the search for the right test endpoints and tools for extrapolation of test endpoints to an assessment endpoint is one of the main research topics at this moment and; the definition of a framework to couple the assessment and test endpoints of RA and EEM. In Norway several research projects are initiated to deal with this challenge. This will make it possible to optimize risk assessment methodologies and defend minimization of expensive monitoring efforts in the future. In short, a joint development of both risk assessment and monitoring is necessary to promote integrative science (Chapman, 1995).

CONCLUSIONS

As described in this paper, up to now environmental risk assessment and environmental effect monitoring protocols have been essentially developed independently. The limitation of communication between the groups has led to discussions on why one should be preferred over the other.

However this paper indicates that opportunities and possibilities exist to couple ERA and EEM in order to achieve an improved and integrated system for ecosystem management and EIA.

ERA and EEM for the offshore oil and gas industry are different approaches and methodologies with the same objective: giving insight in the level of disturbance of the environment. Both methodologies have their strengths and weaknesses and their own position in the framework of Environmental Impact Assessment. Limited interaction between experts of environmental risk assessment and monitoring have resulted in separate development of both methodologies. However, both ERA and EEM have the possibilities to strengthen each other and support the development of both methodologies when placed in a (regulatory) framework. This framework should be a stepwise approach of both risk assessment and monitoring based on recent developments in the field under initiatives such as OSPAR and others. This could form a basis for the evaluation of environmental impacts. To accomplish this synergism new environmental effect monitoring techniques should be implemented; endpoints of ERA and EEM should be coupled and communication between experts should be stimulated.

To become comparable with environmental monitoring risk assessment should be based on realistic fate and transport models, adequately evaluate combined stressors, account for realistic exposures and address measurable endpoints. In addition monitoring must focus on the right and ecologically relevant endpoints.

REFERENCES

Baird, J.D., T.C.M. Brock, P.C. de ruiter, A.B.A. Boxall, J.M. Culp, P. Eldridge, U. Homen, R.G. Jak, A. Kidd, and T. DeWitt. 2001. "The Foodweb Approach in the Environmental Management of Toxic Substances." In Baird J.D. and G.A. Burton (Eds.), *Ecological Variability: Separating Natural from Anthropogenic Causes of Ecosystem Impairment.*

Cairns, J. and B.R. Niederlehner (Eds.). 1995. *Ecological Toxicity Testing: Scale, Complexity, and Relevance.* Lewis Publishers, Boca Raton, FL.

Chapman, P.M. 1995. "Extrapolating Laboratory Toxicity Results to the Field." *Environmental Toxicology and Chemistry.* 14:927-930.

European Commision (EC). 1998. *Risk Assessment for Contaminated Sites in Europe.* LQM Press, Nottingham, UK.

European Commision (EC). 2003. Technical Guidance Document on Risk Assessment in support of Commission Directive 93/67/EEC on Risk Assessment for New Notified Substances Commission Regulation (EC) No 1488/94 on Risk Assessment for existing substances Directive 98/8/EC of the European Parliament and of the Council concerning the placing of biocidal products on the market. European Commission Joint Research Centre. EUR 20418 EN/2.

Hallers-Tjabbes. C.C. Ten, and J.P. Boon. 1995. "Whelks (*Buccinum undatum* L.), Dogwelks (*Nucella lapidus* L.) and TBT—a Cause for Confusion." *Marine Pollution Bulletin.* 30:674-676.

ICES. 2001. *Report of the Steering Group For a Sea-going Workshop On Pelagic Biological Effects Methods* (SGSEA) ICES CM 2001/E:01 (http://www.niva.no/pelagic/web, Nov 2002).

International Organization for Standardization (ISO). 2002. ISO 17776 *Petroleum and Natural Gas Industries—Offshore Production—Guidelines on Tools and Techniques for Hazard Identification and Risk Assessment.*

Jager, D.T. and C.J.M. Visser. (Eds.) 1994. *Uniform System for the Evaluation of Substances (USES).* RIVM VROM WVC.

Jager, T., T.G. Vermeire, M.G.J. Rikken, and P. van der Poel. 2001. "Opportunities for a Probabilistic Risk Assessment of Chemicals in the European Union." *Chemosphere.* 43(2):257-264.

Johnsen, S., T.K. Frost, M. Hjelsvold, and T.R. Utvik. 2000. *The Environmental Impact Factor—a Proposed Tool for Produced Water Impact Reduction, Management and Regulation.* SPE 61178. Society of Petroleum Engineers.

Karman, C.C., E.A. Vik, H.P.M. Schobben, G.D. Øfjord, and H.P. van Dokkum. 1996. *CHARM III.* Main report. R96/355.

Karman, C.C. and H.G. Reerink. 1998. "Dynamic Assessment of the Ecological Risk of the Discharge of Produced Water from Oil and Gas Producing Platforms." *Journal of Hazardous Materials.* 61:43-51.

Karman C.C. 2000. "The Role of Time in Environmental Risk assessment." *Spill Science and Technology Bulletin.* 6(2):159-164.

MacGarvin, M. 1994. "Precaution, Science and the Sin of Hubris" In T. O'Riordan and J. Cameron, (Eds.), *Interpreting the Precautionary Principle*, pp. 69-101. Earthscan.

Montagna, P.A. and D.E. Harper. 1996. "Benthic infaunal long-term response to offshore production platforms in the Gulf of Mexico." *Canadian Journal of Fisheries and Aquatic Sciences.* 53:2567-2588.

OSPAR. 2003. *Agreements of OSPAR with respect to risk assessment for the marine environment agreement.* OSPAR agreement 20-2003.

OSPAR. 2004. *Draft Revised OSPAR Guidelines for Monitoring the Environmental Impact of Offshore Oil and Gas Activities* (OIC 04/7/1-E).

Patin, S. 1999. *Environmental Impact of the Offshore Oil and Gas Industry.* EcoMonitor Publishng, East Northport, New York.

Posthuma, L., G.W. Suter II, and T.P. Traas (Eds). 2002. *Species Sensitivity Distributions in Ecotoxicology.* Lewis publishers. Boca Raton, FL.

Santillo, D., R.L. Stringer, P.A. Johnston, and J. Tickner. 1998. "The Precautionary Principle: Protecting Against Failures of Scientific Method and Risk Assessment." *Marine Pollution Bulletin.* 36(12):939-950.

Schobben, H.P.M. and M.C.Th. Scholten. 1993. "Probabilistic methods for marine ecological risk assessment." *ICES Journal of Marine Science.* 50:349-358.

Scholten M.C.Th., C.C. Karman, and S. Huwer. 2000. "Ecotoxicological Risk Assessment Related to Chemicals and Pollutants in Offshore Oil Production." *Toxicology Letters.* 112/113:283-288.

Schindler D.W. 1987. "Detecting Ecosystem Responses in Anthropogenic Stress." *Canadian Journal of Fisheries and Aquatic Sciences.* 44(suppl 1):6-25.

Smit, M.G.D., R.G. Jak, K.I.E. Holthaus, and C.C. Karman. 2003. *An Outline of the DREAM Project and Development of the Environmental Impact Factor for Produced Water Discharges.* TNO report R2003/376.

Suter, G.W. 1993. *Ecological Risk Assessment*, Lewis Publishers, Inc., Boca Raton, FL.

Thatcher, M., M. Robson, L.R. Henriquez, and C.C. Karman. 1999. *CHARM: Chemical Hazard Assessment and Risk Management for the Use and Discharge of Chemicals Used Offshore.* User guide V1.0.

Vermeire, T.G., D.T. Jager, B. Bussian, J. Devillers, K. den Haan, B. Hansen, I. Lundberg, H. Niessen, S. Robertson, H. Tyle, and P.T.J. van der Zandt. 1997. "European Union System for the Evaluation of Substances (EUSES). Principles and Structure." *Chemosphere.* 34(8):1823-1836.

USEPA. 1993. *Risk assessment in the Federal Government: Managing the Process*, National Academy Press, Washington DC.

USEPA. 1998. *Guidelines for Ecological Risk Assessment*, Federal Register, Vol. 63, No. 93, May 14, pp. 26846-26924.

A Community-based Approach to Environmental Effects Monitoring in the Beaufort Sea

Cott, K. (cottk@dfo-mpo.gc.ca) (Department of Fisheries and Oceans, Inuvik, NT, Canada), D. Cobb (Department of Fisheries and Oceans, Winnipeg, MB, Canada), and D. Chiperzak (Department of Fisheries and Oceans, Yellowknife, NT, Canada)

ABSTRACT: Renewed interest in oil and natural gas in the Beaufort Sea has led to concerns about environmental effects resulting from the oil and gas activities. While environmental effects monitoring programs are being conducted by industry to meet regulatory requirements, and by government agencies in relation to cumulative impact assessment, nobody feels the importance of understanding the environmental changes resulting from oil and gas activities more than the people that live in coastal communities. With the settlement of land claims in Canada's arctic, a co-management approach to resource allocation and use, research, and monitoring has been established. This new form of governance in the arctic is dependent on a high level of community involvement and consultation. The Tariuq (Ocean) community-based monitoring program was piloted in Tuktoyaktuk and Aklavik in 2001 as one of the initiatives under the Beaufort Sea Integrated Management Planning Initiative. It is consistent with the principles outlined in Canada's *Oceans Act* and *Canada's Oceans Strategy*. The program engages coastal communities in monitoring the health of the Beaufort Sea through their participation in the selection of indicators, conducting of monitoring activities, review of monitoring results, and finally the dissemination of the findings back to the community. Tariuq concurrently, builds capacity within the communities and presents an opportunity to engage and train youth in scientific sampling methods. Community-based monitoring provides the benefits of incorporating traditional knowledge and scientific methodology into a monitoring program. It can also provide a broader spatial and temporal coverage than industry and government monitoring programs.

INTRODUCTION

Any change to an aquatic ecosystem alters the composition of that system and may have potentially significant impact on those who use and depend on the resources provided by that system. Monitoring environmental effects such as those caused by changes in climate or by development, is essential in effectively managing in an ever-changing environment.

In the Inuvialuit Settlement Region (ISR) (Fig. 1) monitoring environmental changes has become a collaborative effort between the Inuvialuit, industry and government organisations. With the settlement of the Inuvialuit Land Claim in 1984, the Inuvialuit Final Agreement (IFA) established a co-management approach to resource allocation and use, research, and monitoring. This new form of governance in the ISR is dependent on a high level of community involvement and consultation. Canada's *Oceans Act* and Oceans Strategy provide integrated management as a tool to manage Canada's oceans resources. With an integrated management framework for guidance and the obligations under the IFA, the Department of Fisheries and Oceans (DFO) in partnership with the Inuvialuit and industry developed the Beaufort Sea Integrated Management Planning Initiative, and subsequently the Tariuq (Ocean) community-based monitoring program. Community-based monitoring is a way of utilising the expertise of scientists and community members for gaining essential

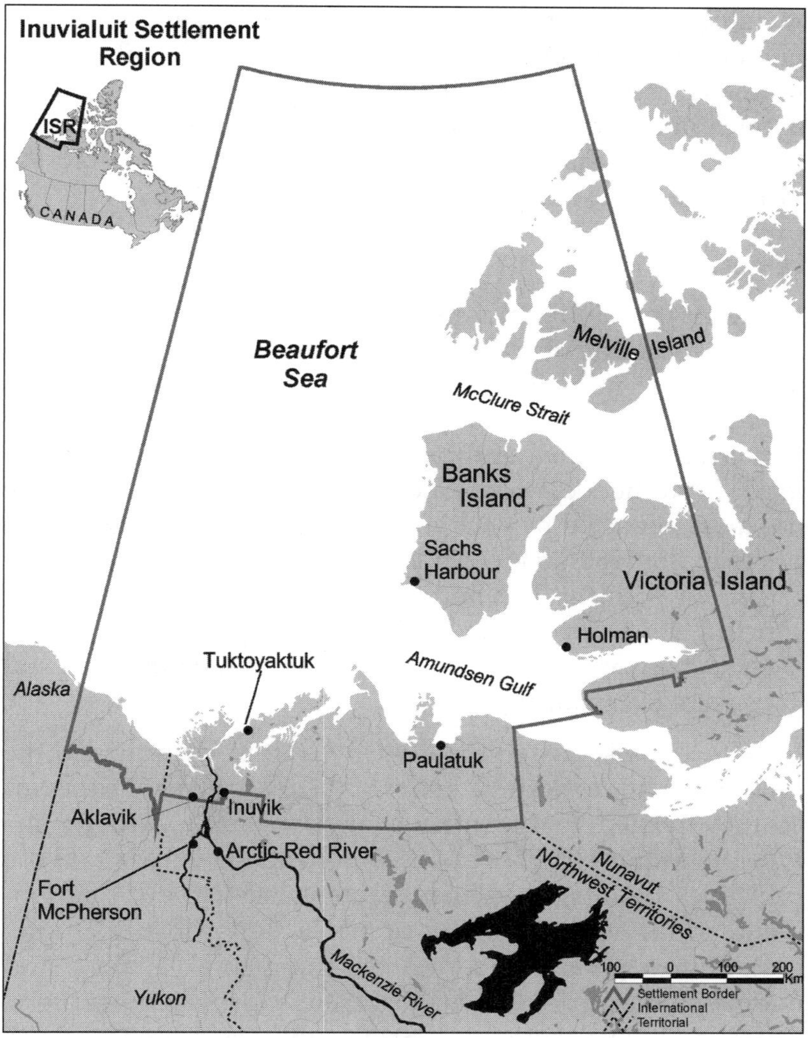

FIGURE 1. Map of the Inuvialuit Settlement Region.

knowledge for ongoing monitoring and management of environmental effects caused by changes due to development, climate, or any other means.

This paper discusses the principles behind developing a community-based monitoring program and how these principles were employed to develop the Tariuq community-based monitoring program in the ISR. How a community-based monitoring program like Tariuq can supplement other environmental effects monitoring programs, the benefits gained by such programs, and overcoming the challenges that arise in developing and maintaining a community-based monitoring program are also explored.

History

In the 1970s, the ISR saw its first oil and gas boom. With the sudden explosion of activity that accompanied the exploration for oil, came numerous scientists, committees, and research initiatives, which worked to fill knowledge gaps about the western arctic. Much of the research was conducted without input from the local population, that is, those who would be the most impacted by changes to the environment. Lower oil prices in the 1980s made extracting the resource less economical, along with the Pipeline Inquiry recommending that the construction of a pipeline be delayed (Berger, 1977), resulted in the activity quickly ending and many of the questions and gaps were left unanswered.

In 1997, Canada passed the *Oceans Act*; this Act provided the tools necessary to address many of the management concerns as they related to the marine environment raised by the Pipeline Inquiry (Berkes et al., 2001). The Act is based on the principles of integrated management, sustainable development, and the precautionary principle, and it identifies integrated management, marine environmental quality, and marine protected areas as management tools. The subsequent, *Canada's Oceans Strategy* (COS) has an overarching goal "to ensure healthy, safe and prosperous oceans for the benefit of current and future generations of Canadians." The COS identifies the need for integration of social, economic, and environmental aspects of decision making through integrated management planning.

Co-management governance and the passage of the *Oceans Act* have spurred new initiatives in the ISR. While regulatory requirements ensure that developers address potential environmental effects caused by development activities, the co-management governance regime as outlined in the IFA, and the management tools of COS, have made community-based monitoring initiatives a practical approach for tracking environmental effects in the Beaufort Sea.

DEVELOPING COMMUNITY-BASED MONITORING PROGRAMS

Establishing a community-based monitoring program requires planning, resources, and patience. The planners should realise that typical community members would not have formal training in scientific collection methods, or necessarily understand the importance of certain protocols. When developing a community-based monitoring program it is essential to start out slowly and with a small scale program; remembering the "keep it simple" mantra. Once a framework for a successful program is developed, it should be easy to expand the program. It is also important to remember that the program must meet the needs of the developer as well as the needs of the community, to ensure community acceptance. Figure 2 demonstrates the five steps followed to develop a successful community-based monitoring program.

Clearly communicating the government or developers objectives for the monitoring program to the members of the community is necessary to set the context for the program and to ensure that community members understand what the environmental implications may be and that the proper people are interested and engaged in the program. The second step, listening to the community's priorities and issues, is essential in selecting indicators that are going to be meaningful to the community as well as to the government or industry, thereby providing value to science, the community and the environment. Thirdly, properly training the monitor is very important in receiving useful data. As most community members will not likely have scientific knowledge or be familiar with sampling techniques and data management, it is essential to take the time to train each monitor. Also, expect that the data from the first few years may be variable and scientifically questionable until the methods are all worked out and standardised. The

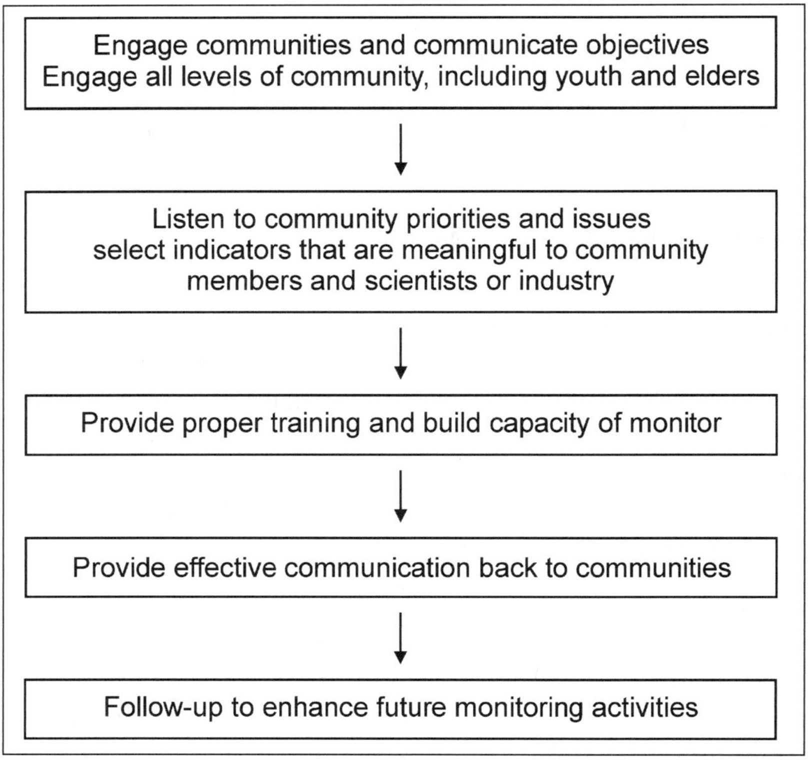

FIGURE 2. Flow Chart of successful community-based monitoring program (adapted from Papst and Cobb, pers comm.).

fourth and fifth steps, communicating results back to communities and following-up to enhance future monitoring activities, are especially important in maintaining a good relationship and improved communication with the community and between the community and scientists, government and industry.

COMMUNITY-BASED MONITORING IN THE ISR

With renewed interest in oil and natural gas in the Beaufort Sea, renewed concerns about environmental effects resulting from the oil and gas activities have arisen in the ISR. The resurgence of activity has brought with it environmental concerns within the communities that depend on the natural resources in the region for maintaining their culture and traditional way of life. Nobody feels the importance of understanding the environmental changes in the Beaufort Sea, which may result from oil and gas activities, more than the people that live in coastal communities in the ISR. This is reflected in the goal for the Tariuq program in Tuktoyaktuk:

> *"The land and water are the foundation of our culture. We can't vocalise it but we live it on a daily basis. If we destroy it, we destroy ourselves. The goal of the Marine Environmental Quality Program in Tuktoyaktuk is to maintain healthy populations of traditional food sources for the benefit of the community and the ISR at large; to gain understanding and share knowledge of the ecosystem so that we can make wise decisions"*

These communities have experienced scientists coming in and out for many years, gathering scientific information and traditional knowledge from the residents. Rarely have the results been brought back to the communities and presented in a meaningful manner. Therefore developing a monitoring program that was scientifically useful as well as meaningful to the communities became a priority for residents in the ISR, and for government and industry agencies that work in these areas.

Community-based monitoring programs have been developing in the ISR since the early 1990s. (Table 1). Community-based monitoring heightens community participation in management decision making and provides communities a sense of ownership over the monitoring program. Community-based monitoring programs involve community members being involved in all aspects of the program from setting the goals, objectives, and priorities, to selecting indicators, conducting sampling, and performing analysis when possible, to the dissemination of information back to the community. Community-based monitoring is not a mechanism to replace all other environmental effects monitoring; for instance, industry will have specific environmental effects monitoring programs related to specific development activities. However, community-based monitoring provides the opportunity to supplement other monitoring programs within the ISR.

TABLE 1. Monitoring Programs in the ISR

ISR Monitoring Programs	Focus	Tariuq (Ocean) Value-Added
Northwest Territories (formerly Mackenzie Valley) Cumulative Impact Monitoring Program (NWTCIMP) and Cumulative Effects Assessment Monitoring Framework (CEAMF)	Looking how the use of land and water and waste deposits affect the environment of the NWT. Monitoring VEC's: water; wildlife & birds; vegetation & fire; Fish; Air & climate; Human health & wellness; Marine life; and Snow/ ice/ permafrost. Identifying Valued Ecosystem Components. Coordinating research and monitoring programs that are already in place.	Have produced MVCIMP-Tariuq database, a searchable database of monitoring programs in Beaufort and Mackenzie Valley. Sharing of information collected increases geographic range of both programs.
EMAN - North	A network of agencies involved in all aspects of monitoring in the north.	Adds a marine component to the ecosystem monitoring.
Arctic Borderlands Ecological Knowledge Co-op	Monitor indicators (75 identified) through interviewing local experts. NWT, Alaska, and northern Yukon involved. Combine observational and traditional knowledge.	Provides long-term scientific data for consideration, as well as building capacity within the communities. Expanded questionnaire to include marine component.
Northern Contaminants Program (DIAND)	Monitoring the levels of contaminants (persistent organic pollutants, heavy metals, and radionuclides) in traditional foods. Ending March 2003.	Tariuq has provided some information from the sampling collected at Tuk solid waste facility.
FJMC Beluga Monitoring Program	Species specific monitoring.	Can enhance information collected by providing long-term and broader information database.
Storm Surge Monitoring Program (NRCAN, Parks, ANWR)	Monitoring of storm surges in Beaufort Sea. Data used in NRCAN coastal erosion model.	Tariuq has contributed to this effort by placing data loggers at the sampling sites each year.
DFO Fisheries Management Index gill netting	Site specific studies related to locally important species, and aimed at setting management plans.	Complimentary by expanding geographic scope.
Other DFO community-based monitoring programs (Char, Beluga, seal)	Species specific monitoring.	Can enhance information collected by providing long-term and broader information database.
Regulatory monitoring	Short-term project specific monitoring occurs on a project by project basis.	Tariuq can provide long-term monitoring that can be used as show changes from natural and anthropogenic stresses.
Nation Program of Action	DOE/DFO land based activities on marine environment.	One of the northern programs that fits under the NPA umbrella.

Tariuq (Ocean)

As part of the implementation of Oceans Act programs in the ISR, DFO initiated the Tariuq (Ocean) community-based monitoring program (The word tariuq means "salty water" or "ocean" in Inuvialuktun. Ocean is used in brackets to alleviate confusion, as tariuq is also the word used for table salt). This program is based on the principles of integrated management, in that it offers flexibility to accommodate all users and cover all concerns; inclusiveness of communities, government, and industry. The program is conducted in an atmosphere with shared responsibility, and is leading towards ecosystem based management. The Tariuq (Ocean) Monitoring Program will eventually contribute to a framework of a marine environmental quality program for the Beaufort Sea.

The Tariuq (Ocean) Monitoring Program of the ISR is being driven by community working groups, and has been active in Tuktoyaktuk and Aklavik since early 2000. The project was initiated in 2000 with community workshops to discuss marine environmental health held in Tuktoyaktuk and Aklavik (Chiperzak and Cockney, 2000a; 2000b). The intent of these workshops was to engage community members in each respective community, to identify goals and objectives of a community-based marine environmental quality monitoring program, and document community concerns.

Following the initial workshops in each community, a marine indicator workshop bringing together members from both communities, was held in Aklavik in 2001 (Eddy 2001). The objectives of the workshop were to move forward on recommendations and goals established by the communities of Tuktoyaktuk and Aklavik, to review and update community concerns, and to discuss and select some initial indicators. The participants included representatives from each community, including youth and elders, as well as government and monitoring experts. Following the workshops, working groups were established in each community. The community working groups include representatives from the hunter and trapper, elder and youth committees, DFO, and depending on the community, either the local community corporation, or Gwich'in Renewable Resource Council.

The Tariuq (Ocean) participants opted to initially monitor fish abundance and health, and water temperature. The monitoring will indicate changes over time and will generate information on contaminants and overall health of the fish. The information collected will provide a baseline on coastal marine and anadromous fish species, and contribute to a better understanding of the marine ecosystem in the Beaufort Sea. As the program develops and expands to include more geographic locations and/or more parameters, the value will be heightened. Currently, the Tariuq (Ocean) working groups, the Beaufort Sea Integrated Management Planning Initiative working group, and DFO are considering expanding the Tariuq (Ocean) Monitoring Program to cover the area in the Beaufort Sea currently being assessed as a potential Marine Protected Area. The information currently being collected supplements the information being collected in other monitoring programs in the region (Table 1).

Monitors were selected in each community by the community working groups and trained by DFO. The monitors selected were able to integrate local and traditional knowledge, and scientific methods into the monitoring program. Each monitor selected a youth from the community as a partner, thereby transferring the knowledge and giving young people in the community training and perhaps a career direction. Fish were sampled using standard fishing techniques. Index gillnets were used, allowing fish community information and catch per unit effort to be calculated. Aging structures were collected, and muscle and liver tissues collected and provided to DFO scientists to be analysed for contaminants (mercury and PAH) and vitamins for various species at each location.

After the first season of monitoring, working group meetings were held in each community to bring the information that was collected back to the participants and discuss what worked and what could be improved. Also, expanding the program to include another monitoring session was agreed upon. With the information collected in the three sampling periods in the first two years, a report was drafted and presented back to the working groups. The working groups elected to maintain the monitoring for another season, and then have a workshop to discuss further expansion of the monitoring efforts after three years of data has been collected and analysed.

Benefits

By integrating the benefits provided by local and centralised management techniques into co-management, significant value is added. Both local and scientific knowledge and expertise are integrated in a way that is beneficial for scientists, government, communities, industry, and ultimately the environment. Using a community-based monitoring program to track environmental changes builds community capacity, aids in resource management, and can help address cumulative effects more effectively.

By training monitors and youth in the communities in scientific data collection methods, the community is gaining value. Value is gained within the community via the information learned as a result of the monitoring. Value is also gained by increasing the community capacity to provide scientific information as well as traditional ecological knowledge to the monitoring program. By community members participating in the monitoring program, the information learned is passed on to the community in a way that is meaningful to members of the community. By participating in the development of the monitoring program, the monitoring results become more relevant to the community members, and community members gain an increased understanding of the causes and effects of environmental changes. This participation by community members may also facilitate support by the community at large for management measures which may be proposed or implemented based on the monitoring results. As the Tariuq program develops, a potential future benefit would be to involve the schools in the community in the data management and analysis phase of the project.

By providing the training in the communities, governments, scientists, and industries can also gain value through enhanced monitoring programs. The temporal and spatial coverage of any community-based monitoring program is enhanced by the monitors being close to, or at the monitoring locations on a year round basis. Local monitors can also provide some experience and local knowledge about the land and water that could otherwise take years to learn, thereby improving operational components such as site selection for sampling locations. Observation of trends can be realised through prolonged monitoring programs, however where the monitor has lived in the area for decades these trends can be identified much sooner. Significant events are also more likely to get recorded, such as a large storm or a strong northerly wind that may have prevented fish from using a specific area at a specific time. These observations can provide significant value to any monitoring program, and the results of long-term monitoring can be incorporated into new projects in the planning phases, thereby reducing preparation costs, and providing a more complete picture of the environment that the project will be within.

A community-based monitoring program also provides value to government and industry beyond the actual information collected. Communication channels are opened between the communities and government and industry representatives. In the ISR, where traditionally scientists and industries have come and gone, leaving with scientific and traditional information, and providing little long term benefit back to the area, these lines of communication are important in developing a trusting relationship. An enhanced collaborative spirit is gained by working together to identify areas of concern and working collaboratively to fill knowledge and information gaps. Through their organisations, communities in the ISR play a major role in the decision making regarding development and renewable resource management. Involvement by communities in monitoring will improve the understanding of potential impacts and remediation that will contribute to informed management decisions for both current and future projects.

Addressing cumulative environmental effects is critical to determining what the ultimate long-term changes in the environment will be from multiple developments. Some government departments have recognised that a segregated approach to environmental effects monitoring does not give a realistic assessment of cumulative effects of the combined activities of several companies working in the same region. Monitoring programs are often short-lived and very isolated geographically, as well as project specific. While project-specific monitoring is designed to meet regulatory requirements, or fill data gaps and determine prolonged project related effects, some environmental changes can be left undetected through this type of monitoring.

With a community-based monitoring program that consistently collects baseline information over a long period of time, short-term project specific monitoring programs can have a base of information for comparison. With multiple activities in an area, monitoring projects can easily start to overlap and become redundant. Table 1 shows how the Tariuq (Ocean) Monitoring Program coordinates with other monitoring programs in the ISR and how an integrated approach to coordinating monitoring programs can eliminate redundancy.

Challenges

While community-based monitoring programs can provide increased value to environmental effects monitoring through capacity building, effective resource management, and addressing cumulative effects, there are also some challenges to developing and sustaining a community-based monitoring program. Some of the challenges faced by community-based monitoring programs are developing a program that meets the needs of individual communities as well as government and industry, addressing the limited capacity both within communities and governments, and gaining acceptance of the data collected by scientific and management bodies.

Meeting the needs of multiple stakeholders is challenging in any forum. Developing a monitoring program that covers the interests of individual communities as well as the developers and/or government agencies can provide additional challenges to planners. This is especially evident if the program is intended to cover a large area and multiple interests. To obtain the benefits of a community-based program, it is essential to recognise community needs in the program, therefore to create a complete monitoring program, it is often necessary to link community-based monitoring programs with other programs and types of sampling.

A limited capacity within communities and governments contributes to the need for multiple monitoring programs in an area to effectively address environmental effects of developments. While individual communities can offer different resources for monitoring programs, not all communities can offer all the resources that a monitoring program may need. Also, maintaining trained reliable monitors can challenge planners of a community-based monitoring program, especially in small, remote communities. Ensuring the guaranteed long-term funding required for monitoring programs through the governments is often difficult. Therefore, the need to supplement the community-based monitoring program with additional programs is highlighted.

Finally, ensuring that the information collected is scientifically defensible and will be accepted by scientific and management bodies is important in ensuring the effectiveness of the monitoring program. If the information can't be used in decision-making, the value of the program is significantly reduced. To ensure quality information is collected, proper training of monitors is required as well as the planners' availability to provide advice and direction to the monitors to ensure proper protocols are followed.

CONCLUSIONS

Community-based monitoring can be a valuable tool. Supporting a community-based monitoring program builds capacity within a community, enhancing community ties with the environment in a scientific as well as traditional way. Community-based monitoring can also aid in collecting long term information for resource management as well as information for addressing cumulative environmental effects. It can help foster strong ties between local communities, government, and industry. The challenges of community-based monitoring programs can be overcome by ensuring a long-term commitment to the program, taking the time to properly develop the program so that it meets multiple stakeholder needs, and properly training monitors in scientific data collection techniques. Ensuring that community needs are met, and that the information collected is brought back to the community members in a meaningful way, will strengthen communication and ties between stakeholders. Community-based monitoring cannot replace all other types of monitoring. There needs to be diversification as well as coordination of monitoring programs in any given area to determine short and long-term environmental effects caused by human activities or natural events. The values gained far outweigh the challenges faced in developing a community-based monitoring program, as has been shown by the Tariuq (Ocean) Monitoring Program in the ISR.

ACKNOWLEDGEMENTS

The authors would like to thank Marlene Bailey for her continued administrative and logistical help with the Tariuq (Ocean) Monitoring Program, as well as the Tariuq working group members and monitors for helping to create a successful program.

REFERENCES

Berger, T.R. 1977. *Northern Frontier, Northern Homeland*, Vol. II. Ottawa: Supply and Services Canada.

Berkes, F., J. Mathias, M. Kislalioglu, and H. Fast. 2001. "The Canadian Arctic and the *Oceans Act*: The Development of Participatory Environmental Research and Management." *Ocean & Coastal Management*. 44:451-469.

Chiperzak, D. and M. Cockney. 2000a. *Results from the Aklavik Community Marine Ecosystem Health (MEH) Workshop, March 6-7, 2000*. DFO Oceans report.

Chiperzak, D. and M. Cockney. 2000b. *Results from the Tuktoyaktuk Community Marine Ecosystem Health (MEH) Workshop, January 25-26, 2000*. DFO Oceans report.

Eddy, S. 2001. *Tuktoyaktuk and Aklavik Tariuq (Ocean) Community-based Monitoring Program—Results from the First Indicators workshop June 28, 2001, Aklavik, NWT*. DFO Oceans report.

Hydrocarbons and Chemicals: Potential Effects and Monitoring in the Deep-Sea

Skadsheim, A. (Arnfinn.Skadsheim@rf.no), J.F. Børseth, A. Bjørnstad, L. Camus, E. Aas, and S. Sanni (RF-Akvamiljø, Randaberg, Norway)

ABSTRACT: For hydrocarbon exploration and production, the inaccessibility of the deep-sea, the low temperature, the high pressure, and the poorly described, but highly diverse biological communities constitute special environmental challenges. For most marine organisms, there is little or no information available on the toxicity of hydrocarbons or any chemical in the deep-sea. Although technical developments have reduced or removed discharges with hydrocarbons and other potentially damaging chemicals by transport to land or re-injection, accidental leaks and spills may occur. Moreover, we may benefit from knowledge-based options for assessing mitigation alternatives and for considering how much an effluent must be cleaned before it possibly can be discharged. Preliminary studies have indicated that high pressure may alter the partitioning of some hydrocarbons between oil, organisms, and water and hence influence effects in biota. Special adaptations in organisms to deep-sea life may possibly also make them vulnerable to industrial activities. It is thus uncertain to what extent knowledge gained in shallow water can be directly transferred to deep-sea environment for risk assessment and environmental monitoring. Recent environmental law regulations emphasize the importance of biological effects. To meet the management needs of the oil and gas industry for deep-sea exploration and production, we are developing an environmental effect and risk estimation methodology for decision-making, impact assessment, and monitoring organised within the Norwegian Deepwater Programme (NDP).

INTRODUCTION

Over the past three decades, oil and gas activity in the North Sea has progressed towards the deeper waters off the Atlantic shelf margin and into the Arctic Ocean. A similar development has occurred off the east coast of Canada. The similar oceanic characteristics and organism assemblages shared by these two very biologically productive regions, suggest that methods used in hydrocarbon exploration and production, and for environmental assessment of these activities, may have much in common, and that future coordination may be beneficial.

Our main focus in the Norwegian Deepwater Programme (NDP) is on discharges from exploratory drilling and various production processes. The NDP addresses methods and information gathering on how to assess the potential for environmental impact, to develop Environmental Effects Monitoring (EEM) methods, and how to make a system for management decisions. Although technical developments have reduced or removed discharges containing hydrocarbons and other potentially damaging chemicals through transport to land or through re-injection, accidental leaks and spills may occur. Moreover, and beyond the scope of NDP, work should include the development of a knowledge-based option to assess how much an effluent must be cleaned before its effects on organisms are so negligible that discharge is acceptable (i.e. a politically-acceptable no effect).

At present, exploration and production of petroleum hydrocarbons are occurring in many places on the slope, from the continents towards the deep-sea. These activities are in the process of passing 2000 m depth. The deep-sea differs from shallow water in depth, pressure, temperature, and darkness. The hostile and inaccessible nature of this environment, combined with a lack of knowledge of physical, chemical, and biological parameters, will influence approaches to assess and monitor the potential for environmental impact.

The approaches presented below have been identified based on our experience and dialogue with operators in the Norwegian offshore sector, and on adaptations to the new guidelines for Environmental Risk Assessment (ERA) developed by the Oslo and Paris Commission for the Protection of the Marine Environment of the North-East Atlantic (OSPAR) (Anonymous, 2003). One important outcome from OSPAR is that increased attention must be paid to the measurements and assessment of the potential for biological effects caused by activities and discharges.

DEVELOPING DEEP-SEA RISK ASSESSMENT

ERA and Decision Making

A methodology is needed to be able to assess environmental risks and make appropriate management decisions. The same general risk assessment methods as those that were developed for present use in Norwegian waters are expected to be applicable. Several systems for risk assessment have been developed in the recent years, and other systems/models are currently being development in parallel to this project. The most comprehensive ongoing work is the Environmental Risk and Management System (ERMS) (Smith et al., 2004). It builds on the models Dose-Related Exposure Assessment Model (DREAM) (Reed and Hetland, 2002), Environmental Impact Factor (EIF) (Johnsen et al., 2000; Frost and Johnsen, 2004), and the DeepBlow model (Johansen, 2000) as the main ones. A model overview has been presented by Melbye (2004). A system has not been chosen at this stage, because the priority is to first gather data and develop knowledge in good accordance with the general internationally accepted principles of risk assessment. By the time the project enters its last phase, and results from ongoing projects are available, an assessment and recommendation will be made of which system/model is the best suited for implementation of a deep-sea risk assessment methodology.

Guidelines Provided by Definitions

The general principles of chemical exposure and ERA are defined as follows (Anonymous, 2003):

- "Environmental Risk Assessment" is the determination of the relationship between the predicted exposure and adverse effects in four major steps: hazard identification, dose-response assessment, exposure assessment and risk characterisation.
- "Hazard identification" is the definition of the adverse effects that a substance has an inherent capacity to cause.
- "Dose (concentration) – response (effect) assessment" is the estimation of the relationship between dose, or level of exposure to a substance, and the incidence and severity of effect.
- "Exposure assessment" is the determination of the emissions (discharges), pathways, and rates of movement of a substance and its transformation or degradation in order to estimate the concentration/doses to which human populations or environmental compartments are or may be exposed.
- "Risk characterisation" is the estimation of the incidence and severity of the adverse effects likely to occur in a human population or environmental compartment due to actual or predicted exposure to a substance, and may include "risk estimation", i.e. the quantification of that likelihood.

To be recognised by the appropriate international conventions, the risk assessment methodology to be developed must have the capacity to support these four major steps above. We acknowledge a demand for effect measurements and pursue the cost-effective nature of biomarkers (Depledge, 1994).

Depledge defines them as: "Biochemical, cellular, physiological, or behavioural variations that can be measured in tissue or body fluid samples or at the level of whole organisms that provide evidence of exposure to and/or effects of, one or more chemical pollutants (and/or radiation)". These biomarkers may be obtained over the wide response range in many types of organisms and they meet many demands at deep-sea ERA and EEM. Added benefits may be improved pointers to the development of more environmentally friendly chemicals, mitigation technologies, and operational procedures.

Initial Tiers and EEM

The large operational, technical, and limited knowledge challenges of deep-sea operations have lead us to conclude that we must develop a tiered approach to ERA, but one, which from the start, provides guidelines to EEM methodology. We propose a combination of chemical and biological effects, especially biomarkers. Such analyses can be conducted in subsequent tiers with increasing demands to data quality. When and how to decide depends on the conclusion strength, stated demands, and cost of implementing an early tier solution versus the outcome from later tiers. Late tier results and decisions may be cheaper, but require more documentation. The effect measurements we prioritise are mainly biomarkers as defined by Depledge (1994). The combination of chemical and biological effect measurements is increasingly applied in Rapid Assessment of Marine Pollution (RAMP) (Grøsvik et al., 2000; Depledge, 2000; Handy et al., 2003; Moore et al., 2004). RAMP facilitates successive tiers and enables accounting for site-specific demands. See also the more comprehensive approach of Hylland et al. (2004). The paper of Handy et al. (2003) places biomarkers in the most up to date regulatory context.

Initially, a suite of chemical and biological measurements is selected, which are rapid, inexpensive, easy to use, contain ecologically relevant information, and are useful to environmental management. The rationale is that chemistry alone presents limited ecological significance, whilst ecological data alone leaves few clues to cause and effect relationships. Combining them provides additional insight, and allows for a gradual increase in sophistication. Biomarker responses often persist long after a transient exposure to a contaminant that has degraded beyond detection (Handy et al. 2003).

SCALING AND LIMITATIONS IN THE DEEP-SEA

The paragraphs in this section describe some deep-sea characteristics to illustrate the scales of some physical events and biological characteristics that may influence deep-sea ERA and EEM. Definitions above are from Lincoln et al. (1982). The deep-sea may be defined as below 1000 m depth and permanently unlit by sunlight. Major divisions from land to the deepest parts are the shelf, the continental slope, the continental rise, the abyssal plain and the hadal zone, or deep-sea trenches. The shelf is the shallow section from land down to 200 m depth. The slope is the subsequent steep decline (average 4°, maximum 20°) down to 1000 m. The slope is followed by the continental rise. It is the gently sloping (average 1°) seabed down to 4000 m where it grades into the abyssal plains. The slopes and rises cover 15% and the abyssal plains 50% of the earth's surface (Gage and Tyler, 1991; Angel, 1997). The abyssal plains are found at somewhat over 4000 m average depths and often defined as covering depths to 6000 m. The trenches are the steepest and deepest parts of the ocean below 6000 m.

The deep-sea vary in both macro- and micro-ways with respect to factors that may influence ERA and EEM, and with respect to the patchiness of occurrence of organisms. Some factors that illustrate this complexity are:

- The oceanic waters may be stratified into three to four main water layers, which may influence the three dimensional spreading of discharges.
- The main new sediment input occurs from the continents and along the continental margins. Sediment that slides on the slopes may reach the abyssal plains.
- Plant and animal production contributes more to the sediment input off the shelf and slope (Gage and Tyler 1991).

- Physical events may influence the fate of discharges and data from monitoring, including internal waves in density boundary layers, gyres from the tides, shallow water gyres and main ocean currents breaking at sub-sea continental protrusions, seasonal and more frequent variations in main weather and current systems (Gage, 1997; Bett, 2001).
- Natural micro-scale disturbances may be involved in maintaining deep-sea benthic diversity (Gage, 1997), and the significance of biological interactions is unknown, but debated (Smith et al., 1998a).

The diversity of at least the larger pelagic (free water masses confined) deep-sea organisms, as exemplified by the fishes, seems to be low and may be related to the absence of many distribution barriers and widely distributed populations (Haedrich, 1997; Gage and Tyler, 1991). This may possibly be an advantage with respect to local impact, but long-lived adults, very fragmentary knowledge about reproductive biology, and how far species move when they rely on very distant food signals, suggest that precautionary approaches may be wise. If over long time periods (decades), a large proportion of a population become attracted to exploration and production sites and increase their potential for impact, there may be reason for concern.

The species diversity in the deep-sea benthos is surprisingly high. It is possibly comparable to that of coral reefs and tropical rain forests (Gage 1997). The biomass is low and many organisms are tiny compared to the shallows (Hessler and Sanders, 1967; Smith et al., 1998a). Food availability is often low. Many species live for several decades to hundreds of years. However, there are also instances of adaptations to seasonally-fluctuating food supplies, longevities, and biological adaptations comparable to the shallows (Gage and Tyler, 1991; Thurston et al., 1998). Natural disturbances may, therefore, not be readily separated from anthropogenic ones.

Many offshore benthic impact assessments and monitoring programmes include expensive benthic community descriptions due to their high ecological relevance. Shallow water benthos contains well-described species, where generally their lifespan is just a fraction of the life of an oil field. Approaching the deep-sea, the lifespan increases up to several times that of an oil field, and the proportion of unknown species also increases. The problem is enhanced by an increasing lack of taxonomists (see also Hartley, 2004). The ecological relevance of benthic community monitoring is partly based on the assumption that if the diversity is maintained, it indicates that the generations succeeded to reproduce as normal, and hence all life stages tolerated or have not suffered damage that influences community structure. However, failed recruitment is not so easy to detect when the adults live longer than the oil field. Damage may thus emerge decades past a field closure.

For organisms living at or in the seabed, the isolation of species and evolution of endemic (native and restricted to a region) forms and the type of organisms dominating the more local benthic communities has been related to topographic barriers on the large scale (Starmans, 1997; Schnack, 1998). However, endemism in the deep-sea will also apply down to small geographic scales as, for example, single sub-sea canyons and seamounts and ridges (Creasy and Rogers, 1999; Etter et al., 1999). Baseline studies along the continental slope have shown that special aggregations of organisms or communities of special concern may be found. Recent examples are the *Lophelia* coral reefs (Mortensen et al., 2001), Darwin mounds (Masson et al., 2003), mud volcanoes (Dimitrov, 2003), and pock marks (Hovland and Judd, 1988). Organism aggregations with special adaptations are associated with vents and seeps (Gage and Tyler, 1991; Smith et al., 1998b). The deep-sea does, therefore, contain many unique organism assemblages relevant for single oil and gas field ERA and EEM.

Various types of natural hydrocarbon sources may contribute to the hydrocarbon signature at sites in the sea. The deep-sea poses a special challenge due to the difficulty of describing and assessing the various contributions as observed even in the shallows (Boehm et al., 2002; Short and Heinz, 2003). Contributions from natural sources, especially to the sediments, may not readily be separated from industrial discharges and other anthropogenic sources unless a balanced understanding is achieved between site-specific hydrocarbon signatures. It is a demand to develop a statistical hydrocarbon source-mixing model, which allocates source contributions from prospective sources by finding the mixture that best explains the composition found in the marine sediments. There are serious challenges for the source-mixing model approach. Sources that are potentially important may not be included for consideration, because samples characterising these sources are unavailable. Inclusion of numerous sources may lead to spurious allocations. In addition, it is critical to describe and base the model on sufficient numbers of the hydrocarbon analytes that provide sufficiently high discrimination.

In the deep-sea the special microbial capability to degrade petroleum hydrocarbons may possibly be comparable to that in shallow waters. Little is known (Rosenberg, 1992), and degradation will be studied in the NDP programme.

STUDIES LEADING TO NDP

Bioavailability and Uptake

In some pilot studies, we have found that the bioavailability of some hydrocarbons possibly may be modified at high pressure in the way that larger molecular sized hydrocarbons may concentrate more in non-polar fractions, like organic tissues, as pressure increases. Altered bioavailability may alter toxicity. In such cases, even the simplest acute toxicity data may not hold for the deep-sea. In addition, for many of the most common organic industrial chemicals, even the simplest EC- or LC_{50} values are lacking for many entire marine phyla (Leung et al., 2001).

Potential Concern for Small Organisms

The small body size of most benthic deep-sea organisms may possibly imply an unfavourable bias for exposure to hydrocarbons, chemicals, and degradation products liable to bioconcentrate. At similar exposures to the same water accommodated fraction (WAF) of a topped crude oil, Skadsheim et al. (2000)

found, at atmospheric pressure, and for three shallow water organisms, that the smallest organism took up more of the largest PAH molecules than the largest organism (Fig.1). The bias may partly be caused by a better degradation capacity, especially for large PAH molecules, in bigger and more complex organisms. However, the influence of the very hydrophobic surface combined with a large surface area to body volume ratio in small aqueous organisms has not been properly dealt with. This may cause an extra bias augmented by pressure

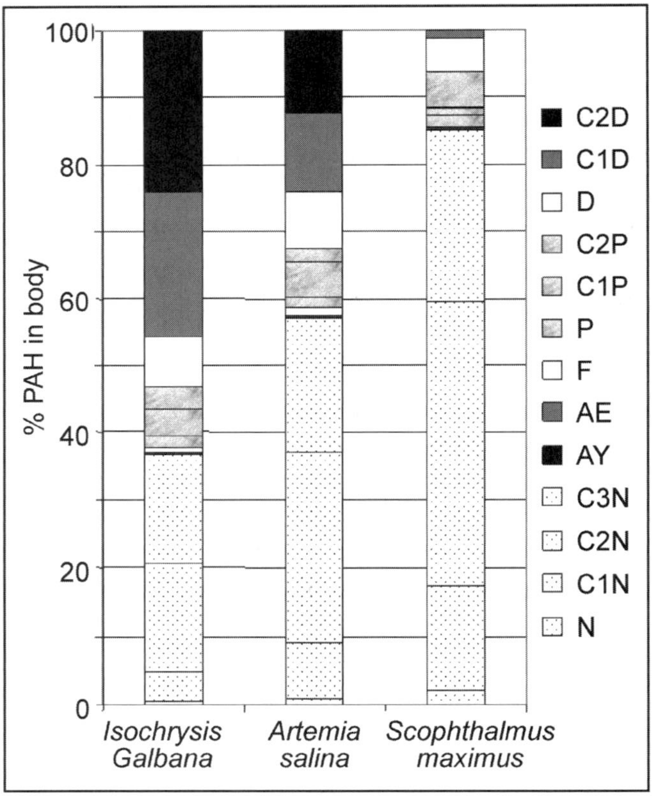

FIGURE 1. PAH in different sized organisms exposed for 24 h at surface pressure. Algae were about 3 x 4 μm, crustaceans were on average 3 mm long, and fish were on average 0.84 g and 2–3 cm long. PAH species stacked in the figure correspond to the order in the legend. N is naphthalenes and C1N to C3N describe the degree of alkylation. AY is acenaphtylene, AE is acenaphthene, F is fluorene, P is phenanthrene and its alkylated forms, and D is dibenzothiophene and its alkylated forms.

towards increased uptake of the least water soluble molecules, which for PAH, would be those of largest molecular size and most chronic effect concern (Meador et al., 1995; Skadsheim et al., 2002).

Chemistry and Biomarkers

The combination of chemical and biomarker measurements have been pursued, for example in a baseline study conducted at 450–1000 m depth in the Faeroe Bank and Faeroe-Shetland Channels and north of UK towards Iceland in the Atlantic (Grøsvik et al., 2000). Baseline levels of metals, total hydrocarbon, and PAH were recorded in fish tissue and sediment samples; metals were also measured in fish bile, and selected biomarkers analysed in the flatfish, Greenland halibut (*Reinhardtius hippoglossoides*) in conjunction with the opening of the 1st Faroese licence round. We measured the detoxifying enzyme cytochrome P450 1A (Cyp1A) induction in the liver and the frequency of DNA adducts in the same tissue. Levels of vitellogenin (Vtg) and zona radiate protein (Zrp) were measured in the blood. Gill and liver tissues were preserved and mounted in paraffin for a reference histology bank. The area was characterised as clean and at normal background values. The data also indicate that oil exploration in the UK area, which abuts the Faroes area and has included drilling since 1972 and production since 1998 (Anonymous, 2001), has not influenced the Faroe area.

Baseline sampling of fish and invertebrates off Angola (Aass et al., 2001) and off Spitzbergen (Aas et al., 2003) at depths below 1000 m have addressed how various biomarker measurements may be obtained, and how the effect expression is recorded with techniques developed for shallow water conditions. The work has also included a PhD (Camus, 2002) and a post doc by Camus on biomarker expressions in the deep-sea scavenging amphipod, *Eurythenes gryllus*. We recapitulate some new results to illustrate that basic knowledge about the natural biomarker expressions is required, and that adaptations to deep-sea life may change toxicity responses compared to those recorded in the shallows. The low metabolism of *E. gryllus*, compared to shallow water amphipods (*Anonyx nugax*) altered the expression level of some enzymes involved in detoxification (Fig. 2). Two were significantly reduced while the third, and probably the one most involved in the basic initial detoxification steps, was kept at the level of shallow water organisms.

Eurythenes gryllus was sampled at 1000 m depth North of Spitzbergen. They were placed in plastic buckets containing either 2 litres of water (control) or 2 litres of copper solution ($CuCl_2$) at 100 μg/l Cu for 7 days at atmospheric pressure. After exposure, hemolymph and digestive gland samples were frozen in liquid nitrogen. TOSC was measured according to Regoli and Winston (1999) on the cytosolic fraction of the

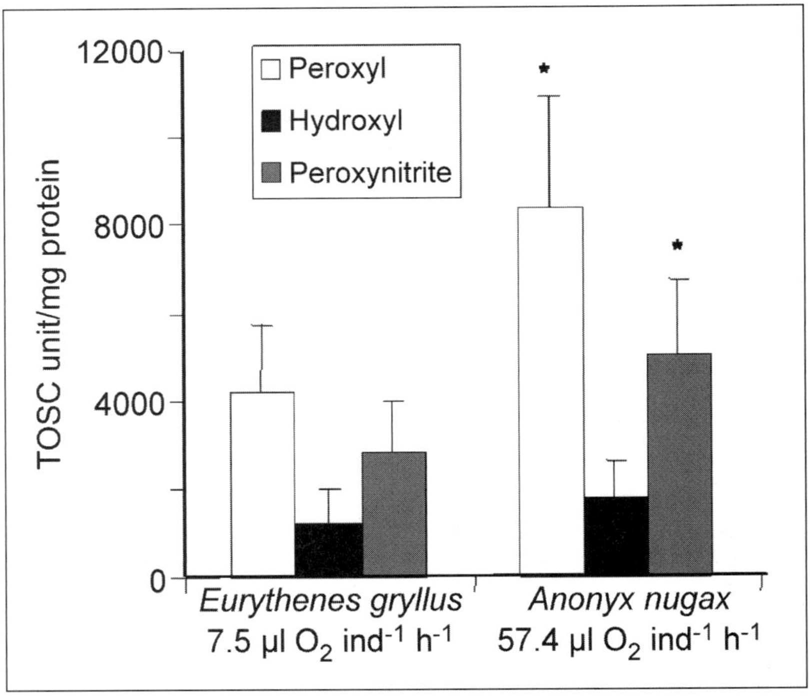

FIGURE 2. Total oxyradical scavenging capacity in the digestive gland of two scavenging crustacean amphipods, the deep-sea species, *E. gryllus*, and the shallow water species, *A. nugax*. Bars show 1 standard deviation; n = 4. * = significant difference between species. Respiration rates are given below the names.

digestive gland and the cell-free hemolymph (Camus, in prep.). The TOSC response differed between body fluids (Fig. 3). The TOSC induction in the digestive gland indicated that enzymes associated with the metabolism of glutathione were activated. Decreased TOSC in the hemolymph could correspond to a metabolism capacity related depletion stress of low molecular weight oxygen scavengers; for example, glutathione can bind directly copper (Regoli and Winston, 1999).

We have also conducted various caging experiments in a Norwegian fjord down to 700 m depth, studying trap solutions, hydrocarbon uptake, and biomarker expressions. Preliminary results are promising for the development of deep-sea ERA and EEM.

We have conducted pilot high pressure studies on methodology and exposure. Mussels (*Mytilus edulis*) and horse mussels (*Modiolus modiolus*) were subjected to 200 bar (2000 m depth) and a crab (*Hyas araneus*) was subjected to 150 bar. Some bivalves were exposed to water soluble fractions of a North Sea crude oil by means of a generator column. Body burden data on PAH and biomarker responses suggested a feasible

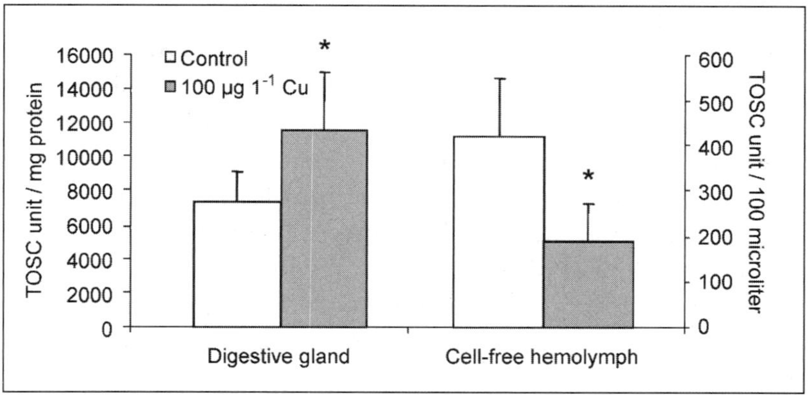

FIGURE 3. Total Oxyradical Scavenging Capacity (TOSC) toward peroxyl measured in the cytosolic fraction of the digestive gland (n = 5 in control and treated) and the cell-free hemolymph (n = 4 in control and treated) of *Eurythenes gryllus* exposed to 100 μg/l for 7 days in static water kept at 0–1°C. Data are expressed as mean and standard deviation. Asterisk indicates significant difference ($P < 0.05$) following Dunnett's test.

design. The organisms survived for months after some hours pressure exposures. PAH uptake in Semi-Permeable Membrane Devices (SPMD) was recorded after flow-through exposure to an oil droplet emulsion sized similar to that of produced water (mean droplet diameter 10 μm). Mussels and SPMD are widely used to monitor hydrocarbon and organohalogen exposure in the water. The tests indicated that the methodology might hold for bivalves and SPMD down to 2000 m depth. In the deep-sea organisms living below 2000 m there may be special biochemical adaptations to life (Somero 1992a, b), and shallow water organisms can probably not be deployed. Angel (1997) suggests that in fish the 2500–2700 m depth range may pose physiological limitations for adapting to increasing depth. We plan experiments with deep-sea species. A joint international approach to initiate recommendations on methods and species or groups of organisms for deep-sea ERA and EEM would probably be a very useful move.

KEY THEMES WITHIN THE NDP PROGRAMME

The seven main themes in the ongoing Norwegian Deepwater Programme (NDP) are summarised below. For methodological reasons and governmental priorities, focus is initially on the water phase, but we have identified the water-sediment interphase and upper sediment layer as key topics that should be addressed. Existing shallow water information and methodology will be assessed for the best possible direct application, but we consider it important to account for limitations set by low temperature and high pressure. Although it may appear overly confident at present, we have tentatively proposed that from the beginning, data should be gathered and made available for risk estimates and assessments (RA), and not only for hazard assessments (HA). HA is the simplest method for decision-making, but it carries an inherent increased probability of misguided conclusions compared to RA. Simplifications in HA for treating fate, degradation, bioavailability, how effects are accounted for, and the incorporation of high safety factors, are not readily comparable across the aforementioned parameters. RA better covers these aspects, but RA faces the challenge of increased information demand, which is not readily met when decision makers need a rapid decision regarding ongoing operations.

Establish Background

We propose to conduct an initial literature study to gather fate and ecotoxicological data. The data will be used to reveal main knowledge gaps and guide the more detailed development of the model framework. Further, it will establish a prioritised list of information for risk and impact systems; assess the potential for novel links between bioavailability, uptake and effect expression in biota, and initiate the creation of a database. At present, it seems like no toxicity tests have been conducted with deep-sea organisms. A pausing state of art, since the habitat covers 2/3 of the earth and is a final sink for many discharged compounds or substances.

Fate—Physical

The fate descriptions will be based on experience gained from the DeepSpill experiment (Johansen et al., 2001) and interpretations resulting in the DeepBlow model (Johansen, 2000). Activities will include collecting literature data on gas bubble size and the potential for hydrate formation to improve equations in, and thereafter verify predictions by the DeepBlow model with field data, and developing guidelines for incorporation of plume fate from bottom discharges in a deep-sea environmental ERA system. In addition, the use of the refined DeepBlow model can be used to predict far-field plume interactions with turbulence, and to predict interactions between discharges and the seabed (Fig. 4).

Fate—Bioavailability

Conducting experiments to examine mainly PAH release from oil films and oil droplets, and making predictions of pressure-dependent uptake in standard monitoring objects (SPMDs and bivalves) can help generate an understanding of PAH availability for water and sediment bound discharges, and

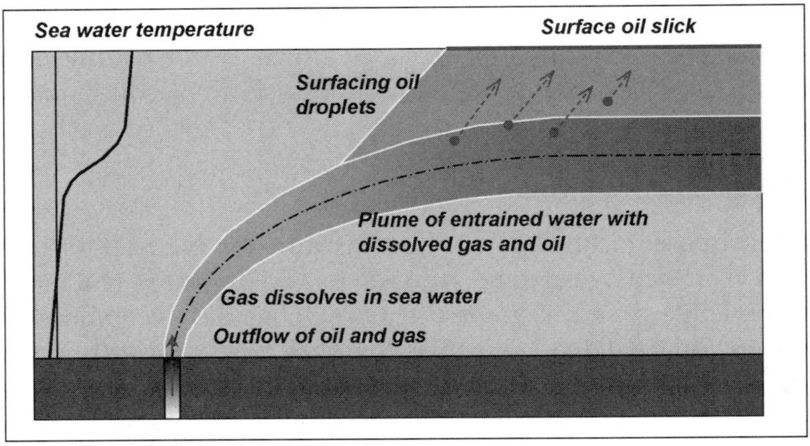

FIGURE 4. Probable behaviour of seabed blowouts in the deep-sea (Johansen, 2000).

assess the general applicability of the n-octanol/water partition coefficient in relation to increasing pressure.

Biodegradation

By comparing degradation by naturally-occurring shallow water and deep-sea bacteria in experiments conducted at surface and high pressure, we may determine the influence of pressure on microbial processes and formation rates of degradation products, their toxicity, and liability for accumulation in sediments and biota. Beyond this, we can assess the suitability of data for RA, define data demands and initiate the creation of a data base. Currently, a postdoctoral study on degradation of hydrocarbons and production chemicals plus estimates on recovery rates is being conducted by R. Kommedal at Stavanger University College.

Valuable Ecological Components (VECs) and Target Resources (TRs)

We use the terms as defined by Sørgård et al. (1995) and Jødestøl et al. (1996) on behalf of the Norwegian Oil Industry Association (OLF). The approach connects the chemical and biomarker measurements to the population, community, and higher levels of biological organisation. Tasks will include 1) elucidating behavioural and toxic responses in the deep and cold water coral, *Lophelia pertusa,* exposed to dissolved and particulate matter (see Roberts and Anderson, 2002), 2) initiating the development of a system for identifying and describing the vulnerability of VECs and TRs. Examples of these are vent and seep communities, including Darwin mounds and pock marks, and, the potential for other unique faunal assemblages, fish stocks, and sand, gravel, mineral or other resources requiring attention, and 3) establishing links to fate events and proposing test organisms and schemes for uptake and effects in organisms defined as potentially vulnerable.

Uptake in Organisms and Effects

Activities within this theme of NDP include 1) examining how exposure and dose and effect responses in shallow water organisms correspond to deep-sea conditions (i.e. how much information can be transferred directly?), 2) developing a method and data basis whereby uptake and effects in deep-sea species may be obtained and compared to existing data, 3) examining hydrocarbon uptake and biomarker responses in selected invertebrates, notably the utility of baseline biomarker measurements in *Lophelia,* and 4) assessing the suitability of biomarkers and key defined species for deep-sea field monitoring, 5) place data applicable for RA in a database, 6) establishing relationships between biomarkers and classical (Darwinian) fitness and effect parameters like growth, reproductive output, and the survival of various life stages. This last task is challenging at any depth because it is still unknown to what degree species lack adaptability in response to environmental forcing (Klerks and Levinton, 1989). A conservative approach may be to treat the ability as limited in ERA models and decision making.

Risk System Evaluation

The last phase of the project includes an evaluation study to identify and recommend the most suitable system or compilation of sub-units for implementation of the developed deep-sea ERA methodology, and the establishment of links between

ERA and EEM. However, we recommend that the deep-sea ERA development should be kept open for some time as it may be premature to recommend any specific model system.

ACKNOWLEDGEMENTS

The preparation of the manuscript and presentation was covered by RF-Rogaland Research. The contributions from L. Camus were covered by the Norwegian Research Council under the program "Marine Resource, Environment and Management", project no146478/120 and by RF-Akvamiljø. J.P. Hartley, an anonymous referee, and Jeff Short are acknowledged for comments that improved the manuscript.

REFERENCES

Aas, E, B. Liewenborg, B.E. Grøsvik, L. Camus, G. Jonsson, J.F. Børseth, and L. Balk. 2003. "DNA Adduct Levels in Fish from Pristine Areas are not Detectable or Low When Analysed Using the Nuclease PI Version of the 32P-postlabelling Technique." *Biomarkers.* 8:445-460.

Aas, E., I.C. Taban, K.B. Øysæd, G. Jonson, L. Balk, B. Liewenborg, S. Torgrimsen, A. Nævdal and J.F. Børseth. 2001. "*Biomarker Analyses of Deep-sea Fish—ZaïAngo Biol2 Survey.*" Akvamiljø. Internal Report. AM 2001/006. 38 pp. RF-Akvamiljø.

Anonymous. 2001. The 2001 Digital Brown Book. Development of UK Oil and Gas Resources. *Department by Trade and Industry, UK. Data by Design Ltd, UK.*

Anonymous. 2003. European Chemical Bureau. Environmental Risk Assessment. European Commission—Joint Research Centre, *EUR 20418 EN/2—TGD.* Part II, Chapter 3. 337 pages.

Angel, M.V. 1997. "What is the Deep-sea?" In D.J. Randall and A.P. Farrell (Eds.), *Deep-sea Fishes*, pp.1-41. Academic Press. San Diego.

Bett, B.J. 2001. "UK Atlantic Margin Environmental Survey: Introduction and Overview of Bathyal Benthic Ecology." *Continental Shelf Research.* 21:917-956.

Boehm, P.D., W.A. Burns, D.S. Page, A.E. Bence, P.J. Mankeiwicz, J.S. Brown, and G.S. Douglas. 2002. "Total Organic Carbon, an Important Tool in a Holistic Approach to Hydrocarbon Source Fingerprinting." *Environmental Forensics.* 3(¾):243-250.

Camus, L. 2002. "Biomarkers Relevant to Oil and Gas Activities in Low Temperature Marine Ecosystems." Ph.D. Thesis. University of Plymouth, UK.

Creasy, S.S. and A.D. Rogers. 1999. "Population Genetics of Bathyal and Abyssal Organisms." *Advances in Marine Biology*. 35:1-151.

Depledge, M.H. 1994. "The Rationale Basis for the Use of Biomarkers as Ecotoxicological Tools." In M.C. Fossi and C. Leonzio (Eds.), *Non-destructive Biomarkers in Vertebrates*. pp. 271-295. Lewis Publishers. Boca Raton.

Depledge, M.H. 2000. "Rapid Assessment of Marine Pollution (RAMP)." In B.B. Que, M.H. Depledge, T.V. Hong, and N.H. Phuong (Eds.), *Proceedings of the Vietnam-U.K. Joint Workshop on Marine Pollution Assessment, Hanoi, 8 June 2000.*

Dimitrov, L.I. 2003. "Mud Volcanoes—A Significant Source of Atmospheric Methane." *Geo-Marine Letters*. 23:155-161.

Etter, R.J., M.A. Rex, M.C. Chase, and J.M. Quattro. 1999. "A Genetic Dimension to Deep-sea Biodiversity." *Deep-sea Research Part I—Oceanographic Research Papers*. 46(6):1095-1099.

Frost, T.K. and S. Johnsen. 2004. "Development and Field Validation of the EIF as a Produced Water Discharge Management Tool" (abstract only). *Workshop on the Offshore Oil and Gas Environmental Effects Monitoring, Bedford Institute of Oceanography, Dartmouth, Nova Scotia, May 26–30, 2003*, Environmental Science Research Fund Report, in press.

Gage, J.D. 1997. "High Benthic Diversity in Deep-sea Sediments: the Importance of Hydrodynamics." In R.F.G. Ormond, J.D. Gage, and M.V. Angel. (Eds.), *Marine Biodiversity: Patterns and Processes,* pp. 148-177. Cambridge University Press, Cambridge.

Gage, J.A. and P.A. Tyler (Eds.). 1991. *Deep-sea Biology: A Natural History of Organisms at the Deep-sea Floor.* Cambridge University Press, Cambridge.

Grøsvik, B.E, A. Bjørnstad, A. Nævdal, S. Westerlund, and E. Aas. 2000. *Hydrocarbon Baseline Study.* RF 2000/221. 81 pp. Report RF-Rogaland Research, Stavanger, Norway.

Haedrich, R.L. 1997. "Distribution and Population Ecology." In D.J. Randall and A. P. Farrell (Eds.), *Deep-sea Fishes*, pp. 69-114. Academic Press. San Diego.

Handy, R.D., T.S. Galloway, and M.H. Depledge. 2003. "A Proposal for the Use of Biomarkers for the Assessment of Chronic Pollution and In Regulatory Toxicology." *Ecotoxicology.* 12:331-343.

Hartley, J.P. 2004. "The Challenge of Deep Water Oilfield Environmental Monitoring West of Britain." (abstract only). *Workshop on the Offshore Oil and Gas Environmental Effects Monitoring, Bedford Institute of Oceanography, Dartmouth, Nova Scotia, May 26-30, 2003*, Environmental Science Research Fund Report. In press.

Hessler, R.R. and H.L. Sanders. 1967. "Faunal Diversity in the Deep-sea." *Deep-Sea Research.* 14:65-78.

Hovland, M. and A. G. Judd. 1988. *Seabed Pockmarks and Seepages. Impacts on Biology, Geology and Environment.* Graham and Trotman, London. 293 pp.

Hylland, K., G. Becker, J. Klungsøyr, T. Lang, A. McIntosch, B. Serigstad, J.E. Thain, K.V. Thomas, T.I.R. Utvik, D. Vethaak and W. Wosniok. 2004. "Biological Effects in Pelagic Ecosystems—An Overview of the ICES BECPELAG Workshop." (abstract only). *Workshop on the Offshore Oil and Gas Environmental Effects Monitoring, Bedford Institute of Oceanography, Dartmouth, Nova Scotia, May 26-30, 2003*, Environmental Science Research Fund Report. In press.

Johansen, Ø. 2000. "DeepBlow—A Lagrangian Plume Model for Deep Water Blowouts." *Spill Science and Technology Bulletin.* 6:103-111.

Johansen, Ø., H. Rye, A.G. Melbye, H.V. Jensen, B. Serigstad, and T. Knutsen. 2001. *DeepSpill JIP—Experimental Discharges of Gas and Oil at Helland Hansen June 2000.* Technical Report. STF 66 F01082, SINTEF Applied Chemistry, 159 pp. Trondheim, Norway.

Johnsen, S., T.K. Frost, M. Hjelsvold, and T.I. Røe Utvik. 2000. *The Environmental Impact Factor—a Proposed Tool for Produced Water Impact Reduction, Management and Regulation.* SPE-paper 61178, presented at the SPE Conference in Stavanger 26-28th June 2000.

Jødestøl, K.A., E. Sørgård, E. Hoell, and B. Fredheim. 1996. *Metode for Miljørettet Risikoanalyse (MIRA): Metodebeskrivelse.* (in Norwegian). DNV Report 95-3562. 75 pp.

Klerks, P.L. and J.S. Levinton. 1986. "Rapid Evolution of Metal Resistance in a Benthic Oligochaete Inhabiting a Metal-polluted Site." *Biological Bulletin.* 176:135 -141.

Leung, K.M.Y., D. Morritt, J.R. Wheeler, P. Whitehouse, N. Sorokin, R. Toy, M. Holt, and M. Crane. 2001. "Can Saltwater Toxicity be

Predicted from Freshwater Data?" *Marine Pollution Bulletin.* 42(11):1007-1013.

Lincoln, R.L., G.A. Boxall, and P.F. Clark (Eds.). 1982. *A Dictionary of Ecology, Evolution and Systematics.* Cambridge University Press.

Masson, D.G., B.J. Bett, D.S.M. Billet, C.L. Jacobs, A.J. Wheeler, and R.B. Wynn. 2003. "The Origin of Deep-water, Coral-topped Mounds in the Northern Rockall Trough, Northeast Atlantic." *Marine Geology.* 194(3-4):159-180.

Meador, J.P., J.E. Stein, W.L. Reichert, and U. Varanasi. 1995. "Bioaccumulation of Polycyclic Aromatic Hydrocarbons by Marine Organisms." *Reviews of Environmental Contamination and Toxicology.* 143:79-166.

Melbye, A. 2004. "Ongoing Monitoring Programs Relevant for the Offshore Oil and Gas Industry in Norway, Emphasizing Methods and Trends Seen in the Perspective of the Development of Hydrocarbon and Produced Water Discharges." (abstract only). *Workshop on the Offshore Oil and Gas Environmental Effects Monitoring, Bedford Institute of Oceanography, Dartmouth, Nova Scotia, May 26-30, 2003,* Environmental Science Research Fund Report. In press.

Moore, M.N., M.H. Depledge, J.W. Readman, and D.R.P. Leonard. 2004. "An Integrated Biomarker-based Strategy for Ecotoxicological Evaluation of Risk in Environmental Management." *Mutation Research.* 552:247-268.

Mortensen, P.B., M. Hovland, J.H. Fosså, and D.M. Furevik. 2001. "Distribution, Abundance and Size of *Lophelia pertusa* Coral Reefs in Mid-Norway in Relation To Seabed Characteristics." *Journal of the Marine Biological Association of the United Kingdom.* 81:581-597.

Reed, M. and B. Hetland. 2002. "DREAM: A Dose-related Exposure Assessment Model Technical Description of Physical-Chemical Fates Components." In *Proceedings of 2002 SPE Conference, Kuala Lumpur.* SPE 73856.

Regoli F. and G.W. Winston. 1999. "Quantification of Total Oxidant Scavenging Capacity of Antioxidants for Peroxynitrite, Peroxyl Radicals and Hydroxyl Radicals." *Toxicology and Applied Pharmacology.* 156:96-105.

Roberts, J.M. and R.M. Anderson. 2002. "A New Laboratory Method for Monitoring Deep-water Coral Polyp Behaviour." *Hydrobiologia.* 471:143-148.

Rosenberg, E. 1992. "The Hydrocarbon-oxidizing Bacteria." In A. Balows, H.G. Truper, M. Dworkin, and W. Hardner (Eds.), *The Procaryotes*, pp. 446-459. Vol. 19. Springer.

Schnack, K. 1998. "Macrofaunal Community Patterns at the Continental Margin of East Greenland." *Reports on Polar Research (Berichte zur Polarforshung)*. (294).

Short, J.W. and R.A. Heintz. 2003. "Normal Alkanes and the Unresolved Complex Mixture as Diagnostic Indicators of Hydrocarbon Source Contributions to Marine Sediments of the Northern Gulf of Alaska." In *Proceedings of the Twenty-sixth Arctic and Marine Oilspill Program (AMOP) Technical Seminar*, pp.155-168 (June 10-12, 2003, Victoria, British Columbia; Environment Canada, Ottawa, Ontario, Canada) AMOP conference.

Skadsheim, A., T. Baussant, R.K. Bechmann, A. Bjørnstad, B. Gaudebert, C. Labes-Carrier, I.C. Jensen, and G. Jonsson. 2000. "Density and Size Dependent Uptake of Various PAC in Organisms of a Model Food Chain." *Polycyclic Aromatic Compounds*. 18:161-175.

Skadsheim, A., G. Kjeilen, A. Melbye, H. Rye, and E. Aas. 2002. *Consequences of Regular Discharges to the Sea—Regional Environmental Impact Assessment, Norwegian Sea. Konsekvenser av regulære utslipp til sjø—RKU Norskehavet.* (in Norwegian). Rapport RF-2002/081. 71 pp. Stavanger, Norway.

Smith, C.R., L.A. Levin, and L.S. Mullineaux. 1998a. "Deep-sea Biodiversity: a Tribute to Robert R. Hessler." *Deep-Sea Research Part II. Topical Studies in Oceanography*. 45(1-11).

Smith, C.R., L.S. Mullineaux, and L.A. Levin. 1998b. "Deep-sea Biodiversity: a Compilation of Recent Advances in Honor of Robert R. Hessler." *Deep-Sea Research Part II. Topical Studies in Oceanography*. 45(1-3).

Smit, M.G.D., R.G. Jak and J.A. van Dalfsen. 2004. "A Concept for the Environmental Impact Factor for Drilling Discharges." (abstract only). *Workshop on the Offshore Oil and Gas Environmental Effects Monitoring, Bedford Institute of Oceanography, Dartmouth, Nova Scotia, May 26-30, 2003*, Environmental Science Research Fund Report. In press.

Somero, G.N. 1992a. "Adaptations to High Hydrostatic-pressure." *Annual Review of Physiology*. 54:557-577.

Somero, G.N. 1992b. "Biochemical Ecology of Deep-sea Animals." *Experentia*. 48(6):537-543.

Starmans, A. 1997. "Comparative Studies on the Ecology and Biodiversity of the Arctic and Antarctic Mega-epibenthos." *Reports on Polar Research (Berichte zur Polarforschung).* (250).

Sørgård, E., K.A. Jødestøl, E. Hoell, and B. Fredheim. 1995. *Metode for Miljørettet Risikoanalyse (MIRA): Grunnlagsrapport.* DNV Report 95-3563. 75 pp.

Thurston, M.H., A.L. Rice, and B.J. Bett. 1998. "Latitudinal Variation in Invertebrate Megafaunal Abundance and Biomass in the North Atlantic Ocean Abyss." *Deep-sea Research Part II—Topical Studies in Oceanography*

Prioritising Environmental Effects Monitoring (EEM) Programs: A Risk-Based Strategy

Sadiq, R. (rehan.sadiq@nrc-cnrc.gc.ca) (Institute for Research in Construction, National Research Council, Ottawa, ON, Canada), B. Veitch, T. Husain, and N. Bose (Faculty of Engineering and Applied Science, Memorial University of Newfoundland, St. John's, NL, Canada)

ABSTRACT: Waste discharges to oceans and rivers deteriorate ambient water quality. Environmental effects monitoring (EEM) programs are designed to evaluate the adverse effects of undesirable substances at various locations in the receiving water body. Population-based and exposure concentration-based approaches are used for the design of EEM programs. The first approach prioritises the EEM program using the density or sensitivity of receptors (ecological entities), whereas the second approach prioritises the EEM program using the intensity of stressors (pollutant concentrations). The population and concentration-based methods focus on frequency (probability of occurrence) and damage potential (hazard), respectively. Adverse effects can be measured only if receptors are exposed to certain pollutant concentrations. The existing approaches do not consider both factors at a time. The proposed approach ranks the EEM program based on the ecological risk posed to ecological entities for controlled discharges or accidental spills. The risk-based method can be helpful for designing and prioritising economical EEM programs. The proposed method is demonstrated with the help of a hypothetical case study.

INTRODUCTION

Environmental effects monitoring (EEM) programs are designed to efficiently evaluate potential ecological impacts caused by controlled and/or accidental discharges into the marine environment. The following are some of the major objectives of EEM programs, which should help to:

- meet regulatory requirements;
- reduce risks to ecological health;
- improve water quality;
- increase public confidence and perception about environment safety; and
- promote reducing the use of hazardous chemicals in treatment.

Generally, EEM programs do not necessarily collect data from the locations of maximum ecological risk. Typical EEM programs can only capture the worst conditions either from the point of view of the sensitivity and density of ecological entities, or the highest exposure concentrations (Baldauf et al., 2002). In this study, the former approach is referred to as population-based and the later approach is referred to as concentration-based. Lawrence (1976) defines risk as a measure of probability and severity of negative adverse effects. Therefore, risk refers to the joint probabilities of an occurrence of an event (concentration-based) and its consequences (population-based). Both existing approaches lack in addressing the risk posed to the ecological entities in totality. A risk-based approach is required, which could consider both factors in the prioritisation of EEM programs.

Baldauf et al. (2001, 2002) recommended air quality monitoring networks based on the highest expected human health risk in order to maximise the potential of obtaining a representative air quality measurement. This methodology was established using the ISC3 dispersion model for ground level concentrations of air pollutants. Baldauf et al. (2001; 2002) used risk assessment techniques for designing ambient air quality monitoring networks to evaluate human health effects from exposures to airborne contaminants. Recently, Khan and Sadiq (2004) proposed a fuzzy synthetic evaluation technique for risk-based monitoring of air pollutants.

The offshore activities by petroleum industries generate several waste streams including drilling waste in the exploration and development phases, and produced water in the production phase. Discharges from deck drainage, kitchen wastes, and wastewaters from washrooms also contribute to marine pollution. These wastes are discharged into the marine environment after treatment. These discharges may contain harmful pollutants even after treatment, which affect the ecological entities. In addition to controlled discharges, accidental spills from loading tanks and offshore storage facilities contribute to the deterioration of the ambient water quality. Stationary and mobile sampling devices are used for monitoring fate and transport of pollutants. The sampling locations are generally designed around the disposal site at varying distances using modelling results. In this study, we propose a new methodology to prioritise, or rank, the EEM locations based on ecological risk. The methodology is demonstrated with the help of a hypothetical example.

METHODS

Ecological Risk-based Approach

Actual exposure concentration (*AEC*) of a pollutant and its threshold value (a concentration to which an ecological entity responds) are important for designing EEM programs. The no-effect concentration (*NEC*), the highest concentration at which no adverse effects are noted, is used as a toxicity threshold for the ecological entity. The *NEC* is compared with the *AEC* to determine the hazard/risk quotient (*RQ*). A risk quotient of more than 1 implies that there is a risk of adverse effects.

Exposure Concentration

Generally, wastewater from a single source contains a variety of pollutants, denoted below by x, in the effluent. Each ecological entity, y, responds differently to these pollutants. The concentration of pollutant, x, varies over the region based on physical and environmental conditions, and is referred to as predicted (total) environmental concentration (*PEC*). The actual exposure concentration of a pollutant, x, to ecological entity, y, in a given or neighbourhood grid, i, can be calculated as given below:

$$AEC_{i,x,y} = PEC_{i,x,(y)} \cdot BAF_{(i),x,y} \cdot p_{i,(x),y} \cdot PAF_{i,(x),y} \qquad (1)$$

where

$x =$	Pollutants of concern (1, 2,....., n).
$y =$	Ecological entities under consideration (1, 2, ..., λ).
$i =$	Network neighbourhood grids (1, 2, ..., m).
$AEC_{i,x,y}$ =	Actual (estimated) exposure concentration of pollutant, x, for an ecological entity, y, in a grid, i (μg/l).
PEC_1 =	Predicted (total) environmental concentration of pollutant, x, in a grid, i (μg/l).
$BAF_{(i),x,y}$ =	Bioavailability factor of pollutant, x, for an ecological entity, y.
$p_{i,(x),y}$ =	Probability of exposure to pollutant, x, for an ecological entity, y, in a grid, i.
$PAF_{i,(x),y}$ =	Population adjustment fraction for an ecological entity, y, in a grid, i.
$(\cdot)$ =	Parenthesis implies that the parameter is not depending on that specific factor (i.e. i, x, y).

The probability of exposure p is given by:

$$p_{i,(x),y} = \left(\frac{ET \cdot EF \cdot ED}{AT \cdot 8760} \right)_{i,(x),y} \qquad (2)$$

where

ET = Exposure time (hour/day); based on daily movement rates.

EF = Exposure frequency (day/year); based on seasonal migration rates.

ED = Exposure duration (year); based on the life of a project (discharged wastewater), maximum value will be equal to the life of an ecological entity.

AT = Average life of an ecological entity (year).

8760 = A conversion factor (hour/year).

The probability of exposure, p, determines the fraction of time over the life span of an ecological entity that it is exposed to a pollutant due to specific discharges. In this study, the regular daily movements and seasonal migration of the ecological entities to other neighbourhoods from the neighbourhood grid, i, are considered in the evaluation of probability of exposure.

The population adjustment fraction (PAF) is defined as the ratio of an ecological entity, y, in a grid, i, to the total population of ecological entities in that area. It can be estimated by:

$$PAF_{i,(x),y} = \frac{\textit{Population of ecological entity } y \textit{ in a grid } i}{\sum_{i=1}^{m} \textit{Population of ecological entity } y} \tag{3}$$

The bioavailability factor (BAF) is the available fraction of the pollutant, x, to species y. Different ecological entities exposed to the same pollutant absorb it differently depending on its physico-chemical characteristics, and the intake ability of each entity. Generally, the dissolved fraction of the pollutants is used as BAF.

Risk quotient (*RQ*)

The response of the system (based on no-effect concentration) can be estimated from $NOEC$ (no observed effect concentration) and LC_{50} (lethal concentration at which 50% of the specific population group will die). Sadiq (2001) and Sadiq et al.

(2003) developed a methodology for converting toxicity data (*NOEC*, LC_{50}) into *NEC* values. After estimating *NEC* for a given ecological entity, the *RQ* can be determined by:

$$RQ_{i,x,y} = \frac{AEC_{i,x,y}}{NEC_{(i),x,y}} \qquad (4)$$

where

$AEC_{i,x,y}$ = Actual exposure concentration of pollutant, x, to ecological entity, y, in a grid, i (μg/l).

$NEC_{(i),x,y}$ = Predicted no-effect concentration of ecological entity, y, for pollutant, x, (μg/l).

The $RQ_{i,x,y}$ is a surrogate measure of risk posed to a ecological entity, y, in a grid, i, due to exposure of pollutant, x. The risk quotient is useful in explaining the relative risk and can be related to absolute risk. The risk quotient can be aggregated for different pollutants and different ecological entities in a neighbourhood grid, i, to predict overall risk. The CHARM model is used extensively in the North Sea as a tool for determining ecological risk of offshore discharges, which recommends a lognormal relationship for converting risk quotients to actual risk. The discussion on this topic is beyond the scope of this paper and interested readers are referred to Sadiq (2001) for more detail.

Weighting Scheme

The risk quotient can be aggregated by simple addition, but risks posed by different pollutants vary in nature and intensity (Sadiq, 2001). Similarly, different ecological entities (sensitive species, commercial fish, and important species in the food web) have different levels of significance.

The analytical hierarchy process (AHP) is the most widely used technique for multiple-criteria analyses. The AHP develops a linear additive model, but, in its standard format, uses procedures for deriving the weights achieved by pair-wise comparisons between criteria or attributes (Ziara et al., 2002).

The pair-wise comparisons of the attributes are arranged into a reciprocal matrix (Saaty, 1996). The pair-wise comparison of the criteria in the AHP method generates a set of matrices. There are a number of ways to derive the vector of priorities (weights), but emphasis on consistency of the matrix leads to the largest or principal eigen value method. The solution is obtained by raising the matrix to a sufficiently large power, then summing over the rows and normalising to obtain the priority vector, *W*. The process is stopped when the difference between the components of the priority vector obtained at the zth power and the $(z+1)^{st}$ power is less than some predefined small value. The vector of priorities is the derived scale associated with the matrix of comparisons.

Saaty (2001) recommended an easy way to get an approximation to the priorities by normalising the geometric means of the rows. The results coincide with the eigen vector for $n \leq 3$. A second method to obtain an approximation is by normalising the elements in each column of the judgement matrix and then averaging over each row. The geometric mean method was used in this study for simplification. The details of this method can be found in Saaty (2001).

Aggregating Risk Quotients

The risk quotient can be aggregated for all pollutants $x = 1, 2, \ldots, n$ for a given ecological entity, y, in a grid, i, by the following relationship:

$$RQ_{i,n,y} = \sum_{x=1}^{n} RQ_{i,x,y} \cdot w_x \qquad (5)$$

where

$RQ_{i,n,y}$ = Final risk quotient due to all pollutants of concern ($x = 1, 2, \ldots, n$).

w_x = Weights estimated for different pollutants using AHP by establishing priority matrix ($\Sigma w_x = 1.0$).

After aggregating the risk quotient over a number of pollutants for an ecological entity, the risk quotient can be aggregated for different ecological entities based on their relative importance assigned by decision-makers.

$$RQ_{i,n,\lambda} = \sum_{y=1}^{\lambda} RQ_{i,n,y} \cdot w_y \qquad (6)$$

where

$RQ_{i,n,l}$ = Final risk quotient due to all ecological entities of concern (y = 1, 2, ..., λ).

w_y = Weights estimated for different ecological entities using AHP by establishing priority matrix (Σw_y = 1.0).

The risk quotient $RQ_{i,n,l}$, determines the risk posed to "all" ecological entities (y = 1, 2, ..., λ) due to "all" pollutants (x = 1,2, ..., n) in a given grid i. The highest value of the final risk quotient in any grid implies that it has the highest rank or priority for the EEM program. Risk quotient values of more than 1.0 represent the probability of adverse effects. If all risk quotient values are less than 1.0, it does not mean that an EEM program is not required. To avoid this confusion, the final risk quotient values can be converted into risk scores (*RS*), which represent the normalised relative risk in different neighbourhoods. The risk scores values are normalised over a scale of 100.

$$RS_i = \left(\frac{RQ_{i,n,\lambda}}{\sum_{i=1}^{m} RQ_{i,n,\lambda}} \right) \cdot 100 \qquad (7)$$

The EEM locations based on estimated risk score can now be ranked by:

$$Rank_i = Rank\ (RS_i)_{i\Sigma m} \qquad (8)$$

The developed methodology was applied to a hypothetical case study in the following section.

APPLICATION OF PROPOSED METHODOLOGY

The proposed risk-based methodology was applied to a hypothetical case study. Figure 1 shows the discharge from a disposal outlet of an offshore oil facility. The total monitoring area was divided into 10 neighbourhood grids ($i = 1, 2, \ldots, m$; $m = 10$) for simplicity. In reality, the monitoring networks can range from a few to hundreds. The area (or volume) of neighbouring grids are not necessarily equal. Three pollutants of concern ($x = 1, 2, \ldots., n$; $n = 3$) were identified in the discharge. As the ecological community consists of thousands of organisms, three entities were considered further based on their commercial importance, sensitivity, and importance in the food web ($y = 1, 2, \ldots., \lambda$; $\lambda = 3$).

A dispersion model (any appropriate long-range calibrated fate and transport model can be used for this purpose) was used to estimate the predicted environmental concentrations in various neighbourhood grids. The dispersion modelling results may be uncertain due to the highly variable environmental conditions, and other model input parameters. In this study, the uncertainties in the model (and scenario) and the input parameters were not considered, but the model can be adapted to work in probabilistic or possibilistic mode. An average concentration (of three pollutants) in each neighbourhood grid was used as a representative predicted environmental concentration (Table 1).

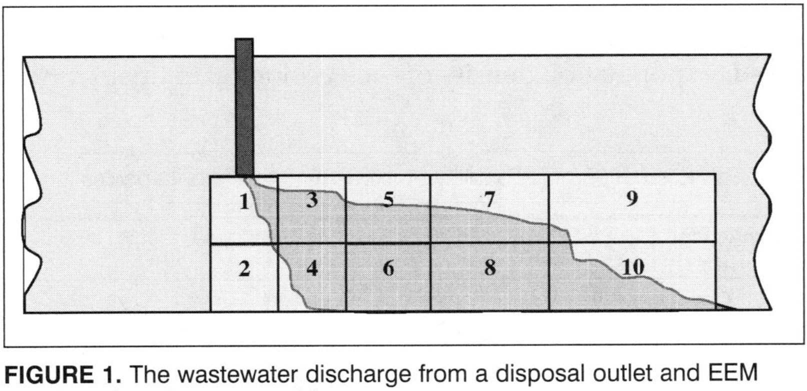

FIGURE 1. The wastewater discharge from a disposal outlet and EEM locations (not drawn to scale).

TABLE 1. Average predicted environmental concentration (μg/l) in various neighbourhood networks.

Neighbourhood Grid (*i*)	$PEC_{i,1,(y)}$	$PEC_{i,2,(y)}$	$PEC_{i,3,(y)}$
1	8.00	0.20	15.00
2	7.50	0.18	12.00
3	7.00	0.15	12.00
4	5.00	0.14	8.00
5	4.90	0.12	6.00
6	3.50	0.10	4.00
7	3.70	0.08	5.00
8	3.00	0.05	3.00
9	3.50	0.05	2.20
10	2.50	0.10	1.50

The probability of exposure, p, for each ecological entity, y, is estimated using equation 2. To estimate p, the exposure parameters ET, ED, EF, and AT are used as reported in Table 2. This information was based on daily movement and seasonal migration rates of different ecological entities. In this example, it is assumed that the regular daily movements and seasonal migration of the species to other neighbourhoods from the neighbourhood grid i, is minimal, and do not significantly affect the probability of exposure, p. The estimated average life of the project is assumed to be 1.5 years, and the average lives of the ecological entities $y = 1$, 2 and 3 are 2 years, 9 months, and 6 months, respectively. As the latter two are shorter than the life of the project, their exposure duration will be limited to their average lives.

The population adjustment fractions, PAF, were estimated based on population density of each ecological entity in differ-

TABLE 2. Estimation of probability of exposure for different species

Ecological Entities	ET	EF	ED	AT	$p_{i,(x),y}$
(y)	(hour/day)	(day/year)	(year)	(year)	
1	20	200	1.50	2.00	0.34
2	18	150	0.75	0.75	0.31
3	20	250	0.50	0.50	0.57

ent neighbourhoods (equation 3). The sum of *PAF* for each ecological entity should be 1.0. Table 3 summarises the *PAF* distributed for all three ecological entities in 10 neighbourhood grids. The bioavailable fractions, *BAF*, of each pollutant, *x*, for different ecological entities are reported in Table 4. Highly stable compounds have a very low *BAF* value (≈ 0), but some compounds are easily dissolvable in water and their *BAF* can be as high as 1.0. The estimated no-effect concentration for different pollutants can be highly variable for the same ecological entity, *y*. Similarly, different ecological entities respond differently to the same pollutant. As mentioned before, these values can be estimated from acute and chronic toxicity parameters LC_{50} and *NOEC*, respectively. Table 5 summarises the *NEC* values estimated in this example.

The risk quotient for each pollutant, x, for ecological entities, *y*, in a grid, *i*, is calculated using equations 1 and 4. After

TABLE 3. Population adjustment fractions for three species in different neighbourhood grids.

Neigbourhood Grid (*i*)	$PAF_{i,(x),1}$	$PAF_{i,(x),2}$	$PAF_{i,(x),3}$
1	0.05	0.03	0.05
2	0.05	0.04	0.10
3	0.10	0.15	0.05
4	0.10	0.10	0.15
5	0.10	0.10	0.08
6	0.20	0.20	0.20
7	0.10	0.05	0.10
8	0.15	0.15	0.25
9	0.05	0.08	0.02
10	0.10	0.10	0.00
=	1.00	1.00	1.00

TABLE 4. Bioavailable fraction for three pollutants for three different ecological entities.

$BAF_{(i),x,y}$	$x = 1$	$x = 2$	$x = 3$
$y = 1$	0.20	0.50	0.20
$y = 2$	0.30	0.40	0.30
$y = 3$	0.50	0.60	0.50

TABLE 5. No-effect concentration for three pollutants for three different ecological entities.

$NEC_{(i),x,y}$	$x = 1$	$x = 2$	$x = 3$
$y = 1$	0.230	0.050	0.900
$y = 2$	0.110	0.025	0.400
$y = 3$	0.050	0.010	0.250

estimating $RQ_{i,x,y}$, these values are grouped for all pollutants for an ecological entity, y, using equation 5. The weights for different pollutants (w_x) are estimated using the AHP technique as described before. Table 6 summarises the weights for different pollutants.

The estimated risk quotients for three species are given in Table 7 for different neighbourhood grids. Generally, these risk quotients are integrated by simple addition. We also applied a weighting scheme for different species using AHP to incorporate the relative significance of each ecological entity in the estimation of risk (Table 6).

The final risk quotients representative of various neighbourhood grids are given in Table 8. All risk quotient values were less than 1.0, but they ranged from 0.08 to 0.98. To prioritise the monitoring locations, we converted them into risk scores. These represent the relative risk values in various grids (e.g. risk to ecological entities is approximately double in grid 4 than in grid 1). The highest risk score obtained was in grid 4, and the lowest risk score was in grid 10. Based on the risk scores, network neighbourhood grids were ranked.

TABLE 6. Estimated weights for various pollutants and species using AHP.

W_x		W_y	
$w_{x=1}$	0.54	$w_{y=1}$	0.20
$w_{x=2}$	0.30	$w_{y=2}$	0.49
$w_{x=3}$	0.16	$w_{y=3}$	0.31
$\sum w_x =$	1.00	$\sum w_y =$	1.00

TABLE 7. Final risk quotients for different pollutants in a given neighbourhood grid, *i*.

Neigbourhood Grid (*i*)	$RQ_{i,n,1}$	$RQ_{i,n,2}$	$RQ_{i,n,3}$
1	0.08	0.13	1.42
2	0.08	0.16	2.63
3	0.14	0.58	1.23
4	0.10	0.28	2.65
5	0.10	0.26	1.35
6	0.14	0.38	2.41
7	0.07	0.10	1.28
8	0.09	0.23	2.52
9	0.03	0.14	0.23
10	0.05	0.13	0.00

TABLE 8. Final risk quotients for different pollutants in a given neighbourhood grid, *i*.

Neigbourhood Grid (*i*)	$RQ_{i,n,}$	RS_i (%)
1	0.52	8.42
2	0.91	14.64
3	0.69	11.12
4	0.98	15.70
5	0.57	9.10
6	0.96	15.45
7	0.46	7.37
8	0.91	14.64
9	0.15	2.34
10	0.08	1.22

For comparison, population and concentration-based rankings were also performed. The ranking orders can be determined using equations 9 and 10 for concentration and population-based approaches, respectively.

$$(Rank_i)_{concentration} = Rank\left(\sum_{x=1}^{n} PEC_{i,x,(y)} \cdot w_x\right)_{i\in m} \tag{9}$$

$$(Rank_i)_{population} = Rank\left(\sum_{y=1}^{\lambda} PAF_{i,(x),y} \cdot w_y\right)_{i\in m} \tag{10}$$

The concentration-based ranking order generally follows the direction of the plume over a period of time. If EEM locations are designed just based on concentration levels, the mean annual wave and current directions can be used. In case of the population-based approach, sensitive and densely populated communities are identified in that area to rank the monitoring locations. The final ranking orders are shown in Table 9 for three methods.

TABLE 9. Final ranking order of EEM neighbourhoods using three approaches.

Neighbourhood Grid (*i*)	Risk-based	Concentration-based	Population-based
1	7	1	10
2	3	2	8
3	5	3	4
4	1	4	3
5	6	5	5
6	2	7	1
7	8	6	6
8	4	8	2
9	9	9	9
10	10	10	7

Correlations between the risk-based and other two methods were estimated. The coefficients of determination were 0.42 (concentration-based) and 0.13 (population-based), both of which are poor, as expected. The concentration-based method focussed on hazard only, whereas the population-based method focussed only on probability of occurrence of adverse effects. The risk-based methodology captures both hazard and probability of occurrence. Validation and calibration of a time dependent version of this model can be built into new and existing EEM programs.

DISCUSSION AND CONCLUSIONS

A new approach for prioritising the EEM program is proposed, which uses risk as the ranking criteria. The risk-based approach considers adverse effects on a representative ecological community using exposure concentration and the toxicity threshold level. The risk is evaluated in surrogate terms as risk quotient, which is the ratio of exposure concentration and a predefined threshold value. The concept of risk quotient is the reverse of factor of safety (ratio of strength to applied load), used in reliability and structural engineering. The structure is considered safe if the ratio is more than 1, but ecological entities are considered safe if the risk quotient is less than 1.

Generally, the EEM locations are not representative of the point of maximum ecological risk. Typical EEM programs can capture the worst conditions either from the point of view of the sensitivity and density of ecological entities (population-based approach) or the highest exposure concentrations (concentration-based approach). Both existing approaches lack in addressing the risk posed to the ecological entities in totality. A risk-based approach uses both sources of information.

In case of non-point pollution sources, a population-based approach might prove more useful. Similarly, for those studies that are focusing on dispersion characteristics of pollutants, a concentration-based approach might be more relevant. But, the risk-based approach provides a more comprehensive framework for prioritising EEM programs. Sensitivity mapping of areas with high ecological relevance in combination with exposure modelling can guide in implementing a risk-based approach. Though a risk-based approach is data demanding, when compared to the two existing approaches, it may prove economical in the long term, such as by avoiding environmental degradation, saving ecological entities, and providing efficient monitoring.

The risk-based approach is valid for controlled discharges, as well as in cases of accidental spills in large water bodies. Indeed, it can be applied to other discharge-effects monitoring designs, including, for example, air pollution and human health effects. The uncertainties related to scenario and model selection, and input parameters were not addressed in this paper. Monte Carlo simulations and fuzzy-based methods can be used to quantify uncertainty in the results.

REFERENCES

Baldauf, R.W., D.D. Lane, and G.A. Marotz. 2001. "Ambient Air Quality Monitoring Network Design for Assessing Human Health Impacts from Exposures to Air Borne Contaminants." *Environmental Monitoring and Assessment*. 66:63-76.

Baldauf, R.W., D.D. Lane, G.A. Marotz, H.W. Barkman, and T. Pierce. 2002. "Application of a Risk Assessment Approach to Designing Ambient Air Quality Monitoring Networks for Evaluating Noncancer Health Impacts." *Environmental Monitoring and Assessment*. 78:213-227.

Khan, F.I. and R. Sadiq. 2004. "Risk-based Prioritization of Air Pollution Monitoring Using Fuzzy Synthetic Evaluation Technique." *Environmental Monitoring and Assessment*. In press.

Lawrence, W.W. 1976. *Of Acceptable Risk*. William Kaufmann, Los Altos, CA.

Saaty, T. L. 1996. *Decision Making With Dependence and Feedback: The Analytic Hierarchy Process*. RWS, Pittsburgh.

Saaty, T. L. 2001. "How to Make a Decision." In T. L. Saaty, and L. G. Vargas, (Eds.), *Models, Methods, Concepts and Applications of the Analytic Hierarchy Process*. Kluwer International Series.

Sadiq, R. 2001. *Drilling waste discharges in the marine environment: a risk based decision methodology*, Ph.D. dissertation, Memorial University of Newfoundland, St. John's, NL, Canada.

Sadiq, R., T. Husain, B. Veitch, and N. Bose. 2003. "Risk Management of Drilling Waste Disposal in the Marine Environment—A Holistic Approach." *Oceanic Engineering International*. 7(1):1-22.

Ziara, M., K. Nigim, A. Enshassi, and B.M. Ayyub. 2002. "Strategic Implementation of Infrastructure Priority Projects: Case Study in Palestine, ASCE." *Journal of Infrastructure Systems*. 8(1):2-11.

Environmental Modelling of Produced Water Dispersion with Implications for Environmental Effects Monitoring Design

Berry, J.A. (jody_berry@hotmail.com) (School for Resource and Environmental Studies, Dalhousie University, Halifax, Nova Scotia, Canada) and P.G. Wells (Environment Canada, Halifax, Nova Scotia, Canada and School for Resource and Environmental Studies)

ABSTRACT: Produced water is the largest waste discharge from the production phase of oil and gas well development. Produced water is a mixture of reservoir formation water and production chemicals from the separation processes, which form a wastewater stream for production platforms. This chemical mixture has a number of toxic components at low concentration, ranging from heavy metals to soluble hydrocarbons. Ecological risk assessment is the quantitative estimation of risk from human activities for a local ecosystem. In this study, analysis of potential environmental effects from produced water was developed using an integrated modelling approach in the context of the ecological risk assessment framework (EPA, 1998). A hydrodynamic dispersion model was used to describe the wastewater plume in the ambient oceanic environment of Sable Island Bank, NS. A fugacity-based model was used to describe the likely plume partitioning in the local environmental media of water, suspended sediment, biota, and sediment. Results from the integrated modelling showed that chronic no-effect concentration (NOECs) for soluble benzene and naphthalene components occurred at a distance of 1.0 meter from the discharge point. The partitioning modelling indicated that low persistence was expected, due to rapid dilution caused by tidal currents for the Sable Island Bank system. The quantitative ecological risk assessment suggested that the risks for adverse environmental effects to both the water column and benthic layer from the low weight, aqueous soluble, hydrocarbon component of produced water on Sable Island Bank are low. This study shows, for the first time, the value to environmental effects monitoring (EEM) design and evaluation of modelling joint toxicity risk quotients.

INTRODUCTION

With the development of Canada's offshore oil and gas reserves, large volumes of produced water (PW) may be discharged into the ocean (Gordon et al., 2000). A production well draws the three reservoir fluids, gas, oil and water, to the surface. The reservoir fluid is separated into the primary phases of gas, oil, and water by the offshore production facility. There are two disposal methods for the water phase (produced water); reinjection into the reservoir, and discharge into the marine environment. Currently in waters off Nova Scotia, Canada, all PW is released into the marine environment after treatment to remove particulate oil. This study will examine PW discharge by the Thebaud production facility on Sable Island Bank to illustrate that environmental modelling can be used to facilitate a continuous improvement program for environmental effects monitoring.

The chemical composition of PW is complex (Neff, 2002). The PW constituents are a mixture of particulate oil (in droplet form), dissolved oil, organic acids, phenols, metals, production chemicals, and radioactive elements. The physical state of PW is also complex. There are true solutions, suspensions, emulsions, particulates, and particle-absorbed mixtures. The complex state of PW adds to the difficulty of assessing environmental fate and risk (Tibbetts, 1992). The amount of

PW discharge is a time-dependent process that is a function of reservoir properties. In offshore oil wells, 98% of produced fluids can be in the water phase (Neff, 2002). Therefore, fate analysis must include an estimate of the temporal production of PW for estimation of risk to the local ecosystem.

Ecological Risk Assessment (ERA) is conducted to translate scientific data into meaningful information about the risk of human activities to the site-specific environment. The purpose of the ERA framework is to allow risk managers to make informed environmental decisions (EPA, 1998). The ERA framework of the Environmental Protection Agency (EPA) was chosen as the analysis method for this study because of the need to differentiate between perceived and real risks for produced water discharge. The Department of Fisheries and Oceans, Environment Canada, and others have indicated that the ERA framework is an appropriate method to examine the management of the various PW components (Gordon et al., 2000; Ray, 2000; Wells, 2000). The ERA framework can be a very useful framework for the implementation of EEM design as will be shown in this study. Exposure analysis can be used with EEM to set up a continuous improvement plan, whereby, the EEM data improves the knowledge of exposure and effects as seen in Figure 1.

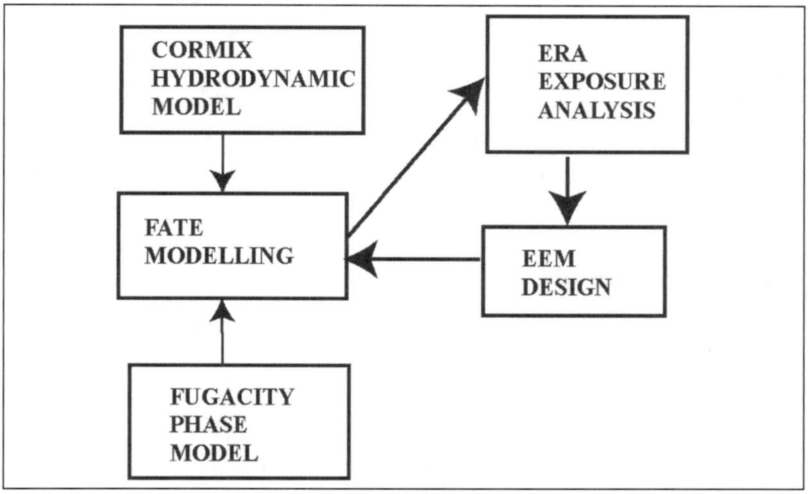

FIGURE 1. Linkages between environmental effects monitoring (EEM), exposure analysis, and the integrated modelling procedure.

PROBLEM FORMULATION AND METHODS FOR ANALYSIS

The risk assessment process includes the use of measurement, testing, and mathematical modelling to quantify the relationship between the initiating event and possible adverse effects (Suter, 1993). The ERA process is used to translate dose-response data into a risk assessment by application of the response information to local site-specific pathways and biota. There are three phases to the ecological risk assessment (ERA) process (Fig. 2). The first phase is problem formulation. Problem formulation is used to evaluate a preliminary hypothesis about possible ecological effects from human activities (EPA, 1998). The second phase of assessment is the analysis phase that considers the two primary components of risk assessment—exposure and effects. The third component of the ERA process is the characterisation of the ecological response and the quantitative estimation of risk.

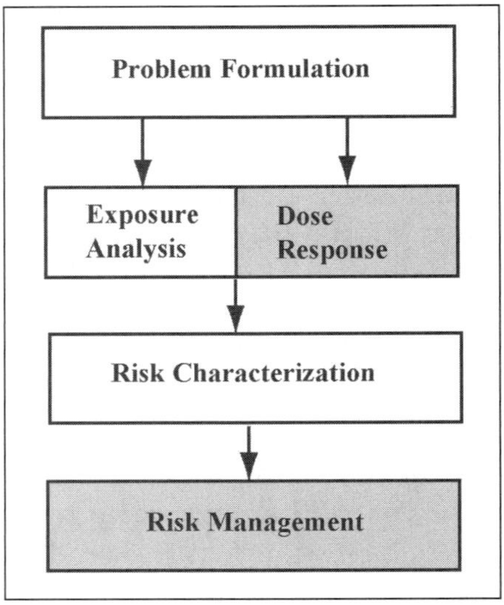

FIGURE 2. Ecological risk assessment framework (as per EPA 1998, and Landis and Yu, 1999). White blocks indicate components addressed in the present study.

Problem Formulation

Problem formulation is used to develop ecologically relevant endpoints that help to sustain the natural structure, function, and biodiversity of the relevant ecosystem (EPA, 1998); in this case, the Thebaud production facility on the Sable Island Bank (SIB) ecosystem. Careful choice of appropriate endpoints facilitates a framework for measuring the site-specific, stress-response relationships (EPA, 1998). Two elements are required to define an assessment endpoint. The first is the identification of the specific valued ecological entity, and the second is a characteristic of the ecological entity that is important to protect or is potentially at risk (EPA, 1998). For PW, the offshore oil separation processes typically removes the particulate oil to 10–20 ppm (Berry, 2003). Therefore, the soluble hydrocarbon component of PW is thought to be one of the main contributors to toxicity and risk (Neff, 2002). The identification of components of concern is shown in Table 1. Of particular interest at lower stressor levels are "no-effect" scenarios (or no observable effect concentrations, the NOECs). Such no-effect levels are defined as stressor levels that indicate no significant adverse effects compared to controls under experimental conditions.

TABLE 1. Typical concentrations of soluble oil components in produced water and estimated risk quotients using experimentally derived NOEC data (from Neff, 2002; Jacobs, 1992; Utvik, 1999a). NA denotes that data are not available; NOEC is the no observable effect concentration derived from EPA chronic exposure data; RQ is the risk quotient or PW concentration divided by NOEC.

Group	Chemical	PW Concentrations (μg/l)	NOEC (μg/l)	RQ
Aromatic	Benzene	80 - 4300	700	6.1
Hydrocarbons	Toluene	80 - 2560	5000	0.5
	Ethylbenzene	20 - 100	NA	NA
	Xylene	13 - 720	NA	NA
Polynuclear	Naphthalene	5.3 - 1000	620	1.6
Aromatic	Anthracene	0 - 21.7	710	0.03
Hydrocarbons	Phenanthrene	0 - 30	4.6	6.5

The risk quotient (RQ) in Table 1 shows that at the highest exposure concentrations, effects could be caused from PW discharges. A RQ that is greater than 1 indicates that an examination of the site-specific pathways is necessary (Landis and Yu, 1999). Toxicity studies on various marine organisms have shown toxic effects from BTEX and PAH compounds at concentrations above 700 μg/l and 620 μg/l (Table 1). Studies have examined laboratory exposure concentrations and found food chain effects in phytoplankton for naphthalene at concentrations of 0.03 ppb (Brakstad et al., 1996). Unicellular microorganisms constitute at least 50 % of the total marine biomass and dominate with respect to volume and metabolic activities (Mann, 2000). Endpoints for the study were chosen to be the plankton (including microorganisms, phytoplankton, and zooplankton) that inhabits the Sable Island Bank ecosystem exposed to the soluble hydrocarbons of naphthalene and benzene.

Analysis Plan

Analysis in the EPA framework is used to characterise site-specific exposure for the ecosystem that is being examined. Exposure is defined as the contact between a stressor and a receptor (EPA, 1998). The objective of analysis is to describe exposure in terms of intensity, space, and time in units that utilise the dose-response analysis. Two primary considerations characterise exposure:

1. The location of a source and the environmental media that receive the stressors.
2. The spatial and temporal distribution of the stressor in the environment, which is accomplished by careful examination of the potential environmental pathways of transport, such as air currents, water currents, density forces, particle forces, the food web, and possible transformation of compounds.

In this study, pathways that affected the persistence and fate of the PW components were examined (Berry, 2003). Physical and chemical oceanographic pathways need to be described to examine dilution and potential compartments of the SIB water column affected by PW dispersion. This was accomplished using CORMIX, an open channel model (Tyler et al., 1997), to examine the plume dispersion for discharge conditions.

Two water column density conditions were examined (unstratified and stratified) using CORMIX. An unstratified water column occurs during the months of October to May, while stratified water column and slower currents occurs during the months of June to September (Hannah et al., 2001). The PW effluent plume will be driven by its initial density (buoyancy flux) in one of two directions, either towards the seafloor or towards the sea surface (Brandsma et al., 1992). Initial descent of PW effluent is a function of the velocity vector of the effluent in relation to current velocity (Brandsma and Smith, 1996). The expansion and subsequent dilution of the PW plume based on density and velocity differences between the effluent and the ocean is called dispersion. Dispersion established estimates of time and environmental compartment volumes for fugacity modelling.

Fugacity modelling (Mackay, 2001 and earlier papers) was used to model the potential chemical pathways for the SIB ecosystem. Fugacity is a physical-chemical property that can be used to mathematically describe the rates that chemicals diffuse or are transported between environmental phases. The chemical transfer rates are expressed as a proportion of fugacity differences between source and destination phases with physical transport equations describing the rates that environmental media interact (Mackay, 2001). Fugacity modelling determined the environmental concentrations of the selected hydrocarbons and their fate after reactions of oxidation, degradation, and photolysis.

Narcosis has been found to be the common mode of toxic action for many organic chemicals. In particular, many studies have concluded that hydrocarbons under static exposure conditions are narcotic (most recently, Rogerson et al., 1983; Abernethy et al., 1986; Abernethy et al., 1988; McCarty et al., 1992; Neff, 2002). These studies reached a number of conclusions:

1. BTEX and PAHs have a common mode of action, narcosis, as do many of the constituents of PW.
2. Since the mode of action is non-selective, there is no influence of molecular reactions, and toxicity is controlled by organism-water partitioning, which is a reflection of the aqueous solubility.
3. No particular single hydrocarbon or group of hydrocarbons can be regarded as the primary toxicants in aqueous solutions of oils, hence the primary focus should be on the mass of the organic molecules or hydrocarbons for the purposes of exposure assessment in ERA.

From these conclusions, an analysis plan was formalised to use benzene and naphthalene to model the risk of total BTEX and PAHs levels measured at offshore production facilities (Table 2). The methods for examining the risk to the SIB ecosystem are described by equation 1 for joint toxicity (adapted from Christensen and Chen, 1991) and equation 2 for risk estimation (adapted from Landis and Yu, 1999). Equation 1 describes the difference between a toxicological response for a chemical acting alone or in a mixture as:

$$\frac{c_1}{C_1} + \frac{c_2}{C_2} + \ldots \frac{c_n}{C_n} = 1.0 \qquad (1)$$

where c_i is the mixture concentration of chemical i that gives a biological response, and C_i is the concentration of chemical i when acting alone that gives a biological response. If the overall ratio of a mixture is greater than the response of the constituents, Equation 1 will result in values greater than 1.0. The work above indicates that the expected ratio for soluble low weight hydrocarbons is 1.0 or additive.

The risk quotient (RQ) was calculated as:

$$RQ = \frac{PW_{\mathrm{EC}}}{EC_{\mathrm{KE}}} \qquad (2)$$

where PW_{EC} is the produced water concentration and EC_{KE} is the known effect threshold concentration.

Maximum concentrations of total BTEX and total PAHs in Table 2 were used as concentrations of benzene and naphthalene. Benzene and naphthalene were examined independently and summed for an estimate of joint toxic action, using EPA chronic toxicity (NOEC) water quality limits (Table 1). The NOEC data was also compared to toxicity data for benzene and naphthalene on plankton (Berry, 2003).

The results from the physical and chemical modelling were combined to create an estimate of risk for the SIB ecosystem. The risk quotient profile (RQP) applies the spatial concentrations to the dose-response effects data in the form of a risk

TABLE 2. Sample concentrations for soluble oil constituents from three production locations (Neff, 2002; Utvik, 1999; Smith et al., 1998).

Compound	Gulf of Mexico (μg/l)	Indonesia (μg/l)	North Sea (μg/l)
Benzene	440 - 2800	84 - 2300	4300
Toluene	340 - 1700	89 - 800	2560
Ethylbenzene	26 - 100	26 - 56	400
Xylenes	160 - 700	130 - 480	430
Total BTEX	966 - 5300	329 - 3636	7690
Naphthalene	5 - 90	45 - 150	1070
Phenanthrene	1 - 8	9 - 29	71
Fluorene	3	3 - 5	3 - 15
Total PAH	100	180	1156

quotient. PW volumes are time dependent and have been seen to increase during the production process (Utvik, 1999; Neff, 2002). Two rates of production were examined. The present rate of PW production at the Thebaud facility was assessed using Canada-Nova Scotia Offshore Petroleum Board's activity reports (Berry, 2003), and used to assess the present risk of PW. Future risk from PW was examined using the maximum platform capacity. The present and maximum production volumes at Thebaud production facility are 135.6 m^3/day and 211.7 m^3/day respectively. The present and maximum rates were used for a temporal estimation of risk from produced water on SIB.

RESULTS

The CORMIX modelling conducted for Thebaud platform used data from Coastal Oceans Inc. to characterise the water column structure and density (Sable Offshore Energy Inc., unpublished data). The data was checked against the published values of Hannah et al. (2001). The velocity of the stratified water column was assumed to be 0.1 m/s and unstratified water column velocity was assumed to be 0.15 m/s (Hannah et al., 2001). The CORMIX program used the production flow rates and water column structure to describe the portions of the SIB water column that were affected by the PW plume in both the stratified and unstratified density conditions. Results are shown in Figures 3 and 4 for the two water column density conditions. CORMIX was also used to examine the advection residence time of the PW constituents of benzene and naphthalene for fugacity modelling, and determined that the PW plume stayed in the 1 km by 1 km system for 3 hours.

In Figure 3(a), the PW plume density is greater than the ambient water column, and the plume descends through the water column to the benthic region. The dilution in Figure 3(b) shows that the rate is proportional to the expansion of the PW plume through the unstratified water column. Figure 4(a) shows that the PW plume is entrapped by the stratified water

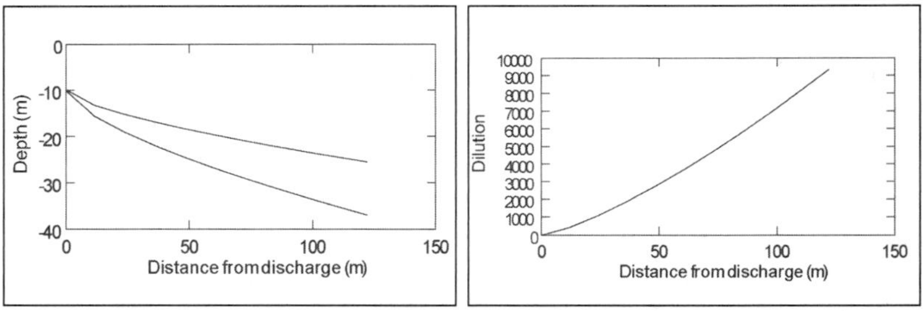

(a) Plume dispersion from discharge (b) Plume dilution from discharge

FIGURE 3. Unstratified water column dispersion of produced water—Thebaud Platform on Sable Island Bank.

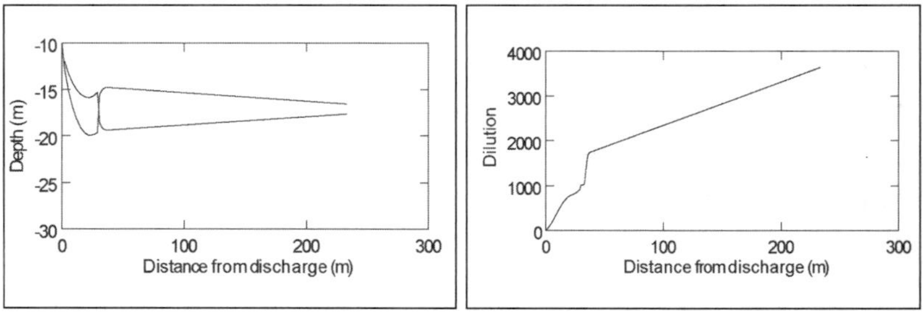

(a) Plume dispersion from discharge (b) Plume dilution from discharge

FIGURE 4. Stratified water column dispersion of produced water—Thebaud Platform on Sable Island Bank.

column structure. This occurs at much shallower depth than the benthic region (16 m vs. 40 m). Figure 4(b) shows that the initial dilution is more rapid before the entrapment process. However, after entrapment occurs, dilution is slower as compared to the unstratified curve (Fig. 3b).

Fugacity modelling results are presented in Tables 3 and 4 for benzene and naphthalene, respectively. The mass flow rates for benzene and naphthalene were calculated using the total BTEX and PAH concentrations (Table 2) combined with the

TABLE 3. Summary of multimedia partitioning of benzene for water column conditions on the Scotian Shelf, using the fugacity level III fate analysis. Z values denote the capacity of a phase for a chemical.

Mass Flow Rate		0.0423 kg/hr		0.066 kg/hr	
Environmental Compartment and Phase	Z Value (mol/m³.Pa)	Amount (kg)	Conc. (μg/g)	Amount (kg)	Conc. (μg/g)
Water: Bulk	1.80E-03	0.125065	3.38E-06	0.195137	5.28E-06
Water	1.80E-03	0.125058	3.38E-06	0.195126	5.28E-06
Sus. Particles	2.98E-02	6.23E-06	3.74E-05	9.71E-06	5.84E-05
Fish	1.21E-02	8.43E-07	2.28E-05	1.32E-06	3.56E-05
Sediment: Bulk	6.44E-03	3.78E-04	4.11E-06	5.89E-04	6.41E-06
Water	1.80E-03	4.21E-05	2.11E-06	6.58E-05	3.29E-06
Solid	9.54E-03	3.36E-04	4.67E-06	5.24E-04	7.28E-06

TABLE 4. Summary of multimedia partitioning of naphthalene for water column conditions on the Scotian Shelf, using the fugacity level III fate analysis. Z values denote the capacity of a phase for a chemical.

Mass Flow Rate		0.00678 kg/hr			0.0106 kg/hr		
Environmental Compartment and Phase	Z Value (mol/m³.Pa)	Amount (kg)	Conc. (μg/g)	RQ	Amount (kg)	Conc. (μg/g)	RQ
Water: Bulk	2.33E-02	2.01E-02	5.43E-07		3.14E-02	8.49E-07	
Water	2.33E-02	2.00E-02	5.42E-07		3.13E-02	8.48E-07	
Sus. Particles	6.7047	1.73E-05	1.04E-04	0.00065	2.71E-05	1.63E-04	0.00102
Fish	2.72549	2.35E-06	6.36E-05		3.67E-06	9.94E-05	
Sediment: Bulk	1.2966	1.48E-04	1.61E-06	1.00E-05	2.31E-04	2.51E-06	1.60E-05
Water	2.33E-02	1.06E-06	5.30E-08		1.65E-06	8.28E-08	
Solid	2.1455	1.47E-04	2.04E-06		2.29E-04	3.18E-06	

flow rates of the production platform to give a mass flow rate. The plume concentrations were checked to see if an additional modelling step was required to describe the benthic region (i.e. sediment plume interaction) (Berry, 2003). For the prevailing depth of 36 meters (i.e. Thebaud platform), the multimedia diffusion parameters were used to describe the water column contaminant transport to the benthic region. The resulting sediment concentrations were then compared to EPA effects range low (ERL) limits for naphthalene to develop the risk quotient in Table 4 for sediment and suspended sediment (Neff, 2002). Suspended sediment concentrations in the water column were assumed to be 3 ppm, based on the work by Muschenheim et al. (1995). The fugacity modelling was conducted for a 1 km by 1 km region that has been documented to be the region of platform effects (Neff, 2002). The physical chemical properties for benzene and naphthalene were obtained from Mackay et al. (2000). The temperature used to model the SIB system was 20°C. This is a conservative estimate and should lead to greater persistence than at the actual water temperature (5–10°C). Results from the fugacity level III analysis were that 98.6% of the chemical mass was removed from the system by advection in the water phase. 1.2% of the chemical mass was removed by loss mechanisms including degradation, oxidation, and photolysis.

The partitioning results indicate that a majority of benzene and naphthalene chemical mass is advected through the local ecosystem, therefore dispersion will control the risk for the SIB ecosystem. Total BTEX and PAH concentrations from Table 2 were applied to the CORMIX dilution curves, and a plot of the joint toxic action risk can be made using equations 1 and 2. This is similar to the RQ presented in Table 1, except the risk quotient profile (RQP) applies the CORMIX dilution curve to give a site-specific spatial estimate of joint toxic risk from the discharge point. The risk quotient results are illustrated for joint toxic action of benzene and naphthalene for both stratified and unstratified water column density in Figures 5 and 6. The plotted risk quotients in Figures 5 and 6 indicated that in the water column, for present and future activities, there is little change in the anticipated chronic response. The dilution

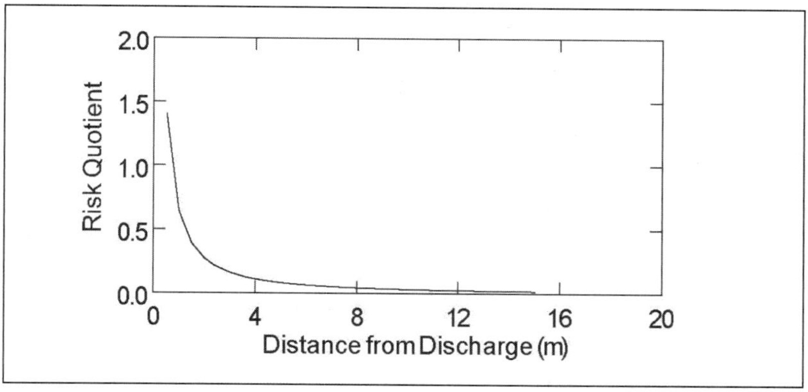

FIGURE 5. Risk quotient for the joint toxic action of benzene and naphthalene for unstratified water column conditions on SIB.

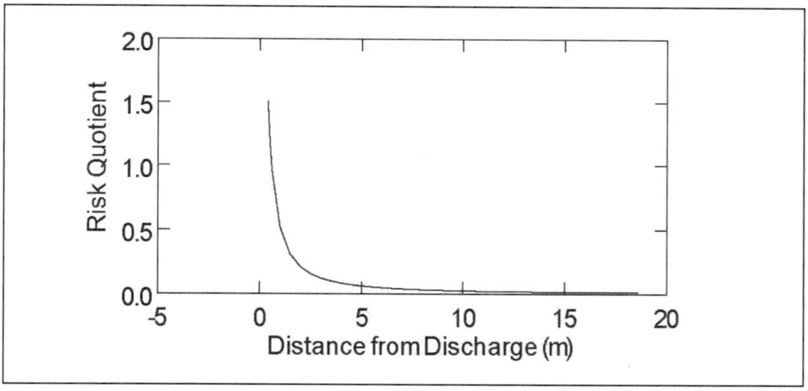

FIGURE 6. Risk quotient for the joint toxic action of benzene and naphthalene for stratified conditions on SIB.

of the PW plume is rapid and NOECs are reached for the joint toxic action within 1 meter of the discharge point.

Environmental effects monitoring (EEM) needs to consider two regions for studies. The first region, defined using an estimate of the distance to reach NOEC concentrations, can be considered the toxic region of study, and has been analysed above. The distance to NOEC (1 m) indicates that chronic toxic effects should be observed in a small region (Figs. 5 and 6). The second region is defined by the distance from the discharge point to background concentrations. The estimation of where the PW plume reaches background concentrations established conservative limits for significant monitoring studies. Once the plume reached background concentrations, it was impossible to distinguish between influences from the ambient environment and the point of PW discharge (Law et al., 1997). The distance to background concentrations was based on drinking water guideline values of 10 μg/l for naphthalene and 5 μg/l for benzene, which were presented by Law et al. (1997). The concentration ratios of the PW plume discharge to background concentrations are plotted against distance from discharge in Figure 7. Figure 7 shows the influence of stratification on dilution concentrations, with dilution being more rapid in the stratified situation up to the 15-meter point (Fig. 7). However, dilution is observed to be much slower after 15 meters of discharge due to diffusion occurring only in the horizontal axis.

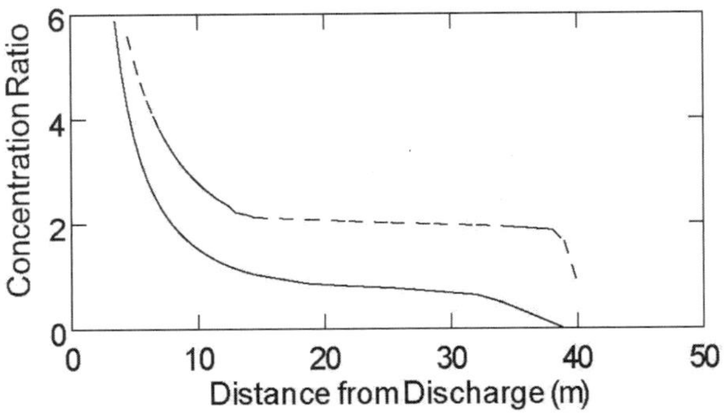

FIGURE 7. Distance to background concentrations of benzene for a stratified water column. The solid black line is based on the present discharge rates, while the dashed black line represents the scenario at the future rate of discharge.

DISCUSSION

The examination of toxic effects from produced water-soluble components benzene and naphthalene indicate that the predicted area of impact (risk) around the Thebaud platform is small. A more detailed analysis is presented in Berry (2003). The dilution of PW components occurred at a distance of 1 meter for both individual and joint toxicity. The parameters that control entrainment of the PW plume were more important than water column stratification for the occurrence of toxic effects. The methodology for joint assessment presented in this study could be used to complete an assessment of the risk for the entire PW mixture, but this was not in the scope of the present study.

The effects of PW dispersion are predicted to be confined to the local water column at the point of discharge and there is a low risk of contamination of benthic communities from PW discharge (Table 4). However, as the depth of the water column changes, the interpretation of risk from PW components will change. For deeper water columns, a majority of the chemicals will sorb to the suspended sediment and biota present in the water column. For shallower depths (less than 40 meters), contamination of the benthic boundary layer is more likely (Berry, 2003). This could lead to problems in shallow slow moving bodies of water, but is highly unlikely for the tidal, current, and mixing conditions that govern Sable Island Bank (Hannah et al., 2001).

It is clear from this study that the exposure of the water column around the Thebaud platform from hydrocarbons (BTEX and PAH) is minimal; concentrations anticipated to cause toxicity only occurring within a meter or so of the PW discharge point. Hence, exhaustive and expensive effects monitoring (i.e. EEM) of the water column at distances more than 100 meters from the PW discharge point is considered unnecessary.

This study, while only considering two categories of hydrocarbons and not the complex PW mixture in total, supports the role of fate modelling and exposure analysis in designing and implementing EEM studies near oil and gas platforms. This study shows, for the first time, the value to EEM of modelling joint toxicity risk quotients. The study also shows in the context of Sable Island Bank there is a need for very confined (limited) sampling near each platform or discharge point. While it is acknowledged that this study was limited by the toxicity data available for PW (i.e. Sable platforms' PW was not sampled), the maximum documented concentration ranges for other locations were used to minimise this limitation. In further studies, samples of PW should be collected for toxicity analysis and a measurement of specific gravity and temperature of the effluent should be recorded. Samples should be collected both before and after discharge for chemical and toxicity analysis to verify the risk estimates in this study. Data from such sampling could be used for further iterations of the models, as in this study, to test the conclusions of limited effects of PW discharges on SIB.

ACKNOWLEDGEMENTS

We would like to thank Kerr McGee, the Department of Fisheries and Oceans, the Panel for Energy Research and Development (PERD), and Environment Canada for the funding that enabled this study. We would like to thank members of the thesis committee Ray Cote, Cal Ross, Dr. Kee Muschenheim, and Andre d'Entremont. The feedback from the committee refined the study and helped immensely. We would also like to thank both Drs. Ken Lee and Charles Hannah for answering many questions during the research.

REFERENCES

Abernethy, S.G., A.M. Bobra, W.Y. Shiu, P.G. Wells, and D. Mackay. 1986. "Acute Lethal Toxicity of Hydrocarbons and Chlorinated Hydrocarbons to Two Planktonic Crustaceans: The Key Role of Organism-water Partitioning." *Aquatic Toxicology*. 8:163-174.

Abernethy, S.G., D. Mackay, and L.S. McCarty. 1988. "'Volume Fraction' Correlation for Narcosis in Aquatic Organisms the Key Role of Partitioning." *Environmental Toxicology and Chemistry*. 7:469-481.

Berry, J.A. 2003. "Integrated Modelling for Exposure Assessment—a Methodology Applied to Assessing the Impact of Produced Water on Sable Bank (Scotian Shelf, Canada)." Masters of Environmental Studies Thesis. School for Resource and Environmental Studies. Dalhousie University, Halifax, Nova Scotia, Canada. 122p.

Brakstad, O.G., A.J. Olsen, T. Nordtug, T.K. Frost, T. Aunnaas, and S. Johnsen. 1996. "Uptake and Degradation of Discharged Produced Water Components in Marine Microorganisms." *Society of Petroleum Engineers*. 35910.

Brandsma, M.G. and J.P. Smith. 1996. "Dispersion Modeling Perspectives on the Environmental Fate of Produced Water Discharges." In *Produced Water Technological/Environmental Issues and Solutions 2*, pp. 215-225. Environmental Science. Vol 52. Plenum Press, New York.

Brandsma, M.G., J.P. Smith, J.E. O'Reilly, R.C. Ayers, and A.L. Holmquist. 1992. "Modeling Offshore Discharges of Produced Water." pp 59-73. In *Produced Water Technological/Environmental Issues and Solutions.* Environmental Science. Vol 46. Plenum Press, New York.

Christensen, E.R. and C.Y. Chen. 1991. "Modelling of Combined Toxic Effects of Chemicals." *Toxic Substances Journal.* 11:1-63.

Gordon, D.C., L.D. Griffiths, G.V. Hurley, A.L. Muecke, D.K. Muschenheim, P.G. Wells. (Eds.) 2000. *Understanding the Environmental Effects of Offshore Hydrocarbon Development.* Canadian Technical Report of Fisheries and Aquatic Sciences, 2311. 82p.

Hannah, C.G., J.A. Shore, and J.W. Loder. 2001. "Seasonal Circulation on the Western and Central Scotian Shelf." *Journal of Physical Oceanography.* 31:591-615.

Jacobs, R.P.W.M., R.O.H. Grant, J. Kwant, J.M. Marquenie, and E. Mentzer. 1992. "The Composition of Produced Water from Shell Operated Oil and Gas Production in the North Sea." pp. 13-23. In *Produced Water Technological/Environmental Issues and Solutions. Environmental Science.* Vol 46. Plenum Press, New York.

Landis, W. and M. Yu, 1999. *Introduction to Environmental Toxicology Impacts of Chemicals upon Ecological Systems.* 2nd Edition. Lewis Publishers, Boca Ranton, FL. 328p.

Law R.J., V.J. Dawes, R.J. Woodhead, and P. Matthiessen. 1997. "Polycyclic Aromatic Hydrocarbons (PAH) in Seawater around England and Wales". *Marine Pollution Bulletin.* 34: 306-322.

Mackay, D. 2001. *Multimedia Environmental Models—the Fugacity Approach.* 2nd Edition. Lewis Publishers. CRC Press, Boca Raton, FL. 261p.

Mackay, D., W.Y. Shiu, and K.C. Ma. 2000. *Physical-Chemical Properties and Environmental Fate and Degradation Handbook.* CRCnetBASE 2000, Chapman & Hall CRCnetBASE, CRC Press LLC., Boca Raton, FL. (CD-ROM.)

Mann, K.H. 2000. *Ecology of Coastal Waters with Implications for Management.* 2nd Edition. Blackwell Science. Malden, Massachusetts. 406p.

McCarty, L.S., D. Mackay, A.D. Smith, G.W. Ozburn, and D.G. Dixon. 1992. "Residue-based Interpretation of Toxicity and Bioconcentration Qsars from Aquatic Bioassyays: Neutral Narcosis Organics." *Environmental Toxicology and Chemistry*. 11:917-930.

Muschenheim D.K., T.G. Milligan, and D.C. Gordon. 1995. *New Technology and Suggested Methodologies for Monitoring Particulate Wastes Discharged from Offshore Oil and Gas Drilling Platforms and their Effects on the Benthic Boundary Layer Environment*. Canadian Technical Report of Fisheries and Aquatic Sciences 2049. Fisheries and Oceans Canada, Ottawa, ON. 55p.

Neff, J.M. 2002. *Bioaccumulation in Marine Organisms. Effects of Contaminants from Oil Well Produced Water*. Elsevier Science ltd. Oxford, UK. 452p.

Ray, J. 2000. "The Ecological Framework for Environmental Effects Monitoring a Perspective From Outside the Region". In *Understanding the Environmental Effects of Offshore Hydrocarbon Development*. pp. 45-53. Canadian Technical Report of Fisheries and Aquatic Sciences 2311.

Rogerson, A., W.Y. Shiu, G.L. Huang, and D. Mackay. 1983. "Determination and Interpretation of Hydrocarbon Toxicity to Ciliate Protozoa." *Aquatic Toxicology*. 3:215-228

Smith, J.P., A.O. Tyler, and Z.A. Sabeur. 1998. "Ecotoxicological Assessment of Produced Waters in Indonesia". *Environmental Toxicology and Water Quality*. 13(4):323-336.

Suter, G.W. 1993. *Ecological Risk Assessment*. Lewis Publishers. Chelsea, Michigan. 538p.

Tibbetts, P.J.C. 1992. "A Comprehensive Determination of Produced Water Composition." In *Produced Water Technological/Environmental Issues and Solutions*. Environmental Science Research. Vol. 46. Plenum Press. New York.

Tyler, A.O., M.C. Rymell, and J.P. Smith. 1997. "Facility-Specific Models as an Essential Aid to Determining Predicted Environmental Concentrations (PECs) for Production Water Risk Analysis." *Society of Petroleum Engineers*. 37868

Utvik, T. 1999. "Chemical Characteristics of Produced Water from Four Offshore Oil Production Platforms in the North Sea." *Chemosphere*. 39:2593-2606.

Wells, P.G. 2000. "Understanding the Environmental Effects of Offshore Hydrocarbon Developments: Workshop Summary and Reflections." In *Understanding the Environmental Effects of Offshore Hydrocarbon Development*. pp. 76-78. Canadian Technical Report of Fisheries and Aquatic Sciences. 2311.

WEB RFERENCES

EPA. 1998. Guidelines for Ecological Risk Assessment. Published on May 14, 1998, Federal Register 63(93):26846-26924. U.S. Environmental Protection Agency, Washington, D.C. http://cfpub.epa.gov/bncea/cfm/recordisplay.cfm

The Role of Mass Balance Models in Offshore Environmental Effects Monitoring

Warren, C.S. and Donald Mackay (dmackay@trentu.ca)
(Canadian Environmental Modelling Centre, Trent University, Peterborough, ON, Canada)

ABSTRACT: It is suggested that mass balance models can play a valuable role in assessing the impacts of offshore oil and gas operations. These models attempt to describe the fate of hydrocarbons released into the marine environment during production or from accidental spills. Traditionally, trajectory models have provided valuable forecasts of the movement of oil, but the focus here is on models which assess transport and transformation processes and seek to establish the masses of hydrocarbon components in various states and subject to various fates including evaporation, degradation, dissolution, dispersion, uptake by biota, association with solid substrates and present as surface slicks. Average concentration ranges can be estimated and compared with monitoring data. Estimates can be made of the nature and extent of contamination of biota. The merits of remedial actions such as the use of chemical dispersants can then be better evaluated. It is concluded that the availability of even an approximate model can be valuable by providing the operators, regulatory agencies and other stakeholders with the likely "big picture" of the impacts of oil on the physical and biological components of the potentially or actually impacted marine ecosystem.

INTRODUCTION

Offshore oil and gas exploration and production have become increasingly important to Atlantic Canada. Several fields are currently in operation on the Scotian Shelf as well as on the Grand Banks of Newfoundland. With these production facilities comes the concern for potential detrimental effects on the marine ecosystem. Comprehensive monitoring programs encompassing measurements in the physical (abiotic) and biotic communities are used to detect chemical concentrations and potential effects to the local ecosystem. While programs of this nature are essential for indicating potential effect concentrations, a carefully developed and validated fate model can be complementary to this endeavour and help alleviate some of the concerns and data gaps associated with monitoring.

Monitoring programs are usually performed on a large scale incorporating a large number of personnel, equipment and ocean vessels. These can be very expensive. For offshore oil operations, monitoring programs are carried out on the high seas, and thus can be dangerous for those involved. In many analytical measurement programs it can be difficult to obtain relevant data that can be compared to previous studies. Precautions have to be taken to ensure tests are carried out by consistent methods at corresponding spatial and temporal frameworks. Detection limits for some contaminants can also be problematic when analysing samples. Furthermore, when performing monitoring programs there is always a risk of acquiring large quantities of under-evaluated data. Data are also often misinterpreted without having an appreciation of the "bigger picture".

Here we assert that a combination of modelling and monitoring can be very effective in enhancing environmental effects monitoring.

MODEL BENEFITS

Incorporating a modelling effort into a monitoring program can help reduce some of the concerns of cost, associated dangers, and analytical challenges. Furthermore, models can help to focus monitoring programs to avoid the collection of large quantities of unusable or uninterpretable data. By formulating the monitoring questions into a model, the need for specific data is established and important processes with potential data gaps may be identified. In these situations, monitoring programs can be focused to acquire data that is needed to produce realistic models with limited uncertainty. Once concentrations have been determined for the physical environment (e.g. water column, sediment bed) links can be made to concentrations in biota and ultimately to effects. It is also possible to build in system food webs to explain the movement of chemicals through food webs.

Models are particularly beneficial when applied to many different chemicals such as hydrocarbons, which form distinct homologous series. Once a single chemical has been evaluated in the model, the user can use appropriate physical-chemical property data to extrapolate to different hydrocarbons and other important chemicals, further reducing monitoring needs. One of the main features of mass balance models is their spatial and temporal predictive ability. Monitoring programs are usually performed over radial transects from the contaminant source at defined time periods for a number of years. This indicates spatial and temporal trends. A model is able to incorporate this information as well as document potential concentrations observed at decreased or increased emissions from spills or effluents (e.g. process waters). Using the model, remedial actions can be explored in order to determine optimal effectiveness.

It is important to remember that modelling exercises are only as effective or realistic as the monitoring data that is used to evaluate them. As a result, it is important to maintain a close relationship between the two and continuously compare modelling results to monitoring data. During this process the aim is to achieve a satisfactory level of agreement. The model is formulated based on knowledge of what is believed to occur in the system. Therefore, when model results and monitoring data agree this suggests that an adequate understanding of the processes exists. When there is not good agreement, this may be a result of gaps in the knowledge applied in the model, or the data may be in error, or both.

MODEL LIMITATIONS

For complex systems with multifaceted contaminant release scenarios, it is often necessary to use spatially and temporally explicit models. This is the case for oil, which changes its composition and transport behaviour the longer it is exposed to the environment. However, if monitoring data are not available to track the changes of the oil, it becomes difficult to evaluate the model and be certain of results. Furthermore, as models become spatially complex it can be difficult to interpret results and assess the appropriateness of the model constructs. In many cases, it can also be difficult to obtain reliable input parameters to run a model, which is only intensified by any increases in model complexity. This can be the case for food web models in which a good understanding of the dynamics of the food web is essential to the credibility of the model.

MODEL AVAILABILITY

There is a wide variety of models available for application to marine and fresh water systems such as estuaries, lakes, and rivers. In most cases, different types of models are designed to apply to different aspects of the environment. Typically, fate and transport models, or abiotic models, look at the movement and degradation of contaminants in a system, while separate bioaccumulation and biouptake models are used to assess contaminant exposure experienced by biota. Furthermore, effects modelling involves the comparison of contaminant concentrations in biota to known effect concentrations. The incorporation of these three modelling components allows the user to assess the release of contaminants in the environment and the subsequent effects on biota. Monitoring data can be incorporated as input to the biouptake and effects models or results from the abiotic model can be used for the biouptake model and subsequent predictions used for the effects modelling. The extent to which monitoring data are used is dependent on its availability as well as the users' intensions for the modelling exercise.

In a review by Paquin et al. (2003), a number of mass balance models are discussed that can be used to examine contaminant fate and transport in aquatic systems, such as, WQAM (Water Quality Assessment Methodology), CTAP (Chemical Transport and Analysis Program) and WASP5 (Water Quality Analysis Simulation Program-5), just to name a few. Other models have also been developed to address offshore marine systems, which incorporate the abiotic, biouptake, and effects modelling components. Applied Science Associated (ASA) has developed models dealing with issues related to oil production, such as OilToxEx, which has been used for assessing toxicity and exposure resulting from oil spills (French-McCay, 2002). This model is part of a larger ASA project, SIMAP (Spill Impact Model Application Package) that incorporates a suite of models focused on oil spill fate and biological effects. Another example is that of Cranford et al. (2003), in which a series of different models were used to examine the impact of drilling wastes on scallops in the North Atlantic Ocean.

The above are just a few of the different models that may be used to examine oil transport and effects in the environment. There is merit in having several different models that address the same contaminant and environmental scenarios. Different models can provide different insights into the importance of fate processes and help the user avoid reliance on one model in all circumstances. In this paper we present one such approach of incorporating the abiotic, biouptake, and effects modelling components, employing the fugacity based QWASI (Quantitative Water Air Sediment Interaction) and food web models. One advantage to fugacity modelling is the ease at which abiotic and biotic models can be linked. Furthermore, bioavailability is automatically taken into account in fugacity models.

ABIOTIC MODELLING

We suggest the use of the Quantitative Water Air Sediment Interaction (QWASI) model for the development of an offshore model for predicting concentrations of petroleum hydrocarbons and related chemicals in the water column and sediment bed of systems of concern near production facilities. This model was developed by Mackay et al. (1983) and has since been used for the study of contaminants in a variety of aquatic systems (Diamond et al., 1996; Lun et al., 1998; Warren et al., 2002). QWASI incorporates a defined volume of water interacting with a defined volume of sediment and also with the atmosphere. To simplify the model equations, the water and sediment compartments are treated as being well mixed, thus a single concentration prevails in each. The QWASI unit in Figure 1, which shows the fate and transport processes treated, illustrates a representative segment. All significant processes are quantified, as follows:

- Emission into the water columns from production or spills
- Inflow from adjacent segments of both water and suspended sediments
- Input from the atmosphere
- Volatilization to the atmosphere
- Degradation in the water column and sediment
- Water to sediment transport by deposition and diffusion
- Sediment to water transport by resuspension and diffusion
- Sediment burial
- Outflow in water and suspended sediments to adjacent segments

The QWASI model can be formulated as a single segment or with multiple segments. As a single segment, the model can only provide a single concentration in the water and the sediment. This provides a basic representation of the system, which does not provide information on the extent of contaminant plume development but can give an idea of

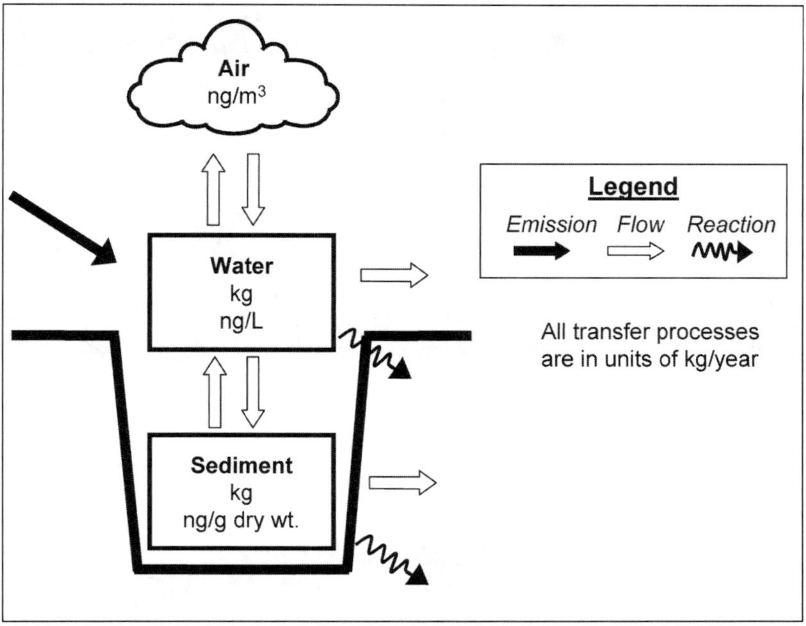

FIGURE 1. Mass balance diagram for a QWASI model unit.

important processes and "ballpark" concentrations without extensive data demand. Figure 2 is a schematic of a one-segment oil exploration effects QWASI environment.

The more spatially resolved multiple segment model provides a description of regions of higher concentrations and potential problem areas. It can be formulated to examine differences along the horizontal transect (Fig. 3A) as well as the vertical water column and sediment bed (Fig. 3B).

The entire multiple segment model is a 3-dimensional matrix of individual QWASI unit environments, as shown in Figure 4. While this model has more extensive data requirements, it has the ability to provide a more accurate description of chemicals in the defined system.

The QWASI model provides a description of contaminant concentrations in each of the water and sediment compartments defined for the system. Dissolved and particulate sorbed

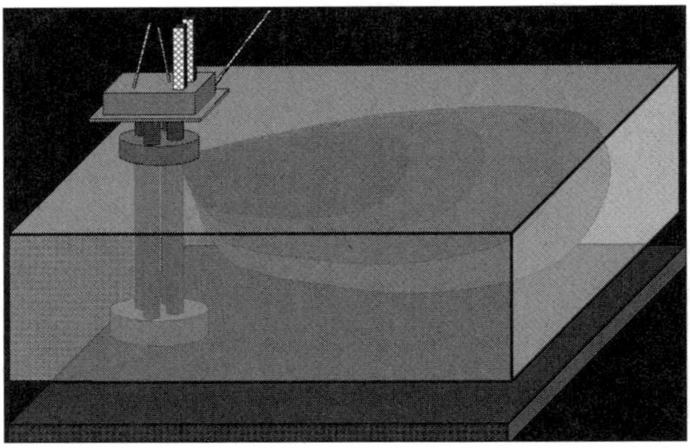

FIGURE 2. One segment representation of QWASI application to oil exploration effects.

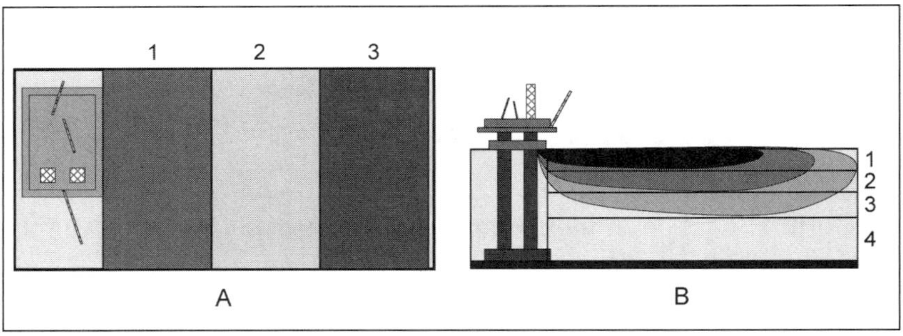

FIGURE 3. Multiple segment representation of QWASI application to oil exploration effects. **A)** Horizontal resolution; **B)** Vertical resolution

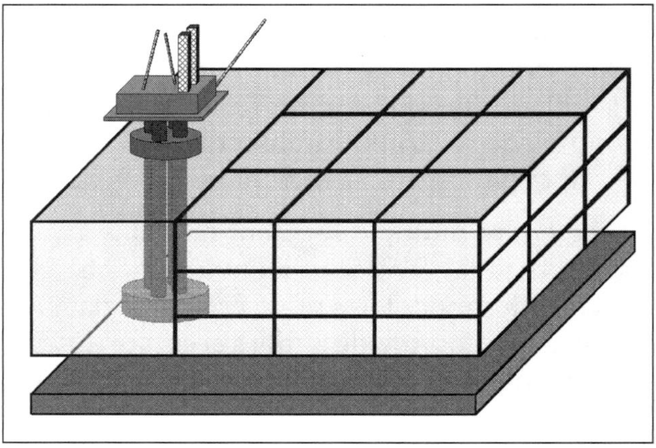

FIGURE 4. Multiple-segment QWASI for application to oil exploration effects.

chemical are considered by the model. As previously mentioned, once this model has been run, comparison is made to monitoring data and it is hoped that both data sets are mutually consistent. This abiotic model application provides the basis for a bioaccumulation and effects evaluation.

BIOUPTAKE MODELLING

Chemicals in the environment are not generally considered contaminants until they are shown to cause some detrimental effect on biotic organisms. As a result contaminant monitoring and effects assessments often focus on living organisms, particularly those most susceptible to contaminants (e.g. benthic invertebrates). There is also concern for other organisms, such as commercial fish and crustacean species and larger organisms existing higher in the food chain. Of special interest are chemicals that have the ability to bioaccumulate and biomagnify. As with abiotic monitoring, biotic monitoring experiences the same limitations. Again, modelling can offer a complementary role.

Bioaccumulation models have been developed, which explain the important processes occurring during this phenomenon. One such model is the Fish model developed by Clark et al. (1990). Since this time, other studies, such as Fraser (2002) have examined the processes of bioaccumulation. Figure 5 is a diagram describing the processes that are quantified in the Fish model.

Loss processes considered in the Fish model include metabolism, egestion, and respiration. Growth dilution can also be

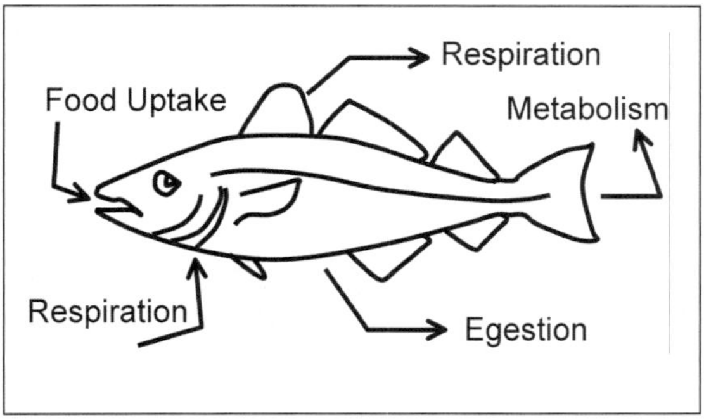

FIGURE 5. Uptake and loss processes considered in the Fish model.

included. In the model, the contaminant concentration and fugacity in water, obtained from an abiotic model or monitoring data, is used to determine the uptake by respiration. Food uptake can be estimated or measured directly from food sources, or obtained from modelled food webs, which usually rely on concentrations (measured or modelled) from the sediment bed. The basic units of a food web are benthic organisms, which are in interaction with the sediments, and pelagic zooplankton, which approach equilibrium with the water column.

Food web models, such as Campfens and Mackay (1997) and Morrison et al. (1997), can be used to examine the process of biomagnification up the food chain. Hickie et al. (2000) have applied such models to marine mammals, which are of particular concern as these species are located towards the top of the food chain. Figure 6 is a diagram of an example food web. As shown in this figure, the modelling process for food webs can become difficult and data intensive, however, it can also provide a more accurate picture of processes in the biotic environment.

Again, the biotic modelling results must be evaluated against monitoring data in a check for consistency. Agreement is

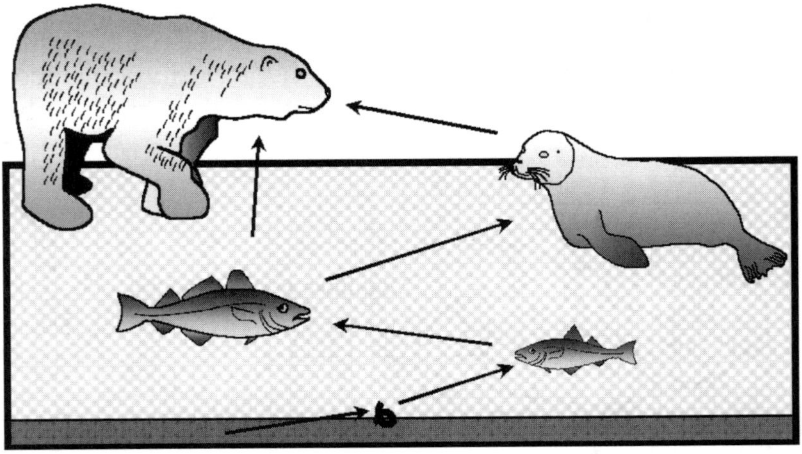

FIGURE 6. Food web diagram.

important to ensure there is an adequate understanding of the processes occurring in the system.

EFFECTS MODELLING

The modelling process can be applied from chemical fate in the abiotic environment to effects on organisms. Effects modelling can be carried out by two approaches:

1. External exposure or dose
2. Internal or delivered concentration or dose

For the external exposure approach, effects are determined by comparing water or sediment concentrations to some external criterion of effect (e.g. an aquatic LC_{50}). Biouptake modelling is not required. The method of internal or delivered concentration involves comparing biouptake modelling results with an internal criterion, such as a critical body residue. While this exercise requires more modelling and is more data demanding it has the potential to provide a better defined and more accurate picture of the contaminant level in an organism in terms of concentrations that will likely cause detrimental effects. In any case, both approaches to the effects modelling can be carried out for purposes of consistency.

MODELLING APPROACH

Each of the model types suggested above can be used individually, depending on data availability and the intensions of the user. As a coherent series, for modelling a system from contaminant input to biological effects, the abiotic model is first used, from which water and sediment concentrations are predicted and evaluated against monitoring data. These results are then used as inputs to the biouptake model in which results are again evaluated with monitoring values. Using the biouptake results, effects modelling can be carried out. At each point along this series of models, monitoring data can be used as the inputs to a model rather than results obtained from the preceding model. Figure 7 is a diagram describing the links between the individual models.

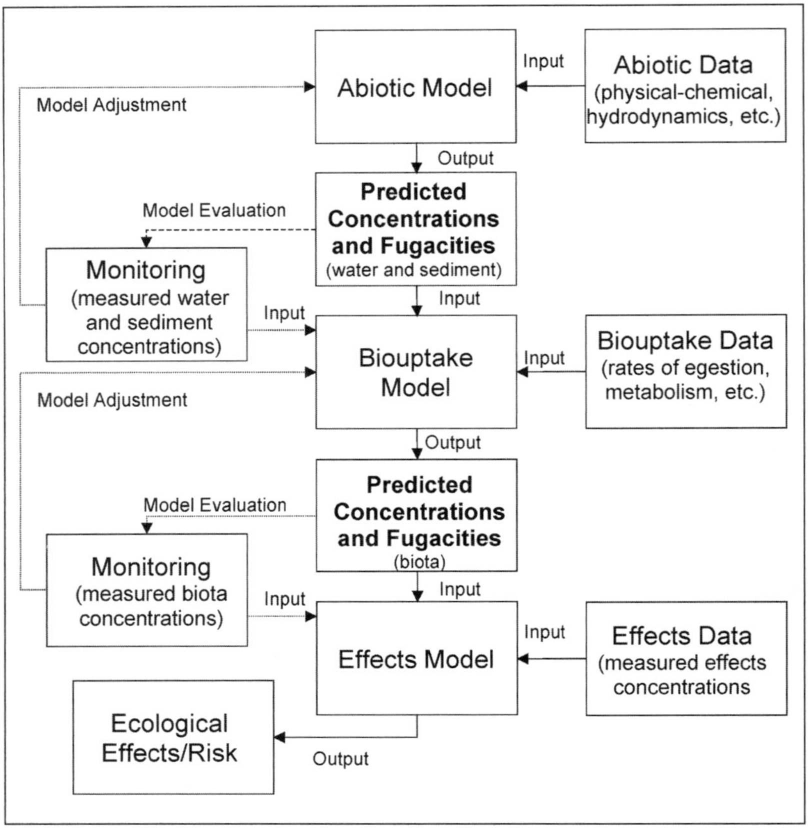

FIGURE 7. Linkage between models, monitoring, and input data.

CONCLUSIONS

In this paper, we have presented a framework for the complementary application of modelling to environmental effects monitoring programs. Using this approach, both abiotic and biotic fate of chemicals can be determined for a system from chronic as well as episodic sources. Modelling provides a means to develop and test new ideas for data interpretation as well as helping to focus monitoring programs, possibly reducing the extent and expense of monitoring needs. The forecasting or predictive ability of models is also beneficial for estimating potential effects resulting from increased contaminant production, accidental spills, as well as a variety of remedial options. It is important when developing a modelling approach to strive for simplicity. Model complexity can be increased as needed, to the limits of the input data available. Furthermore, with any modelling exercise, it is essential to seek consistency between model and monitoring results. We thus suggest that environmental effects monitoring programs be designed to incorporate a modest modelling component which has the potential to help establish the "big picture" of the entire sequence of discharges, environmental fate, bio-uptake, and effects, thus leading to more enlightened and economic management of offshore operations.

ACKOWLEDGEMENTS

We thank the Natural Science and Engineering Research Council of Canada (NSERC) and the Department of Fisheries and Oceans (DFO) for scholarship support and research funding, as well as the consortium of industrial partners that support research at the Canadian Environmental Modelling Centre, www.trentu.ca/cemc.

REFERENCES

Diamond, M.L., D. Mackay, D.J. Poulton, and F.A. Stride. 1996. "Assessing Chemical Behavior and Developing Remedial Actions Using a Mass Balance Model of Chemical Fate in the Bay of Quinte." *Water Research* 30:405-421.

Campfens, J. and Mackay D. 1997. "Fugacity-Based Model of PCB Bioaccumulation in Complex Aquatic Food Webs." *Environmental Science and Technology* 31:577-583.

Clark, K.E., Gobas, F.A.P.C. and Mackay, D. 1990. "Model of Organic Chemical Uptake and Clearance by Fish from Food and Water." *Environmental Science and Technology* 24:1203-1213.

Cranford, P.J., D.C. Gordon, C.G. Hannah, J.W. Loder, T.G. Milligan, D.K. Muschenheim, and Y. Shen. 2003. "Modelling Potential Effects of Petroleum Exploration Drilling on Northeastern Georges Bank Scallop Stocks." *Ecological Modelling*. 166:19-39.

Fraser, A. 2002. "Modeling Bioaccumulation of Organic Chemicals in Complex Aquatic Food Webs which Include Birds and Mammals

Feeding from the Aquatic Environment." M.Sc. Thesis, Trent University, Peterborough, Ontario, Canada.

French-McCay, D.P. 2002. "Development and Application of an Oil Toxicity and Exposure Model, OilToxEx." *Environmental Toxicology and Chemistry.* 21:2080-2094.

Lun, R., K. Lee, L. De Marco, C. Nalewajko, and D. Mackay. 1998. "A Model of the Fate of Polycyclic Aromatic Hydrocarbons in the Saguenay Fjord." *Environmental Toxicology and Chemistry.* 17:333-341.

Mackay, D., M. Joy, and S. Paterson. 1983. "Quantitative Water, Air, Sediment Interaction (QWASI) Fugacity Model for Describing the Fate of Chemicals in Lakes." *Chemosphere.* 12:981-997.

Morrison, H.A., F.A.P.C. Gobas, R. Lazar, D.M. Whittle, and G.D. Haffner. 1997. "Development and Verification of a Benthic/Pelagic Food Web Bioaccumulation Model for PCB Congeners in Western Lake Erie." *Environmental Science and Technology.* 31:3267-3273.

Paquin, P.R., K. Farley, R.C. Santore, C.D. Kavvadas, K.G. Mooney, R.P. Winfield, K-B. Wu, and D.M. Di Toro. 2003. *Metals in Aquatic Systems: A Review of Exposure, Bioaccumulation, and Toxicity Models.* Society of Environmental Toxicology and Chemistry (SETAC), Pensacola, FL.

Warren, C.S., D. Mackay, N.P. Bahadur, G.B. Boocock. 2002. "A Suite of Multi-Segment Fugacity Models Describing the Fate of Organic Contaminants in Aquatic Systems: Application to the Rihand Reservoir, India." *Water Research.* 36:4341-4355.

Assessing the Environmental Fate and Behavior of Oil Discharges in the Marine Ecosystem Using the Fugacity Model

Khan, I. (ibrahim.khan@dal.ca) and M.R. Islam (Dalhousie University, Halifax, NS, Canada)

ABSTRACT: Among the pollutants released in the ocean, oil has been identified as a major source of concern due to the volume discharged and potential for toxicity. The major sources of oil discharges are run-off, oil spills, natural seepage, blowouts, vessel operations, leaky submerged pipelines, and offshore oil and gas operations. An accurate or reliable assessment of the effects of oil on the marine environment is one of the vital issues involved in making environmental management decisions. This research focuses on the environmental fate and likely consequences of toxic oil compounds as considered from the marine management perspective. Predictive models have now been produced for the assessment of risk associated with oil spills. Following the evaluation of existing popular models, the fugacity model was selected for a simulation exercise in this study. To provide a better perspective on the potential impacts of oil released into the environment, the likely behaviours of the oils are analysed and presented pictorially. Two toxic compounds of oil, benzo[a]pyrene and pyrene, were tested using the fugacity model. Their transformation in the air, water, soil, and sediment, and accumulation in fish and suspended particles was determined for environmental conditions similar to that found in Canada's Atlantic region. In analysing the results and considering the limitations of the fugacity model, the use of an additional food chain based model component was proposed as a means to overcome its shortcomings. This approach improves our understanding of various oil spills issues and provides non-technical personnel with a decision-making and marine resources management tool.

INTRODUCTION

Pollution is a major problem in marine ecosystems and that has been steadily increasing due to numerous land and ocean-based activities. Common anthropogenic marine pollution sources are shipping and boating, mining, land-based industries, municipal wastes, and offshore oil and gas operations. Of all pollutants released in the ocean, oil occupies first place in both volume and toxicity. Due to high market demand, global offshore operations have increased. Most recently, this has included the east coast of Canada. With recent advancements in technology, oil and gas operations now extend to more than 3000 meters in deep-sea areas (Pinder, 2001), and to new places that were previously considered unproductive. As a consequence, concerns over the potential hazards from oil and other pollutants released into marine ecosystems have increased. Accordingly, present levels of knowledge, technology, and the tools for understanding the consequences of oil pollution need to be improved.

Oil can cause direct physical damage to wildlife, fisheries, aquaculture, and important habitats, such as fish habitats, mangroves, shores, and coral reefs. It also has the potential for toxicological effects on exposed organisms (GESAMP, 1993; Wells et al., 1995; Gulec and Holdway, 1999; Moles, 1999; Mitchell and Holdway, 2000; Holdway, 2002; Kingston, 2002). The effect level varies with species and may range from acute toxic effects that kill the exposed organism within short time periods to chronic sub-lethal effects that are longer lasting over the lifespan of the organism and/or its future generations. The severity of the effects of oil depends on its different hydrocarbon compound concentrations, exposure time, the type and composition of oil, and surrounding environmental parameters. When oil is released in the marine environment, it immediately starts to spread and change through dissolution, emulsification, oxidation and destruction, sedimentation, and microbial degradation (Patin, 1999; Kingston, 2002).

The effect of oil in the marine ecosystem has been extensively researched to assess its toxicity, weathering, degradation, and bioremediation (Patin, 1999; Holdway, 2002; Kingston, 2002). These types of studies have been conducted through direct experimental measurements, as well as through environmental models. Considering the dynamic nature of the marine ecosystem, environmental modelling is a suitable tool for studying the fate and effects of toxic compounds. A clear understanding of the impacts of oil releases into marine ecosystems requires an accurate assessment of the magnitude, spatial extent, duration of exposure, and interactions with that environment. In this research, steps were taken to ensure that the primary factors controlling the impact and behaviour of oil in the marine environment were included in the modelling effort. Application of the output from the models will aid marine mangers, policy makers, and other non-technical personnel in their overall understanding of oil fate and effects in the marine environment.

Oil in the Marine Environment

Petroleum products are some of the most toxic elements released in the marine environment. Annually, more than 1300 thousand tonnes of oils enter the marine environment worldwide (National Research Council, 2003). The maximum average worldwide discharge estimates are provided in Figure 1. In previous decades, there were greater amounts of oil discharged in the marine environment (GESAMP, 1993). However, the volume has been significantly reduced by the introduction of new legislation and advances in technology

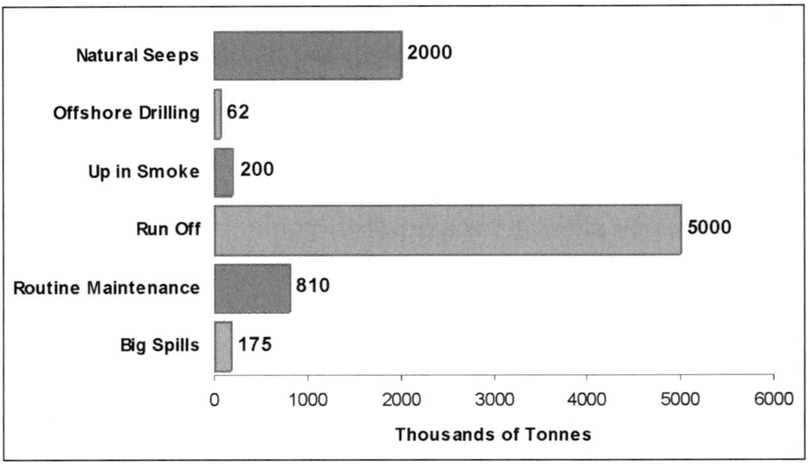

FIGURE 1. Average yearly (1990–1999) worldwide release of oils in the marine environment from different sources (maximum estimated amount) (from Committee on Oil in the Sea, 2003).

and operational practices by industry. The main sources of oil entering the ocean are from natural seeps, offshore extraction of oil and gas, land-based and river runoff, big spills, and operational discharges from vessels (National Research Council, 2003; GESAMP, 1993). Although accidental oil spills attract more media attention, hundreds of millions of gallons of oil quietly enter the marine ecosystem, mostly from non-accidental sources.

Behaviour of Oil

Petroleum oils include all liquids derived from crude oil. Crude oil itself is mainly complex mixtures of hydrocarbons and hydrocarbon-like chemicals that include atoms of nitrogen, oxygen, or sulphur. It also includes some metals such as vanadium and mercury (GESAMP, 1993; Wells et al., 1995). Crude oil consists of hundreds of different hydrocarbons and other non-carbon compounds. Some may contain 30% or more of aromatic hydrocarbons while others have virtually no aromatic hydrocarbons (Patin, 1999). Some aspects of the toxicity of petroleum oils depend on these chemical compositions

(GESAMP, 1993). According to specific gravity, oils are categorised into light, average, and heavy. Most of the oils are lighter than water.

As soon as oil is spilled or released into the marine environment, the chemical composition of the oil starts to change. This is known as the weathering process. The rate of change varies with surrounding environmental conditions as well as initial composition. The weathering process takes one or more years to complete. There are different steps in the weathering process, such as spreading, evaporation, dissolution, dispersion, emulsification, sedimentation, photolysis, and bioaccumulation (Fig. 2). Released or spilled oils spread over the surface of water as a slick, and the tides, waves, currents, and winds greatly accelerate spreading. Within the first few days, the volatile components of crude oils evaporate. Generally, 30 to 70% of oils might evaporate, depending on their specific gravity. For example, lighter oil has a greater evaporation rate (Patin, 1999; National Research Council, 2003). Some of the components of evaporated oils are oxidised with the UV radiation of sunlight and produce toxic acidic and phenolic compounds, which are highly toxic. However, it is reported that

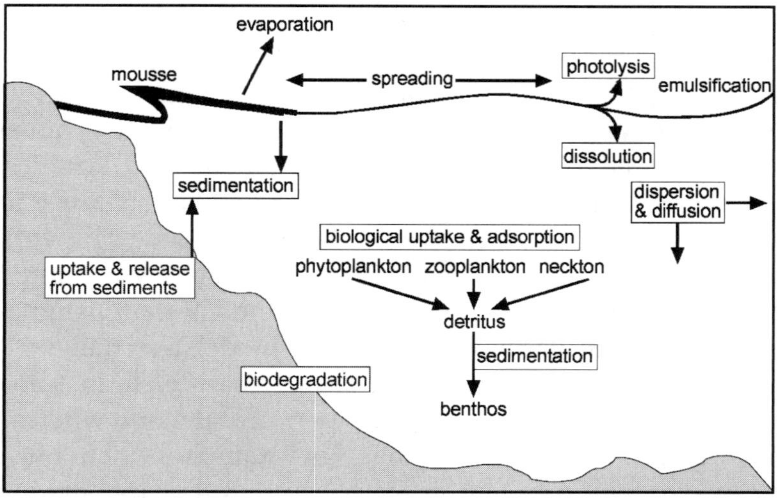

FIGURE 2. Oil weathering pathways in the marine environment.

their concentration is relatively low and their impact is not ecologically significant (Kingston, 2002).

Sea waves, currents, and upwelling help form a water-in-oil emulsion where water droplets are incorporated in oils. This process generates "chocolate mousse" or "mousse". Emulsification greatly changes the behaviour of the oil that it turns into a denser and semi-solid material. Generally, 3 to 15% of crude oil dissolves in seawater. This soluble portion of the oil is light in weight. However, it is relatively toxic for aquatic organisms. Due to rough sea conditions, oil goes through a mixing process known as dispersion. Dispersion takes place vertically, as well as horizontally. This process basically removes the oil by natural means. For example, the dispersed oil, in the form of oil droplets, is bioaccumulated by different microorganisms. These droplets attach to suspended solids drop down to the sea bottom, and then go through microbial degradation process.

Marine Environmental Modelling

Modelling is gaining popularity in recent years. It provides opportunities to understand the complexity of study areas. Government and law enforcement agencies have been using models to assess mechanisms for toxic chemical management (Park et al., 2000). The main advantages of models are their low cost and convenience in operating. The main cost is in the development of the model. Once it is developed, the operating cost is minimal. The cost of a full-scale experiment is much higher. It needs experts, research equipment, and facilities. Another advantage of modelling is that it is easier to use to study any site. This overcomes the obstacles posed by very harsh environments and the difficulty associated with limited access to the study area. By introducing the different parameters of that particular environment in a model, a virtual environment can be created and studied. Just as well, in some cases, it is hard to conduct actual experimental work, whereas a model can work around this. For example, when huge amounts of toxic compounds are released in the large marine environment, using environmental models, this can be easily analysed. Also, predicating long-term future effects cannot be studied in advance; by using models these can be easily estimated or predicted.

Modelling the marine environment is different from any lentic water environment because of dynamic characteristics, such as tides, currents, winds, water-depths, upwelling, species diversity, and food chain dynamics that involve many trophic levels within the ocean water column. It is also necessary to estimate the levels and impacts of the toxic compounds in organisms.

There are numerous models for the prediction of oil spills and oil related hazards. Some of them are: AES OIL SPILL model, ASA PC OIL SPILL model, ASA COASTAL SEA model, CHEMSPL, CHI/VKI model, OIL IN ICE model, PC Oil Slick simulation, S.L. ROSS oil spill model, SHORT TERM USGS model, SIMAP, and NRDAM/CME (GESAMP, 1993; MaCay and Payne, 2001; French, 1998, 2001). These models mainly address physical transportation and damage estimation. In general, most of the existing models poorly address and estimate how oil impacts marine organisms, the food chain, or biological processes. Also, very limited work has been done to estimate the effects of oil or other wastes at the species level.

APPLICATION OF THE FUGACITY MODEL

Fugacity modelling is considered one of the best ways to simulate the fate and behaviour of chemicals. Fugacity is a measure of the tendency of a toxic chemical to escape from one phase to another. For example, spilled oil flees from water to the atmosphere or sediments or into the tissues of aquatic organisms. This model is a valuable tool for understanding how contaminants migrate into different phases. The Level III Fugacity-based Environmental Model (CEMC, 2003) was used in this research to predict the fate and transport of compounds of crude oil by incorporating not only the single medium fate and transport but the inter-medium transfers as well. The present study method is one of the robust models. It was developed based on the work of Mackay and Paterson (1983), Mackay (2001), and Webster et al. (2003).

Environmental Scenario

In this study, the model was run using a set of environmental conditions similar to that found in Atlantic Canada. The physical size (primary compartment) of the spill areas were hypothetically given, but all other environmental data/properties required for this study, such as organic carbon fraction, lipid fraction, volume fraction, and transport velocities, were derived from the CEMC (2003) database, which is representative of the Atlantic maritime region of Canada. Table 1 provides the environmental properties of the studied area.

The hypothetical spilled area was 100 km^2 and 10% of the area was considered as water bodies. The area was characterised as a coastal zone by providing a 30 km-long coastline, 2 km in width, and 100 m water depth. The characteristics of the coastline were kept similar to the average Atlantic Canada coastal environment. The bulk compartments of this studied area were divided into water, soil, sediment, groundwater, coastal water, and terrestrial plants. There were 10 sub-compartments, namely, air vapour, air particles, water particles, biota, soil air, soil water, soil solids, pore water, and sediment solids. The total volume and densities of these compartments are shown in Table 1.

The temperature data of the proposed study area, including year round mean (5.5°C), winter mean (-4.5°C), and summer mean (17.3°C) are based on actual recorded values for Atlantic maritime Canada. In the model, input organic carbon fraction value was used for the particles in water, soil solids, and sediment solids. The lipid fraction value was used for the fish and vegetation. Other environmental data used in the model included:

- water surface area,
- total volume depth of active layer of sediment,
- particle concentrations in the inflow water,
- aerosols in air,
- volume fraction of solids in surface sediment,
- particle densities (suspended sediment, sediment solids, aerosols),

TABLE 1. Environmental properties of the study area.

Physical Parameters		**Bulk Compartments**		
Total Surface Area (km^2)	100		Volume (m^3)	Density (kg/m^3)
Surface Covered by Water (%)	10	Air	20000000000	1.268369
Average Air Height (km)	2	Water	200000000	1000.007
Average Water Depth (m)	20	Soil	9000000	1500.254
Average Soil Depth (cm)	10	Sediment	100000	1420
Average Sediment Depth (cm)	1	Groundwater	-	1000
Length of Coastline Km	30	Coastal Water	6000000000	1.268369
Average Depth of Coast (m)	100	Terrestrial Plants	-	900
Average Width of Coast (km)	2	**Sub-Compartments**		
Temperature Conditions (°C)	5.5	Air Vapour	20000000000	1.268369
Year round mean		Air Particles	4	2400
Winter mean	-4.5	Water	200000000	1000
Summer mean	17.3	Water Particles	1000	2400
Organic Carbons (g/g)		Biota	200	1000
Particles in Water	0.2	Soil Air	1800000	1.268
Soil Solids	0.02	Soil Water	2700000	1000
Sediment Solids	0.04	Soil Solids	4500000	2400
Fish Lipid	0.048	Pore Water	70000	1000
Vegetation Lipid	0.01	Sediment Solids	30000	2400
Transport Velocities (m/h)				
U1 Air side air-water MTC	3			
U2 Water side air-water MTC	0.03			
U3 Rain rate	0.000141			
U4 Aerosol dry deposition velocity	10.8			
U5 Soil air phase diffusion MTC	0.04			
U6 Soil water phase diffusion MTC	0.00001			
U7 Soil air boundary layer MTC	1			
U8 Sediment-water MTC	0.0001			
U9 Sediment deposition velocity	.000000046			
U10 Sediment resuspen. velocity	.000000011			
U11 Soil water runoff rate	.0000564			
U12 Soil solids runoff rate	.0000000282			
U13 Sediment Burial	3.5E-08			
U14 Diffusion to Higher Altitudes	0.01			
U15 Leaching from Soil	.0000141			
Scavenging ratio (unitless)				
Rain	200000			
Snow	1000000			

- organic carbon fraction,
- water inflow and outflow data,
- sediment deposition,
- re-suspension and burial rates,
- atmospheric deposition parameters (aerosol dry deposition velocity, scavenging ratio, rain rate), and
- mass transfer coefficients (volatilization—air side and water side, sediment-water diffusion)

Transport velocities used in the model were taken as actual values of this region. In this study, they are mass transfer coefficients (MTC), flows per unit area, or ratios of diffusivity to path length, and have the dimensions of velocity (m/h). As a first approximation, transport velocities are independent of the chemical, although they do contain diffusion rates that are somewhat chemical dependent. The chemical-to-chemical variability of diffusivity is, however, usually negligible when compared with the variability in Z values, where, Z is the fugacity capacity (mol/m^3-Pa) of pollutants in each compartment. In this simulation, deposition was treated as three parallel processes: 1) dissolution in rain water, 2) wet deposition of particles by rain with a default scavenging ratio of 0.2×10^6 or snow with a scavenging ratio of 10^6, and 3) dry deposition of particles with a default deposition velocity of 10.8 m/h (0.3 cm/s) (CEMC, 2003).

Chemical Properties of Tested Oil Compounds

Two major compounds of crude oil that are of known environmental concern, benzo[a]pyrene and pyrene, were selected for this simulation study running through the Fugacity-based Environmental Model Level III (Mackay, 2001; Webster et al., 2003). These chemicals are considered as the environmental indicators of oil in the sediments. It is also reported that these chemicals are available in the sediments even 10 years after of the oil spill (Bence and Burns, 1995). Pyrene has four benzene rings and benzo[a]pyrene has five carbon rings. Benzo[a]-pyrene is a polycyclic aromatic hydrocarbon which is a potent human carcinogen.

In this simulation study, the chemical properties of these selected chemicals, such as molecular mass, $LogK_{ow}$, $LogK_{oa}$, K_{aw}, water solubility, vapour pressure, melting point, Henery's Law constant, reaction half-lives (air, water, soil, and sediment), and partition coefficient for all the sub-compartments were required as model input. The chemical properties of benzo[a]pyrene and pyrene were collected from the Mackay et al. (2000) and are shown in Table 2. The half-lives of both chemicals are similar. The half-life value for air, water,

TABLE 2. Chemical properties of benzo[a]pyrene and pyrene (from Mackay et al. (2000) and CEMC (2003).

	Benzo[a]pyrene		Pyrene	
Molar Mass (g/mol)	252.3		202.3	
Data Temperature (°C)	25		25	
$LogK_{ow}$	6.04		5.18	
Vapour Pressure (VP) (Pa)	.0000007		.0000006	
Melting Point (°C)	175		156	
Fugacity Ratio	.0276		.0506	
Sub-cooled Liquid VP (Pa)	.0000254		.00118	
Water Solubility (g/m^3)	.00380		.132	
Henry's Law Constant ($Pa.m^3/mol$)	.0465		.0919	
Partition Coefficients	dimensionless	l/kg	dimensionless	l/kg
Octanol-Water (K_{ow})	.00000109		.0000151	
Organic Carbon-Water (K_{oc})		.0000449		.00062
Air-Water (K_{aw})	.0000191		.000371	
Soil Solids-Water	.000215	.00899	.00297	.00124
Sediment Solids-Water	.000431	.000179	.00595	.00248
Suspended Particles-Water	.0000215	.000899	.000297	.000124
Fish-Water	.000526	.000526	.00726	.00726
Aerosol-Air	.0000000000236		.0000000506	

soil, and sediment are 170, 1,700, 17,000, and 55,000 hours, respectively. The detailed chemical data of these two oil compounds are given in Table 2.

MODELLING RESULTS AND DISCUSSION

Simulations were run for benzo[a]pyrene and pyrene in the hypothetical environment using the model, while the physical dimensions of the study area was pre-defined as a 30 km length of coastline, 2 km in width with a mean depth of 100 m, and all other environmental properties were kept similar to the Atlantic maritime Canada (Table 1).

For the simulation, it was estimated that annually 1,275 kg of benzo[a]pyrene was emitted in 100 km^2 study area from different sources. The emission rate of this compound in air, water, soil and sediment were decided as 300, 450, 500 and 25 kg/yr, respectively. The inflow rate in air and water were 200 and 2.73 kg/hr, respectively. The inflow concentrations were 2 ng/m^3 and 3 ng/l in the air and water, respectively. The result of the benzo[a]pyrene simulation is illustrated in Figure 3. In the case of the pyrene simulation, it was decided that

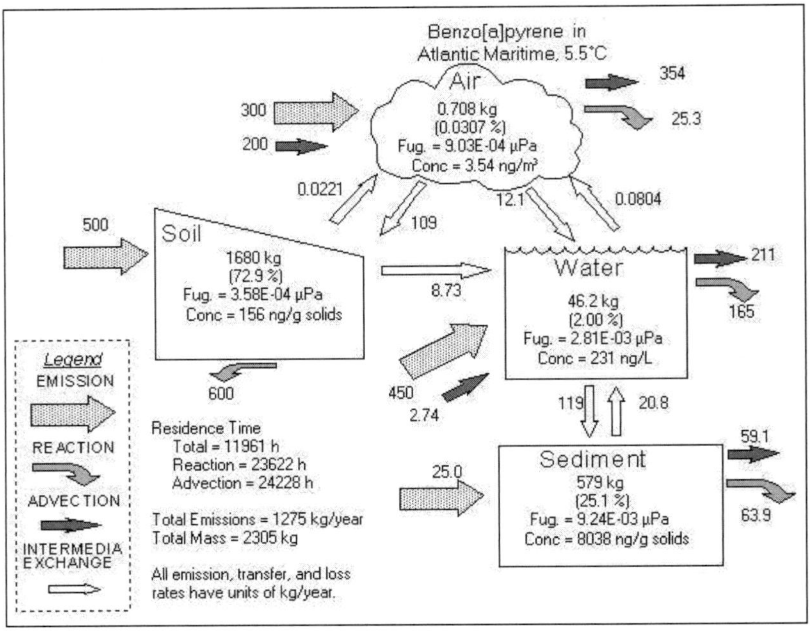

FIGURE 3. Diagram of fugacity model output for benzo[a]pyrene.

2,068 kg was emitted in the study area every year. The emissions in air, water, soil, and sediments were 98, 1,800, 120, and 50 kg, respectively. In this case, it was estimated that most of emission took place in the water area. The emission result of pyrene is shown in Figure 4.

The main compartments of the study area include soil, water, sediments, and air. Different types of arrow symbols in Figures 3 and 4 represent the emission, reaction, advection, and intermedia exchanges of the chemical among the compartments. Inside the shapes, the value of residing amount and percentage, fugacity value, and concentrations are shown. These pictorial presentations provide the comparative scenario of all compartments. This may be helpful to decide which compartment will require more attention in case of a real accident. In case of other existing models (non-fugacity models), it just assesses the effects level within a particular compartment rather than this whole picture (Mackay, 1985).

Based on the emission rate, inflow concentration, and other physiochemical parameters, the total amount of benzo[a]-

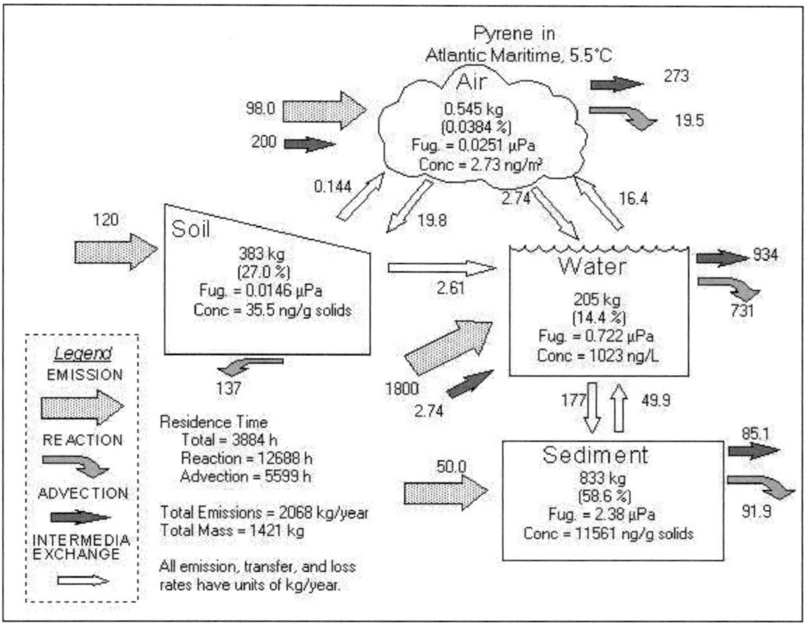

FIGURE 4. Diagram of fugacity model output for pyrene.

pyrene found in the whole system was 2,305 kg. The total overall residence time of this chemical is 11,961 hours. The total amount of benzo[a]pyrene in the system is higher than the yearly emitted amount (1,275 kg). It is due to a longer residence time of this chemical in the study area. It shows that the total residence time in the system is 11,961 hours, which is higher than the emission time (8,760 hours). In this simulation test, the chemical entering the system is staying longer than a year. In other words, one can state that the emissions are effectively retained within the environment. This finding supports the work of Bence and Burns (1995). They reported that benzo[a]pyrene is found in the sediments of the spill site 10 years after the accident.

A total of 1,421 kg pyrene is found in the whole system, whereas initially, 2068 kg pyrene was emitted. The total overall residence time of this chemical was calculated to be 3,884 hours (Fig. 4), which is less than the emitting time (8,760 hours). In the simulation test of pyrene, more was emitted than for benzo[a]pyrene, but the total amount of pyrene found was less than benzo[a]pyrene. This may be due to the difference in emission pattern. In case of benzo[a]pyrene, most of the chemical was emitted in the soil compartment; whereas most of the pyrene was emitted in the water compartments (Figs. 3 and 4). However, the two also have different chemical properties.

The detailed results of the benzo[a]pyrene and pyrene are shown in Table 3. It shows the fugacity, amount of chemical, and concentration for the all compartments and sub-compartments. In case of benzo[a]pyrene, the greatest amount of chemical among all compartments is 1,679.61 kg found in soil, followed by sediments (578.7 kg), and water (46.16 kg). The amount of chemical in the air is very low, at only 0.7 kg. The amount of chemical is also determined for each sub-compartment. In the bulk water compartment, for example, total chemical is 46.16 kg. Bulk water compartment has three sub-compartments, which are suspended particles, fish, and water (only). The benzo[a]pyrene concentration in these three compartments are 29.32, 1.42, and 15.41 kg, respectively (Table 3). An interesting observation in the water compartment is that most of the chemical (63%) of this compartment is accumu-

TABLE 3. Simulation results for benzo[a]pyrene and pyrene.

	Benzo[a]pyrene			Pyrene		
	Fugacity (Pa)	Amount (kg)	Conc. (g/m³)	Fugacity (Pa)	Amount (kg)	Conc. (g/m³)
Air: Bulk	9.03E-10	7.08E-1	3.54E-09	2.51E-08	5.45E-01	2.73E-09
Air Vapour	9.03E-10	1.97E-02	9.84E-11	2.51E-08	4.38E-01	2.19E-09
Aerosol	9.03E-10	6.88E-01	172.1142	2.51E-08	1.06 E-01	26.73778
Water: Bulk	2.81E-09	4.61E+01	2.31E-04	7.22E-07	2.04 E+02	1.02E-03
Water	2.81E-09	1.54E+01	7.71E-05	7.22E-07	1.60E+02	8.02E-04
Susp. Particles	2.81E-09	2.93E+01	2.93E+01	7.22E-07	4.21E+01	4.21E+01
Fish	2.81E-09	1.42E+00	7.13E+00	7.22E-07	2.05E+00	1.02E+01
Soil: Bulk	3.58E-10	1.6E+03	1.86E_01	1.46E-08	3.83E+02	4.26E-02
Air	3.58E-10	7.01E-08	3.89E-11	1.46E-08	2.29E-06	1.27E-09
Water	3.58E-10	2.65E-02	9.81E-06	1.46E-08	4.38E-02	1.62E-05
Solid	3.58E-10	1.67E+3	0.373242	1.46E-08	3.83E+02	8.52E-02
Sediment: Bulk	9.24E-09	5.78E+02	5.78E+00	2.38E-06	8.32E+02	8.32E+00
Water	9.24E-09	1.77E-02	2.54E-04	2.38E-06	1.84E-01	2.64E-03
Solid	9.24E-09	5.78E+02	19.29018	2.38E-06	8.32E+02	2.77E+01
Groundwater	1.63E-13	-	4.47E-09	4.80E-11	-	5.34E-08
Coastal Water	1.82E-11	-	3.99E-07	1.99E-09	-	1.77E-06
Terrestrial Plants	9.03E-10	-	4.78E-01	2.51E-08	-	7.43E-02

lated in suspended particles. The suspended particles of water are mainly phytoplankton and zooplankton or silt. In the region of study, the silt concentration is comparatively lower because there is very little silt discharge from the river network. Therefore, the suspended particles in the water are mainly plankton communities.

Based on the result of this study, we can say that in case of a real accident, oil spills may severely damage the plankton community, which is the primary food source of the marine ecosystem. As these primary producers are at the base of the food pyramid, detrimental impacts on them may result in a decline of overall productivity of the ecosystem. This is an important issue that deserves detailed studies to quantify the extent to which it would affect the food chain. In this simulation study, the effects on different trophic levels of the food chain could not be determined. The concentration of benzo[a]pyrene in fish is poorly estimated because the size of the fish population, their migration and movement, and other biological data were not considered. This is a major drawback of this model. However, other existing models that are commonly used in oil spill estimation, such as OIL IN ICE model, PC Oil Slick simulation, SIMAP, and NRDAM/CME (French, 1998; 2001; McCay and Payne, 2001; McCay, 2003), do not consider the food chain-based assessment.

Fugacity (Pa) value of benzo[a]pyrene and pyrene for all environmental compartments are shown in Table 3. It was also found that the fugacity value was much higher in the sediment and air than any other compartment. The fugacity of pyrene was also estimated through the model for all compartments and sub-compartments. The highest concentration was found in sediments, with the next highest concentration in soil compartments. The fugacity value in the air was negligible.

The transport rates for spilled compounds are calculated and shown in Table 4. The highest transport rate for both of the oil compounds studied has been found to be between air and soil. In the case of pyrene, water to air transportation rate is higher than that of benzo[a]pyrene. In case of a real spill accident, this transport rate can identify which compartment is adversely

TABLE 4. Calculated intermedia transport and individual process D values.

Intermedia Transport (kg/h)	Benzo[a]pyrene	Pyrene
Air to water	12.09228	2.742805
Air to soil	108.6537	19.79159
Water to air	.00804	16.42926
Water to sediments	118.8296	176.8371
Soil to air	0.022074	0.143905
Soil to water	8.734469	2.614213
Individual Process D Values (mol/Path)		
Air-water diffusion (air-side)	12949.48	12949.48
Air-water diffusion (water-side)	32600000	1649157
Air-water diffusion (overall)	12944.34	12848.59
Rain dissolution to water	153344	7751.038
Rain dissolution to soil	1380097	69759.34
Aerosol deposition to water (dry)	1630865	11379.17
Aerosol deposition to water (wet)	4258369	29712.28
Aerosol deposition to water (total)	5889234	41091.46
Aerosol deposition to soil (dry)	14700000	102412.6
Aerosol deposition to soil (wet)	38300000	267410.6
Aerosol deposition to soil (total)	53000000	369823.1
Soil to water runoff (water)	552038.6	27903.74
Soil to water runoff (solids)	10500000	73270.49
Soil-air diffusion (air-phase)	1553.937	1553.937
Soil-air diffusion (water-phase)	97879.18	4947.471
Soil-air diffusion (boundary layer)	38848.43	38848.43
Soil-air diffusion(overall)	27934.46	5569.358
Water-sediment diffusion	108754.6	5497.19
Water to sediment deposition	19000000	132799.2
Sediment to water resuspension	910264.1	6351.264

affected, and will help in the development of a management plan. If we know in which compartment the spill has occurred, it is easier to determine the effect level by observing the transport rates. For example, if a spill occurred on the soil it is less detrimental than a spill in water, because the transport rate between oil and other media is less than that of the water to media. Considering the transport rate managers can take the precautionary measure to determine vulnerability of marine zones.

The individual process diffusive transfer between phases are calculated from quantities, such as the inter-phase transfer areas, mass transfer coefficients, uptake and release rates or times of chemicals into phases such as fish or sediments, and Z values. The calculated values of D for both of the studied chemicals are shown in the Table 4. The values for air-water diffusion, rain dissolution to water and soil, aerosol deposition to water, aerosol deposition to soil, soil to water runoff, soil air diffusion, and water sediment diffusion are calculated and shown in Table 4.

PROPOSED FOOD CHAIN MODEL

In this study, it was found that the fugacity model can be used as a screening tool to determine the fate of hydrocarbons or any other contaminants in large environments. This model identifies the quantitative data of oil compounds in each of the environmental compartments, such as soil, water, suspended particles, fish tissue, sediment, aerosol and air. This model can not clearly identify the impact of spilled compounds on the multitude of aquatic species that may be present. There are many models that have been used to study the impacts of oil spills, but very few are appropriate for the marine environment, because they do not consider the transformation of oil at the organism level. It is, therefore, necessary to develop a suitable model that will address the transformation of toxic compounds at different trophic levels.

A proposed integrated food chain model is shown in Figure 5. In this model, the fate model is integrated with the food chain component. This model mainly focuses on the water compartment and its inhabitants. The bulk water compartment is subdivided to other sub-compartments. These sub-compartments are similar to different components of a food chain, such as primary producers (phytoplankton), primary consumers (zooplankton), secondary consumers (fish, squid) and higher trophic level consumers (large fish, marine mammals). The interrelationships between the sub-compartments (food chain components) are shown using the arrow symbols (Fig. 5). By estimating the different partition coefficients such as water-

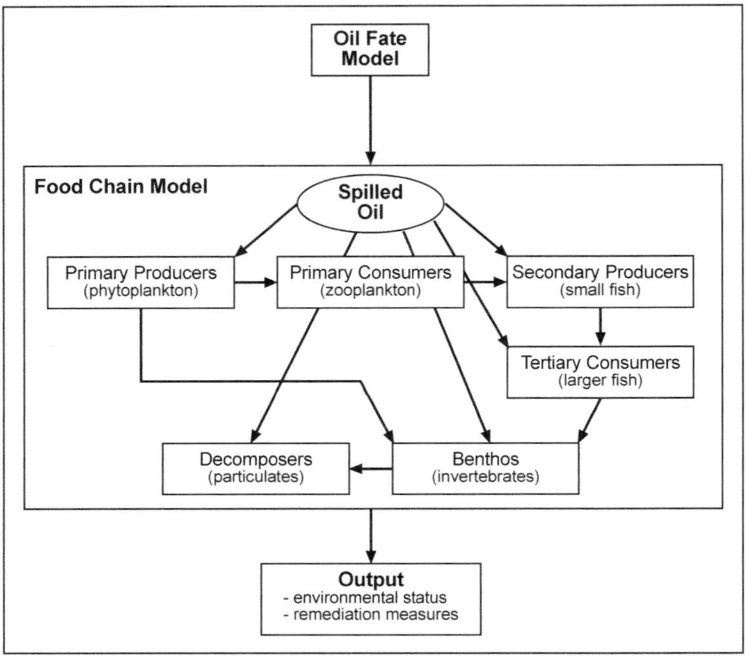

FIGURE 5. Schematic diagram of a proposed food chain model.

phytoplankton, phytoplankton-zooplankton, zooplankton-small fish, small fish-nekton, benthos-sediments, and debris-phytoplankton, we will be able to clearly assess the probable impacts of oil on food chains as a whole. The proposed model is the integration of the fate and food chain models, which will identify the environmental and biological effects. At present, two or more different models are required for estimating these effects; however the proposed food chain model can assess these effects as single integrated model.

CONCLUSIONS

It can be concluded from this study that an application of the fugacity model may provide a quantitative estimate for spilled toxic compounds, such as benzo[a]pyrene and pyrene in a large environment. While other models simply provide a collection of disparate quantities, using this fugacity model, the detailed behaviour of these chemicals can be understood. Similar to other environmental models, this model not only estimates the concentration of released compounds, but also estimates their fates, transformation, and behaviour. This model even attempts to reveal the amount of released toxic compounds transfused into organisms, such as fish tissue. However, this model can not estimate the detailed impacts on other species. The integrated food chain model is proposed to overcome these limitations.

REFERENCES

Bence, A.E. and W.A. Burns. 1995. "Fingerprinting Hydrocarbons in the Biological Resources of the Exxon Valdez Spill Area." In Peter G. Wells, James N. Butler, and Jane S. Hughes (Eds.), *Exxon Valdez Oil Spill: Fate and Effects in Alaskan Waters, ASTM STP 1219*, American Society for Testing and Materials, Philadelphia, 1995.

Canadian Environmental Modelling Centre (CEMC). 2003. "Level III Version 2.70 Released March 2002", Trent University, Peterborough, Ontario, Canada.

French, D.P. 1998. "Evolution of Oil Trajectory, Fate and Impact Assessment Models." *Oil Spill 98 Conference, 1998*. Wessex Institute of Technology, Ashurst Lodge, Ashurst, Southampton, United Kingdom

French, D. 2001. "Modelling Oil Spills Impacts." In R. Garcia-Martinez and C. Brebbia (Eds.), *State of the Art in Oil Modelling and Processes*. WIT Press, Ashurst Lodge, Ashurst, Southampton, UK, 2001.

GESAMP (IMP/FAO/UNESCO/WMO/WHO/IAEA/UN/UNEP Joint group of Experts on the Scientific Aspects of Marine Pollution). 1993. *Impact of Oil and Related Chemicals and Wastes on the Marine Environment*. Report Studies. GESAMP No. 50 189pp.

Gulec, I. and D.A. Holdway. 1999. "Toxicity of Laboratory Burned Oil to the Amphipod *Allorchestes compressa* and the Snail *Polinices conicus*". *Spill Science Technology Bulletin* 5(2):135-139.

Holdway, D.A. 2002. "The Acute and Chronic Effects of Wastes Associated with Offshore Oil and Gas Production on Temperature and

Tropical Marine Ecological Process." *Marine Pollution Bulletin* 44:185-203.

Kingston, P.F. 2002. "Long-term Environmental Impact of Oil Spills." *Spill Science and Technology Bulletin* 7(1-1): 53-61

Mackay, D. 1979. "Finding Fugacity Feasible." *Environmental Science and Technology*. 13:1218.

Mackay, D. 1985. "The Fugacity Concept in Environmental Modelling" In O. Hutzinger (Ed.), *The Handbook of Environmental Chemistry*, Volume 2/Pat C. 120-140.

Mackay, D. 2001. "Multimedia Environmental Models: The Fugacity Approach - Second Edition", Lewis Publishers, Boca Raton. pp. 201-213.

Mackay, D., M. Joy, and S. Paterson. 1983. "A Quantitative Water, Air, Sediment Interaction (QWASI) Fugacity Model for Describing the Fate of Chemicals in Lakes." *Chemosphere*. 12:981-997.

Mackay, D., W.Y. Shiu, and K.C. Ma. 2000. "Physical-Chemical Properties and Environmental Fate and Degradation Handbook." CRC-netBASE 2000, Chapman & Hall CRCnetBASE, CRC Press LLC., Boca Raton, FL. (CD-ROM.)

McCay, D.P.F. and J.R. Payne. 2001. "Modelling of Oil Fate and Water Concentrations With and Without Application of Dispersants". *Proceedings of the 24th Arctic and Marine Oilspill (AMOP) Technical Seminar*, Edmonton, Alberta, Canada, June 12-14, 2001, Environment Canada, pp. 611-645.

McCay, D.F., 2003, "Development and Application of Damage Assessment Modelling: Example for the North Cape Oil Spill." *Marine Pollution Bulletin*. 47:341-359.

Mitchell, F.M. and D.A. Holdway. 2000. "The Acute and Chronic Toxicity of the Dispersants Corexit 9500 and 9527, Water Accommodated Fraction (WAF) of Crude Oil, and Dispersant Enhance WAF (DEWAF) to *Hydra viridissima* (Green Hydra)." *Water Research*. 34(1):343-348.

Moles, A. 1999. "Parasitism, Feeding Rate, and Hydrocarbon Uptake of Pink Shrimp *Pandalus borealis* Fed a Crude Oil Contaminated Diet." *Environmental Contamination Toxicology*. 62:259-265.

National Research Council. 2003. *Oil in the Sea III: Inputs, Fates, and Effects*. National Research Council. The National Academies Press, Washington, D.C.

Park, M.V., A.G. Gillett, and J.R. Burema. 2000. "A GIS-framework for Modelling Environmental Fate of Chernobyl-Derived Radiocesium." *4th International Conference on Integrating GIS and Environmental Modelling (GIS/EM4): Problems, Prospects and Research Needs*. Banff, Alberta, Canada, September 2-8, 2000.

Patin, S. 1999. *Environmental Impact of the Offshore Oil and Gas Industry*. EcoMonitor Publishing, East Northport, New York. 425 pp.

Pinder, D., 2001. "Offshore Oil and Gas: Global Resource Knowledge and Technological Change." *Ocean and Coastal Management*. 44:579-600.

Webster, E., D. Mackay, A. Di Guardo, D. Kane, and D. Woodfine. 2003. "Regional Differences in Chemical Fate Model Outcomes". *Chemosphere*. Accepted for publication, August 2003.

Wells, P.G., J.N. Buttler, and J.S. Hughes. 1995. "Introduction, Overview, Issues," In Peter G. Wells, James N. Butler, and Jane S. Hughes (Eds.), *Exxon Valdez Oil Spill: Fate and Effects in Alaskan Waters, ASTM STP 1219*, American Society for Testing and Materials, Philadelphia, 1995.

Evaluation of Offshore Drilling Cuttings Management Technologies Using Multicriteria Decision Making

Thanyamanta, W. (woraka@engr.mun.ca), K. Hawboldt, T. Husain, N. Bose, and B. Veitch, (Faculty of Engineering and Applied Science, Memorial University, St. John's, NL, Canada)

ABSTRACT: Due to increasingly stringent environmental regulations, and the high risk and cost of ship-to-shore treatment alternatives, on-site cost-effective solutions to manage drill cuttings for offshore oil drilling operations are of interest. Existing and emerging drilling cuttings management technologies differ from each other in technical, environmental, and economic factors. In this study, multicriteria decision making was applied to assess eight cuttings management alternatives. The criteria include one threshold criterion, conformity with regulations, and four major categories of decision making criteria: technical feasibility, rig compatibility, environmental impacts, and costs. The alternatives evaluated include technologies currently in use offshore and those used onshore, but with potential for offshore applications. The total score for each option was then calculated using a weighted additive overall value model. From the evaluation, the three optimum drilling cuttings management technologies are the vertical centrifuge, horizontal centrifuge, and re-injection.

INTRODUCTION

In offshore oil and gas operations, drilling fluids and drilling cuttings are the major waste generated from drilling activities and may pose environmental impacts when not properly managed. The main effects of cuttings discharges include benthic smothering, sediment alteration, toxic threats, bioaccumulation, and anoxic conditions in the sediment. The level of impacts depends mainly on the types and properties of drilling fluids retained on the cuttings. There are basically three types of drilling fluids: water-based, oil-based, and synthetic-based. The different drilling fluids provide different drilling performance and the fluids' chemical properties lead to different environmental impacts once discharged into the ocean. Therefore, the management of cuttings contaminated with these types of fluids is strictly controlled by regulations and varies depending on the types of the fluids used.

Synthetic-based fluid (SBF) cuttings, which are the focus of this study, may be allowed to be discharged after sufficient treatment, or release to the environment may be prohibited. Current management technologies for SBF drilling cuttings include landfarming, stabilisation/solidification, onsite mechanical treatment and subsurface re-injection (USEPA, 2000). This paper presents an evaluation of these technologies using a practical and comprehensive method to select the most suitable cuttings management technology for ongoing offshore operations. The technologies being evaluated are intended to treat the cuttings to the level that is the lowest possible for the technologies at the time of publication. Treating the cuttings onsite is believed to help minimise transportation accidents and toxicity effects during handling of cuttings. Onsite treatment may also make the cuttings acceptable for discharge overboard depending on the local regulations. The evaluation was conducted using a deterministic multicriteria decision making model that compares the selected management options based on various factors including technical feasibility, rig compatibility, environmental and safety, and cost aspects.

IDENTIFYING CRITERIA AND EVALUATION MEASURES

The first step in the multicriteria decision making process is to identify the evaluation criteria following the value-focused thinking approach proposed by Keeney (1992). The criteria are the controlling factors under which the drilling cuttings management options will be scored in the evaluation. The criteria were chosen to accurately reflect the issues important in drilling cuttings management. In this study, criteria were divided into two major types: a threshold criterion and decision making criteria. To select the optimum cuttings management technologies to be used offshore and avoid transportation of cuttings wastes for on-land disposal, the conformity to discharge regulations was used as the threshold criterion to screen out inappropriate options. The technology-based discharge standard of 6.9% retention of base fluid, based on USEPA's cuttings discharge standards and the Offshore Waste Treatment Guidelines (USEPA, 2000; NEB et al., 2002), were used in the screening. The options which were not able to meet the regulatory discharge standard were rejected and not considered further. The decision making criteria are used to evaluate and compare options. These criteria provide trade-offs of important factors influencing the decision making. For drilling cuttings management, the decision making criteria were divided into four major categories: technical feasibility, rig compatibility, environmental and safety, and costs. The criteria hierarchy is shown in Figure 1. The numbers in brackets represent the weighting factors, which will be described later in this paper. Under each criterion, an evaluation measure, which is a scale used to measure each option's attainment, was identified. The options were then rated according to the evaluation measures. The values or ratings obtained for an option can be either quantitative or qualitative.

IDENTIFYING ALTERNATIVES

For a technology to be selected as an option in the evaluation, it is necessary for it to meet the threshold criterion. These require the technology either to be able to reduce the amount of retained base fluids on discharged cuttings to lower than

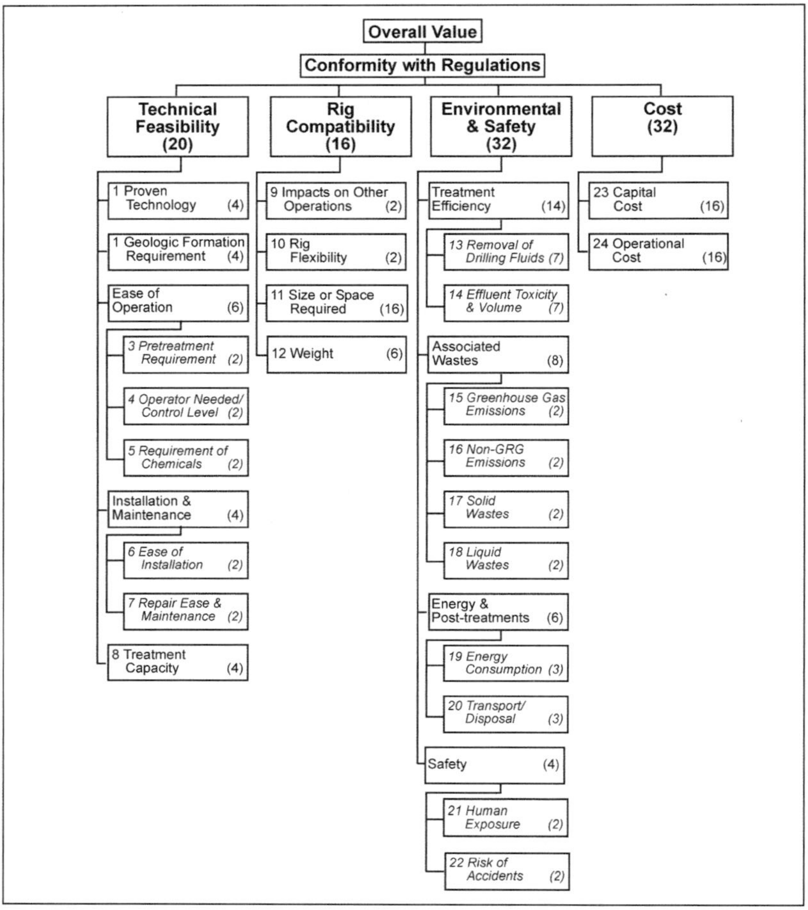

FIGURE 1. Decision making criteria hierarchy and weighting factors (values in parenthesis).

6.9%, or not involve ocean discharge. In addition, the selected technologies must either be currently in use offshore, or have potential to be used offshore. According to these requirements, technologies using various types of treatment or disposal mechanisms including mechanical treatment, thermal treatment, chemical treatment, biological treatment, and re-injection were selected. The technologies include the vertical centrifuge, horizontal centrifuge, thermal desorption, incineration, grinding, stabilisation, biological treatment in a bio-reactor, and re-injection. Descriptions of the technologies can be found in many sources, such as CAPP (2001), Cripps et al.

(1998), and elsewhere. These are proven treatment and disposal technologies that are able to meet the regulatory standard. The vertical centrifuge, horizontal centrifuge, and re-injection have been previously used offshore, while the other technologies are current land-based.

DATA GATHERING

The data to perform the evaluation were collected from several sources, including reports, journal papers, and industry personnel surveys (e.g. Cripps et al., 1998, UKOOA, 2000, and companies' brochures). As part of the industry research, two sets of questionnaires, one for suppliers and the other for users of drilling fluids, were distributed. The questionnaires were directed towards the criteria outlined. Fourteen copies of the suppliers' questionnaires were distributed to companies supplying drilling cuttings treatment and disposal systems and six were completed and returned. The user's questionnaires included similar questions regarding cuttings management technologies currently in use, as well as preferences in technologies to deal with drilling cuttings. One copy of the user's questionnaire was given out and completed.

ASSIGNING WEIGHTS

In this study, the direct weighting method was used to assign weights to the criteria (Parnell et al., 1999). Using this method, weights that represent the relative importance of the criteria are directly assigned to the criteria (Parnell et al., 1999). A total weight of 100 was distributed to all the criteria. The assigned weights are shown as the number in brackets of Figure 1. These weights were subjectively assigned based partly on the results from the user's questionnaire. Relatively higher weights were given to the criteria representing the most important properties the user would consider in choosing a treatment system for drilling cuttings. These criteria include treatment capacity, size, weight, treatment efficiency, and so on. Therefore, relying in part on the user's opinion, this weight assigning process may give more importance to the practicability of the evaluated systems than to the environmental aspects. However, as this concern is realised, equal importance was given to technical, environmental, and cost criteria in this evaluation. As shown in Figure 1, weights of 20, 16, 32, and 32 were assigned to the technical feasibility, rig compatibility, environmental and safety, and costs, respectively. The sum of the weights of technical feasibility and rig compatibility is 36, which is almost equal to the weights assigned to the other two categories of the criteria. Moreover, as this is a subjective process, sensitivity analysis was also conducted to determine the effect of the assigned criteria weights on the results of the evaluation.

The four major categories of criteria were divided into sub-criteria. These sub-criteria were grouped for simplicity in assigning weights. Appropriate weights were given to each group according to their relative importance and the sub-criteria in the same group were assigned equal weights. The weights of the sub-criteria in each group add up to the weight of the upper-level criteria and the weights of the criteria in the same level add up to 100. For example, as shown in Figure 1, three sub-criteria were grouped into the higher level criterion, ease of operation. The ease of operation criterion was assigned the weight of six and each of the three criteria under the ease of operation criterion was assigned the equal weight of two.

SCORING

The scoring schemes used in the evaluation were divided into two types according to whether the data were quantitative or qualitative.

Quantitative Scheme

Quantitative data such as weight and capacity of management systems were normalised to interval scaled values before being used in the evaluation. This was done in order that the values for different criteria conform to an equivalent scale. The normalisation was conducted using linear value functions. The quantitative data may have values that are either increasing (a higher value indicates a higher ranking) or decreasing (a lower value indicates a higher ranking). Therefore, a negative sign was assigned to the scores with decreasing values. This results in two different ranges of normalised scores, 0 to 10 for increasing values and -10 to 0 for decreasing values.

The general equation used to normalise quantitative values is:

$$Score\ (in\ 0\text{ - }10\ range) = \frac{\pm\left(Quantitative\ data\ value\text{ - }Minimum\ Value\right)\times 10}{Maximum\ Value\text{ - }Minimum\ Value} \quad (1)$$

Qualitative Scheme

Where quantitative data were not available, subjective rankings were used to measure or evaluate the option. In order to compare options, the subjective rankings were converted into numbers within the range of 0 to 10 using conversion charts. The conversion charts (Fig. 2) were divided and marked with the numbers from 0 to 10 where the higher value represents a preferred characteristic. As seen in Figure 2, the technology that has been previously used offshore is preferred and assigned the score of nine. Those which have only been used onshore for treating drilling cuttings were assigned the lower score of three.

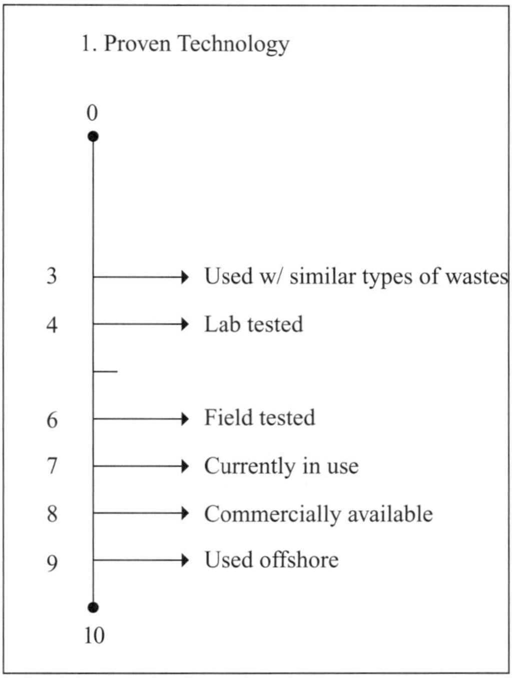

FIGURE 2. Conversion chart for the proven technology and availability criterion.

OVERALL VALUE MODEL

In this study, the widely used Additive Value Model was selected to be used as the overall value model because of its simplicity and robustness (Hobbs and Meier, 2000). This model provides a single index that can be used to compare alternatives.

The overall value for each cuttings management option can be calculated by using the equation (modified from Parnell et al., 1999):

$$Overall\ Value, V(X) = \sum_{i=1}^{n} W_i X_i \quad (2)$$

where i is the criterion identification number, X_i represents the score under the ith criterion, W_i is the corresponding weighting factor, and n is the total number of criteria

With the weights assigned and the ranges of the scores as mentioned in the preceding section, the overall values may range from -560 to 440. The minimum overall value of -560 is obtained by the worst option with the lowest scores (0 or -10) under all of the criteria while the maximum overall value (440) is the overall value of the ideal option obtaining the highest scores (0 or 10) for all criteria.

RESULTS OF THE EVALUATION

In this section, drilling cuttings management options were ranked and compared using overall values. According to the calculated overall values for each cuttings management option, the three best alternatives were the vertical centrifuge, horizontal centrifuge, and re-injection, with overall values of 291, 285, and 182 respectively. Therefore, based on the overall values alone, the three options were considered the optimum alternatives for drilling cuttings management under the established set of criteria. These options are also the only technologies currently used offshore. The bioreactor, which ranks fourth in this evaluation, obtained a slightly lower overall value than re-injection. This means that even though this option has never been used offshore, it is one of the most promising options for offshore applications, although how its potential might be realised in practice is a challenge. The results are shown in Figure 3 below.

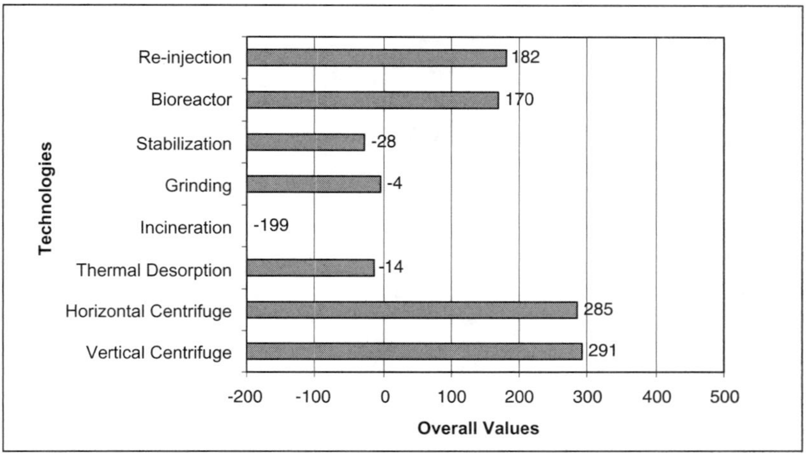

FIGURE 3. Overall values for the alternative drilling cuttings management technologies.

Figure 4 presents a comparison of the total scores of each major category of the criteria. Centrifuges scored the best under technical feasibility. This could have been due to their high capacity to process the waste and the relative ease of their operation and maintenance. In addition, these are well known technologies and have been previously used offshore. Under rig compatibility, centrifuges (vertical and horizontal) also attained the highest scores, mainly because of their compact size and weight. Again, previous offshore application enhanced the high rig compatibility. Re-injection was assigned the highest score under the environmental and safety category because of the reduction of toxic substances and future liability. Under the cost criteria, centrifuges scored the highest.

To compare the cost-effectiveness of each option, the relationship between offshore applicability and costs was considered. The offshore applicability was defined as the applicability of a treatment option to be used offshore with respect to technical

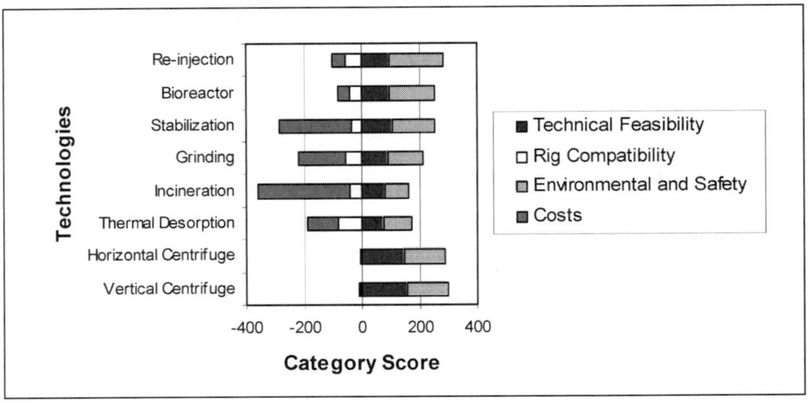

FIGURE 4. Category scores of each technology option.

feasibility, rig compatibility, and environmental and safety. The sum of the three category scores under technical feasibility, rig compatibility, and environmental impacts and safety, were used to represent the applicability score. The values were then plotted against cost category scores of the evaluated options as shown in Figure 5 to outline the performance of each management option compared with their costs. From the figure, the two types of centrifuges (vertical and horizontal) provided high values of offshore applicability with the lowest costs (least negative and nearly zero scores). Re-injection, despite its slightly higher costs than the bioreactor, showed significantly better offshore applicability. Therefore, re-injection was the third most cost-effective option for this comparison.

DISCUSSION AND CONCLUSIONS

In this evaluation, drilling cuttings management options for offshore drilling operations were compared based on technical feasibility, rig compatibility, environmental and safety considerations, and costs. The results of the evaluation using multicriteria decision making shows that the optimum onsite management technologies for drilling cuttings are the vertical centrifuge, horizontal centrifuge, and re-injection.

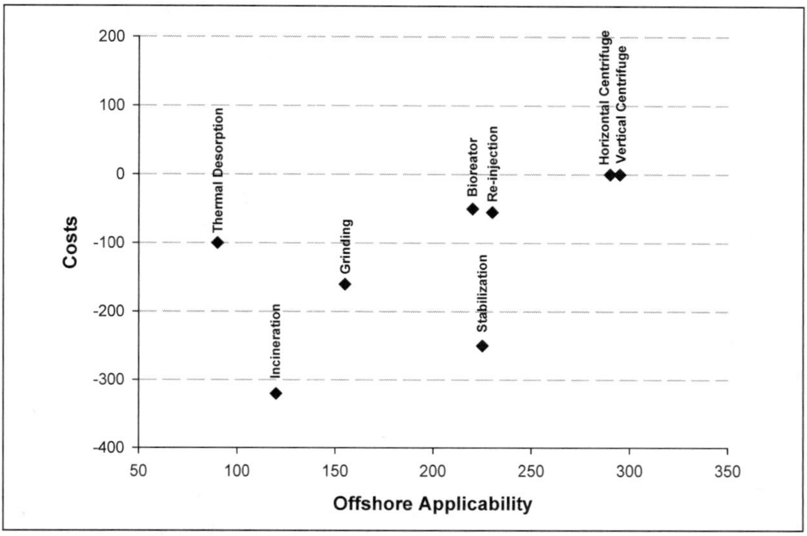

FIGURE 5. Costs versus offshore applicability.

These technologies are the only technologies widely used in current offshore operations. The bioreactor, which is in fourth place, is considered the most promising onshore technology for offshore applications, although how it could be effectively deployed offshore is not clear.

The results of the evaluation were limited by many factors, including availability of data, number of criteria, and differences in the local discharge standards that were used as threshold criteria. In a real situation, selecting a technology to deal with contaminated drilling cuttings depends on many site specific issues. As this evaluation provides only a generic overview of preferred types of drilling cuttings management technologies, the details of the evaluation, such as the set of criteria and the input data, could be modified on a case-by-case basis.

In addition, the validity of the evaluation results is influenced by many parameters, including the availability of data and the distribution of weights. Therefore, uncertainty and sensitivity analyses were conducted and will be reported elsewhere to verify the results of the evaluation.

Briefly, the centrifuges and re-injection scored the best when uncertainty values were considered along with the overall scores, and the uncertainty associated with these options were relatively low. The bioreactor ranked fourth with larger associated uncertainties due to lack of information on its size, weight, and energy consumption. This means that the bioreactor might rank better or worse if more accurate data was available. For example, to provide for a sufficient rate of treatment, the size of a bioreactor may have to be very large as bioremediation is a very slow process. However, the sizes of bioreactors vary considerably, so an assumed number with a high associated uncertainty value was assigned in this evaluation.

The effects of changes to the criteria weights on the ranks of the options were observed through a sensitivity analysis where a criterion or a group of criteria was assigned weight(s) that were decreased by up to 75% (0.25 times as much as the original weight) and/or increased by up to 150% (2.50 times as much as the original weight). It was shown that the results of the evaluation were not significantly sensitive to changes in criteria weight distribution. Therefore, it can be concluded that the best ranked options are significantly superior to others in this set of evaluated technologies.

ACKNOWLEDGEMENTS

The authors would like to thank the Natural Sciences and Engineering Research Council of Canada (NSERC) for financial support through the Strategic Project: Offshore Environmental Engineering using Autonomous Underwater Vehicles.

REFERENCES

Canadian Association of Petroleum Producers (CAPP). 2001. *Offshore Drilling Waste Management Review*. Technical Report 2001-0007.

Cripps, S.J., G. Picken, J.P. Aabel, O.K. Andersen, C. Heyworth, M. Jakobsen, R. Kristiansen, C. Marken, J.E. Paulsen, D. Shaw, A. Annand, T.G. Jacobsen, and I.B. Henriksen. 1998. *Disposal of Oil-based*

Cuttings. Prepared for the Norwegian Oil Industry Association (OLF) by Rogaland Research (RF). RF-98/097, Vers. 3.

Hobbs, B.F. and P. Meier. 2000. *International Series in Operations Research & Management Science. Energy Decisions and the Environment. A Guide to the Use of Multicriteria Methods.* (F.S. Hillier, Series Ed.) Kluwer Academic Publishers, Massachusetts, USA.

Keeney, R.L. 1992. *Value-Focused Thinking.* Harvard University Press, Massachusetts, USA.

National Energy Board, Canada–Newfoundland Offshore Petroleum Board, and Canada–Nova Scotia Offshore Petroleum Board. 2002. *Offshore Waste treatment Guidelines.*

Parnell, G.S., J.A. Jackson, J.M. Jr. Kloeber, and R.F. Deckro. 1999. *Improving DOE Environmental Management: Using CERCLA-Based Decision Analysis for Remedial Alternative Evaluation in the RI/FS Process.* VCU-MAS-99-1.

United Kingdom Offshore Operators Association (UKOOA). 2000. *Task 7.2: Techniques for the Offshore Treatment and Disposal of Retrieved Drill Cutting Piles.* Final report prepared by Environment Resources Management, 17 Jan 2000.

U.S. Environmental Protection Agency. 2000. *Development Document for Final Effluent Limitations Guidelines and Standards for Synthetic-Based Drilling Fluids and other Non-Aqueous Drilling Fluids in the Oil and Gas Extraction Point Source Category.* EPA-821-B-00-013.

A Plan for Implementing Environmental Effects Monitoring During Oil Spills from Sable Offshore Energy Project Installations

Trudel, K. (KenTrudel@slross.com)
(S.L. Ross Environmental Research Ltd., Ottawa, ON Canada)

ABSTRACT: This paper describes a three-tiered system for implementing spill-related environmental effects monitoring (EEM), in which the level of monitoring is linked to the type and size of spill and the magnitude of environmental damage. It describes the system using monitoring of colonial nesting marine birds and commercial fishery species to illustrate the application. Monitoring activities are divided into three tiers, each successive tier gathering more and better data than the previous one. The rationale for the tiers, as well as the criteria for escalating from one monitoring tier to the next is discussed.

INTRODUCTION

This paper describes a system for implementing environmental effects monitoring (EEM) for marine oil spills on Canada's Atlantic Coast. Marine oil spills can be large (tens of thousands of tonnes of oil) and destructive. Environmental contamination can occur over large areas with major impacts on local oil-sensitive living habitats (e.g. coastal marsh), wildlife populations, and local commercial fisheries. However, most spills are a few litres to a few barrels of oil, and cause little damage; oil slicks cover only a few hundred square meters, dissipate within hours, few if any wildlife are injured, and there is no impact on commercial fishing. When spills occur, EEM is needed for a variety of purposes, but monitoring can be costly, both in financial terms and in terms of demands on manpower and logistic resources. For EEM to be practical, it is important to match the level of monitoring effort to the level of environmental risk or damage.

The need for tiered implementation of oil spill EEM was identified during a recent review of EEM plans for spills from offshore oil production facilities in eastern Canada. Over the period of 1999 to 2002, spills from all offshore production in Canada occurred at a rate of 20 to 50 spills per year and virtually all of these were limited to a few litres to a tonne of oil (CNOPB, 2002; CNSOPB, 2003), and probably caused little damage. Government regulators were interested in having some monitoring conducted for these spills. However, the oil spill EEM plans for most offshore facilities had been developed to meet the demands of large spills (thousands of tonnes of oil) and would be extremely costly to implement. Both government and industry expressed an interest in developing a monitoring implementation system that linked monitoring effort to the size of the spill and the level of damage that might occur. Under Sable Offshore Energy Incorporated (SOEI) sponsorship, a three-tiered system for implementing spill-EEM was developed (SOEI, 2002) in which the level of monitoring is linked to several criteria: the type and size of spill, the spatial extent of the zone of contamination, and the type and magnitude of observed environmental damage.

For purposes of this paper, the implementation system is illustrated using two versions of the plan, a simple plan for marine birds, and a more complex plan for commercial fisheries. The bird plan is an example of a simple plan, aimed solely at quantifying the level of mortality in stakeholder-specified populations of valued ecosystem components (VECs), and tracking population recovery. This is done for purposes of verifying the prediction of environmental impact assessment models used in project permitting. The commercial fishery plan is more complex and sophisticated. Its purpose is to provide the information needed to manage the closure and re-opening of commercial fisheries during a spill. Monitoring must quantify, with precision, concentrations, geographic extent, and persistence of spill-related hydrocarbon contamination in environmental compartments (water, sediments), and off-flavours and tissue contamination (for hazardous chemicals) in commercial fishery species in the vicinity of a spill site. The plan contains devices that allow monitoring to achieve the objectives of closing fisheries, while at the same time, minimizing both the size of the area closed and the duration of the closure.

At present, there are no standards or guidelines in Canada regarding environmental effects monitoring for marine oil spills. However, general guidance for the work was obtained from two sources. The National EEM Office (Environment Canada) provides general guidance for EEM for industrial installations (EC, 1997; EC, 1998), and plans have been developed for EEM of routine operations at offshore oil production sites in Canadian waters (e.g. Jacques Whitford Environmental Limited (JWEL), 2000).

MONITORING STRATEGY

General Approach

The objective of EEM for oil spills is to gather information concerning damage caused by a spill. For marine bird populations, monitoring quantifies population damage and post-spill recovery in stakeholder-specified species and stocks. For commercial fisheries, monitoring quantifies the concentration, geographic extent, and persistence of contamination by hydrocarbons in fish tissues and environmental compartments. Monitoring provides the information needed to manage the impact of the spill on local commercial fisheries (e.g. closure and re-opening of fisheries near the spill site).

In general, EEM for marine spills should address at least four questions: 1) Have significant damages and/or exposures to hazardous discharges occurred? 2) Are the observed damages spill-related, or from another cause? 3) What are the severity and magnitude of the damages? and 4) Are impacts greater than predicted during the planning stages of the project? Since the spilled oil and the damage they cause are transient and many of the signs are detectible for only days to weeks, two other questions should be added: 5) How long do the exposures persist? and 6) How quickly do the populations recover, and does recovery occur at the expected rate?

In general, three methods have been employed in assessing the damage caused by spills in the past: 1) quantifying damage directly by counting oiled birds, or measuring contamination in fish tissues, 2) inferring damage indirectly by measuring populations' sizes before and after a spill incident, 3) inferring the proportion of the target population that has been exposed to the oil spill by mapping the spatial distribution of the oil slick and then overlaying this, with a map of the local spatial distribution of the target wildlife populations.

In this context, the term "small spill" refers not only to the size of the spill, but also to its impact or potential impact. Since impact depends on factors such as oil properties, wind, and susceptibility of the VECs in question, these conditions must also be considered when defining "small". In the present application, the definition of "small spill" varies according to the sensitivity of the receiving environment.

Levels of Monitoring Effort

The objectives of monitoring may vary from one group of valued ecosystem components (VECs) to another, but in general, the levels of monitoring effort is the same for all groups, and as follows:

Monitoring Level 1 is for small spills that dissipate naturally near the spill site. These spills may cause no damage, or no more damage than is usually caused on a daily basis by the presence of the platform. Monitoring activities should involve limited, general observations of the spill, the slick (location, dimension, percent coverage, appearance/thickness), and affected biota (e.g. oiled or injured birds) from platforms-of-opportunity, including vessels or aircraft that are already on-scene.

Monitoring Level 2 is for larger spills that: a) persist beyond the navigation safety zone surrounding the production or drilling platform; and therefore b) may cause damage over a larger area than that routinely impacted by the production platform. Monitoring activities include Level 1 monitoring, as well as: a) surveillance of contamination in specific environment compartments (e.g. slick, water column), and b) measuring specific types of damage to target VECs (e.g. counts of oiled birds at shorelines, dead or distressed fish, screening of fish tissues for hydrocarbon contamination).

Monitoring Level 3 is for spills that clearly are causing damage to wildlife or commercial fisheries. Monitoring includes Levels 1 and 2 monitoring, as well as: a) surveying the amount or spatial extent of specific types of damage (e.g. mapping the spatial extent of contamination in fish tissues), b) using more sophisticated analytical procedures needed to address specific endpoints (e.g. using GC-MS to quantify contamination of fish tissues with specific hydrocarbons, rather than simply using fluorescence methods to screen for the presence of hydrocarbons), c) using spill finger-printing methods to determine whether or not all damage is linked to the spill, and d) conducting post-spill monitoring to assess declining levels of environmental or tissue contamination and recovery of impacted wildlife populations.

STRATEGIES FOR SPECIFIC TARGET GROUPS

The following information provides an overview of the practical issues for consideration when assessing damage to marine bird populations, and the contamination of commercial fishery stocks. For details of the approach and methods, see the operator's spill response plan (SOEI, 2002).

Colonial Nesting Sea Birds

The east coast of Canada supports large numbers of seabirds. These include both locally-breeding species and seasonal migrants, all of which are of interest to stakeholders. The objectives of monitoring seabirds are to determine: whether a spill has caused mortalities, and if so, the numbers killed; that the observed mortalities have been caused by the spill; whether the impacts "predicted" in environmental impact statements are consistent with the "actual" impact of the spill; and whether damaged populations recover as expected. It is important to note that in areas near petroleum operations, impacts of routine operations are measured, and baseline values for monitoring parameters are verified on an on-going basis through regular (annual or biannual) monitoring activities. The monitoring activities are conducted as part of the operators' routine operational EEM programs. This work includes studies of local breeding VEC populations, such as the work of Horne and Shepard (1999), and seasonal rates of wildlife stranding on local shorelines, such as work by Lucas (2001).

Monitoring activities are summarized in Table 1, and the decision-making process is described in Figure 1. Monitoring levels and strategies for implementing them are as follows:

Monitoring Level 1. Monitoring is undertaken in all spills and seeks to: 1) determine the size of the spill, 2) assess the persistence of the slick, and 3) quantify bird oiling and mortalities near the spill site. Trained personnel that operate from "platforms-of-opportunity", and already at the platform, make these assessments, using simple visual methods. For most small, non-persistent spills, Level 1 monitoring may be all that is required.

Monitoring Level 2. Monitoring effort escalates to Level 2 in spills where the oil slick persists and moves more than 1 km from the spill site, and therefore, may result in oiling and mortality to a significant proportion of target populations. The objectives are to quantify slick persistence and local abundance of target species and compare this estimate with the predicted impact for a spill of this volume, which is based on

TABLE 1. Monitoring activities for assessing damage to colonial nesting birds.

No.	Activity[1]	Description
1	Level 1 assessment of spill behaviour[2]	Estimate volume of oil spilled using various methods. Assess location, size, and thickness of slick visually by trained personnel at scene.
2	Level 1 computer simulation of impact	Predict the number of birds killed based on computer simulation using: 1) oil fate and impact model, 2) actual spill conditions, 3) real-time environmental conditions, and 4) historical information on vulnerability of the target species to local spills (e.g. seasonality, foraging habits).
3	Level 1 assessment of bird impacts	Trained personnel at scene assess oiled birds visually.
4	Level 1 collection of samples oil spilled oil	Take and preserve samples of oil from the source and the slick for future source identification.
5	Level 2 assessment of slick behaviour	Visual or airborne remote sensing (ARS) assessment of location, size, and persistence of slick made by specialized personnel (refer to strategies and methods for oil slicks)
6	Level 2 aerial surveys of actual bird abundance near spill site during spill	Visual airborne surveys of bird distributions in areas contaminated by the oil slick. Resulting information on bird distributions compared with slick distributions to yield numbers if individuals at risk.
7	Level 2 ship surveys of actual bird abundance near spill site during spill	Visual ship-based surveys of bird distribution in areas contaminated by the oil slick. Resulting information on bird distributions compared with slick distributions to yield numbers of individuals at risk.
8	Level 2 oiled bird surveys on nearby shorelines [3]	Systematic shoreline surveys of nearby shorelines to determine the numbers and condition of birds of target species washed ashore. Results used in models to back-calculate total numbers killed, by species.
9	Level 2 assessments of populations immediately following the spill.	The numbers of breeding adults is measured during the first nesting season following the spill in order to provide a third, independent estimate of damage.
10	Level 3 fingerprinting of oils from oiled birds for source verification	Fingerprinting of oil taken from birds encountered in oiled-bird surveys in order to link oil to spill.
11	Level 3 autopsy of oiled birds in shoreline survey	Autopsies conducted to determine whether or not oiling might have been cause of death/distress.
12	Level 3 measurement of population size in the years following the spill.	If the population size is significantly reduced in the year following the spill, the population is monitored for a number of years in order to track recovery.

[1] Samples of spilled oil are collected regularly in all spills, as are metocean information including wind speed and direction, wave height, air and sea temperature, and current speed and direction. [2] Under low-visibility conditions airborne remote sensing may be used to assess slicks. [3] In general, using oiled bird stranding rates on beaches to estimate of offshore bird mortality suffers from sources of inaccuracy, because most birds oiled at sea do reach shore. In the present context, however, location-specific models have been developed and calibrated in order to improve accuracy (e.g. Weise and Jones, 2001).

environmental impact statement predictions. Specialized personnel and more rigorous methods (e.g. real-time aerial surveys) are used to quantify the real-time fate of oil slicks, the pelagic distributions of seabirds near the spill, and the oiled bird strandings on nearby shorelines. In the present application, Level 2 monitoring is undertaken: 1) in all batch spills of crude oils where slicks persist farther than 1 km from the site, 2) in all batch spills occurring within 3 km of a shoreline, or within the foraging range of occupied marine bird breeding

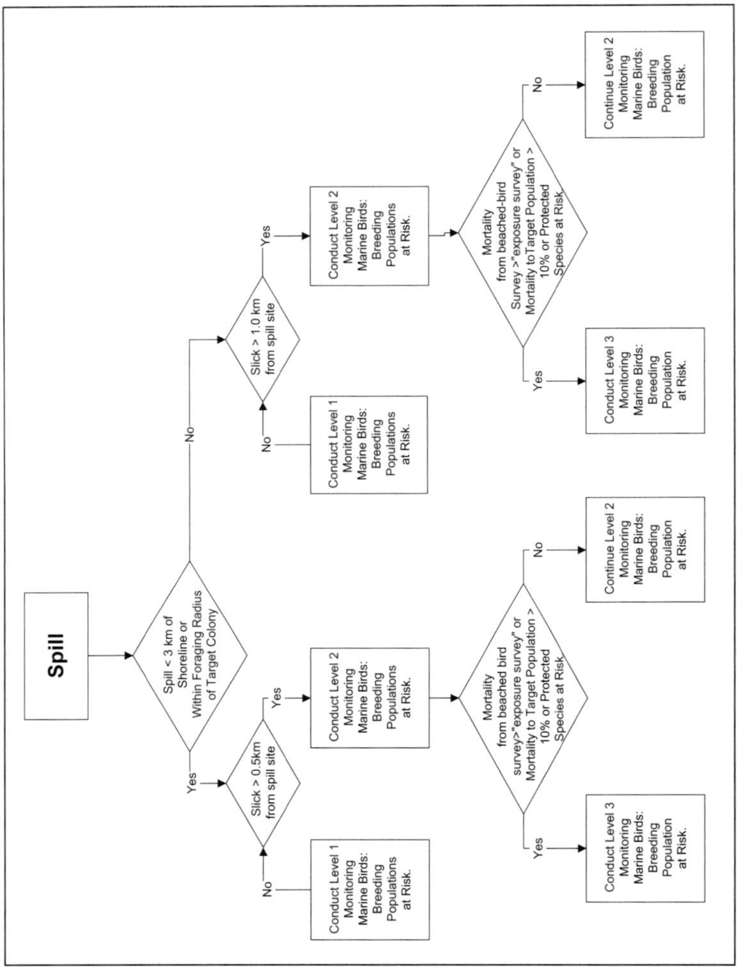

FIGURE 1. Decision tree for EEM: marine birds.

colonies where slicks persist beyond 0.5 km from the spill site, or 3) in continuous spills larger than 0.1 cubic meters of oil per day lasting longer than six hours.

Monitoring Level 3. Monitoring effort escalates to Level 3 when the spill is clearly doing significant damage to seabird populations. The objectives of monitoring are: 1) to verify that all of the observed bird oilings and mortalities are linked to the spill and 2) to track the recovery of the damaged population(s). Level 3 monitoring involves the extensive use of analytical procedures, such as the fingerprinting of oil samples, and post-spill surveillance of impacted populations. Level 3 monitoring is undertaken: 1) if "observed" impacts greatly exceed "predicted" levels, 2) if "observed' impacts exceed the annual variation in population size, or 3) if endangered or threatened species are at risk.

The rationale behind escalating monitoring from Level 2 to Level 3 when damage exceeds predicted levels is as follows. One basic objective of all EEM programs for offshore petroleum operations in Canadian waters is to verify that the actual impacts are no greater than those predicted in the environmental impact statements (EIS) prepared during project planning and permitting. It is implicit that if actual damage from an operation greatly exceeds predicted levels, then action may be required to reduce environmental disruption to acceptable levels. If, during a Level 2 spill, the damage greatly exceeds that predicted by models used in the EIS process, stakeholders may conclude that the EIS predictions have underestimated the real risks to birds. They may further demand action to identify the cause of the unpredicted impact and to reduce risks to acceptable levels in future operations.

Commercial Fishery Species (Contamination and Off-flavour in Fish Tissue)

When oil spills occur near commercial fishing grounds, regulatory agencies may close local fisheries temporarily until the spill is controlled and any environmental and tissue contamination with hydrocarbons in commercial species has declined to acceptable levels (e.g. Mauseth and Martin, 1997). EEM provides information upon which decisions regarding the closing and re-opening of fisheries are made. In recent years, closure and re-opening decisions have focused on two measures of contamination: off-flavours and the presence of health-related compounds in fish tissues, such as benzopyrene (Mauseth and Martin, 1997; Gilroy, 2000). At present, there are no Canadian standards or guidelines for petroleum spill-related closures. However, during recent North American oil spills, decisions regarding fishery closure and re-opening have been made based on criteria such as: whether or not the source of the spill has been controlled; presence of visible oil slicks; and presence of oil contamination in fish tissues (Mauseth and Martin, 1997, Mearnes and Yender, 1997; Gilroy, 2000). The present approach was developed based on the spill experience as reported by these aforementioned authors. The approach applies to pelagic, demersal, and benthic species. Monitoring fish tissues serves a number of purposes. It verifies whether or not significant tissue contamination has taken place. It determines the spatial extent of fishery contamination. It assesses whether any contamination is linked to the spill. If tissue contamination has taken place, monitoring provides a means to measure the change in tissue contamination with time through the course of the spill incident.

Monitoring activities are summarized in Table 2. Levels of monitoring activity and strategies for implementing them are as outlined below (see also Figs. 2a, b, c).

Monitoring Level 1. Level 1 monitoring is undertaken in all spills in order to assess and verify the size and persistence of the spill and to identify fish mortalities near the spill site. Trained personnel, already at the scene, conduct monitoring. Monitors use simple visual methods and operate from "platforms of opportunity". For small spills only Level 1 monitoring may be required.

Monitoring Level 2. Monitoring effort escalates to Level 2 in larger spills when spill volume exceeds preset volume thresholds, and there is the potential for tissue contamination in commercial fishery species. Monitoring objectives are to determine whether fish tissues have become contaminated with hydrocarbons above baseline levels, and to compare the

TABLE 2. Monitoring activities for assessing contamination and off-flavours in commercial fishery stocks.

No.	Monitoring Activity	Description
1	Level 1 assessment of spill volume and slick behaviour	Estimate volume of oil spilled using various methods. Assess location, size, and thickness of slick visually by trained personnel at scene.
2	Computer simulation of impact	Estimate the spatial extent and persistence of environmental contamination and tissue contamination in specific target species.
3	Level 2 assessment of spill behaviour	Assess slick persistence and movement using visual/instrumental methods by trained expert from aircraft.
4	Level 2 assessment of hydrocarbons in water column	Assess concentration and distribution of spill-related hydrocarbons in surface waters and at depth by in-situ fluorescence.
5	Level 2 assessment of seabed sediments for hydrocarbon contamination	Assess distribution of hydrocarbons in flocculent layer and seabed sediments by screening method (see SOEI, 2002).
6	Level 2 assessment of target species for tissue contamination	Assess presence of contamination of target fishery species in contaminated area, by sampling, screening for tissue contamination and raw-odour grading.
7	Level 3 source verification of contaminants in fish tissue and environment	Conduct fingerprinting of hydrocarbons in extracts of fish tissues for source verification.
8	Level 3 quantification of specific health-related PAHs in fish tissue	Quantify health-related PAH compounds in extracts of contaminated fish tissue.
9	Level 3 assessment of spatial extent of tissue contamination	Sample stocks of target species to assess spatial extent of tissue contamination. Analysis is by hydrocarbons screening and taste-testing methods first, followed by PAH testing on positive samples.
10	Level 3 assessment of post-spill persistence of environmental and tissue contamination	After the spill source is contained, assess persistence of slicks, in-water contamination, and contamination in fish tissues.

"observed" magnitude of contamination with that predicted in the EIS. Baseline levels of contamination are determined from results of routine operational monitoring that is conducted periodically for all offshore petroleum operations in Canadian waters. The spatial extent of oil slicks, water column, and seabed contamination are measured to define the area where fish are exposed to hydrocarbons. Fishery species are collected periodically from this defined area after the spill has been controlled and surface oil collected. Edible tissues are analyzed to determine whether they are contaminated. Level 2 monitoring is undertaken: 1) in continuous spills of more than 0.1 cubic meter of oil per day lasting longer than six hours, and 2) in any batch spill in which the in-water hydrocarbon exposure concentrations might cause significant contamination of fish tissues at distances beyond safety zone around the offshore installation.

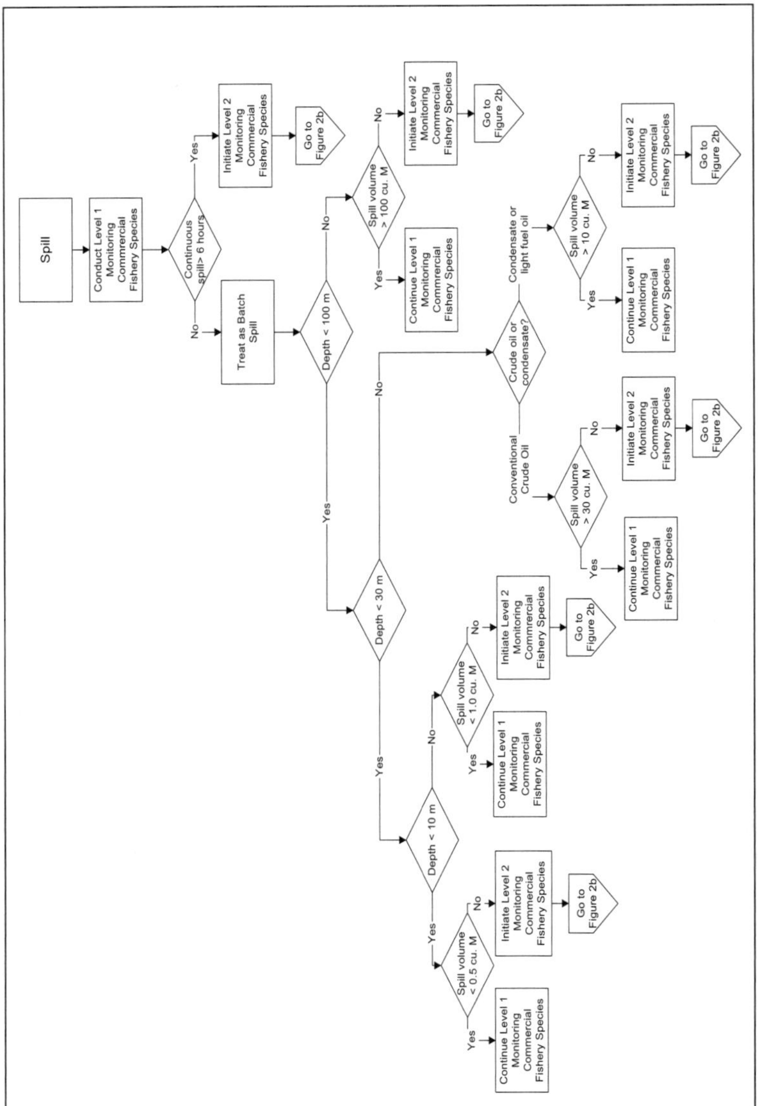

FIGURE 2a. Decision tree for initiating EEM for commercial fishery species.

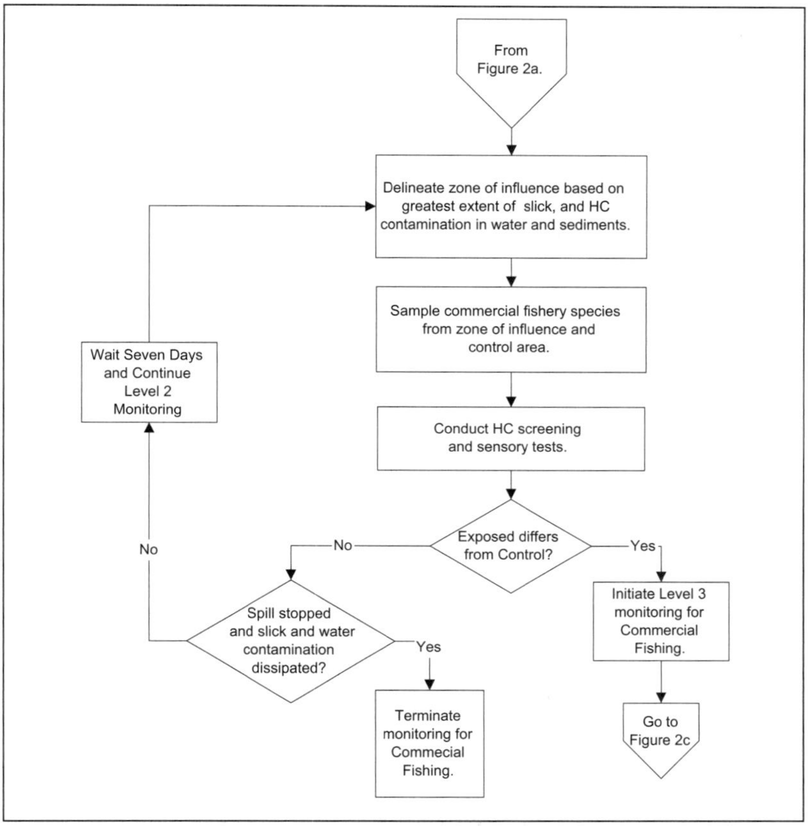

FIGURE 2b. Decision tree for initiating EEM for commercial fishery species cont'd.

Monitoring Level 3. Monitoring escalates to Level 3 if significant contamination of fishery products is detected. The objectives are to: 1) verify the source of contamination, 2) determine whether tissue contamination exceeds acceptable levels, 3) delimit the spatial extent of tissue contamination, and 4) track its decline following the spill. Level 3 monitoring is undertaken when tissues of fish within the zone of contamination differ significantly from background, either for taste or level of petroleum-derived hydrocarbons, as measured by a screening test. In Level 3 monitoring, the spatial extent of tissue contamination is determined by sampling along transects radiating from the spill source (see SOEI, 2002). Levels of contamination

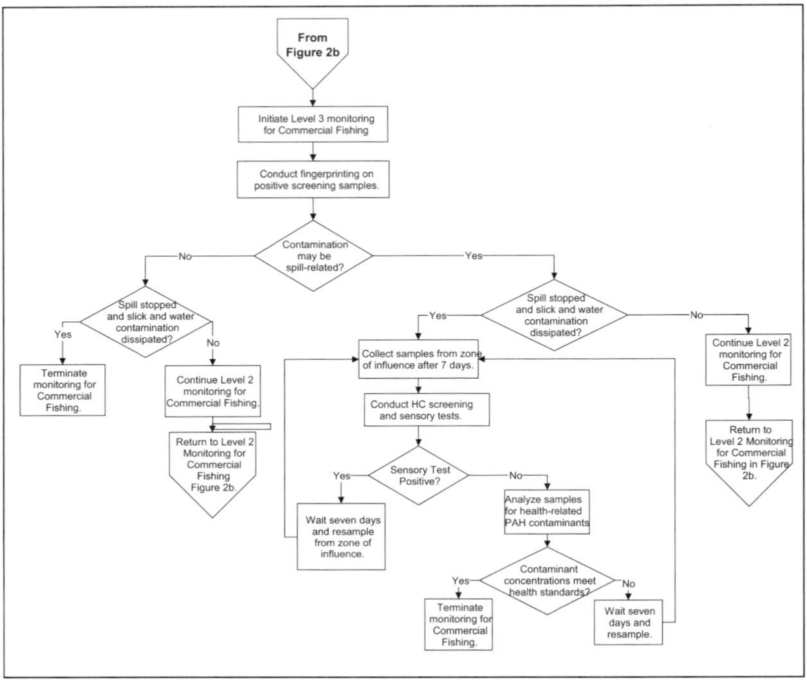

FIGURE 2c. Decision tree for initiating EEM for commercial fishery species cont'd.

in fish tissues are established definitively by a) taste-panel testing and b) quantitative chemical analysis of tissues for a certain health-related polycyclic aromatic hydrocarbons (PAH) (see SOEI, 2002). Once the spill source is controlled, environmental and tissue samples are collected at regular intervals to track the decline in contamination to background or acceptable levels.

DISCUSSION

The criteria described above for escalating from one monitoring level to the next are workable under most spill conditions. One exception may be the transition from Level 1 to Level 2 in the fish contamination protocol. Ideally, the Level 1 to 2 escalation should be based on results of measurements made in Level 1. However, visual observations cannot assess contamination in the water column or in fish tissues. Fortunately, in spills of medium to heavy density oils, slick persistence and movement can be used as an indirect indicator of in-water contamination. The risk of in-water hydrocarbon concentrations to fisheries and slick persistence are functions of spill size, oil type, and sea state. Numerical modelling suggests that spills of medium and heavy crude oils disperse little at low to moderate sea states, so their slicks should persist, and in-water hydrocarbon concentrations should be low. It follows that if a spill of medium density oil at low or moderate sea state produces a small (< few hundreds of meters in diameter), non-persistent slick (dissipates < one km from spill site), it probably indicates that the spill is small, and probably does not exceed the pre-established thresholds for escalating to a higher level of monitoring. On the other hand, if a spill produces a much larger (many km^2), and more persistent slick (persists for many km), it suggests that the spill itself is large and escalation to Level 2 monitoring is justified. Under these conditions, Level 1 monitoring is not longer adequate, and direct measurement of in-water oil concentrations is needed to assess risks of fish contamination. At very high sea states, slick size is unreliable as an indicator of spill size and in-water oil exposure, so operators must use other approaches to decide about a Level 1 to 2 transition.

When non-persistent oils, such as gasoline or condensates are spilled, slick persistence is less useful in estimating in-water oil contamination. Numerical modelling suggests that light fuels and condensates disperse quickly under average wind-wave conditions, with slicks of even larger spills (many tens of m^3 of oil) dissipating within a few hundreds of meters of the spill. Unfortunately, the in-water contamination generated by these spills may be sufficient to contaminate fish tissues several kilometres, or more, from the spill site. There appears to be at least two alternative approaches for the Level 1 to 2 transition, for these lighter oils. A conservative approach is to conduct Level 2 monitoring for all spills larger than a few litres. However, this might pose unwarranted safety risks for operators and would prove costly. A more realistic alternative may be to establish, by consultation among stakeholders, a risk-based standard, a critical spill size above which, spills are assumed to cause unacceptably high exposures to fishery species at some specified distance from the spill site. The critical spill size might vary with wind speed and water depth. Ideally, the standard should be based on practical spill experience, but computer simulations might be used until adequate spill experience is gained.

ACKNOWLEDGEMENTS

This project was funded by Sable Offshore Energy Incorporated (SOEI). The author wishes to thank Michael Coolen, Geoffrey Hurley, and Cal Ross for their advice and guidance during the work. The author also wishes to acknowledge the support of the Canadian Association of Petroleum Producers (CAPP) for initiating the project.

REFERENCES

Canada–Nova Scotia Offshore Petroleum Board (CNSOPB). 2003. *Environmental Spills.* Canada–Nova Scotia Offshore Petroleum Board, Halifax, NS.

Canada–Newfoundland Offshore Petroleum Board (CNOPB). 2002. *Spill Frequency and Volume Annual Summary.* Canada–Newfoundland Offshore Petroleum Board, St. John's, NF.

Environment Canada. 1997. *Aquatic Environmental Effects Monitoring Requirements.* Environmental Canada. 1997. EEM/1997/1 (Revised EPS 1/RM/18).

Environment Canada. 1998. *Pulp and Paper Technical Guidance for Aquatic Environmental Effects Monitoring.* 1998. EEM/1998/1.

Gilroy, D.J. 2000. "Derivation of Shellfish Harvest Reopening Criteria Following the New Carissa Oil Spill in Coos Bay, Oregon." *Journal of Toxicology and Environmental Health.* 60(5):317-329.

Horne, A.G. and P. Shepard. 1999. *Sable Island Tern Project: 1998 Final Report.* Unpublished Report Prepared for Sable Offshore Energy, Incorporated (SOEI).

Jacques Whitford Environmental Limited. 1999. *Environmental Effects Monitoring Program.* Prepared for Petro-Canada Offshore Development and Operations, St. John's, NF.

Lucas, Z. 2001. *Beached Bird Surveys on Sable Island, 2001.* Unpublished Report to Sable Offshore Energy Incorporated, 2 pp.

Mauseth, G.S. and C.A. Martin. 1997. "Closing and Reopening Fisheries Following Oil Spills: Three Different Case Studies with Similar Problems." *Arctic and Marine Oilspill Program (AMOP).* 20:1283-1303.

Mearnes, A.J. and R. Yender. 1997. "Workshop on Managing Seafood Problems during the Response Phase of an Oil Spill." *Arctic and Marine Oilspill Program (AMOP).* 20:203-214.

Sable Offshore Energy Incorporated (SOEI). 2002. *Environmental Effects Monitoring Plan for Marine Spills.* Prepared by S.L. Ross Environmental Research Ltd. for Sable Offshore Energy Project, Sable Document Number: 10-E-04-0007-B1, August 2002.

Weise, F. and I. Jones. 2001. "Experimental Support for a New Drift Block Design to Assess Seabird Mortality from Oil Pollution." *Auk.* 118:1062-1068.

PART 2

Approaches and Technologies

Quantifying Fine-grained Drill Waste in Scotian Shelf Suspended Sediments

Milligan, T.G. (milligant@mar.dfo-mpo.gc.ca), (Department of Fisheries and Oceans, Dartmouth, NS Canada) T. Tedford (University of British Columbia, Vancouver, BC, Canada), D.K. Muschenheim (Acadia Centre for Estuarine Research, Acadia University Wolfville, NS Canada), and C.G. Hannah (Department of Fisheries and Oceans, Dartmouth, NS Canada)

ABSTRACT: Laboratory studies have shown that elevated concentrations of bentonite and barite, the two major constituents of water based drilling muds, can affect the growth of suspension feeding organisms. The disaggregated inorganic grain size distribution of bentonite is unusual in that it is dominated by the smallest grain sizes, giving it a negative slope between 0.87 and 2.4 mm. This distribution is in marked contrast to that of naturally occurring suspensions on continental shelves, which have positive slopes in this region of the curve. Using the shape of the bentonite size distribution to identify drilling wastes, samples collected near an operational drilling platform on the Scotian Shelf showed elevated concentrations of bentonite in the surface plume and in the benthic boundary layer. Flocculation was identified as the mechanism for the rapid settling of drill wastes containing bentonite to the near bed region. Based on these findings, models were developed to predict the concentration of fine-grained drill wastes in the benthic boundary layer. To provide a method of quantifying bentonite in suspension and to verify model predictions, a method to estimate the concentration of drill waste using the disaggregated inorganic grain size of suspended sediment samples was developed.

INTRODUCTION

Standard offshore drilling practices in Canadian jurisdiction allow for the unregulated release of water based drilling muds during both exploration and production phases. The two major constituents of this waste stream are bentonite clay, which is used to control the rheology of the drilling fluid, and barite, which is primarily a weighting agent. Bentonite and barite have been shown to have a negative effect on scallop growth above certain concentrations (Cranford and Gordon, 1992). As a result of these and other potential environmental impacts related to offshore oil development, regular environmental effects monitoring (EEM) is required to identify and minimize zones of impact.

To ensure compliance with environmental regulations, operators conduct sediment and water column sampling to evaluate the levels of drill waste present. Whereas barite can be identified using standard chemical analyses, bentonite can not easily be chemically distinguished from other naturally occurring clays. Both regulators and operators require a means to determine the concentration of drill wastes in the water column and in the near bottom region. A need also exists to validate the results provided by a benthic boundary layer transport model (bblt) for drill waste dispersion that has been used to predict drill waste distributions (MacLaren Plansearch, 1997; Thompson et al., 2000) and potential sublethal effects on sea scallops *Placopecten magellanicus* (Gordon et al., 2000; Cranford et al., 2003). Early predictions of drill waste dispersion using bblt have not been supported by EEM results on the Scotian Shelf, which found lower than predicted concentrations. More recent simulations focussed on the dispersion of barite are more consistent with EEM results (Hannah et al., 2003). A means to differentiate between the transport pathways of bentonite and barite is required to aid in the interpretation of EEM data on the Scotian Shelf (Hannah et al., 2003). The method presented here uses the difference between the disaggregated inorganic grain size (DIGS) of bentonite and that of naturally occurring particulate material to calculate the amount of bentonite in suspension and to investigate the fate of drill wastes.

Flocculation incorporates material from suspension, both organic and inorganic, into rapidly sinking aggregates in an unbiased manner (Kranck, 1980; Hill et al., 2000). Within each floc, the disaggregated inorganic grain size (the size distribution of individual inorganic grains after the removal of organic material) will be the same as that of the parent suspension (Kranck, 1980). Mechanical weathering of siliciclastic rocks results in the formation of straight DIGS distributions, with regularly varying concentrations in each size class when plotted as log concentration vs. log diameter (Fig. 1; Bennett, 1936; Barndorff-Nielsen, 1977; Bagnold and Barndorff-Nielsen, 1980; Kranck and Milligan, 1985).

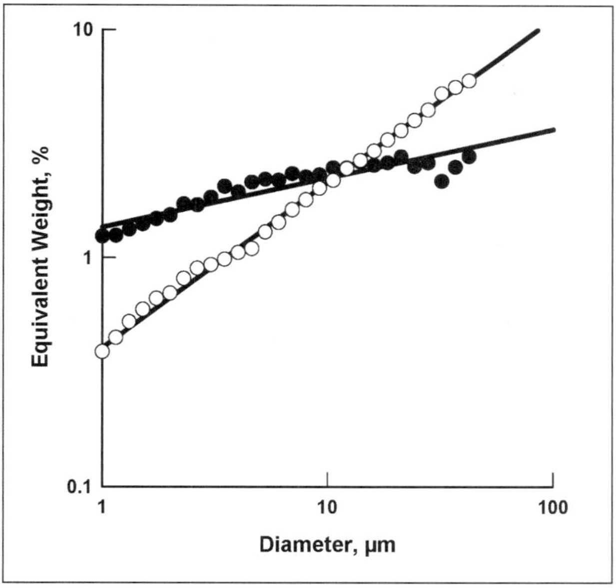

FIGURE 1. Disaggregated inorganic grain size of tills derived primarily from granite (open circles) and slate (filled circles) showing straight-line distributions created from mechanical weathering by glaciers. The source slope is the slope of the regression line and is dependent on the type of rock.

The slope of the distribution is dependent on the type of material being weathered, hence more resistant types of rock such as granite will produce steeper slopes (Bagnold and Barndorff-Nielsen, 1980; Kranck and Milligan, 1985). These straight-line, source distributions are ubiquitous in rivers (Kranck and Milligan, 1983) and in the ocean (Kranck and Milligan, 1992; Stevens et al., 1996). The DIGS distributions of suspended and bottom sediments normally consist of a straight line tail of fine grained material derived from the source material and a region of coarser material with varying degrees of sorting (Kranck and Milligan, 1985; Kranck et al., 1996). Barite, which is produced from crushing and then screening to provide a specified median size, has a positive source slope slightly greater than that normally found in suspension (Fig. 2). In marked contrast, the DIGS for bentonite has a highly negative source slope close to -2 (Fig. 2). It is this distinctive slope that forms the basis for determining the presence of bentonite in the water column and near bottom.

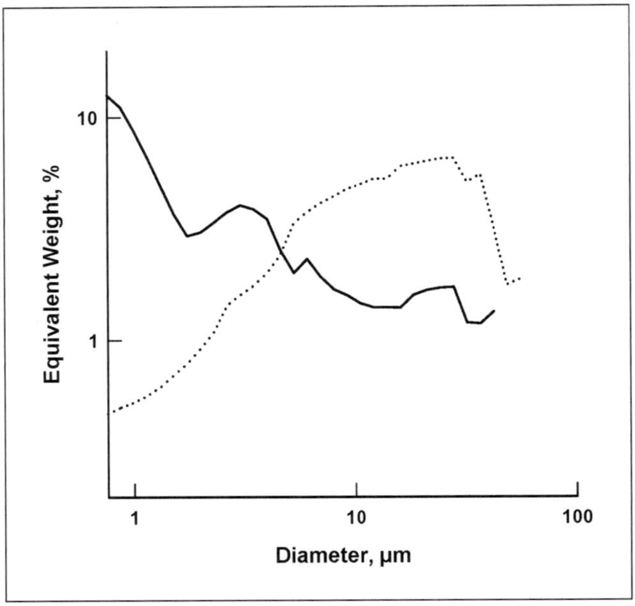

FIGURE 2. DIGS of bentonite (solid line) and barite (dotted line) used in formulation of drilling mud.

Using a Coulter Counter to analyze the DIGS distribution of material in suspension, Muschenheim and Milligan (1996) found evidence for the flocculation and deposition of drill waste at an actively drilling platform on the Scotian Shelf off of eastern Canada in 1991 and 1993 (Fig. 3). The jackup platform, *Rowan Gorilla III,* was at the time being operated by LASMO Nova Scotia Ltd. for production drilling at the Cohasset and Panuke fields. Pay zones in these fields were at fairly shallow depths and the wells were drilled with 'gel' (bentonite) mud containing only small amounts of barite. Thus the particulate composition of the drilling fluid—and the discharged waste stream—was primarily bentonite. The presence of drilling waste was identified in the discharge plume and in samples collected from the benthic boundary layer (BBL) near the rig using the unique grain size signature of bentonite (Fig. 2). The area affected by the deposition of flocculated drill mud was determined using image analysis of sea floor video that recorded regions of bottom covered in a dark mat of flocculated material and that sampling revealed to contain bentonite (Fig. 4). Missing from that study was a method to determine the quantity of bentonite in the samples collected.

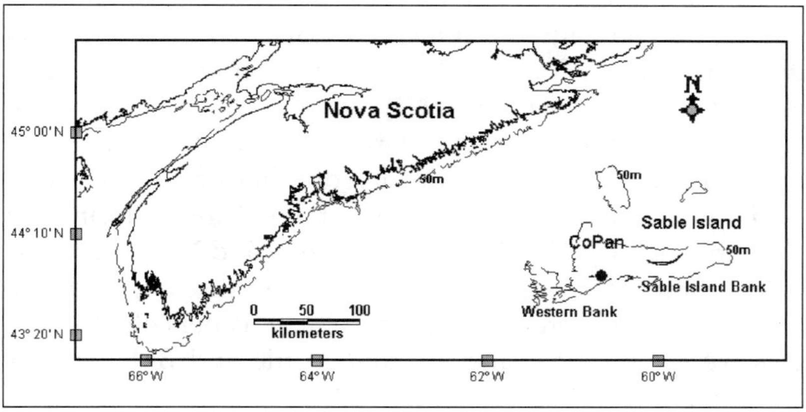

FIGURE 3. Map showing location of Sable Island and Western Banks and position of Panuke and Cohasset (CoPan) oil production sites.

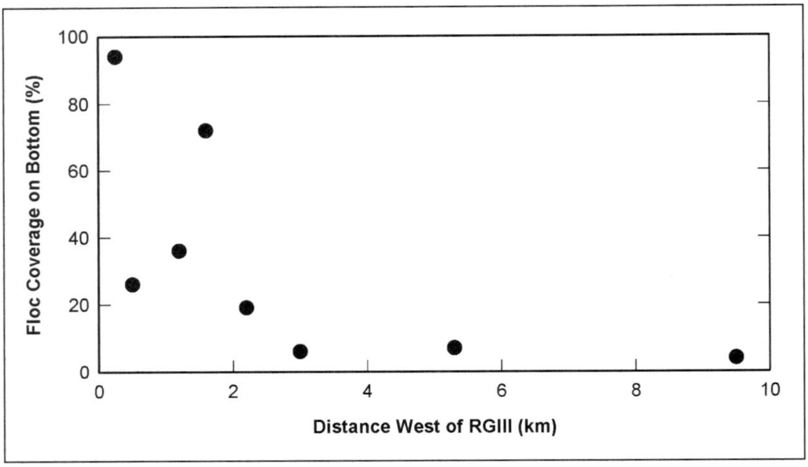

FIGURE 4. Percent area of the seabed covered by flocculated material in September 1993. Areas were determined by image analysis of bottom video. Disaggregated inorganic grain size analysis of samples from the benthic boundary layer at these stations revealed that there was bentonite present at stations out to 3 km (Muschenheim and Milligan, 1996).

MATERIALS AND METHODS

Over a period of 5 years, suspended sediment samples were collected in the vicinity of the jack-up rig *Rowan Gorilla III* (RGIII) located at the Panuke and later the Cohasset formations (CoPan) on the Scotian Shelf in water depths from 30–50 m (Fig. 3). The bottom sediment at CoPan consists of fine to medium sand with extremely low concentrations of silt and clay (Muschenheim et al., 1995). In July 1991, samples were collected in the drill mud discharge plume and from the benthic boundary layer (BBL) along the drift line of the plume created by the RGIII while drilling at the Panuke formation (Muschenheim et al., 1995; Muschenheim and Milligan, 1996). At this time, water based drill muds were discharged above the water surface, creating a visible plume. Water column samples were obtained using a Niskin bottle and BBL samples were collected with a specially designed near bottom sampler (BOSS) that simultaneously collected samples from 7 depths within the bottom 0.5 m (Muschenheim et al., 1995). In subsequent years, samples were collected at stations along transects running E–W and N–S from the platform at the Cohasset site and were not specifically targeted at the discharge plume, as wastes were no longer being discharged at the surface.

The primary source of data for this report was collected in September 1993, seven months into a planned 11-month period of drilling. During this period, drill wastes were being discharged just below the surface. At the transect stations water column samples were obtained at approximately 1, 10, 20, 30 m below the surface and an additional sample within 3 m of the bottom. In addition, results from a 24 h anchor station 0.3 km N of the RGIII are reported. At this station, water column and BOSS samples were collected every 1.5 h along with CTD, optical backscatter, and other ancillary data. Niskin bottles were taken at the same depths as the transect stations. BOSS samples in the BBL were taken at 0.05, 0.10, 0.15, 0.25, 0.35, 0.45, and 0.5 mab (metres above bottom). To get enough material from the BOSS samples for analysis, the samples from 0.05 to 0.25 mab and from 0.35 to 0.5 mab were combined and assigned heights of 0.15 mab and 0.40 mab, respectively. To monitor currents, a single S4 electromagnetic current meter was placed 1 m above bottom, 3 km NE of the RGIII for the duration of the sampling effort.

In April 1994, sampling was limited to video observations of the bottom and water column sampling along the same 1993 orthogonal transects extended out to 20 km from the RGIII located at Cohasset. No samples were taken in the BBL. Drilling at Cohasset had ceased 4 months prior to the cruise, and discharges from the site were limited to produced water. In September 1996, in addition to stations near CoPan, a number of stations were occupied in the haddock nursery area of Western Bank, an area of the Scotian Shelf closed to both fishing and oil and gas development. The water depth on Western Bank was near 70 m and the bottom sediment type was gravel and small cobble rather than well sorted sand.

Water samples were filtered onto 0.8 μm AAWP Millipore cellulose acetate filters to determine total suspended particulate matter (SPM) using standard gravimetric analysis. The DIGS for the filtered samples was then determined following the protocols described in Milligan and Kranck (1991). Sheldon (1972) showed that the retention efficiencies of 0.8 μm depth filters exceeded that of 0.4 μm membrane filters while allowing a greater volume to be filtered. Prior to particle size analysis on a Coulter Multisizer IIe, organic material was removed using low temperature plasma ashing and hydrogen peroxide digestion. The remaining inorganic fraction was resuspended in 1% NaCl and then disaggregated using a sapphire tipped ultrasonic probe before counting. Size analysis was performed with a 30 μm and 200 μm aperture tube giving a size range from 0.63 to 100 μm. Results were plotted as log concentration expressed as parts per million (ppm) vs. diameter divided into 32 logarithmically equal size classes (channels) over a size range from 0.63–64 μm. Plotting the DIGS spectra as log volume vs. log diameter preserves the slopes of the distributions over a wide range of concentrations (Kranck and Milligan, 1991). The total concentration of inorganic material in the samples, expressed in ppm, was calculated from the area under the DIGS curve.

The method for determining the concentration of bentonite is based on its unusual grain size distribution, which differs greatly from that of naturally occurring inorganic mineral suspensions found on continental shelves (Fig. 2). The presence of bentonite alters the normally positive source slope of the suspended sediments found on both Western and Sable Banks (Figs. 5, 6). Samples collected at Panuke in 1991 clearly show bentonite enrichment in the fine end of the distribution when compared to the DIGS of bentonite (Fig. 6). Using the difference between the negatively skewed bentonite distribution and the straight line distribution of the background inorganic grains, the amount of bentonite in each sample was determined by subtracting the background size spectra for uncontaminated samples from that of samples containing bentonite. For each DIGS distribution, the concentration of bentonite was determined for the area under the curve between 0.87 μm and 2.4 μm, the region represented by the source slope (Fig. 7).

Data below 0.87 μm were discarded, as this region of the curve is frequently associated with spurious noise counts generated

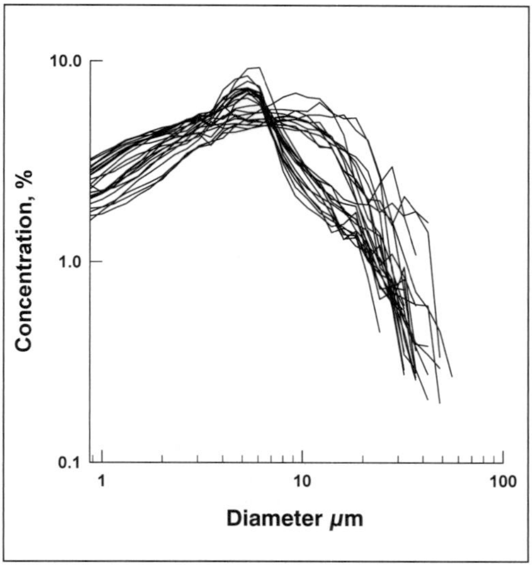

FIGURE 5. DIGS distributions for samples collected from Western Bank in September 1996 showing common source slope from 0.87 to 2.0 μm. A diatom peak at 6–8 μm occurs in samples collected near the surface and increasing maximal size in found in samples collected near bottom. Note there is no change is the source slope between 0.87 and 2.4 μm.

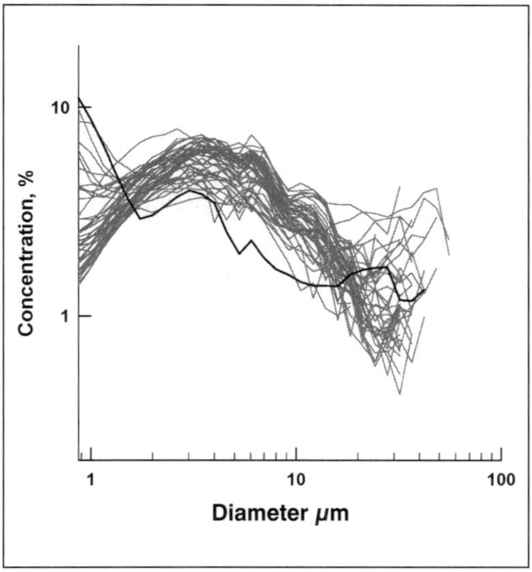

FiIGURE 6. DIGS for samples collected near the RGIII in June 1991. Spectra have been normalized to 100% to illustrate similarity between curves and effect of addition of bentonite. The DIGS of bentonite (black) is shown for comparison.

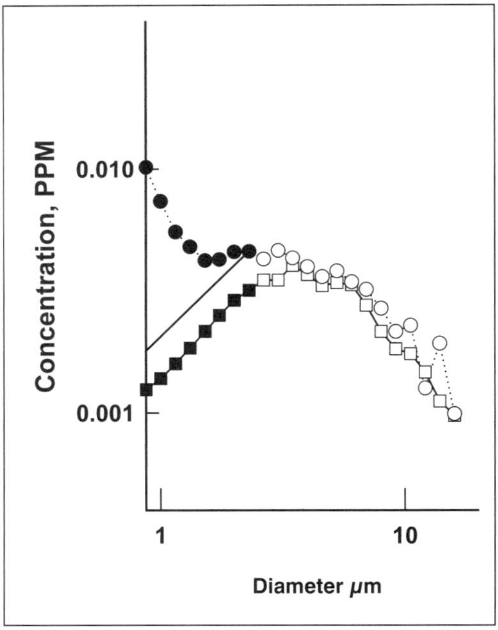

FIGURE 7. DIGS plots of suspended sediment near the RGIII showing uncontaminated (squares) and bentonite enriched (circles) spectra. Solid squares (channels 1–8) are used to determine the standard curve, which is then scaled so that the value in channel 8 is equal to that of the enriched size distribution (black line). The concentration of bentonite is calculated by subtracting the standard curve from the enriched curve.

during the counting process (Milligan and Kranck, 1991). A standard curve representing the background suspended DIGS from 0.87 to 2.4 μm was calculated using the first 8 channels of the DIGS distributions from all samples with positive source slopes collected during each sampling period and/or study area. To define the standard curve, the slope was determined for the first 8 points in each sample using linear regression of the logarithms of the concentrations (Table 1). The slope of the standard curve for each group of samples was calculated by taking the average of the regression slopes that were positive and had r^2 values greater than 0.95. To account for a variation in total concentration, the standard curve was scaled so that the values in channel 8 (2.4 μm) of the standard curve and sample of interest were equal (Fig. 7). Using the scaling factor and the slope of the standard curve, the concentrations for channels 1–7 of the standard curve were then calculated. To estimate the concentration of bentonite in suspension the area under the curve, defined by the difference between the

TABLE 1. Results of regression analysis to determine source slope for standard curve using channels 1–8 for samples collected at CoPan site and on Western Bank. Number of samples with positive source slopes and $r^2 > 0.95$ (n) and total number of samples collected are indicated.

Year	Month	Area	Discharging	Mean Slope	STD	n	Total	% wt Bentonite
1991	July	Panuke	Yes	1.1	0.2	22	42	48
1993	Sept	Cohasset	Yes	0.61	0.12	147	303	51
1994	April	Cohasset	No	0.85	0.11	43	49	12
1996	Sept	Western Bank	No	0.57	0.11	24	24	0
1996	Sept	Cohasset	No	0.48	0.11	33	39	15

standard curve and the sample curve, was determined by subtraction. Negative values, where the concentration in the standard curve was greater than the sample curve, were not included in the summation for total concentration of bentonite.

Results from the curve fitting and subtraction were expressed as a volume of bentonite in ppm between 0.87 and 2.4 μm. This value was then used to determine an equivalent weight per volume concentration using a specific gravity of 2500 kg/m^3. Reported values for the specific gravity for bentonite range from 2400–2600 kg/m^3. Contours of bentonite concentration over time and depth for the 1993 anchor station were determined use a kriging algorithm (Surfer, v 7.0, Golden Software, Golden CO).

RESULTS

Total suspended particulate matter (SPM) values on Sable Bank remote from the RGIII and on Western Bank were low except in the BBL during periods when sand was resuspended from the bottom. Background water column values ranged from 0.3 to 5.5 mg/l in the vicinity of the RGIII and exceeded 2000 mg/l in the BBL when sand was in suspension. On Western Bank, SPM values were similar with suspension values between 0.3 and 5 mg/l and a maximum near bed value of 1000 mg/l. Low DIGS volumes were found in the water column at both sites. Inorganic volume concentrations on Sable Bank were between 0.01 and 0.15 ppm with the highest values occurring nearest the bottom. BBL inorganic volumes sometimes reached 1400 ppm during resuspension events. Water column values for Western Bank were similar to those found on Sable Bank, with a maximum of 0.2 ppm in the deepest sample. BBL values were of the same order as those on Sable.

The background DIGS from all sampling areas and times were characterized by fine-grained tails with slightly positive source slopes (Figs 5, 6). Most had rounded distributions with poorly developed modes. A coarsening in the DIGS was seen in the near bottom samples, and during the Western Bank sampling, a diatom peak was present at 10 μm in the surface samples (Fig. 5). During any one sampling period, the source slopes of the samples not containing bentonite were constant (Table 1). All samples from Western Bank had positive source slopes with a mean of 0.57 and standard deviation of 0.11. Regressions on Coulter channels 1–7 for these samples all fell within the 95% confidence limit indicating that throughout this area, the size distribution of the fine grained inorganic fraction was constant. No samples collected on Western Bank had negative source slopes indicating that drill wastes were not present in this area.

In 1991 during drilling at Panuke, several surface and BBL samples had negative source slopes, indicating the presence of bentonite (Fig. 6) (Muschenheim and Milligan, 1996). Similar results were obtained in 1993 where the maximum concentration of bentonite in the water column was ~ 0.4 mg/l. The highest values were found at 1 and 40 m depth 250 m E and W of the RGIII and at 40 m depth, 15 km E of the platform. Values in the intermediate water column were lower with a maximum value of 0.03 mg/l. In the BBL, the maximum bentonite concentration was 2.5 mg/l. Maximum values in 1994 did not exceed 0.01 mg/l. In 1996, values in both the water column and BBL did not exceed 0.01 mg/l.

Bentonite concentrations at the anchor station, 300 m N of the RGIII, are shown in a depth-time plot in Figure. 8. The currents during the anchor station measured 3 km to the NE are shown in Figure. 9. The maximum bentonite concentration at

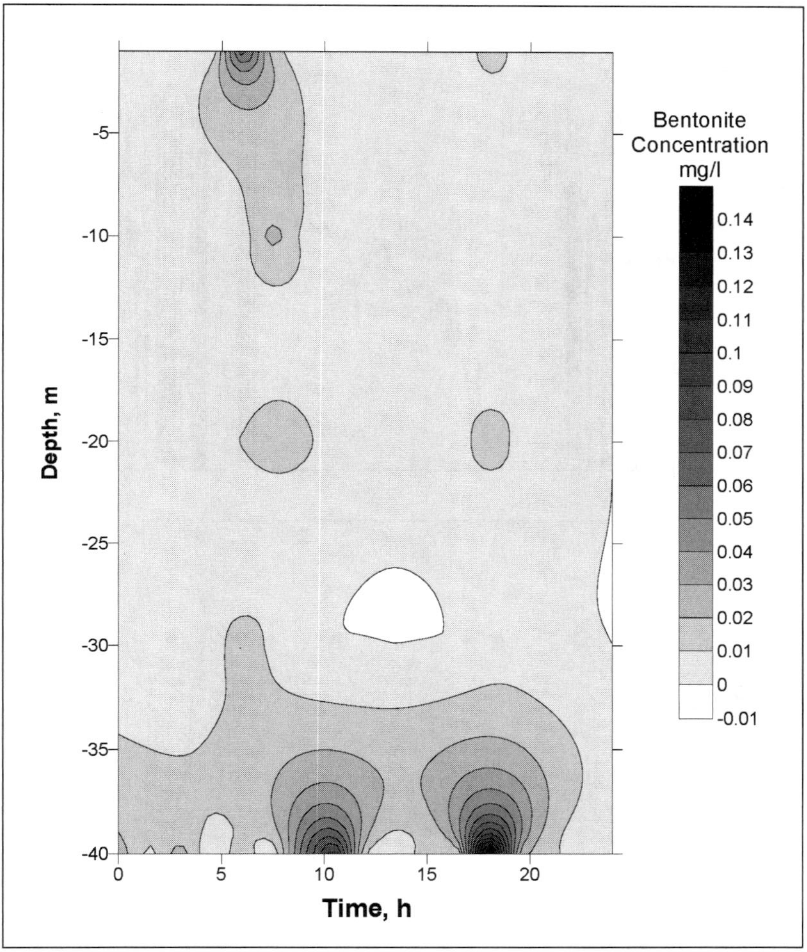

FIGURE 8. Contour plot of bentonite concentration calculated from DIGS of samples collected in the water column and benthic boundary layer over 24 hours. Station was located 300 m N of the RGIII and occupied in September 1993.

the surface (hour 7.5) reaches 0.07 mg/l during a period of sustained currents from the RGIII towards the anchor station (360 degrees/due north). The other local maxima at the surface and mid-depth, with concentrations of 0.02–0.03 mg/l, also occur during times with currents from the south and greater than 0.1 m/s. The largest concentrations in the BBL occurred near hours 10 and 17 and had values of 0.10–0.25 mg/l. The maxima

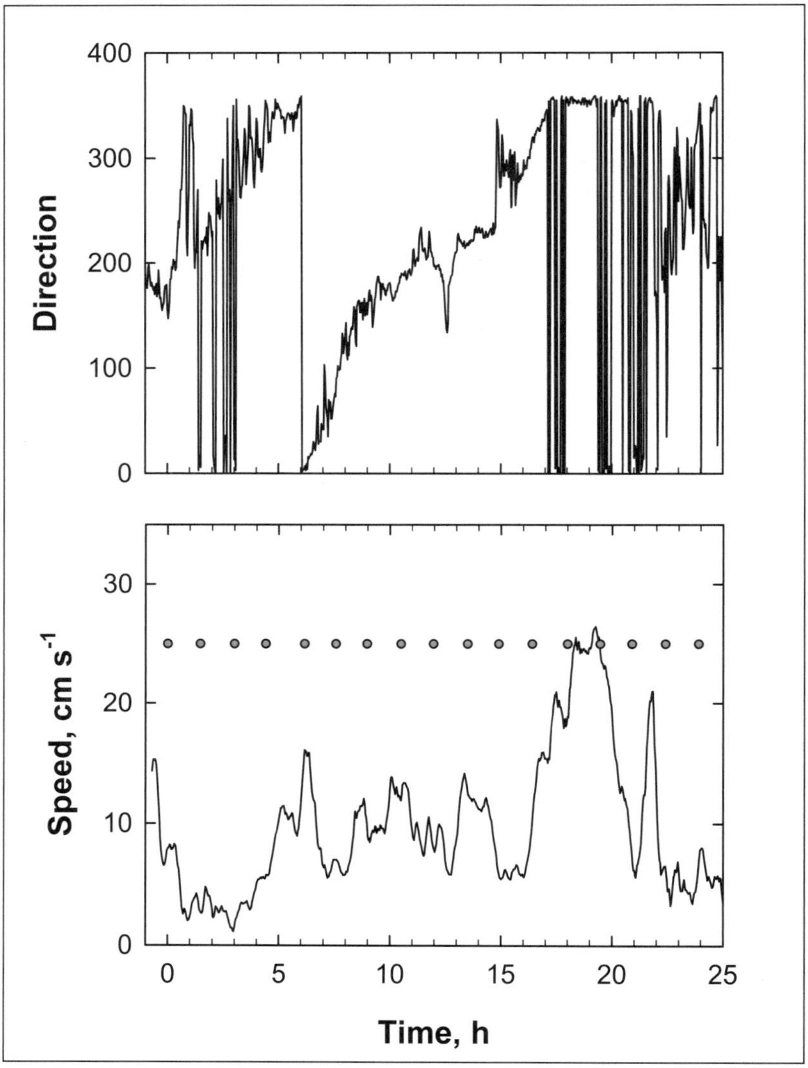

FIGURE 9. S4 current meter record for CoPan anchor station in September 1993 showing current speed and direction (in degrees). Meter was located 1 m above the seabed, 3.0 km NE of the RGIII. Circles indicate sampling times.

were associated with periods of current greater than 0.15 m/s. However, not all events with currents greater than 0.15 m/s lead to elevated concentrations.

DISCUSSION

The similarity in the water column DIGS on Western and Sable Banks, and in other areas of the Scotian Shelf not reported here, demonstrates that the size distribution of the background mineral fraction of the suspended sediment is constant. Within any one sampling period, the background DIGS is sufficiently similar that an average distribution can be determined for the straight-line portion of the curve between 0.87 and 2.4 μm. While concentration and the occurrence of coarse sediment introduced by resuspension or diatom frustules may vary spatially and temporally, the slope of the region of the DIGS curve between 0.87 and 2.4 μm that is defined by the source sediment does not change significantly except in the vicinity of water based drilling waste discharges. Using the difference between the source slope of the background suspended sediment and that of samples enriched with bentonite, it is possible to examine the transport and deposition of fine-grained drill waste at CoPan and to extrapolate those results to other areas of Sable Bank where drilling is taking place.

Consider the variation in bentonite concentration at the anchor station, located 300 m N of the RGIII. In 1993, the discharge point for drill mud was located 10 m below the surface. We interpret the occurrence of a high surface bentonite value at 6 h as a signature of the discharge plume being advected past the anchor station. Increased concentrations at 10 m and 20 m suggest that material is settling from the plume (Figs. 8, 9). The highest concentrations in the BBL are not associated with high surface concentrations supporting the hypothesis that the source of the surface bentonite was the discharge plume.

The two major peaks in the BBL concentrations (10 h and 18 h) correlate with current speeds in excess of 10 cm/s. However, three other periods when current speed exceeded 10 cm/s are associated with minimums in concentration. Three possibilities exist to explain the apparent disconnect between bottom stress and concentration. First, the occurrence of the high concentration of bentonite is the result of settling from the discharge plume and not resuspension. Assuming a settling velocity of 0.1 cm/s for flocculated drill mud, material would reach the bottom in 4000 s in 40 m water. This implies a mean advection speed of 7.5 cm/s which is consistent with the observations from the S4 for the period from 6 h to 10 h. The second hypothesis is resuspension from the bottom. The lack of a simple relationship between resuspension and bottom stress could be due to patchiness in the coverage of observed mats of flocculated material (Fig. 4). Coverage on the bottom was not consistent and varied between 25% and 95% within 2 km of the RGIII. The third possibility is that the erosion of the bottom also suspends large quantities of fine-grained sediment held within the sand bed. In some cases this will contaminate the bentonite signal by introducing fine-grained material which is not part of the standard curve that is in suspension. This likely contributes to the large signal at 10.5 h and 16.5 h when the total SPM in the bottom BOSS sample exceeded 2000 mg/l (Muschenheim et al., 1995). The lack of a clear interpretation of the near bottom concentrations means that caution is called for when interpreting results near the bottom.

To map the horizontal distribution of the bentonite in 1993, we calculated the potential amount of bentonite at each station by summing the total mass in the water column based on the water column samples. The calculation proceeds as follows. For the transect stations the total mass of bentonite was determined by dividing the water column into 5 boxes representing the Niskin bottle depths. Box depths were 1–5 m, 5–15 m, 15–25 m, 25–35 m and 35 m to bottom. In cases where no Niskin sample was taken, the mass was distributed equally between adjacent boxes. Within each box a uniform concentration of bentonite determined from the DIGS curve was assumed. It is important to note that there are no data below 3 mab. Thus, the calculation likely underestimates the total bentonite in the water. The results are expressed as mass per unit area, g/m^2 (Fig. 10).

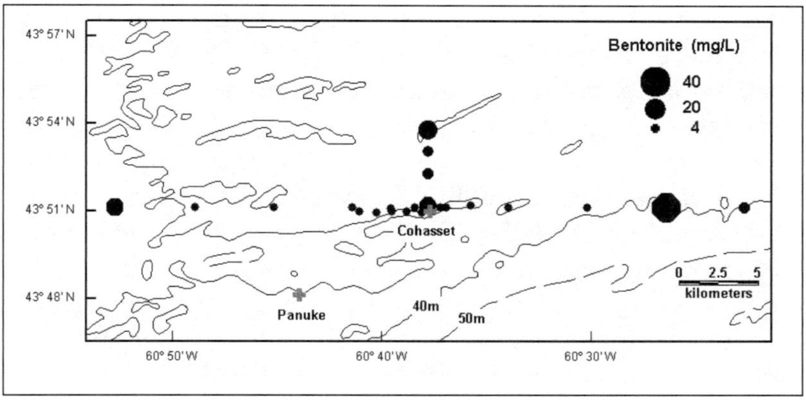

FIGURE 10. Estimated concentrations of bentonite in the bottom 10 cm in September 1993. Values are calculated from the difference between background disaggregated inorganic grains size distributions and bentonite enriched grain size spectra for samples collected at 5 depths in the water column. Estimate assumes total mass in the water column accumulates in the bottom 10 cm.

The greatest amounts of bentonite in 1993 were found very near the RGIII and in areas of increased water depth at a significant distance from the rig (Fig. 10). This pattern does not reflect the pattern of flocculated material observed on the bottom in the video sampling (Fig. 4 and Muschenheim et al., (1995)). Where bottom video images taken along the transect to the west of the rig revealed a dense layer of flocculated material covering up to 95% of the bottom, concentrations determined from bentonite in the water column were relatively low (Fig. 10). This suggests that in the vicinity of the RGIII, floc settling velocities were great enough to overcome resuspension and hence accumulate as a mat on the bottom. In the same manner as algal mats observed on tidal flats, the development of polymeric substances would be expected to increase the cohesivity of the flocs on the bottom near the RGIII, thus increasing the critical erosion shear stress (Tolhurst et al., 1999; Friend et al., 2003). Mats of flocculated material, similar to those observed at CoPan in 1993, have been seen in bottom videos taken at various times and at different locations on the Scotian Shelf during other EEM surveys (S. Belford, pers. com.). The occurrence of mats of flocculated sediment prevents representative sampling of drill wastes in the BBL. Even with specialized sampling equipment such as BOSS, which was designed to sample within the bottom 50 cm, if the bottom stress is low then the mat will conceivably be below the bottom sample port of BOSS located at 5 cm above bottom. EEM sampling on Sable Bank is done with a self-closing Niskin bottle triggered at 0.5 m above the bottom, the lowest point of which is approximately 0.2 m above bottom. Samples collected during periods of low stress could severely underestimate the concentration of drill waste in the BBL. Likewise, if sampling occurs during very high stress periods, bed sediments could mask the bentonite signal.

In April 1994, four months after drilling was completed at CoPan, the bentonite signature in the DIGS spectra had disappeared from the water column samples (Fig. 11) and the estimated bentonite concentrations had fallen to below 0.01 mg/l. During this time, extremely high concentrations of flocs, presumably related to a spring diatom bloom, were in suspension which obscured the view of bottom making it impossible to determine if a dense layer of flocs, similar to that found in 1993, existed near the RGIII (Muschenheim and Milligan, 1995).

While the results presented here do not allow for a rigorous evaluation of the bblt model, they do provide support for conclusions drawn from the modelling results. First, Figure 10 supports the idea that drill mud discharges can be transported significant distances from a rig at detectable concentrations. This is consistent with predictions for Venture, Thebaud, and North Triumph that showed that the zone of influence of transport of flocculated drill mud could extend out to 5–15 km (Maclaren Plansearch 1997; Jacques Whitford 1998). Second, concentrations are consistent with the predictions of 2 mg/l at distances of 5–15 km reported in Maclaren Plansearch (1997), based on bblt simulations that computed near-bottom concentrations averaged over the bottom 50 cm. Computing an observed concentration with all the material in the bottom 50 cm is relevant because in bblt the near bed concentration

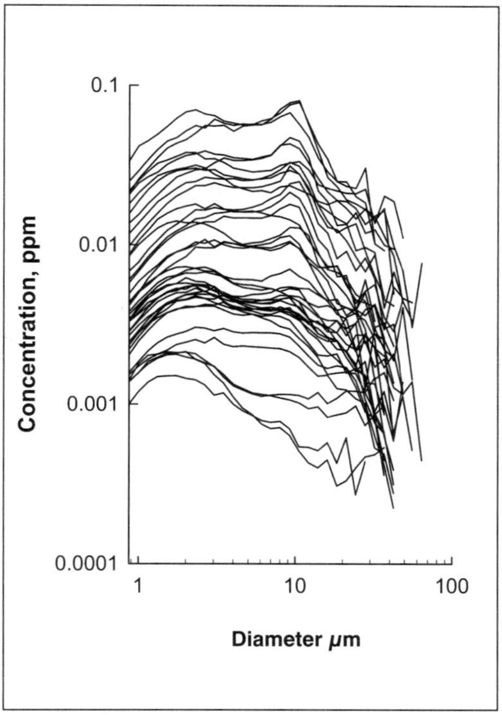

FIGURE 11. DIGS spectra for transect stations carried out near the RGIII in April 1994, four months after completion of drilling. Spectra have not been normalized to illustrate range of concentrations found. Note absence of bentonite enrichment between 0.87 and 2.4 μm. A diatom peak similar to that on Western Bank can be seen at ~10μm.

profile is in equilibrium with the bottom stress. When the currents and thus the bottom stress get small, settling velocities are sufficiently high that the concentration profile collapses and all the material falls to the bottom. In tidally dominated regimes such as Sable Island Bank, the bblt concentration profile collapses every tidal cycle and periodically gives rise to large near-bottom concentrations. Using the same assumptions as bblt, concentrations in the bottom 50 cm in 1993 would be in the range 0.4–8 mg/l. At the order of magnitude level the bblt simulation reported in Maclaren Plansearch (1997) and the 1993 data are in agreement. Both move approximately the right amount of material the right distance.

One drawback to the method presented is its inability to differentiate the barite component. The size distribution of barite is not sufficiently different from that of the background inorganic fraction for it to significantly alter the size distribution of the suspended sediment. Recent coring studies on Sable Bank carried out with a hydraulically damped corer have found that barite is being incorporated in the sand bed to a depth in excess of 30 cm near Thebaud, a gas production site (Muschenheim et al., in press). The persistence of barite in high energy, coarse-grained sediment has also been shown on George's Bank (Neff et al., 1989). Huettel et al. (1996) and Packman et al. (1997) showed that fine particulate material percolates through sandy sediments under tangential flow in a flume. However, Curran et al. (2002), also in a flume study, showed that flocculated drill mud does not become incorporated into a sand bed. DIGS analysis of barite shows that the size distribution has a positive source slope and a well-developed modal peak at about 20 μm giving it a settling velocity on the order of 0.1 cm/s (Fig 2). Several in situ studies of flocs in suspension have found their settling velocity to also be near 0.1 cm/s (Hill et al., 2000). The granular nature of the barite, and its settling velocity very near to that of flocculated drill mud, suggest that the transport pathways for barite might be different than those of bentonite. The appearance of barite at depth in the sandy sediments of Sable Bank suggests that barite is settling with the sand grains and might be subject to transport only during periods when sands are in suspension. Whether or not significant amounts of barite are being sequestered has to be determined.

The apparent difference in transport pathways for bentonite and barite suggest that modelling efforts might require a means to separate water-based drill mud discharges into two components, barite and bentonite. Flocculation has been shown to be a major controlling factor in the transport of surface active contaminants (Milligan and Loring, 1997). Due to the difference in specific surface area for bentonite and barite, this would suggest that other contaminants in the discharge plume would preferentially move with the bentonite fraction. The occurrence of high total mass concentrations of bentonite in areas of lower stress remote from drilling activity could be a pathway for contaminant uptake by benthic organisms. However, it should be pointed out that concentrations overall are low and are subject to frequent resuspension. Four months after drilling was completed at CoPan, there was no evidence of bentonite in suspension (Fig. 11).

CONCLUSION

The disaggregated inorganic grain size of suspended sediment can be used to determine the concentration of bentonite discharged during drilling operations on the Scotian Shelf. The size distributions of samples containing bentonite are sufficiently different from naturally occurring inorganic sediment that the amount of bentonite in suspension can be estimated from the difference in their grain size spectra. This paper reports the first observational estimates of bentonite concentrations associated with drill mud discharges in Canadian waters. The concentrations in both the water column and BBL are small relative to the no effect concentration of 2 mg/l reported for sea scallops by Cranford and Gordon (1992) and Cranford et al. (1999). However, integrated values, assuming all the material in the water column accumulates in the bottom 0.5 m, could exceed effects levels intermittently. Four months after the completion of drilling at Cohasset, bentonite was no longer present in sufficient quantities to be detectable with the proposed method.

Using suspended sediment DIGS, patterns of drill waste dispersion have been determined that can be compared to the results of the bblt model. Simulations of a realistic discharge for North Triumph, also located on Sable Bank, returned values for barite dispersion of the same order as that found for bentonite at the CoPan site. Considering the lack of data for the discharge at CoPan and the difference in physical conditions at the two locations, the predicted and measured concentrations are surprisingly similar. As predicted by the bblt model, measurable quantities of bentonite (drill wastes) can be detected out to 15 km from the source. Both observed and modelled distributions are patchy in space and time.

The formation of mats of flocculated material on the bottom presents a problem for sampling. The work reported here provides no insight into the origin or fate of the mats of flocculated material on the bottom first reported in Muschenheim and Milligan (1996) and observed sporadically since then. Muschenheim and Milligan (1996) reported that the mats contained bentonite but were unable to provide quantitative estimates. The mats were so thin that they were not reliably sampled by the BOSS port 5 cm above the bottom. The difficulty in interpreting the near-bottom concentrations reduces our confidence in the bentonite estimates at times when there is active sand suspension. Future analysis of the grain size distribution data may shed some light on this problem, as might improved sampling methods that would sample the floc layer intact. Routine video observation of the bottom during active discharge is likely required to determine the factors that contribute to its formation and dissipation.

While the distinct size distribution of bentonite allows it to be used as a tracer for drill mud, the DIGS of barite is not discernable from that of the background inorganic fraction. On the other hand, the concentration of barite can be determined chemically but no such test is readily available to differentiate bentonite from other naturally occurring clays in suspension. The granular nature and high specific gravity of barite suggests that its transport pathway might not be with flocs but rather with the coarse silt and sand fractions making it necessary to develop a two component model for drill waste transport. Chemical analysis can be used to determine the concentration of barium but a specific method such as that presented here is required to determine the transport and deposition of bentonite. This is especially important as surface-active contaminants are likely to be transported in flocs along with the bentonite fraction of drill wastes. Understanding the fate of the flocculated material resulting from both solid waste discharges and from produced water, remains an essential factor for determining the significance of interactions between offshore oil and gas development and the environment.

ACKNOWLEDGEMENTS

The effort and enthusiasm of the officers and crews of the MV *Ryan Leet*, CCGS *Hudson* and CSS *Parizeau* are gratefully acknowledged. K. Saunders and A. Prior performed the size analysis on the samples and kept all the data straight over several years and cruises. Ole Mikkelsen and Brent Law helped with earlier versions of the manuscript.

REFERENCES

Bagnold, R.A. and O. Barndorff-Nielsen. 1980. "The Pattern of Natural Size Distributions." *Sedimentology.* 27:199-207.

Barndorff-Nielsen, O. 1977. "Exponentially Decreasing Distributions for the Logarithm of Particle Size." *Procedures of the Royal Society of London. A.* 352:401-419.

Bennett, J.G. 1936. "Broken Coal." *Journal of Industrial Fuel.* 19:22-29.

Cranford, P.J. and D.C. Gordon. 1992. "The Influence of Dilute Clay Suspensions on Sea Scallop (*Placopecten magellanicus*) Feeding Activity and Tissue Growth." *Netherlands Journal of Sea Research.* 30:107-120.

Cranford, P.J., D.C. Gordon, Jr., K. Lee, S.L. Armsworthy, and G.-H. Tremblay. 1999. "Chronic Toxicity and Physical Disturbance Effects of Water- and Oil-based Drilling Fluids and Some Major Constituents on Adult Sea Scallops (*Placopecten magellanicus*)." *Marine Environmental Research.* 48:225-256.

Cranford, P.J., D.C. Gordon, C.G. Hannah, J.W. Loder, T.G. Milligan, D.K. Muschenheim, and Y. Shen. 2003. "Modelling Potential Effects of Petroleum Exploration Drilling on Northeastern Georges Bank Scallop Stocks." *Ecological Modelling.* 1/2:19-39.

Curran, K.J., P.S. Hill, and T.G. Milligan. 2002. "The Role of Particle Aggregation in Size-dependent Deposition of Drill Mud." *Continental Shelf Research.* 22(3):403-414.

Friend, P.L., P. Ciavola, S. Cappucci, and R. Santos. 2003. "Biodependent Bed Parameters as a Proxy Tool for Sediment Stability in Mixed Habitat Intertidal Areas." *Continental Shelf Research.* 23(17-19):1899-1917.

Gordon Jr., D.C., P.J. Cranford, C.G. Hannah, J.W. Loder, T.G. Milligan, D.K. Muschenheim, and Y. Shen. 2000. "Modelling the Transport and Effects on Scallops of Water-based Drilling Mud from Potential Hydrocarbon Exploration on Georges Bank." *Canadian Technical Report of Fisheries and Aquatic Science.* 2317, 115 pp.

Hannah, C.G., A. Drozdowski, D.K. Muschenheim, J.W. Loder, S. Belford, and M. MacNeil. 2003. *Evaluation of Drilling Mud Dispersion Models at SOEI Tier I Sites: Part 1 North Triumph, Fall 1999.* Canadian Technical Report of Hydrography and Ocean Sciences 232, 56 pp. (Available from http://www.dfo-mpo.gc.ca/Library/275999.pdf).

Hill, P.S., T.G. Milligan, and W.R. Geyer. 2000. "Controls on Effective Settling Velocity of Suspended Sediment in the Eel River Flood Plume." *Continental Shelf Research.* 20:2095-2111.

Huettel, M., W. Ziebis, and S. Forsteer. 1996. "Flow-induced Uptake of Particulate Matter in Permeable Sediments." *Limnology and Oceanography.* 41(2):309-322.

Jacques Whitford Environment Limited. 1998. *Offshore Environmental Effects Monitoring Program Baseline and Drilling Phases for Venture, Thebaud and North Triumph.* Report to Canada Nova Scotia Offshore Petroleum Board on Behalf of Sable Offshore Energy Inc. July 29, 1998.

Kranck, K. 1980. "Experiments on the Significance of Flocculation in the Settling Behaviour of Fine Grained Sediment in Still Water. *Canadian Journal of Earth Sciences.* 17:1517-1526.

Kranck, K. and T. Milligan. 1983. "Grain Size Distribution of Inorganic Suspended River Sediment. Transport of Carbon and Minerals." In E.T. Degens (Ed.), *Major World Rivers, Part II.* Mitteilungen aus dem Geologisch-Paleontologisch Institut der Universitat Hamburg, Heft 52.

Kranck, K. and T.G. Milligan. 1985. The Origin of Grain Size Spectra of Suspension Deposited Sediment. *Geomarine Letters.* 5:61-66.

Kranck, K. and T.G. Milligan. 1991. "Grain Size in Oceanography." In J.P. Syvitski, (Ed.), *Theory, Methods, and Applications of Particle Size Analysis,* pp. 332-345. Cambridge University Press, New York.

Kranck, K. and T.G. Milligan. 1992. "Characteristics of Suspended Particles at an 11 Hour Anchor Station in San Francisco Bay, California." *Journal of Geophysical Research.* 97(C7):11373-11382.

Kranck, K., P.C. Smith, and T.G. Milligan. 1996. "Grain Size Characteristics of Fine-Grained Unflocculated Sediments I: 'One-round' Distributions." *Sedimentology.* 43:589-596.

MacLaren Plansearch. 1997. *Physical Fate of Drilling and Production Effluent Discharges and Impact on Marine Environment, Part 1: Drilling Waste Discharges.* Phase B - Impact Assessment Final Report.

Milligan, T.G. and K. Kranck, 1991. Electro-resistance particle size analysers." In J.P. Syvitski, (Ed.), *Theory, Methods, and Applications of Particle Size Analysis,* pp. 109-118. Cambridge University Press, New York.

Milligan, T.G. and D.H. Loring. 1997. "The Effect of Flocculation on the Size Distributions of Bottom Sediment in Coastal Inlets: Implications for Contaminant Transport." *Water, Air, and Soil Pollution.* 99:33-42.

Muschenheim, D.K., T.G. Milligan, and D.C. Gordon Jr. 1995. *New Technology and Suggested Methodologies for Monitoring Particulate Wastes Discharged from Offshore Oil and Gas Drilling Platforms and their Effects on the Benthic Boundary Layer Environment.* Canadian Technical Report of Fisheries and Aquatic Science. 2049, x+55p.

Muschenheim, D.K. and T.G. Milligan. 1996. "Flocculation and Accumulation of Fine Drilling Waste Particles on the Scotian Shelf." *Marine Pollution Bulletin.* 32(10):740-745.

Muschenheim, D.K., T.G. Milligan, B. Law, G. Morton, and P.A. Yeats, In press. *Application of Slow-Coring to Drilling Waste Deposition Studies on Sable Island Bank* Canadian Technical Report of Fisheries and Aquatic Science. 2478: vi + 35.

Neff, J.M., M.H. Bothner, N.J. Maciolek, and J.F. Grassle, 1989. "Impacts of Exploratory Drilling For Oil and Gas on the Benthic Environment of Georges Bank." *Marine Environmental Research.* 27:77-114.

Packman, A.I., N.H. Brooks, and J.J. Morgan. 1997. "Experimental Techniques for Laboratory Investigation of Clay Colloid Transport and Filtration in a Stream with a Sand Bed." *Water, Air, and Soil Pollution.* 99:113-122.

Sheldon, R. 1972. "Size Separation of Marine Seston by Membrane and Glass-fiber Filters." *Limnology and Oceanography.* 17:494-498.

Stevens, R.L., H. Bengtsson, and A. Lepland, 1996. "Textural Provinces and Transport Interpretations with Fine-Grained Sediment in the Skagerrak." *Journal of Sea Research.* 35(1-3):99-110.

Thomson, D.H., R.A. Davis, R. Belore, E. Gonzalez, J. Christian, V.D. Moulton, and R.E. Harris. 2000. *Environmental Assessment of Exploration Drilling Off Nova Scotia. Report prepared for Canada - Nova Scotia Offshore Petroleum Board and Mobil Oil Canada Properties.* Shell Canada Ltd., Imperial Oil Resources Ltd., Gulf Canada Resources Ltd., Chevron Canada Resources, EnCana Petroleum Ltd., Murphy Oil

Company Ltd., and Norsk Hydro Canada Oil & Gas Inc. Report prepared by LGL Ltd. Environmental Research Associates, LGL Report, TA 2281: x+280 p.

Tolhurst, T. J., K. S. Black, S. A. Shayler, S. Mather, I. Black, K. Baker, and D.M. Paterson. 1999. "Measuring the In Situ Erosion Shear Stress of Intertidal Sediments with the Cohesive Strength Meter (CSM)." *Estuarine, Coastal and Shelf Science.* 49(2):281-294.

Transport Properties of Discharged Synthetic-based Drilling Cuttings

Niu, H. (haibo@engr.mun.ca), T. Husain, B. Veitch, N. Bose, and K. Hawboldt (Faculty of Engineering and Applied Science, Memorial University of Newfoundland, St. John's, NL, Canada,)

ABSTRACT: There is concern over the potential environmental effects of offshore discharges of drilling cuttings and synthetic-based drilling fluids (SBFs) from hydrocarbon exploration and production operations. As a result, research is required to understand the transport mechanisms and fate of cuttings in water and sediment once they are discharged. The transport and fate of SBF cuttings in the marine environment is influenced by the settling velocity and flocculation processes. An experimental approach was taken to improve our understanding of these processes. Laboratory experiments were conducted with cutting samples from an oil well on the east coast of Canada. The effects of particle shape and size on the settling mechanism and the effects of salinity, fluid shear, discharge concentration, and oily components on the rate of flocculation and the settling speeds of flocs were studied. A thermally treated sample was also used to study the effects of oily components on flocculation. The settling velocities of coarse cutting particles ($D > 100\ \mu$m) were found to be functions of particle diameter, density, sphericity, fluid properties, and composition. The untreated coarse cuttings particles ($D >$ 100 μm) tended to clump together and settled quickly, while the treated coarse cuttings particles ($D >$ 100 μm) settled as individual particles with relatively low speeds. The flocculation tests demonstrated that steady state median floc size decreased as shear stress and concentration increased, and that cuttings particles flocculate faster in seawater than in freshwater. For the same diameter and salinity, the flocs formed at high fluid shears had higher settling velocites than flocs formed at low shears and concentrations. It was also shown that flocs formed by untreated cuttings settled faster than flocs formed by thermally treated cuttings.

INTRODUCTION

The exploration and extraction of offshore oil and gas from beneath the ocean floor requires the disposal of large quantities of drilling wastes. Drilling waste is rock cuttings and adhering drilling fluids. The drilling fluids, or muds, are a suspension of solids and dissolved materials in a base of water, oil, or other synthetic material. According to the base materials used, different varieties of drilling fluids can be categorized as water-based fluids (WBFs), oil-based fluids (OBFs), or synthetic-based fluids (SBFs).

During drilling operations, the drilling fluid is first pumped from the mud tanks down the hollow drillstring to lubricate the drill bit. After passing through the nozzles of the drill bit, the flowing drilling fluids sweep the crushed rock cuttings from beneath the bit and carry them up the annular space between the drillstring and the hole to the surface. Once the mixture of drilling fluid and cuttings returns to the surface, it is sent to a solids control system to remove the drilling cuttings. After passing through the solids control system, the drilling fluids are recirculated and the drilling cuttings are treated before being discharged into the sea. Once discharged, the dispersion of drilling wastes is controlled by various transport mechanisms, which include flocculation, settling, re-entrainment, and re-suspension. These transport processes are illustrated in Figure 1.

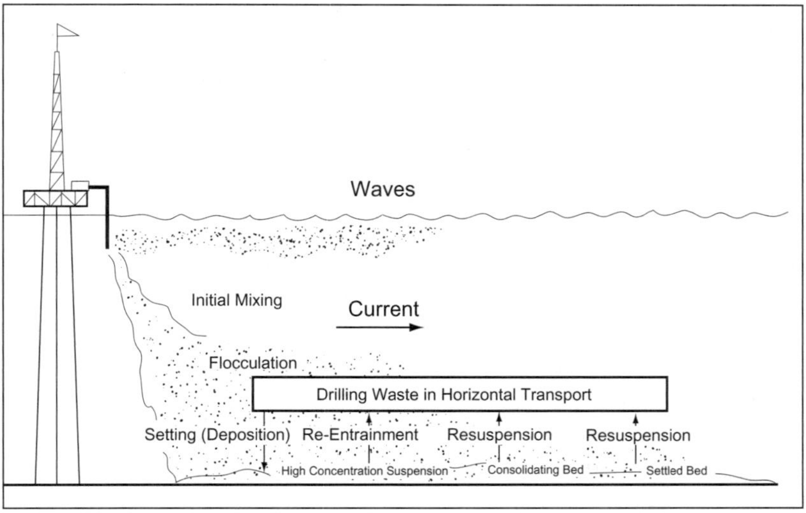

FIGURE 1. Transport processes of discharged drilling wastes.

Most of the waste materials sink quickly to the bottom near the well site. These materials usually accumulate in an area about 40 to 50 m in diameter immediately down current of the well site. The diameter is strongly dependent on the water depth and current speeds at the site (NRC, 1983). In many areas, because of the strong currents, dispersion of the settled materials is rapid, and there is no visible accumulation of cuttings on the sea floor. During the settling process of drilling cuttings, soluble and particulate fluid additives adhering to the cuttings are to some extent washed off into the water column. Due to the effect of turbulence, the fine-grained particles ($D < 100\ \mu$m) within the discharge plume collide with each other and form flocs with relatively larger diameters and faster settling speeds (Kranck et al., 1992; Milligan, 1995). The re-suspended materials are transported away by ocean currents and re-deposited depending on environmental factors (such as water depth, bottom configuration, and energy regime).

The discharge of these waste materials can result in either physical impacts, such as burial or smothering of benthic fauna and flora, or toxic impacts such as reduced growth rate and increased mortality (NRC, 1983). A number of drilling waste transport models have been developed to give operators and regulatory agencies the capability of predicting the fate of drilling discharges under a variety of ocean conditions. Khondaker (2000) has conducted a systematic review of some available models. It was shown by the review that the ultimate accuracy of transport models relies on our knowledge of complex transport processes (such as settling and flocculation) of drilling wastes. Carles and Bryden (1999) have shown that the transport models are very sensitive to settling velocity equations, which in turn depend on cutting characteristics (type, size, shape, etc). The transport models are also very sensitive to the flocculation process as it changes the size distribution, density, structure, and therefore, the settling velocities of drilling wastes.

Due to the fact that flocculation and settling processes are still partially understood, and that they depend on cutting characteristics, it is necessary to study these processes using different types of cuttings to improve the accuracy of prediction. For WBFs, Huang (1992) studied the flocculation due to both fluid shear and differential settling using laboratory flocculators. The settling velocity of flocs was also measured in a settling tube by a double-exposure photographic method. It was shown by Huang's (1992) experiments that the settling velocity of flocs is a function of its size and the conditions at which the floc is formed. Curran et al. (2002) studied the flocculation of WBFs using a recirculating flume channel. For cuttings produced using OBFs, Gerard (1996) tested the flocculation of this type of cuttings in a grid column. The settling velocity of cutting particles was also measured using a sediment balance approach. SBFs are a relatively new generation of drilling fluid that were developed to have the same performance as oil based drilling fluid, but with a lower environmental impact. The composition and structure of cuttings produced by SBFs are quite different from those produced by other types of drilling fluids and therefore their transport properties are different. This study focuses on the understanding of flocculation and settling processes of SBFs produced cuttings to provide input to transport models.

MATERIALS AND METHODS

The drilling cuttings used in this study were obtained from an exploration well on the east coast of Canada. Cuttings from two different depths (formation A and B) were used. The bulk density of Formation A and B are 1900 kg/m^3 and 1833 kg/m^3 respectively. The particle size distributions for the two formations are also different. To study the effects of oily components on flocculation and settling, a portion of the cuttings were dried in an oven at 110°C for 24 hours to remove the oily component and were then prepared (using sieving methods) for the experiments.

The settling velocity experiments of drilling cutting particles with diameters larger than 100 μm was conducted in a 2.5 m high and 0.14 m inner diameter plexiglas settling column. At the start of the settling experiment, the column was filled with either freshwater or artificial seawater depending on the test conditions. A small amount of pre-separated cutting samples was dumped into the column from the top. A digital imaging system was used to record the motion of settling particles. The system, which is shown in Figure 2, consists of a high speed CCD camera with lenses, computer, and illuminating devices. The settling velocity and the sphericity value were obtained by examining the obtained images using image analysing software.

In presenting the experimental data, the projected area equivalent diameter was used to classify particles and to correlate particle size with settling velocity. The particle sphericity was calculated using Wadell's (1933) formula

$$\psi = \frac{d_{cn}}{D_C} \tag{1}$$

where ψ is the approximated degree of sphericity, d_{cn} is the diameter of a circle equal in size to the cross-section area of the particle and D_C is the diameter of the smallest circle that can circumscribe the particle.

The schematic diagram of the flocculation experiment system is also shown in Figure 2. A Phipps and Bird six-paddle stirrer was employed in this study to produce fluid shear. The stirrer

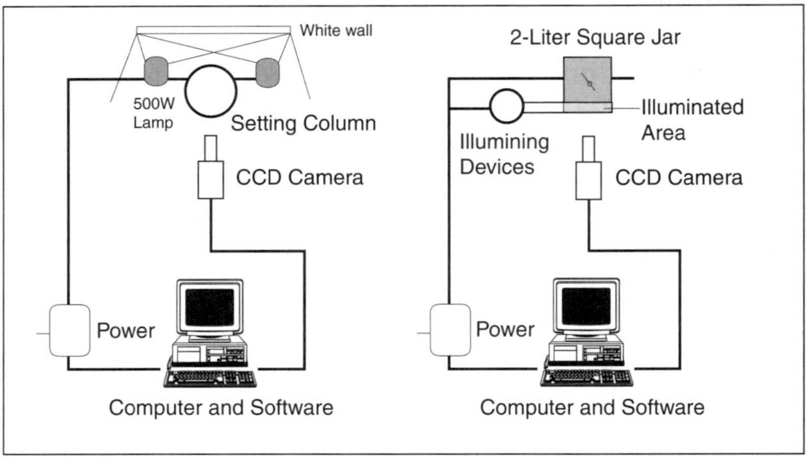

FIGURE 2. System configuration for settling column test (left) and flocculation test (right).

has a rotating speed up to 300 rpm that can generate a wide range of fluid shear. At the start of the experiment, a 2 l sample ($D < 100\ \mu$m) in a square test beaker at certain concentrations (25, 50, 100, and 200 mg/l) was loaded with the stirrer at high speed for 90 seconds to disaggregate the samples and to evenly distribute the sample. The initial particle size distribution was recorded using the high speed CCD camera at this time. The stirrer was then run at a constant speed to produce desired fluid shear values (25, 50, 100, 200 G) and a timer was started. As the experiment proceeded, the images were taken after a certain flocculation time (5, 10, 20, 40, 60, 80, 120 min). The particle size distribution was derived from analysis of the images.

At the end of flocculation tests, the stirrer was stopped. The suspension was given 30 seconds to allow the fast rotation motion to subside. The images were taken at that time to record the settling motion of flocs. The settling velocity of flocs was obtained from the image analysis using the ImageJ software with a user-written plug-in. The images were first calibrated by comparing the pixel size with pre-recorded known distance. The particles (in focus) were then separated from those that were out of focus by a user defined threshold value. The detailed information of each particle and the distribution was then calculated by the software.

RESULTS AND DISCUSSION

Settling velocity results for both untreated and treated cutting particles with diameters larger than 100 μm were obtained under different test conditions. For untreated drilling cutting particles, it was found that they settle as clumps rather than individual particles and the shape effect on settling velocity is not significant. Figure 3 presents the seawater settling test results for formation A with a bulk density of 1900 kg/m^3. The results from the tests of formation A in freshwater and formation B in both seawater and freshwater are similar to formation A in seawater, but with different settling rates due to the variation of water density and cutting bulk density. Unlike untreated cuttings, treated cuttings settle as individual particles. The settling velocity of treated cutting particles is affected by sphericity as well as diameter and density. Through the regression analysis of experimental data, the relationship between particle drag coefficient (C_D) and Reynolds number (*Re*) was established. By introducing the definition of C_D and *Re* into the relationship, settling velocity equations for both treated and untreated drilling cutting particles were obtained and are presented in Equations 2 and 3.

$$u = \frac{-15.36\mu + \sqrt{236.08\mu^2 + 0.78(\rho_s - \rho_f)\rho_f D^3 g}}{\rho_f D} \quad \text{(Untreated)} \quad (2)$$

$$u = \frac{-3b\mu + \sqrt{9b^2\mu^2 + 48a(\rho_s - \rho_f)\rho_f g D^3}}{6a\rho_f D} \quad \text{(Treated)} \quad (3)$$

where u is the settling velocity in m/s, ρ_s and ρ_f are the drilling cutting density and fluid density respectively, in kg/m^3, μ is the dynamic viscosity in kg/ms, D is the particle diameter in m,

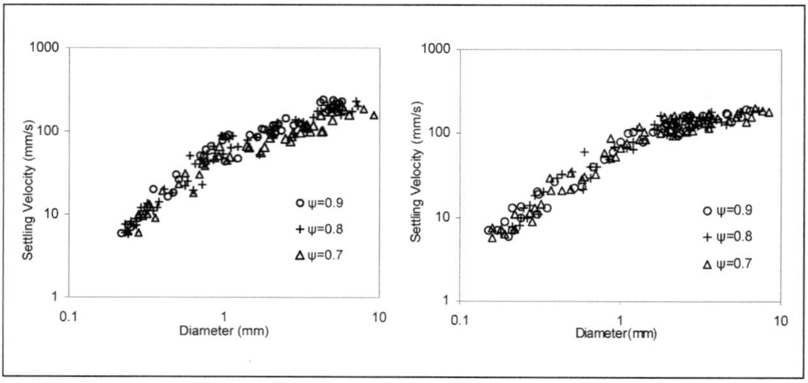

FIGURE 3. Seawater settling velocity data of formation A. Untreated cuttings (left), treated cuttings (right).

and g is the gravitational acceleration in m/s^2. Dimensionless parameters a and b in Equation 3 are defined as:

$$a = -3.44 + 13.85\psi - 10.31\psi^2 \tag{4}$$

$$b = 225.72 - 406.60\psi + 223.37\psi^2 \tag{5}$$

Figure 4 presents the results from Equations 2 and 3 by substituting the cutting densities of 1850 and 2650 kg/m^3, seawater density of 1024.5 kg/m^3, and sphericity of 0.8 (in order to compare the settling velocity of cuttings that are affected by sphericity and cuttings that are not affected by sphericity, a most frequently occurring sphericity value of 0.8 was used). The settling velocity results of OBFs produced cuttings with density of 1850 kg/m^3 from Gerard (1996), quartz sands with density of 2650 kg/m^3 from Sleath (1984), and drilling cuttings (unknown type) with density of 1850 kg/m^3 from Chien (1992) are plotted for reference. It is shown in Figure 4 that treated SBFs cuttings settle faster than untreated cuttings and OBFs cuttings with the same density. The treated cuttings have a settling velocity close to irregular shaped quartz sands. The reason for this is that the structure of treated cuttings is

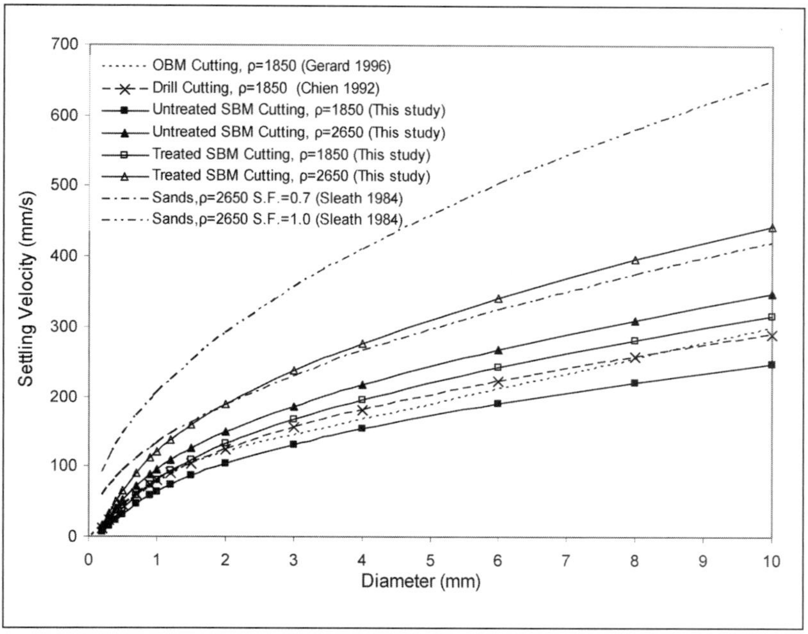

FIGURE 4. Setting velocity for SBFs produced drilling cuttings.

similar to quartz sands and they exist as relatively individual particles. The untreated SBFs and OBFs produced cuttings have a porous structure that makes them have a smaller sphericity value and higher drag forces than the treated cuttings with same density and diameter.

The time variation of median floc diameter was obtained under the combination of fluid shear and solid concentration in both seawater and freshwater. It was shown by the experiments that the median floc size continued to increase as the flocculation proceeded, until it reached steady state (Fig. 5). After this point, the floc median size remained constant. Table 1 shows the time needed to reach steady state under fluid shear of 25 G. Table 2 gives the steady state median floc size under fluid shear of 25 G. Tables 1 and 2 show that an increase in concentration resulted in a decrease in the time to reach steady state, and a decrease in steady state median diameter. It

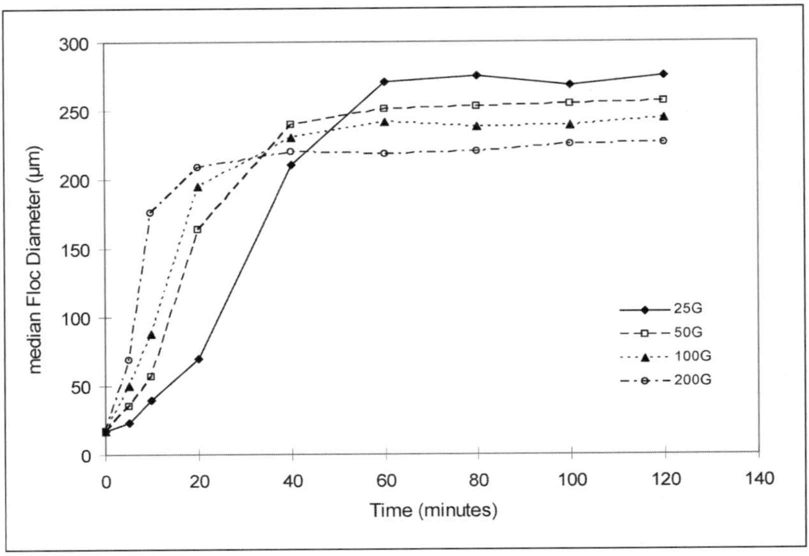

FIGURE 5. Time variation of median floc diameter for untreated cuttings at 25mg/l in seawater.

TABLE 1. Time to reach steady state under fluid shear of 25 G.

Concentration (mg/l)	Time (minutes)		
	Treated/ Seawater	Untreated/ Seawater	Untreated/ Freshwater
25	80	60	100
50	65	50	90
100	50	40	80
200	40	30	60

TABLE 2. Steady state floc median diameter under fluid shear of 25 G.

Concentration (mg/l)	Diameter (μm)		
	Treated/ Seawater	Untreated/ Seawater	Untreated/ Freshwater
25	129.3	273.1	101.5
50	96.3	254.3	94.7
100	61.8	238.6	89.4
200	52.5	223.1	79.9

was also shown by the experiments that the time to reach steady state and steady state median diameter decreased as the fluid shear increased (Fig. 5).

It can be seen from Tables 1 and 2 that untreated cuttings flocculate faster and have larger steady state median diameter than treated cuttings. The reason is that the oily components in untreated cuttings act as adhering agents that help to stick the particles together and enhance the flocculation efficiency. Tables 1 and 2 also show that the cuttings flocculate faster in seawater than in freshwater. The steady state median floc diameters are also larger in seawater than in freshwater. The reason is that most particles in water are charged. The surface charge on a particle and the associated counter-ion charge together form an electrical double-layer. Part of the counter-ion is located close to the particle surface and the remainder is distributed more broadly, which forms a diffuser layer. When two particles approach each other, their diffuse layers overlap and a repulsion force is experienced and it will reduce the collision efficiency. If sufficient salt is added (seawater condition), the particles will be destabilised and the flocculation is enhanced as a result of increased collision efficiency (Gregory, 1989). As a consequence, the time needed to reach steady state is shorter in seawater than in freshwater and the steady state median floc diameter is larger in seawater than in freshwater.

Flocs are fragile and their density cannot be measured directly. Therefore, it is difficult to derive a settling velocity equation in the same way as for solid particles. In most cases, the settling velocity for flocs has been summarised as

$$u = AD^{m} \qquad (6)$$

where A and m are parameters that depend on the type of particles being investigated and the conditions under which the flocs were produced (Burban et al., 1990).

From the experiments, the settling velocity of flocs produced from SBFs cuttings in seawater was obtained as a function of both fluid shear and floc diameter. These relationships are described as

$$u = 2.16\times10^{-5}\, G^{1.42}\, D^{\frac{2.22}{G^{0.35}}} \qquad \text{(Untreated)} \quad (7)$$

$$u = 2.22\times10^{-4}\, G^{1.07}\, D^{\frac{1.84}{G^{0.32}}} \qquad \text{(Treated)} \quad (8)$$

where u is the floc settling velocity in mm/s, G is the fluid shear in s^{-1}, and D is the floc diameter in μm. Particle concentration had no effect on the settling velocity of flocs in this work (Fig. 6), although it is an important factor that affects the flocculation.

The results from Equations 7 and 8 are plotted in Figures 7 and 8. The settling velocity of flocs produced from WBFs by Huang (1992) are also plotted for reference. It can be seen from these figures that the settling velocity increased as the fluid shear increased. The settling velocity of treated SBM flocs are close to the settling velocity of WBM flocs and faster than measured for the untreated cuttings. This indicates that the flocs formed by treated cuttings have a higher density.

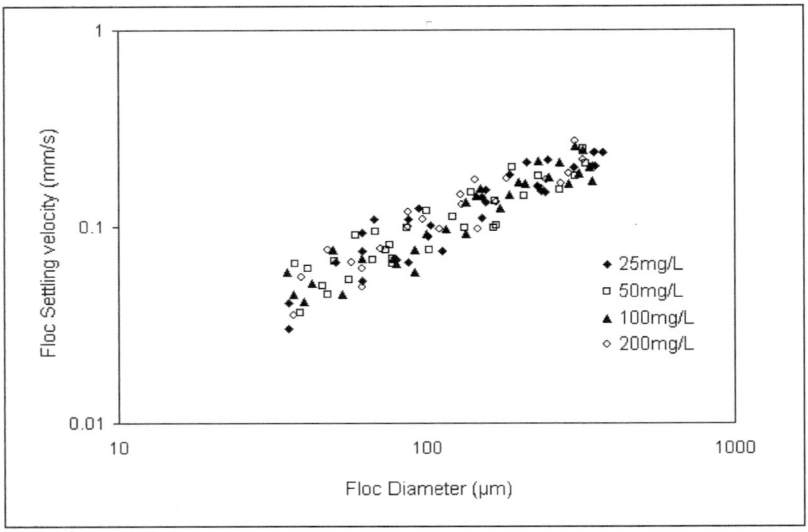

FIGURE 6. Seawater settling velocity of flocs produced from treated cuttings under different concentrations.

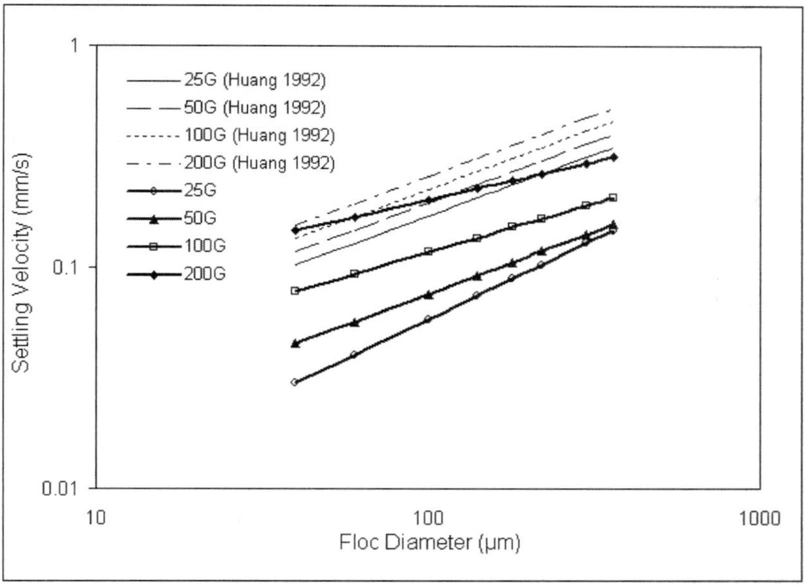

FIGURE 7. Seawater settling velocity of flocs produced from untreated cuttings under different shear rates.

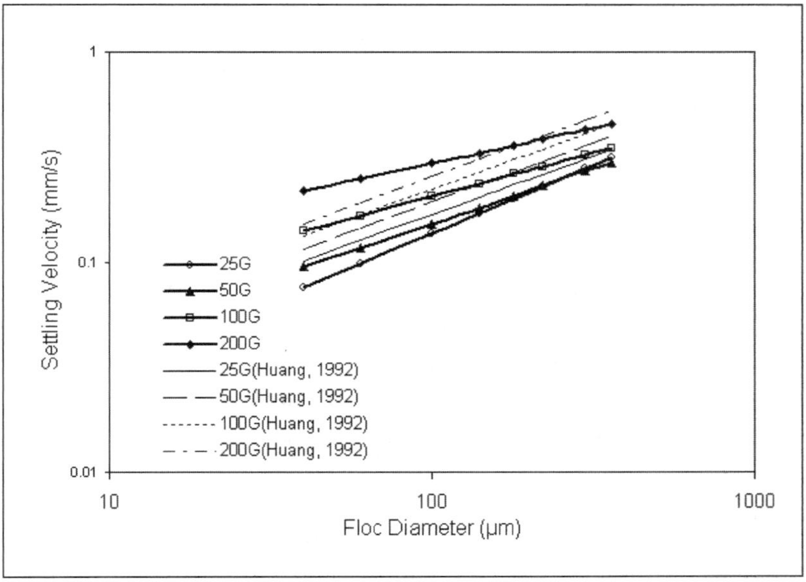

FIGURE 8. Seawater settling velocity of flocs produced from treated cuttings under different shear rates.

The seabed distribution of drilling waste is sensitive to the particle size distribution and settling velocity equations (Bryden and Carles, 1998; Carles and Bryden, 1999). This study shows that the flocculation process changed the original particle size distribution of drilling cuttings. The increased fluid shear allowed the fine-grained particles to collide, adhere together, and form flocs. Because the settling velocity of flocs is larger than the settling velocity of individual particles, particle flocculation is responsible for discharged drilling cuttings reaching the seabed more rapidly than can be achieved without flocculation. The pollutant concentration in sediments at a given distance from the platform will be higher as a result of flocculation than that without flocculation. The thermal treatment changed the particle size distribution because the treatment breaks the larger particles into small particles and the oily components that acted as adhering agents are removed. Therefore, the treated cuttings would be expected to be more dispersive than untreated cuttings (i.e. lower particle concentration at same distance from site of discharge). However, the higher settling velocity (as shown in Figure 4) makes the treated cuttings less dispersive than untreated cuttings (higher sediment concentration at same distance). The ultimate dispersion ability of drilling cuttings has to be determined by considering both the settling velocity and the particle size distribution.

CONCLUSIONS

Experimental investigations of the effects of the shape and oily components on the settling of individual drilling cutting particles with diameters larger than 100 μm were made. The experiments were performed in both seawater and freshwater. The experimental results showed that the settling velocity of untreated drilling cutting particles was a function of particle size, bulk density, and fluid properties. The settling velocity of treated cuttings was a function of particle sphericity, as well as size, bulk density, and fluid properties. It was also demonstrated that the treated cuttings settled faster than untreated cuttings with the same bulk density.

The flocculation experiments were conducted under different combinations of fluid shear, solids concentration, salinity, and quantity of retained oily components. From the experimental results, it was found that the steady state particle size distribution depended on the conditions under which the steady state was approached. The median floc size decreased as the shear stress increased, particles flocculated faster in seawater than in freshwater for the same type of cuttings, and flocs were smaller in freshwater than in seawater for the same type of cuttings. Flocs produced from untreated cuttings flocculated faster than treated cuttings in the same water. Flocs produced from treated cuttings were smaller than untreated cuttings in the same water.

It was shown by the floc settling experiments that the settling velocity of flocs was a function of both floc size and fluid shear. The settling velocity increased as fluid shear increased.

ACKNOWLEDGMENTS

This work was funded by the NSERC Strategic Project entitled "*Offshore Environmental Engineering using Autonomous Underwater Vehicles*".

REFERENCES

Bryden, I. and Carles, L. 1998. "Predicting the Behaviour of Drilled Cuttings Released into Sea Water." *Proceedings of the 8th International Offshore and Polar Engineering Conference*, May 24–29, 1998, Montreal, Canada, 2:301-306.

Burban, P-Y, Y-J, Xu, J. McNeil, and W. Lick. 1990. "Settling Velocity of Flocs in Freshwater and Seawater." *Journal of Geophysical Research*. 95(C10):18213-18220.

Carles, L. and I. Bryden. 1999. "The Sensitivity of a Dispersion Model to Cutting Settling Speeds." *Journal of the Society for Underwater Technology*. 24(1):19-24.

Chien, S-F. "Settling Velocity of Irregularly Shaped Particles." SPE Paper 26121. In *Proceedings of 1994 SPE Annual Technical Conference and Exhibition—Drilling and General Petroleum Engineering*, September 25-28, 1994, New Orleans, Louisiana.

Curran, K.J., P.S. Hill, and T.G. Milligan. 2002. "The Role of Particle Aggregation in Size-dependent Deposition of Drill Mud." *Continental Shelf Research*. 22:405-416.

Gerard, A.L.D. 1996. "Laboratory Investigations on the Fate and Physicochemical Properties of Drilling Cuttings after Discharged into the Sea." In *Physical and Biological Effects of Processed Oily Drill Cuttings*, E&P Forum Report, Paper 3, pp. 16-24.

Gregory, J. 1989. "Fundamentals of Flocculation." *Critical Reviews in Environmental Control*. 19(3):185-230

Huang, H. 1992. "Transport Properties of Drilling Muds and Detroit River Sediments." PhD Dissertation. University of California, Santa Barbara.

Khondaker, A.N. 2000. "Modeling the Fate of Drilling Waste in Marine Environment—An Overview." *Computers and Geosciences*. 26:531-540.

Kranck, K., E. Petticrew, T.G. Milligan, and I.G. Droppo. 1992. "In Situ Particle Size Distribution Resulting from Flocculation." In A. J. Mehta (Eds.), *Nearshore and Estuarine Cohesive Sediment Transport*, pp. 60-74. American Geophysical Union, Washington, DC.

Milligan, T.G. 1995. "An Examination of the Settling Behaviour of a Flocculated Suspension." *Netherland Journal of Sea Research*. 33(2):163-171.

National Research Council (NRC).1983. *Drilling Discharges in the Marine Environment*. National Academy Press, Washington D.C. 1983.

Sleath, J.F.A. 1984. *Sea Bed Mechanics*. John Willey & Sons.

Wadell, H. 1933. "Sphericity and Roundness of Rock Particles." *Journal of Geology*. 41:310-331.

Chronic Effects of Synthetic Drilling Mud on Sea Scallops (*Placopecten magellanicus*)

Armsworthy, S.L. (armsworthys@mar.dfo-mpo.gc.ca), P.J. Cranford, K. Lee, and T. King (Department of Fisheries and Oceans, Dartmouth, NS, Canada)

ABSTRACT: The potential impact of different types and concentrations of used synthetic drilling mud (SBM) formulations and their major components on adult sea scallops (*Placopecten magellanicus*) was assessed in laboratory exposure studies. In addition to monitoring acute responses (% mortality), the test suite included chronic (sublethal) measurements that included changes in somatic and reproductive tissue growth, and feeding rate. Sea scallop mortalities in all treatments (0.07 to 9.55 mg/l) were 13% or lower. Used SBM formulations containing Novaplus® and IPAR-3® base fluids resulted in significant deleterious effects on the final weight of somatic and reproductive tissues (except adductor muscle) at 1.4 and 1.5 mg/l, respectively. No significant effects were detected at the lower concentrations tested. Winter exposure experiments with the Novaplus® mud gave an effects-threshold for all tissues of less than 0.9 mg/l. Scallops exposed to an ester-based mud produced with Petrofree® drilling fluid displayed significant reductions in somatic tissue weight at 0.2 mg/l, but effects on gonad were not detected until the exposure concentration reached 1.4 mg/l. Scallops exposed to 0.3 to 0.7 mg/l suspensions of the major solid components added to the ester-based mud formulation (clay and barite) also displayed significant reductions in tissue dry weight. These results, when combined with observations of significant reductions in scallop clearance rate for all SBM and component particle treatments (> 0.1 mg/l) indicate that the fine particulate fraction of the drilling muds was the primary cause of observed impacts on sea scallop tissue growth.

INTRODUCTION

The rapid growth of offshore oil and gas exploration and production on the eastern Canadian continental shelf has generated the need for both general and region-specific scientific information on the environmental consequences of drilling activities. Offshore activities generate a number of wastes, including discharges from deck drainage, domestic waste, drilling fluids and cuttings, and production water (NEB et al., 2002). Drilling fluids and cuttings are the major waste created during oil and gas exploration and development activities. An important environmental concern is the effect on resident marine organisms of long-term exposure to low concentrations of these discharged materials. In Atlantic Canada, SBMs are currently in use on the Scotian Shelf and Grand Banks, ecosystems highly valued for their commercial fisheries potential.

Synthetic oil-based drilling muds (SBM) have replaced previous oil-based formulations shown to exhibit high chronic toxicity (Cranford et al., 1999), and known to cause significant benthic community impacts at offshore fields (Olsgard and Gray, 1995). The producers of SBMs highlight their lower toxicity values and superior drilling properties relative to water-based mud (WBM) formulations. Prior to implementation of new waste discharge guidelines in 2002, synthetic oil accounted for a large proportion of the total mass of cuttings discharged from the Hibernia development (Cranford et al., 2005). Current waste treatment and discharge guidelines for eastern Canada permit the release of well cuttings containing less than 6.9% synthetic base fluids (NEB et al., 2002). However, this threshold limit is questioned as there is little or no information on the chronic toxicity of SBM formulations to resident species, including benthic organisms of substantial commercial value. As part of a multidisciplinary study designed in part to contribute to the development of scientifically defensible waste disposal regulations, chronic biological effects studies using a variety of SBMs was conducted on the commercially important sea scallop (*Placopecten magellanicus*).

As a result of rapid waste dispersion in the marine environment, suspended drilling waste concentrations greater than 10 mg/l in the water column are only found in close proximity to discharge points (Neff, 1987). For a significant impact to occur at the population level, a long-term biological response would have to be caused by low waste concentrations, a large portion of the population would have to be exposed, and the observed biological effects would have to be relevant to the target species survival, growth, and/or reproduction. The high sensitivity of adult *P. magellanicus* to different types and components of drilling wastes has previously been demonstrated. Scallops are filter-feeders and drilling mud can potentially be detrimental to the health of the organism due to direct toxic effects or by interfering with the energetics of food utilisation, as particulates may become a significant non-nutritional component of their diet. It has been proposed that the transport of used drill mud solids to the benthic environment, their accumulation in the benthic boundary layer (Muschenheim and Milligan, 1996), and subsequent potential resuspension, combined with scallops' limited locomotion, could extend the period and degree of impact to sea scallops (Cranford et al., 2001). Deleterious effects on tissue growth and reproduction from chronic exposure to drilling wastes (including WBM) have been observed within the concentration range of 0.7 to 10.0 mg/l (Cranford and Gordon, 1992; Cranford et al., 1999). Results of model-based assessments of the risk of WBM discharges to extensive Georges Bank scallop stocks indicate that near-bed waste concentrations can exceed no-effects thresholds over a large area under some oceanographic conditions

(Cranford et al, 2003). Thus, routine waste discharges during production drilling activities could result in detectable population level effects on scallops if permitted in higher risk areas.

The objective of this study was to investigate the acute and chronic effects on adult sea scallops of synthetic drilling fluids and their major solid constituents. Waste hazard was assessed from acute and chronic mortalities, effects on somatic and reproductive tissue growth, and changes in feeding physiology. Since pollution stress does not directly affect growth, but rather acts indirectly by influencing physiological processes controlling energy acquisition and utilisation (Bayne and Newell, 1983), feeding activity is considered to be a sensitive sublethal measure of stress. Scallop clearance rate (volume of water cleared of food particles per unit time) has previously been shown to be significantly depressed by suspended WBM concentrations, resulting in reduced growth rate (Cranford et al., 1999). Clearance rate measurements were estimated in this study to make evident the essential role that food supply and food collection activities have on observed growth effects from SBM exposure.

MATERIALS AND METHODS

The experimental protocol used in this study is similar to that described in Cranford and Gordon (1992) and Cranford et al. (1999). Adult sea scallops were collected from Georges and Browns Bank by commercial dragger and transported to the Bedford Institute of Oceanography in Dartmouth, Nova Scotia within 12 h of collection. Scallops having no visible collection injuries were held for one month prior to experiments in tanks receiving flowing unfiltered seawater pumped from a 3 m depth from the Bedford Basin. This period allowed scallops to recover from collection stress and to acclimate to controlled experimental conditions. The acute lethal toxicity of different concentrations and formulations of drilling muds was assessed by observing mortalities in control and experimental populations over a 96-h exposure period. Chronic sublethal toxicity of various concentrations was assessed through long-term growth experiments and by measurement of food acquisition processes.

Long-term Growth Experiments

Four chronic exposure studies were conducted using scallops of a narrow size range (Table 1). The scallops (~ 35 scallops/tank) were randomly distributed among nine fibreglass raceway tanks, each measuring 481 cm long by 37 cm wide, with a stand pipe drain that maintained water depth at 19 cm (total volume = 338 l). The tanks were covered with a plastic-wrapped rigid insulation board to prevent airborne particles from settling in the tanks and to minimise heat exchange and disturbance to the animals. To maintain particles in suspension, each tank was connected to a jet pump that recirculated the seawater, producing a flow velocity of 2.0 cm/sec around the scallops. Unfiltered Bedford Basin seawater from 3 m depth was delivered to each raceway tank after passing through in-line flow meters that maintained flow rate at 10 l/min. Seawater and suspended particles had a residence time in the raceway tanks of 34 min.

Used drilling muds were obtained from production platforms working off Nova Scotia, and included a Novaplus®-based formulation from the Sable Offshore Energy Project, and an IPAR-3® mud from Hibernia. In addition, a low viscosity ester-based drilling fluid was prepared for this study by Baroid/Halliburton. This formulation consisted of a typical suspension of Baroid® (ground barium sulphate), Gelltone II® (bentonite clay) and Suspentone® (undisclosed clay) in Petrofree® ester-based fluid. Concentrated drilling mud stock solutions were delivered through Solvaflex® tubing to raceway tanks via peristaltic pumps. One tank received no drilling mud and served as the control. Drilling mud stock concentrations were pumped into the recirculating pump intake of each raceway so that they were rapidly mixed with seawater to give the desired concentration range (Table 1). Scallops were exposed continuously to the treatments for between 53 and 72 days. Unfortunately, due to an unexpected failure of a seawater valve, Baroid® and Geltone II® treatments were terminated early.

TABLE 1. Synthetic drilling mud and constituent exposures schedules, target concentrations, scallop size range (shell height), and parameters measured.

Exposure Dates	Duration	Treatments (target concentration)	Size Range
Feb. 17 - Apr. 10, 1999	53 days	Novaplus®; Control, 0.1, 0.5, 1.0, 5.0, and 10.0 mg/l	103 ± 2 mm
July 5 - Aug. 30, 1999	55 days	Control Novaplus®; 0.1, 0.5, 1.0, and 5.0 mg/l IPAR-3®; 0.1, 0.5, 1.0, and 5.0 mg/l	90 ± 4 mm
June 12 - Aug. 24, 2000	72 days *37 days 58 days	Control Petrofree®; 0.1, 0.5, 1.0, 5.0, 10 mg/l Baroid®*, Geltone II®, and Suspentone® (0.5 mg/l separate treatments)	103 ± 2 mm
June 25 - Aug. 21, 2001	58 days	Barite; Control, 0.5, 1.0, 5.0 mg/l	107 ± 2 mm

On a daily basis, temperature and salinity measurements were taken, seawater flow rates were checked, stock tanks were refilled when necessary, and expired animals (shell gaping, retracted mantle, and no response to physical stimulation) were removed and the mortality was recorded. Each week, water samples were collected from raceways and the tanks were cleaned (by siphon) of faeces and any other deposited material. Water samples were analysed for suspended particulate material (SPM; mg/l), particulate organic material (POM; mg/l and % of SPM), particle size distribution, and hydrocarbon content. SPM was estimated by passing a known volume of water through a pre-combusted, pre-tared glass fibre filter (Whatman GF/D), and drying to constant weight (± 1 mg) in a 60°C drying oven. POM was then determined as weight loss after ignition (6 h at 450°C). The abundance and size distribution (2 to 40 μm) of drilling mud and seston particles in raceways were determined with a Coulter Counter (model TAII) fitted with a 100 μm aperture tube.

On the initial day of each of chronic exposure experiment, prior to addition of drilling wastes, 25 scallops were randomly selected from the nine raceway tanks and dissected into individual body components (gonad, adductor muscle, digestive gland, and remaining tissue consisting of gills and mantle). The tissues were placed in pre-weighed glass scintillation vials, dried for 5 d at 60°C, and reweighed to (± 0.1 g). Shell height (dorsal hinge to ventral margin) was measured to the nearest mm using callipers. On the median and final days of the growth experiments, where possible, fifteen scallops were selected from each raceway and treated as above for determination of shell height and dry body component weights.

Clearance Rate Estimates

Clearance rate (l/individual/h) was estimated using methods described in Cranford and Gordon (1992). Drilling mud suspensions maintained in raceways during growth experiments were pumped from ~ 2 cm above the bottom of the tank to a series of six flow-through feeding chambers. For each experimental treatment, five scallops were temporarily moved from the appropriate growth tank and placed in individual feeding chambers. The scallops were allowed to acclimate for 20 minutes before water samples were collected at the outlet of each chamber. The acclimation period allowed the scallops to resume normal feeding posture with shell open, mantle inflated, and tentacles extended. The remaining empty feeding chamber served as a reference for estimating the quantity and size distribution of particles removed by the scallops in the five test-chambers. Flow rates through feeding chambers were adjusted for each treatment to allow the scallops to clear a maximum of 30% of the available particles. Water samples collected from the outflow of the reference and test chambers were analysed by Coulter Counter fitted with a 100 μm aperture. Clearance rate was calculated from particle count data according to Cranford and Gordon (1992). The presence or absence of faeces and pseudofeces was noted after water samples were collected. A minimum of 10 independent clearance rate estimates was made for each drilling mud treatment.

Hydrocarbon Analyses

Samples (water and tissues) from each treatment and samples of each drilling mud were recovered on two separate occasions during the long-term growth experiment. Seawater samples (4 litres) were siphoned from the raceways into glass solvent bottles, which were previously cleaned, dried, and rinsed with hexane. Using a glass syringe, 200 μl of surrogates (120 ng/μl n-dodecane-d_{26}, 128 ng/μl n-hexadecane-d_{34}, 101 ng/μl n-tetracosane-d_{50}, 97 ng/μl, hexatricontane-d_{74}, 42 ng/μl hexamethylbenzene, 52 ng/μl 9,10-dimethylanthracene; Cambridge Isotope Laboratories Andover, MA, USA) were added to each sample. Water samples, separated into two lots, were poured into separatory funnels (2 l) and immediately partitioned into dichloromethane (3 $\times$ 50 ml Caledon, Georgetown, ON, Canada, distilled in glass). The combined extracts were concentrated and solvent transferred into a 5 ml vial with hexane (Caledon, Georgetown, ON, Canada, distilled in glass). The vials caps were lined with aluminum foil, sealed with Teflon tape, and frozen for transport and storage prior to analysis by gas chromatography/mass spectrometry (GC/MS).

On each sampling day, five scallops from each treatment and the control were collected and allowed to depurate in unfiltered seawater for a period of 24 h prior to dissection. Composite body component samples of the five scallops (sexes combined) were frozen for transport and storage in glass jars. Wet tissue samples were subjected to 6 M methanol-KOH saponification (Tremblay et al., 1992). The digests were diluted with water and partitioned into hexane. The extracts were purified using a Solid-Phase extraction (VWR-Canlab, Mont-Royal, QC, Canada, cat no. BJ9400) column packed with silica gel (activated 200 °C for 17 hr and deactivated 5% w/v with HPLC grade water; Whatman Laboratory Division, Clifton, NJ, USA, 60A 70-230 mesh ASTM for HPLC, cat no. 4791-010). The column was washed with hexane (Caledon, Georgetown, ON, Canada, distilled in glass) and the sample was applied as a 1 ml extract and PAHs are eluated with 10 mL hexane:dichloromethane (4:1 v/v). Extracts of water, muds, and biological tissues were analysed using high resolution gas chromatography (Agilent 6890 GC) coupled to a mass selective detector (Agilent 5973N) (Willmington, DE, USA) in the selective ion monitoring mode (SIM). The following GC (Ultra-2, 0.17 μm capillary column, Supelco Canada) conditions were used for the analysis: splitless injection 225°C; temperature programs 60 °C hold 2 min ramp at 6 °C/min to 300° C hold 13 min. A duplicate, certified reference material and operational blank was routinely performed with each batch of 10 samples.

Statistical Analysis

For each exposure, final dry body component weights from each waste treatment were compared using a multivariate analysis of covariance (MANCOVA) where shell height was the covariate. Significant (P-value ≤ 0.05) MANCOVA results were followed by post-hoc multiple comparison tests based on the Pillai Trace multivariate test statistic. Planned comparisons were limited to differences in mean values between the control and each waste treatment. Clearance rate estimates for scallops from each treatment were compared using analysis of covariance (ANCOVA) with shell height as the covariate. ANCOVA were followed by Tukey multiple comparison tests to explore difference among treatments. The null hypothesis tested in all cases was that mean values were equal for all treatments. Statistical tests were performed using SYSTAT Version 6.1 (SPSS, Inc., Chicago, ILL).

RESULTS AND DISCUSSION

Exposure Conditions

Temperature, salinity, and concentration of total particulate (SPM) and organic matter (POM) content of suspended matter available to scallops in the raceway tanks are summarised in Table 2. SPM and POM concentrations increased with increasing drilling mud concentrations, however, the organic content of SPM (% POM; scallop food quality) decreased with increasing mud additions (Table 2). The actual treatment concentration was calculated by subtracting mean SPM concentrations for the control (natural particles) from mean concentration in each the drilling waste treatments (Table 2). SPM measurements from each raceway tank confirmed that the actual exposure concentration was close to the target level for many of the treatments, but also show that the intended levels for the 2001 barite exposures were not reached (Table 2). The particle size distribution and volume concentration of drilling mud and natural particles in the raceway tanks were

TABLE 2. Experimental conditions in the raceway tanks during exposures of *P. magellanicus* to additions of used drilling muds and major constituents. T = temperature range, SPM = suspended particulate material (mean ± 2 SE), and POM = particulate organic material (mean ± 2SE).

Exposure Target level; mg/l	T (°C)	SPM ± 2SE (mg/l)	POM ± 2SE (mg/l)	POM ± 2SE (% of SPM)	Average Exposure (mg/l)
Novaplus® 1999 (winter)					
Control	2.3 - 4.7	1.25±0.18	0.59 ± 0.11	47.34 ± 5.36	
0.1		1.32 ± 0.16	0.64 ± 0.08	48.93 ± 3.82	0.1
0.5		1.41 ± 0.19	0.79 ± 0.21	53.74 ± 6.73	0.2
1.0		2.16 ± 0.21	0.90 ± 0.14	41.52 ± 3.18	0.9
5.0		5.71 ± 0.50	1.61 ± 0.20	28.11 ± 1.91	4.5
10.0		10.80 ± 1.02	2.77 ± 0.32	25.76 ± 2.11	9.5
Novaplus® 1999 (summer)					
Control	9.7 - 20.7	1.83 ± 0.47	1.37 ± 0.40	75.53 ± 5.30	
0.1		2.15 ± 0.59	1.56 ± 0.47	72.01 ± 7.26	0.3
0.5		2.86 ± 1.38	1.84 ± 0.86	65.31 ± 2.80	1.0
1.0		3.35 ± 0.80	1.88 ± 0.49	55.40 ± 3.94	1.5
5.0		5.44 ± 1.27	2.38 ± 0.71	43.24 ± 9.24	3.6
IPAR-3 ® 1999 (summer)					
Control	9.7 - 20.7	1.83 ± 0.47	1.37 ± 0.40	75.53 ± 5.30	
0.1		2.51 ± 0.83	1.45 ± 0.43	64.52 ± 7.84	0.7
0.5		3.25 ± 0.64	1.83 ± 0.50	54.30 ± 5.02	1.4
1.0		3.71 ± 0.81	1.92 ± 0.52	50.92 ± 3.55	1.9
5.0		9.98 ± 1.98	3.23 ± 0.89	37.16 ± 2.32	8.2
Petrofree® 2000 (summer)					
Control	11.0 - 21.5	0.97 ± 0.28	0.62 ± 0.14	66.40 ± 5.81	
0.1		1.19 ± 0.24	0.76 ± 0.10	65.50 ± 5.84	0.2
0.5		1.99 ± 0.46	1.32 ± 0.31	63.25 ± 8.25	1.0
1.0		2.32 ± 0.35	1.39 ± 0.32	59.48 ± 12.78	1.4
5.0		4.92 ± 0.88	1.69 ± 0.56	38.42 ± 11.75	4.0
10.0		9.03 ± 2.13	4.52 ± 1.69	51.21 ± 12.85	8.1
0.5 Suspentone®		1.69 ± 0.25	1.04 ± 0.14	66.39 ± 5.47	0.7
0.5 Geltone II®		1.65 ± 0.31	0.99 ± 0.14	62.93 ± 6.64	0.7
0.5 Baroid		1.29 ± 0.31	0.71 ± 0.12	60.40 ± 10.99	0.3
Barite 2001 (summer)					
Control	7.4 - 16.1	0.76 ± 0.17	0.44 ± 0.14	54.59 ± 9.95	
0.5		0.63 ± 0.10	0.32 ± 0.07	52.59 ± 8.65	0
1.0		0.88 ± 0.14	0.43 ± 0.12	39.58 ± 7.71	0.1
5.0		1.33 ± 0.23	0.43 ± 0.13	31.31 ± 6.80	0.4

quantified using Coulter Counter data. Bentonite and barite particles were observed to be similar in size distribution to those shown in Cranford et al. (1999). The IPAR-3® and Petrofree® mud size spectra contained modes at 12 μm (bentonite) and 25 μm (barite), while the Novaplus® treatments were dominated by a fine clay signature at 5 μm.

The total resolved hydrocarbon concentrations in the recovered seawater samples from the experimental raceway tanks increased with the addition of Novaplus® over the concentration range (1 to 10 mg/l) of the study (Table 3). The compositional indices (odd:even carbon alkanes < C_{23}) was similar between that of Novaplus® and the hydrocarbons within the seawater samples (1.24 ± 0.04) spiked with various concentrations of Novaplus®. The compositional indices of the remaining fraction of higher molecular wt alkanes (odd:even carbon alkanes >C_{23}) was much more variable as they were less soluble in water. The concentrations of PAHs and alkylated PAHs suggest that their content in the Novaplus® was very low or that they were or immiscible in water (high molecular weight > 4 rings with $\log K_{ow}$ > 4). Indeed the majority of resolved aromatic compounds were low molecular weight alkylated PAHs with 2–3 ring structures dominating (< 2% of the total resolved organics).

TABLE 3. Hydrocarbon constituents in seawater in the Novaplus® experiment.

	Control (0 mg/l)	0.1 (mg/l)	0.5 (mg/l)	1.0 (mg/l)	5.0 (mg/l)	10.0 (mg/l)
∑ alkanes (ng/l)	6842	4049	7445	15536	80128	157394
∑ odd < C23	1518	1210	3293	6931	44811	84443
∑ odd > C24	3392	1574	1115	2323	639	3104
∑ even < C23	1206	982	2737	5775	34602	69495
∑ even > C24	571	283	300	508	77	352
∑ odd	4910	2784	4409	9253	45450	87547
∑ even	1777	1265	3036	6283	34679	69846
odd:even	2.76	2.20	1.45	1.47	1.31	1.25
odd:even < C23	1.26	1.23	1.20	1.20	1.30	1.22
odd:even > C24	5.94	5.56	3.72	4.57	8.30	8.82
∑ PAHs (ng/l)	11	2.4	46		6.6	1.9
∑ alkylated PAHs	132	95	57	102	96	63
Total resolved organics (ng/l)	6985	4146	7547	15637	80231	157459
∑ 2-3 ring PAHs	139	96	60	102	101	63
∑ 4 ring PAHs	4.4	1.2	42		1.4	2.5
∑ 5 ring PAHs	0.16				0.25	
∑ 6 ring PAHs			0.02		0.49	

A unique trend for hydrocarbons within the seawater samples was observed within the IPAR-3 concentration gradient (Fig. 1). Contrary to expectations based on dilution of the drilling mud, alkane concentrations only increased from the control to 0.5 mg/l after which concentrations declined rapidly. A similar trend was also reported for the PAH fractions despite their low concentrations. This observed difference from Novaplus® may be attributed to product specific differences in chemical behaviour following their release into the environment.

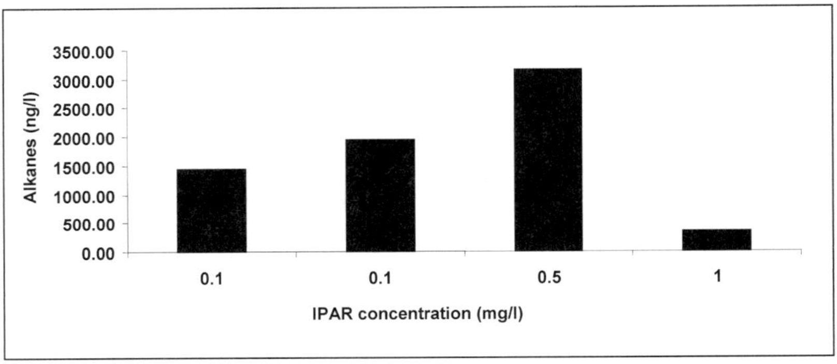

FIGURE 1. Concentration of alkanes in water of the IPAR Experiment.

The concentration of PAH and alkylated PAH within the water samples with additions of the Petrofree® ester-mud drilling fluid formulation were also low with 2–3 ring structure dominating (ranging 1–22% of the total resolved organics). Elevated concentrations of alkanes (primarily high molecular weight alkanes > C24) above the control value were apparent with the addition of 0.5 mg/l SBM (Table 4). Contrary to the Novaplus®, the concentration of alkanes did not increase with the concentration gradient of SBM added.

Based on the results presented, the organics present in the seawater (low molecular weight) and ester-muds introduced to synthetic drilling fluids are bioavailable to marine organisms. However, other factors, such as solubility and biodegradation rates, can alter the organic constituents' distribution after they have been released into the environment.

TABLE 4. Hydrocarbon constituents in seawater receiving Petrofree® ester-mud formulation additions

	Control (0 mg/l)	0.1 (mg/l)	0.5 (mg/l)	1.0 (mg/l)	5.0 (mg/l)	10.0 (mg/l)
∑alkanes (ng·g-1)	10688	8554	31677	19039	36581	25190
∑ odd <C23	3023	3833	2612	3768	5839	3764
∑ odd >C24	3387	1364	14868	7123	14128	10655
∑ even <C23	1297	1259	982	2101	3542	1989
∑ even >C24	1913	618	12550	5574	11600	7978
∑ odd	6410	5197	17480	10891	19967	14419
∑ even	3210	1877	13532	7675	15141	9968
odd:even	1.99	2.77	1.29	1.41	1.32	1.45
odd:even < C23	2.33	3.04	2.66	1.79	1.65	1.89
odd:even > C24	1.77	2.21	1.18	1.28	1.22	1.34
∑ 2-3 ring PAHs	3030	2466	1331	963	1314	1345
∑ 4 ring PAHs	1014	947	505	198	343	311
∑ 5 ring PAHs	155	165	80	40	60	78
∑ 6 ring PAHs	4.15	11	2.83	1.99	2.17	3.86
∑ PAHs (ng/g)	877	900	529	315	457	455
∑ alkylated PAHs (ng/g)	3307	2676	1381	889	1262	1282

Biological Effects

Mortalities of *P. magellanicus* exposed to all waste treatments were generally comparable or were only slightly higher than observed in the control population (Fig. 2). Seasonal variations were observed as the mortality of scallops during the winter experiment (including the control) was higher than observed during summer.

The average dry weight of all tissue components at the end of Novaplus® SBM chronic exposures was significantly affected (mean tissue weight less than control values) at concentrations equal to and greater than 0.9 mg/l (winter exposure) and 1.5 mg/l (summer exposure) (Tables 5 and 6). Treatment effects were also detected by MANCOVA for other drilling fluids at similar concentrations. Significant effects on final weight were observed at concentrations equal to and greater than 1.4 mg/l for IPAR-3®, and greater than 0.2 mg/l for the Petrofree® ester-based mud formulation (Tables 5 and 6). Effects thresholds were not determined for solid drilling mud constituents (only one concentration tested), but significant effects were detected in the 0.7 mg/l Suspentone®, 0.7 mg/l Geltone II®, and 0.3 mg/l Baroid® treatments (Tables 5 and 6).

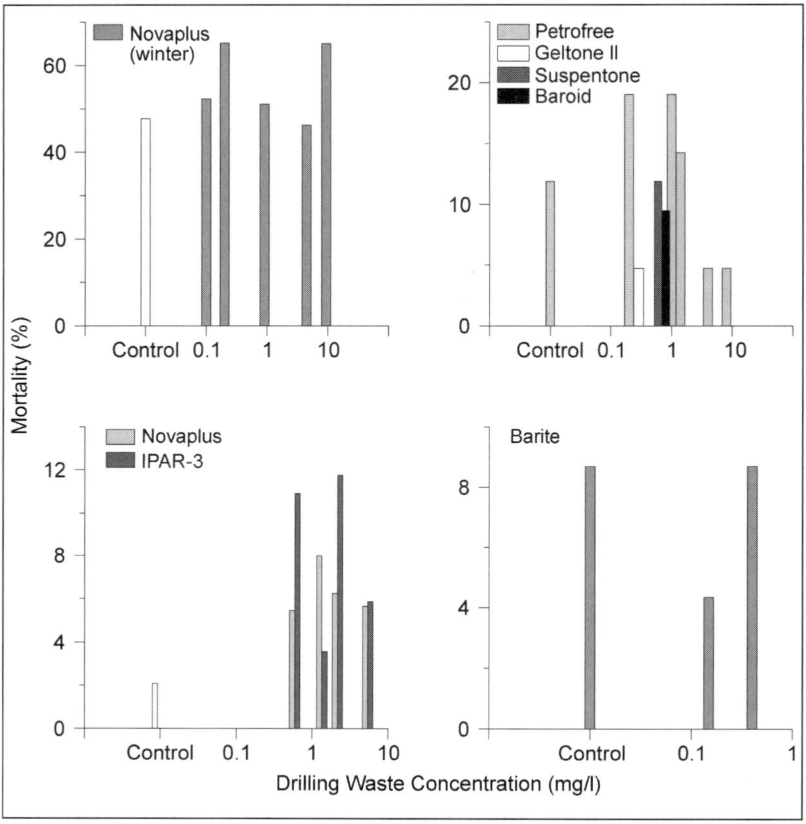

FIGURE 2. Mortality (% of initial abundance) of *P. magellanicus* by the end of chronic exposure to SBM and components.

No overall significant effect on final tissue weight was detected by MANCOVA for barite exposures conducted in 2001 (Tables 5 and 6). Previous chronic exposures of sea scallops to barite suspensions resulted in the complete cessation of gonad growth at the lowest concentration tested (0.5 mg/l; Cranford et al., 1999). This is consistent with the results obtained from the Baroid® treatments (above). Barite concentrations measured in the 2001 experiments were lower than intended, and two of the treatments had mean concentrations lower than measured in the control raceway (Table 2). This may have resulted from an unexpected reduction in flow rates

TABLE 5. Gonad, digestive gland, adductor muscle and remaining tissue dry weights (mean DW ± 2SE) of *P. magellanicus* sampled on the initial and final days of chronic exposure experiments.
*Treatment identifiers used in Table 6.

Treatment (mg/l)	Day	Gonad DW (g)	Adductor DW (g)	Digestive DW (g)	Remaining DW (g)
Novaplus® 1999 (winter)					
Initial	0	0.17 ± 0.04	2.35 ± 0.13	1.02 ± 0.09	1.85 ± 0.07
Final Control (T1*)	53	1.01 ± 0.22	4.02 ± 0.41	2.57 ± 0.41	2.49 ± 0.17
0.1 (T2)	53	0.72 ± 0.35	2.89 ± 0.50	1.69 ± 0.53	2.27 ± 0.15
0.2 (T3)	terminated	-	-	-	-
0.9 (T4)	53	0.24 ± 0.04	2.60 ± 0.35	1.37 ± 0.36	2.08 ± 0.23
4.5 (T5)	53	0.48 ± 0.17	3.10 ± 0.39	1.57 ± 0.35	2.25 ± 0.13
9.5 (T6)	terminated	-	-	-	-
Novaplus® and IPAR-3® 1999 (summer)					
Initial	0	1.28 ± 0.23	4.19 ± 0.21	1.35 ± 0.10	2.13 ± 0.08
Final Control (T1)	55	2.37 ± 0.31	3.88 ± 0.63	1.37 ± 0.17	2.10 ± 0.13
0.3 Novaplus (T2)	55	1.82 ± 0.19	3.44 ± 0.30	1.11 ± 0.14	1.83 ± 0.16
1.0 Novaplus (T3)	55	1.81 ± 0.42	3.75 ± 0.23	1.12 ± 0.12	1.83 ± 0.17
1.5 Novaplus (T4)	55	1.39 ± 0.26	3.24 ± 0.25	0.98 ± 0.09	1.75 ± 0.09
3.6 Novaplus (T5)	55	1.66 ± 0.28	2.99 ± 0.18	0.82 ± 0.09	1.72 ± 0.07
0.7 IPAR-3 (T2)	55	1.82 ± 0.34	3.30 ± 0.22	1.23 ± 0.13	1.85 ± 0.11
1.4 IPAR-3 (T3)	55	1.63 ± 0.26	2.95 ± 0.28	0.89 ± 0.10	1.70 ± 0.11
1.9 IPAR-3 (T4)	55	1.50 ± 0.29	3.05 ± 0.32	0.90 ± 0.10	1.77 ± 0.11
8.2 IPAR-3 (T5)	55	0.44 ± 0.13	2.46 ± 0.20	0.64 ± 0.10	1.45 ± 0.12
Petrofree® 2000 (summer)					
Initial	0	2.07 ± 0.31	3.73 ± 0.26	1.27 ± 0.10	2.42 ± 0.12
Final Control (T1)	72	2.18 ± 0.46	3.72 ± 0.25	1.39 ± 0.10	2.31 ± 0.11
0.2 (T2)	72	1.60 ± 0.44	3.80 ± 0.32	1.33 ± 0.14	1.90 ± 0.07
1.0 (T3)	72	1.09 ± 0.37	3.46 ± 0.39	0.56 ± 0.08	0.30 ± 0.20
1.4 (T4)	72	0.53 ± 0.15	2.62 ± 0.27	0.92 ± 0.07	2.04 ± 0.09
4.0 (T5)	72	0.54 ± 0.14	3.32 ± 0.31	1.01 ± 0.11	2.17 ± 0.11
8.1 (T6)	72	1.84 ± 0.29	3.19 ± 0.21	0.61 ± 0.07	1.44 ± 0.13
0.7 Suspentone (T7)	58	1.16 ± 0.68	3.17 ± 0.44	0.60 ± 0.07	1.80 ± 0.18
0.7 Geltone II (T8)	58	0.65 ± 0.24	3.63 ± 0.41	0.90 ± 0.14	1.56 ± 0.17
0.3 Baroid (T9)	37	1.91 ± 0.24	3.65 ± 0.25	0.81 ± 0.09	1.60 ± 0.22
Barite 2001 (summer)					
Initial	0	1.71 ± 0.69	5.02 ± 0.79	1.63 ± 0.42	2.99 ± 0.49
Final Control (T1)	58	2.39 ± 1.03	5.44 ± 1.22	1.03 ± 0.32	2.60 ± 0.34
0 (T2)	58	2.79 ± 0.77	4.65 ± 0.94	0.91 ± 0.25	2.48 ± 0.30
0 (T3)	58	2.77 ± 0.54	4.84 ± 0.69	0.87 ± 0.19	2.54 ± 0.26
0.4 (T4)	58	2.58 ± 0.53	4.89 ± 0.54	0.94 ± 0.21	2.57 ± 0.18

TABLE 6. Summary table of results from one-way multivariate analysis of covariance (MANCOVA) on sea scallop tissue component dry weight (DW) on the final day of exposure. Summary data and treatment identifiers (T1 to T9) are given in Table 3. The indicated multiple comparison hypothesis tests were performed only after a significant MANCOVA result. *$P \leq 0.05$.

	Multivariate Test Statistics				**Univariate *F*-test *P*-values**			
Waste and Hypothesis	Pillai Trace	*F*-statistic	*df*	*P*	Gonad DW	Adductor DW	Digestive DW	Remaining DW
Novaplus® 1999 (winter)								
All equal	1.042	2.661	12, 60	0.006*	0.001*	0.000*	0.000*	0.001*
T1=T2	0.475	1.812	4, 8	0.220	-	-	-	-
T1=T4	0.828	9.658	4, 8	0.004*	0.000*	0.001*	0.001*	0.001*
T1=T5	0.651	4.661	4, 10	0.022*	0.001*	0.002*	0.001*	0.002*
IPAR-3® 1999 (summer)								
All equal	0.9596	6.228	12, 159	0.000*	0.000*	0.000*	0.000*	0.000*
T1=T2	0.216	1.581	4, 23	0.213	-	-	-	-
T1=T3	0.394	3.740	4, 23	0.017*	0.005*	0.327	0.007*	0.002*
T1=T4	0.442	4.355	4, 23	0.010*	0.005*	0.297	0.003*	0.004*
T1=T5	0.862	4.353	4, 23	0.001*	0.000*	0.135	0.000*	0.000*
Novaplus® 1999 (summer)								
All equal	0.495	2.401	16, 272	0.002*	0.016*	0110	0.000*	0.043*
T1=T2	0.167	1.155	4, 23	0.356	-	-	-	-
T1=T3	0.283	2.267	4, 23	0.093	-	-	-	-
T1=T4	0.442	4.547	4, 23	0.007*	0.001*	0.593	0.035*	0.006*
T1=T5	1.317	7.573	4, 23	0.000*	0.025*	0.202	0.001*	0.000*
Petrofree® and constituents 2000 (summer)								
All equal	1.574	10.868	32, 536	0.000*	0.000*	0.000*	0.000*	0.000*
T1=T2	0.671	12.745	4, 25	0.000*	0.097	0.710	0.473	0.000*
T1=T3	0.954	77.783	4, 15	0.000*	0.081	0.684	0.000*	0.000*
T1=T4	0.821	29.756	4, 26	0.000*	0.000*	0.000*	0.000*	0.024*
T1=T5	0.719	19.815	4, 31	0.000*	0.000*	0.075	0.000*	0.122
T1=T6	0.878	57.570	4, 32	0.000*	0.207	0.003*	0.000*	0.000*
T1=T7	0.878	28.682	4, 16	0.000*	0.104	0.041*	0.000*	0.000*
T1=T8	0.869	38.213	4, 23	0.000*	0.000*	0.717	0.000*	0.000*
T1=T9	0.716	29.645	4, 47	0.000*	0.246	0.899	0.000*	0.000*
Barite 2001 (summer)								
All equal	0.256	1.282	12, 159	0.234	-	-	-	-

within the raceway tanks, which would have allowed suspended barite to adhere to seston and settle out of suspension.

Differences were observed on the relative impact of drilling wastes on different tissue components. Significant treatment effects were detected at relatively low waste concentrations in remaining tissue (gills and mantle) and digestive gland (Table 6). Slightly higher concentrations were generally required to cause a reduction in gonad mass. Adductor muscle weight was less prone to impact, and some wastes did not significantly impact adductor at the highest waste concentrations tested.

Clearance rate measurements for *P. magellanicus* chronically exposed to the various waste treatments are summarised in Figure 3. ANCOVA detected significant effects on clearance rate for all the drilling muds and components studied ($P < 0.001$). The post-hoc multiple comparisons between control and different waste treatments also showed significant effects from all concentrations tested ($P \leq 0.022$), including the lowest (0.1 mg/l Novaplus®). Major solid components in SBM had a similar impact on clearance rate as the whole SBM (Fig. 3), and ANCOVA indicted significant effects of barite and

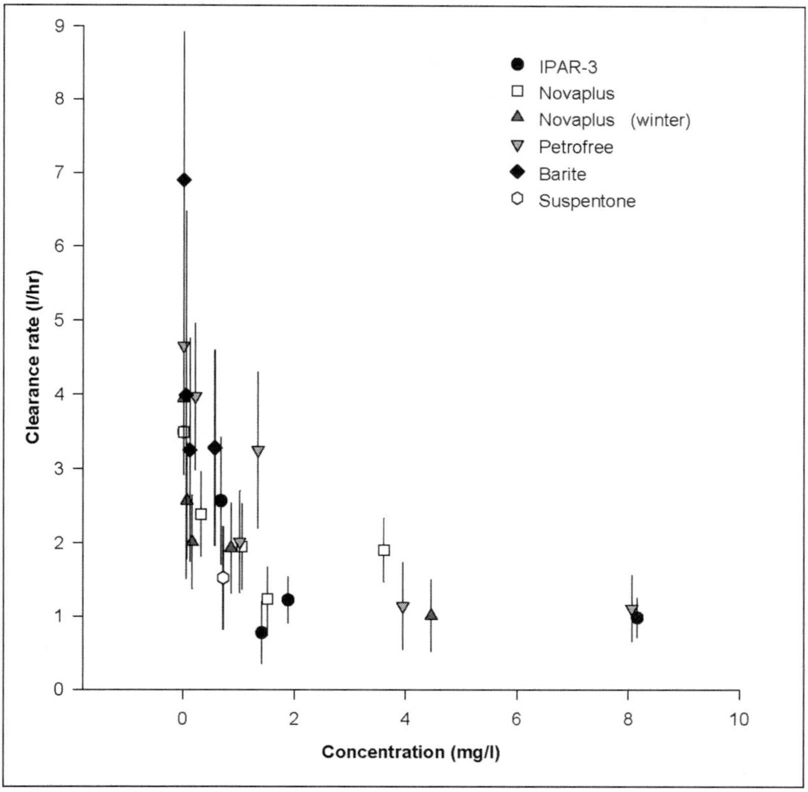

FIGURE 3. Relation between *P. magellanicus* clearance rate (mean ± 2SE) and concentrations of SBM and major constituents.

Suspentone® ($P \leq 0.007$). Cranford and Gordon (1992) and Cranford et al. (1999) previously showed that 1 to 2 mg/l of suspended bentonite and a water-based mud (primarily bentonite) caused similar reductions in sea scallop feeding rate. These results indicate that the solid fraction of the drilling mud is largely responsible for the reduced clearance rates observed for whole SBMs. The reduced feeding rates, combined with reductions in food absorption (Cranford et al., 1999) are therefore believed to be the cause of the observed impacts on somatic and reproductive tissue growth.

Bioaccumulation of Hydrocarbons

Based on the analysis of gonad, digestive gland, adductor muscle, and remaining body tissue, the digestive gland was the primary site for the bioaccumulation of hydrocarbons following exposure to IPAR-3® (Fig. 4). Total resolved alkane concentrations in digestive glands increased relative to the control with the addition of 0.7 mg/l of IPAR-3®, and continued to increase in the 1.4 mg/l treatment before they declined at the higher concentrations (Fig. 4). This trend could result from the inhibitory effects of increasing SBM concentration on scallop clearance rates (above). With increasing mud concentration, the decline in clearance rate would reduce waste ingestion and limit the bioaccumulation of SBM. Bioaccumulation of PAHs in IPAR-3® also occurred primarily in digestive glands (Fig. 4). Highest PAH bioaccumulations were observed in scallops from the 8.2 mg/l treatment, perhaps indicating that PAH uptake was both direct (absorption) and indirect (ingestion). Bivalves such as blue mussels from sites contaminated with PAHs are known to bioaccumulate these compounds (Ernest et al. 1999). The exposure studies with various drilling muds clearly suggest that the discharge of drilling waste in the ocean may impact the biological response and subsequent health of sea scallops. Distinct differences in the organic composition between the IPAR-3® and Novaplus® result in unique chemical profiles for the biological samples that were exposed to the test synthetic drilling fluid in the exposure studies. It will also be dependent on bioavailablity and bioaccumulation of the compounds that are present in the fluids. Bioavailability can

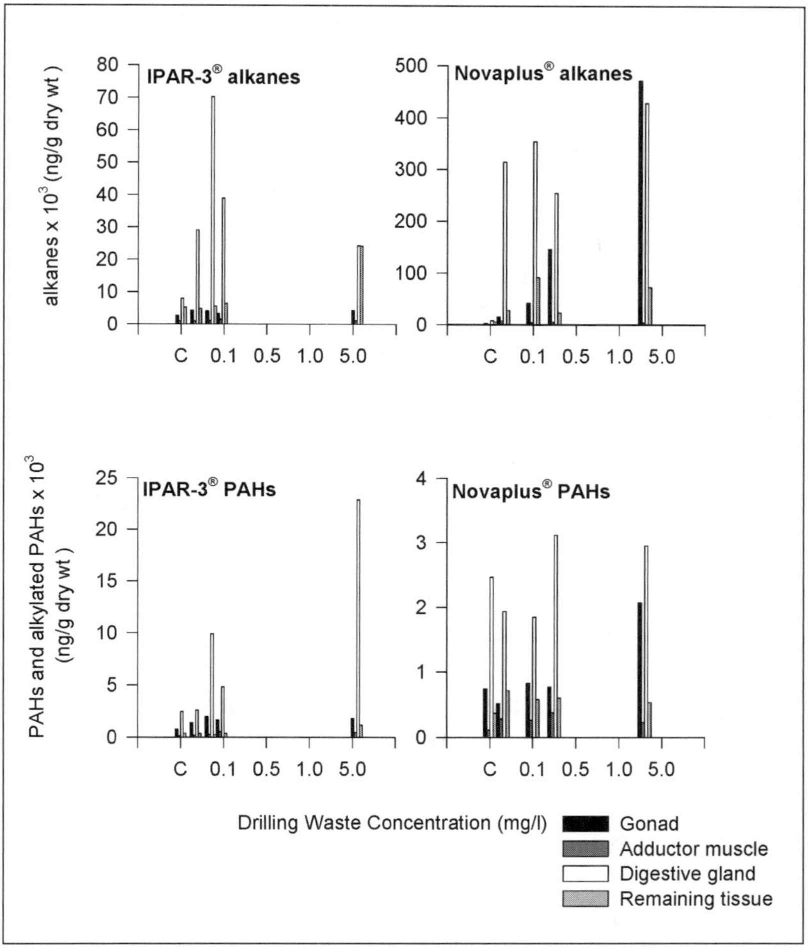

FIGURE 4. Total resolved alkanes and total PAHs and alkylated PAHs in tissues from scallops sampled on the final day of exposure to Novaplus® and IPAR-3® treatments.

be affected by the composition of the particles present in the water column. Some characteristics that effect binding of hydrophobic organic compounds include aromaticity, hydrophobic acid content, and molecular size (McCarthy et al., 1989; Kukkonen and Oikari, 1991; Chin et al., 1997; Haitzer et al., 1999). Differences in physicochemical factors (i.e. lipophilicity, size, and spatial configuration of individual compounds and changing PAH profiles), geochemical factors (e.g. feeding behavior and uptake) will all influence contaminant bioavailability and toxicity (Long et al., 1998).

Significant alkane and PAH uptake was detected in both digestive glands and gonads during the Novaplus® exposures. Although the Novaplus® exposures were generally of lower concentration than the IPAR-3® treatments, bioaccumulations of alkanes were approximately an order of magnitude higher in animals exposed to Novaplus® (Fig. 4). There was clear evidence for a dose-dependent increase in Novaplus® alkane concentrations in the gonads (Fig. 4). This is uncharacteristic for sea scallops, which typically accumulate contaminants in the digestive gland (Cranford et al., 1999, Cranford, 2005). The difference in alkanes uptake between Novaplus® and IPAR-3® are not attributed to physiological differences between scallops from different sources as the studies were conducted concurrently. The dissimilarity between the two tests for alkanes is probably due to differences in the chemical composition of the drilling mud formulations that altered their bioavailability and/or the biological response of the sea scallops. Total PAHs and alkylated PAHs concentrations in scallops were similar for similar concentrations of both SBMs. The alkylated PAHs were highest in concentration with 2–3 ring structures as the dominant compounds.

CONCLUSIONS

The biological effects of drilling wastes have been attributed to: (1) chemical toxicity of specific hydrocarbons in base fluids, metal contaminants associated with barite (metals), and biodegradation products, (2) organic enrichment of sediment that may produce anoxia, and/or (3) physical smothering. Previous observations of growth and clearance rate inhibition in sea scallops during chronic exposure to low-toxicity mineral oil-based muds were attributed to a narcotic effect from base-oil toxicity (Cranford et al., 1999). The carrier-fluid in SBMs is specifically engineered to exhibit low toxicity and their major solid components (bentonite clay and barite) are assumed to be inert (Neff, 1987). Organic enrichment and smothering impacts are typically of low concern owing to the limited geographic extent where such impacts could occur at offshore drilling sites. Results from this and previous studies with scallops show that an additional impact mechanism exists that could have consequences at the population level of some species. Physical effects caused by chronic exposure to environmentally relevant concentrations (< 10 mg/l) of suspended solids found in most drilling muds (e.g. bentonite and barite) are capable of causing a significant impact on growth, reproduction, feeding rates, and food absorption efficiency, at least in *P. magellanicus* (Cranford and Gordon, 1992; Cranford et al., 1999; this study). These results may help to explain why observed impacts on benthic communities have been detected at greater distances from drilling platforms than can be explained by the distribution of hydrocarbon and metal contaminants (Olsgard and Gray, 1995).

The underlying mechanism for changes in feeding behaviour may be related to clay-sized particles interfering with ciliary activity on the feeding structures (Cranford and Gordon, 1992) or physical damage to the gills. SEM observations of the ctenidia of suspension-feeding bivalves have shown gill damage after exposure to barite (Barlow and Kingston, 2001). It appears unlikely that the low concentrations tested in the present study resulted in permanent damage as previous experiments have shown feeding rates to return to natural levels after removal of the waste particles (Cranford et al., 1999).

ACKNOWLEDGEMENTS

The authors wish to thank David Knickle of Adams and Knickle Ltd. of Lunenburg, NS and the crew of the MV *Cashalot I* for providing scallops for this study. Sable Offshore Energy Inc. and MI Drilling Fluids provided the Novaplus® and IPAR-3® drilling muds, and Baroid/Halliburton provided the Petrofree-based drilling mud formulation. We also thank Gilles H. Tremblay and Sylvie St.-Pierre for conducting the hydrocarbon analyses at the Institute Maurice Lamontagne in Mont-Joli, Quebec.

REFERENCES

Barlow, M.J. and P.F. Kingston. 2001. "Observations on the Effects of Barite on the Gill Tissues of the Suspension Feeder *Cerastoderma Edule* (Linne) and the Deposit Feeder Macoma balthica (Linne)." *Marine Pollution Bulletin.* 42(1):71-76.

Bayne, B.L. and R.C. Newell. 1983. "Physiological Energetics of Marine Molluscs". In A.S.M., Saleuddin and K.M. Wilbur, (Eds.), *The Mollusca, Physiology, Part I,* pp. 407-515. Academic Press, New York.

Chin, Y.P., G.R. Aiken, and K.M. Danielsen. 1997. "Binding of Pyrene to Aquatic and Commercial Humic Substances: The Role of Molecular Weight and Aromaticity." *Environmental Science and Technology.* 31(6):1630-1635.

Cranford, P.J., S.L. Armsworthy, S. McGee, and T. King, K. Lee, and G.H. Tremblay. 2005. "Scallops as Sentinel Organisms for Offshore Environmental Effects Monitoring." In: S.L. Armsworthy, P.J. Cranford, and K. Lee (Eds.), *Proceedings of the Offshore Oil and Gas Environmental Effects Monitoring Workshop: Approaches and Technologies, Bedford Institute of Oceanography, May 26–30, 2003.* Battelle Press. In prep.

Cranford, Peter J., Kenneth Lee, John W. Loder, Timothy G. Milligan, D.K. Muschenheim, and Jerry Payne. 2001. *Scientific Considerations and Research Results Relevant to the Review of the 1996 Offshore Waste treatment Guidelines.* Canadian Technical Report of Fisheries and Aquatic Sciences, 2364.

Cranford, P.J. and D.C. Gordon. 1992. "The Influence of Dilute Clay Suspensions on Sea Scallop (*Placopecten magellanicus*) Feeding Activity and Tissue Growth." *Netherlands Journal of Sea Research.* 30:107-120.

Cranford, P.J., D.C. Gordon Jr., C.G. Hannah, J.W. Loder, T.G. Milligan, D.K. Muschenheim, and Y. Shen. 2003. "Modelling Potential Effects of Petroleum Exploration Drilling on Northeastern Georges Bank Scallop Stocks." *Ecological Modelling.* 166:19-39.

Cranford, P.J., D.C. Gordon Jr., K. Lee, S.L. Armsworthy, and G.-H. Tremblay. 1999. "Chronic Toxicity and Physical Disturbance Effects of Water- and Oil-based Drilling Fluids and their Major Constituents on Adult Sea Scallops (*Placopecten magellanicus*)." *Marine Environmental Research.* 48:225-256.

Ernst, B., D. Pither, S. Matthews, and G. Julien. 1999. *Concentration and Distribution of Polycyclic Aromatic Hydrocarbons (PAHs), PCBs, and*

Metals in Sediments, Mussels (Mytilus edulis and Modiolus modiolus), and Lobster (Homarus americanus) from Sydney Harbour, Nova Scotia, Surveillance Report EPS-5-AR-99-7 Atlantic Region.

Haitzer, M., G. Abbt-Braun, W. Traunspurger, and C.E.W. Steinberg. 1999. "Effect of Humic Substances on the Bioconcentration of Polycyclic Aromatic Hydrocarbons: Correlations with Spectroscopic and Chemical Properties of Humic Substances." *Environmental Toxicology and Chemistry.* 18(12):2782-2788.

Kukkonen, J., J.F. McCarthy, and A. Oikari. 1990. "Effects of XAD-8 Fractions of Dissolved Organic Carbon on the Sorption and Bioavailability of Organic Micropollutants." *Archives of Environmental Contaminants and Toxicology.* 19:551-557.

Long, E.R., and R. MacDonald. 1998. "Recommended Uses of Empirically Derived, Sediment Quality Guidelines for Marine and Estuarine Ecosystems." *Human and Ecological Risk Assessment.* 4(5):1019-1039.

McCarthy, J.F., L.E. Roberson, and L.W. Burris. 1989. "Association of Benzo[*a*]pyrene with Dissolved Organic Matter: Prediction of Kdom from Structural and Chemical Properties of the Organic Matter." *Chemosphere.* 19(12):1911-1920.

Muschenheim, D.K., and T.G. Milligan. 1996. "Flocculation and Accumulation of Fine Drilling Waste Particulates on the Scotian Shelf (Canada)". *Marine Pollution Bulletin.* 32:740-745.

National Energy Board (NEB), Canada–Newfoundland Offshore Petroleum Board, and Canada–Nova Scotia Offshore Petroleum Board. 2002. *Offshore Waste Treatment Guidelines.*

Neff, J.M. 1987. "Biological Effects of Drilling Fluids, Drill Cuttings and Produced Water." In Boesch, D.F. and Rabalais, N.N. (Eds.), *Long-term Environmental Effects of Offshore Oil and Gas Development.*

Olsgard, F. and J.S. Gray. 1995. "A Comprehensive Analysis of the Effects of Offshore Oil and Gas Exploration and Production on the Benthic Communities of the Norwegian Continental Shelf." *Marine Ecology Progress Series.* 122:277-306.

Tremblay, G.H., J.N. Gearing, and M. Noël, "Méthodes d'Analyse et Résultats sur les Concentrations en Hydrocarbures Aliphatiques dans les Sédiments du Fleuve Saint-Laurent". *Rapport Statistique Canadien sur l'Hydrographie et les Sciences Océaniques,* Vol. 103 vi + 54 p., 1992.

Scallops as Sentinel Organisms for Off-Shore Environmental Effects Monitoring

Cranford, P.J. (cranfordp@dfo-mpo.gc.ca), S.L. Armsworthy, S. McGee, T. King, K. Lee (Fisheries and Oceans Canada, Dartmouth, NS, Canada), and G.H. Tremblay (Fisheries and Oceans Canada, Mont-Joli, PQ, Canada)

ABSTRACT: Two environmental effects monitoring approaches were tested at the Hibernia offshore oil production field. Both methods use bivalves as sentinel organisms for assessing operational waste bioavailability, and for relating contaminant exposure and dose to the onset of biological effects. Bivalve cages containing caged Icelandic and sea scallops and blue mussels were placed on the seabed at different distances to the Hibernia platform. Lethal and sublethal (shell and tissue growth) impacts and hydrocarbon body burden measurements were used to assess the spatial extent of contamination and benthic impacts from drilling activities. The second monitoring approach utilised an in situ biological effects monitoring system (HABITRAP) that provided daily measurements of drilling waste exposure (barium and hydrocarbon sedimentation rate) and bivalve biodeposition rate, a measure known to be sensitive to the presence of drilling wastes. Additional supporting technologies developed for this study included an inexpensive subsurface float release mechanism for the bivalve cages that prevented mooring loss from ship collisions with surface floats, and a new barium extraction methodology that was developed to improve recovery from sediment containing high barite concentrations. Standard extraction methods were shown to inefficiently extract barite at concentrations greater than 200 μg/g. Observations of bivalve responses to temporal and spatial variations in metal and hydrocarbon contaminants are compared with the results of benthic community analysis.

INTRODUCTION

Offshore oil and gas operations off the eastern coast of Canada have increased within the last decade and are expanding to include shallow coastal and deep slope waters. An ongoing concern related to the development of this industry is that long-term exposure of marine organisms to operational drilling and production wastes may cause ill effects on fish and fish habitat. A commonly held viewpoint is that impacts on commercial fish and shellfish should be restricted to the immediate vicinity of production drilling platforms. However, laboratory and field observations (Reiersen et al., 1989; Gray et al., 1990; Daan et al., 1990; Olsgard and Gray, 1995), and numerical model predictions (Cranford et al., 2003) indicate that important lethal and sublethal impacts may occur at much greater distances than originally envisaged. Uncertainties related to the significance of environmental impacts from operational waste discharges has resulted in the establishment of costly mitigation measures (waste treatment technologies) and environmental effects monitoring (EEM) programs at offshore oil and gas production sites in Canada and elsewhere. To address environmental concerns, a multidisciplinary research program was initiated by the Department of Fisheries and Oceans to study the environmental pathways, transport rate, and ultimate fate and effects of drilling and production wastes.

Although there is no standardised EEM approach for offshore oil and gas operations in Canada, established monitoring programs rely heavily on the analysis of hydrocarbon and metals contaminants in sediments, standard sediment toxicity biotests, and/or benthic community analysis. The rationale for focusing current EEM programs on the benthos is that the bulk of drilling wastes rapidly descend to the seabed and benthic organisms tend to be sessile and integrate the biological effects of contaminants over time. Benthic communities are subject to large natural spatial and temporal variations, making it difficult and costly to attribute community changes to the presence of contaminants. A shortcoming of laboratory toxicity tests is that they generally do not include resident species of concern, including commercial species. Loss of contaminants during sampling, and unrealistic exposure conditions in the laboratory may also lead to the collection of inapplicable results. As a result, considerable scope exists for the development of improved technologies for measuring environmental impacts that also reduce costs to industry.

Toxicity screening of drilling wastes has identified the sea scallop (*Placopecten magellanicus*) as one of the most sensitive species studied (Neff, 1987). Chronic toxicity tests in which scallop cohorts were exposed under environmentally relevant conditions to drilling fluids (muds) showed that low levels (0.05 to 2 mg/l) of all major types of muds (includes water-, low toxicity mineral oil-, synthetic-, and ester-based muds) can significantly impact growth and reproduction (Cranford and Gordon, 1992; Cranford, 1995; Cranford et al., 1999; Armsworthy et al., 2005). Such impacts are considered, from both ecological and fisheries perspectives, to be the most important sublethal effects on adult organisms (Capuzzo, 1988). Fine particles in drilling fluids (bentonite and barite) were the major cause of the observed effects from the water-, synthetic-, and ester-based muds, as a result of their impact on feeding and digestion processes (Armsworthy et al., 2005). Similar physical effects were detected by Barlow and Kingston (2001) in other bivalve species. A modelling study of the potential impact of exploration drilling on a major offshore scallop fishery indicated that under certain hydrographic conditions, the routine discharge of water-based mud may impact scallop growth over an area exceeding 200 km^2 (Cranford et al., 2003).

The high sensitivity of sea scallops to waste drilling mud, combined with their high socioeconomic value, and ability to bioaccumulate contaminants, makes them an ideal candidate as a sentinel species for assessing operational waste bioavailability, and for relating contaminant dose to biological effects (Cranford, 2003). Two EEM approaches are tested and reported that utilise scallops for measuring biological effects from offshore operations. The first consists of a series of bivalve cages used to monitor spatial variations in contaminant bioaccumulation and biological effects. The second approach is based on the time-series sediment trap technique developed by Cranford and Hargrave (1994) for continuously and autonomously monitoring scallop feeding rate, a measure that is closely related to ambient waste concentration and growth rate (Cranford et al., 1999; Armsworthy et al., 2005). This 'in situ biodeposition' method was adapted to provide additional information on contaminant exposure. Supporting data on sediment metal concentrations and benthic communities were also collected at the test site to allow a comparison of EEM methodologies.

MATERIALS AND METHODS

Study Sites and Benthic Sampling

The Hibernia gravity base structure (GBS) is located 315 km east southeast of St. John's, Newfoundland, on the Grand Banks in 80 m of water at 46°45.03′ N, 48° 31.15′ W. A 15-year, 83-well drilling program was initiated by Hibernia Management and Development Company Limited (HMDC) in 1997 and first oil was achieved in the same year. The GBS is capable of drilling two wells at a time with some highly deviated wells expected to exceed 7,000 m horizontal distance. Drilling has been conducted with a combination of water-, silica-, and synthetic-base muds (SBM). Before March 2001, SBM cuttings meeting the regulatory guidelines for oil retention (48 hour rolling average of 15g/100g) were discharged to the marine environment at 60 m depth. After March 2001, SBM cuttings were injected into subsurface formations. Cuttings discharge data for the study period (below) were provided by HMDC.

Sampling stations were established along a northwest transect at 500, 1000, 1500, 2500, 4000, and 6000 m distance to the drilling platform (Fig. 1). This transect was selected based on observations of a relatively uniform sandy seabed in this sector (HMDC, personal communication). Seabed samples were collected on June 16, 2000 at the six sampling stations using a large grab sampler (VideoGrab) deployed from CCGS *Hudson*. Three replicate 0.5 m^2 grab samples were collected at each station. Sediment was subsampled (upper 1 cm) for metals analysis, and immediately screened through a 1 mm mesh. Benthic organisms were fixed in formalin for macrofauna analysis. Where possible, all organisms were identified to the species level.

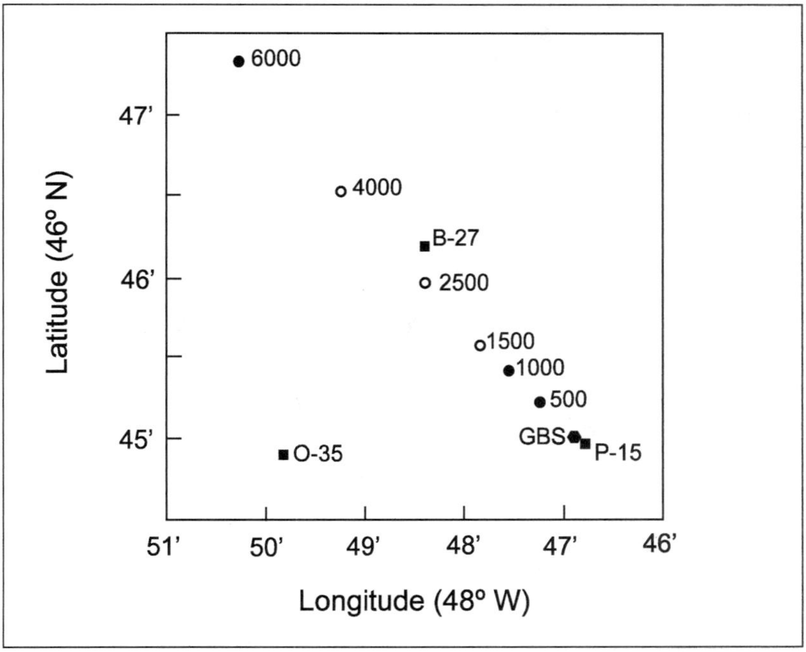

FIGURE 1. Location of bivalve cages and seabed sampling sites (circles) relative to the Hibernia gravity base structure (GBS). Site designations refer to the distance (m) to the GBS. Closed circles indicate sites where sediment traps were deployed along with the bivalve cage. The locations of three abandoned oil wells (P-15, O-35, and B-27) are shown (squares).

Bivalve Cages

Six cages were constructed to hold three bivalve species near the seabed (Fig. 2). *P. magellanicus* were selected for this study because of their known sensitivity to drilling wastes (above), and were collected by commercial dragger from Georges Bank off Nova Scotia. The Icelandic scallop (*Chlamys islandica*) resides in commercial quantities on the Grand Banks, and was collected by commercial dragger from Banquereau Bank, also off Nova Scotia. Blue mussels (*Mytilus edulis*) have been

FIGURE 2. Biological effects monitoring tools tested at Hibernia in 2000. Bivalve cages (top) were deployed on the seabed using an acoustic release (top left photo) and are recovered after the subsurface floats located on the top of the cage are released by a timed mechanism. *Placopecten magellanicus, Chlamys islandica,* and *Mytilus edulis* were held separately in 'pocket' nets (top right) behind a predator mesh on each wing of the cage (top left). Time-series sediment traps (bottom left) were deployed near the seabed with and without adult *C. islandica* (bottom right) to monitor scallop biodeposition and sedimentation rates over consecutive 3-day intervals for a total of 72 days (photos by the author).

widely employed for monitoring contaminant concentrations (e.g. Mussel Watch) and were used in this study to compare interspecific differences in biological effects. The mussels were obtained from a coastal aquaculture site in Ship Harbour, NS.

After holding animals in flow-through tanks for one month to remove animals that expire from collection injuries, they were graded and only a relatively narrow size range of each species was used for the study. *P. magellanicus* and *C. islandica* were graded to between 97 to 102 mm, and 85 to 97 mm shell height, respectively, and *M. edulus* were between 52 and 60 mm shell length. Animals were placed in individual pockets of a net held inside a 1 m x 1 m PVC frame (Fig. 2). These 'bivalve frames' were placed in an insulated fish box with flowing sea-water and transported to Hibernia on CCGS *Hudson*.

The aluminum bivalve cages were designed to be rapidly assembled prior to deployment. Each cage holds three bivalve frames (one containing each species) (Fig. 2). After the bivalve frames and a semi-permeable membrane device (SPMD) were installed in the cage, they were covered by a polyethylene predator mesh (1 cm x 1cm). To avoid loss of cages from the cutting of mooring lines by ship traffic, a subsurface float mechanism was developed that brought a recovery line to the surface at the designated recovery time. This was accomplished using an inexpensive 'burn-wire' programmable release mechanism that was developed for this study. The cages were placed on the seabed at the six sites (Fig. 1) on June 17, 2000, using an acoustic release mechanism. Cages were recovered on September 9, 2000 from CCGS *Teleost*, except for the cage located 4000 m from the GBS. This cage was recovered two weeks early by an industry vessel and is not included in the following analysis.

Initial samples of each species were collected from the holding tank onboard ship on June 15, 2000 by sampling extra bivalve frames that were not placed in the cages. Initial and final samples each consisted of 125 mussels and 35 scallops (each species) for biological analysis and 20 mussels and 5 scallops for chemical analysis. Animals collected for biological analysis were frozen whole for transport while tissues for chemical analysis were immediately dissected prior to freezing and storage in cleaned, certified glass jars with Teflon lined caps (Wheaton). Biological analysis consisted of measurement of shell length (mussels) or height (scallops) to the nearest 0.1 mm with callipers, and measurement of tissue dry weight (80°C). For scallops, tissues were separated into gonad, digestive gland, adductor muscle, and remaining tissues (remaining tissues consisted primarily of the mantle and gills).

Multivariate one-way analysis of variance (MANOVA) was used to test the null hypothesis that mean shell morphology and body component dry weight were equal for initial (June 17, 2000) and final (September 9, 2000) samples, and between the five bivalve cages at the end of the study. For the two scallop species, the five dependent variables were shell height, gonad, adductor, digestive gland, and remaining tissue dry weight. The mussel analysis was limited to shell length and total tissue dry weight. The 6000 m site was used as a reference station for assessing temporal and spatial effects. Prior to analysis, univariate normality and homoscadasticity were confirmed by examining normal probability and residual plots, respectively (Wilkinson et al., 1996). Significant MANOVA results were followed up by multiple comparison tests to discern which bivalve cage sites and dependent variables account for the difference. Tests were conducted using SYSTAT Version 6.1 (SYSTAT, Inc., Evanston, Illinois).

Sedimentation and Biodeposition

Two time-series sediment traps were deployed at each of the 500, 1000, and 6000 m stations (Fig. 1) on June 17, 2000 for measuring the vertical flux of particles, including drilling wastes, and for monitoring biodeposition rate (faeces production) by Icelandic scallops. The 1.5 m high traps were constructed for this study (Fig. 2) and have a 0.5 m^2 aperture area covered by a honeycomb baffle (cell width = 2.5 cm; cell depth = 15.8 cm) that prevents ambient currents from flushing trapped particles out of the funnel. A 2 cm mesh polyethylene cage was attached to the baffle on top of each trap (Fig. 2). One trap at each station contained a cohort of 15 Icelandic scallops

(Fig. 2), while the other trap did not contain animals and served as a control for measuring sedimentation rate. Each trap held 24, 250 ml sample cups containing a dense preservative (0.1% mercuric chloride in 35 ppt NaCl). All sediment traps were programmed to collect material in sample cups over 24 consecutive 72h sampling periods starting at 0000 h on June 20, 2000. The traps were deployed between an anchor (train wheel) and subsurface floats (15 m above trap) such that the top of the trap was located 3 m above the seabed. The traps were retrieved on September 9, 2000 from CCGS *Teleost* after triggering an acoustic release mechanism under the trap. The samples were immediately removed from the traps and the scallops were stored frozen prior to biological analysis (same methods as for caged scallops). Sediment trap samples were homogenised and subsampled for analysis of total dry weight (particles collected on prewashed and tared 25 mm GF/C filters and dried at 60°C), and organic content (weight loss upon ignition at 450°C for 4 h). Additional subsamples from control traps were analysed for metals and hydrocarbons (below).

Average daily egestion rates (*E*; mg dry weight/day) by Icelandic scallops over each three day sampling period was determined as:

$$E \text{ (mg DW/g/d)} = (DW_s - DW_c) \text{ / scallop tissue dry weight / 3 days,}$$

where DW_s is the total dry weight of particulate matter collected in sample cups under the scallop trap, and DW_c is the total dry weight of particles in corresponding cups under the control trap (Cranford and Hargrave, 1994).

Metals Analysis

Concentrations of Ba, Ca, Cd, Cr, Cu, Fe, Mg, Mn, Pb, Sr and Zn in dried surface sediments and control sediment trap subsamples were determined by ICP-ES (Inductively Coupled Plasma Emission Spectrometry). Commonly used methods for extraction of Ba from sediment containing high concentration of barite have been shown to be inefficient (Hartley, 1996). Therefore, a new two-step extraction methodology was developed to improve recovery for barium. Samples of surface sediment collected by VideoGrab were dried and sieved

(1 mm), and 0.5 g was initially digested using 2 ml of concentrated nitric acid. Each sample was evaporated to near dryness and 1 ml of 1:1 $HNO_3/HClO_4$ was added. This solution was heated to fumes of perchloric acid and almost to dryness. After cooling, 5 drops of nitric acid plus 1 ml of hydrochloric acid was added and the solution was warmed for a few minutes. Water (~40 ml) was added and the sample was heated to near boiling for 30 minutes. The solution was cooled and filtered through a 25 mm, 0.45 μm cellulose ester membrane into a 50 ml polypropylene centrifuge tube. The filtrate was then diluted to 50 ml for ICP-ES analysis.

A second extraction procedure was performed on material retained on the cellulose ester filter. The filter was treated with 1 ml of nitric acid, followed by 0.5 ml of the 1:1 $HNO_3/HCLO_4$ mixture. The sample was carefully evaporated to dryness and cooled. Sodium hydroxide (1 ml, 4%) and 3-5 ml of DTPA (2% in dilute NH_3OH) was added and the solution warmed for one hour at approximately 60°C. The solution was diluted to 50 ml in a polypropylene centrifuge tube for ICP-ES analysis. Metal concentrations obtained from both extraction steps were summed. For quality control purposes, two of the sediment samples were prepared and analysed in duplicate along with two analytical reagent blanks. Four reagent blanks along with the blank filters were prepared and analysed concurrently with the samples. In addition three samples each of pure synthesised barium sulfate (Sigma-Aldrich, Oakville, Ontario, Canada) and used drilling mud obtained from the Hibernia platform were prepared and analysed.

The concentration of metals in sediment trap samples was determined using the same methodologies as described above for seabed samples. Particulate material collected in sediment trap cups were homogenized, subsampled quantitatively by pipette, and filtered onto tared 45 mm, 0.45 μm cellulose ester membrane filters. The filters were dried, reweighed to determine subsample mass, and metal concentrations were determined by ICP-ES using the two-step extraction procedure described above.

Hydrocarbon Analysis

The sediment trap samples were decanted to remove the water layer. The remaining sediments were dried, sieved (1 mm mesh sieve) and stored in cleaned glass vials prior to extraction. Dried sediments were mixed with anhydrous sodium sulphate (VWR-Canlab, Mont-Royal, Québec, Canada, 10-60 mesh) and copper (J.T. Baker, Phillipsburg, New Jersey, USA), placed in a double thickness, 33 mm x 94 mm cellulose extraction thimble (Whatman International Ltd., England), spiked with surrogate internal standards (predueterated PAHs: d_{10}-phenanthrene, d_{10}-pyrene, d_{12}-benzo[a]pyrene, d_{12}-benzo[b]fluoranthene, d_{14}-dibenz[a,h]anthracene; Cambridge Isotope Laboratories, Andover, Massachusetts, USA), and covered with clean glass wool (VWR-Canlab). The sediments were extracted with 300 ml glass distilled dichloromethane for 18 hr (modified version of EPA method 3540 C). After extraction, the extracts were concentrated on a Turbo Vap (Zymark) to 1 ml. The extracts were purified using a solid-phase extraction column (VWR-Canlab) packed with silica gel (activated at 200°C for 17 hr and deactivated 5% w/v with HPLC grade water). The column was washed with glass distilled hexane and the sample was applied as a 1 ml extract and PAHs were eluted with 10 ml hexane:dichloromethane (4:1 v/v). The SPMDs that were attached to the bivalve cages were treated similar to sediments with the exception that no anhydrous sodium sulfate was added prior to soxhlet extraction.

The bivalves from the cages were also analysed for hydrocarbon content. Homogenised tissue (20 g), 100 μl of surrogates (400 mg/l predueterared PAHs same as sediments), 45 ml hexane, 90 ml methanol, and 13 ml potassium hydroxide (12 N) were added to a flask. The mixture was refluxed (Liebig Condenser) for 4 hr, cooled to ambient temperature and filtered under vacuum (Whatman 42, 70 mm filters). The filtrate was added to a decantation funnel and 25 ml of hexane was added and mixed for 2 min. The extraction was repeated two times and the hexane extracts were added to a Turbo Vap concentrator tube. The extracts were concentrated to 1.0 ml and purified using a 5% deactivated silica gel column (the same procedure used for the sediment extracts).

Extracts of sediments were analysed using high resolution gas chromatography (Agilent 6890 GC) coupled to a mass selective detector (Agilent 5973N, Willmington, Deleware, USA) in the selective ion monitoring mode. The following gas chromatography (MDN-5S column 30 m x 0.25mm id 0.25 μm film thickness, Supelco Canada) conditions were used: cool on-column injection with oven track mode (track 3°C higher that the oven temperature program); 80°C hold 2 min; temperature ramp at 4°C/min to 280°C held for 10 min. PAHs were obtained from Ultra Scientific, Kingstown, Rhode Island USA; Absolute Standards from Hemden, Connecticut, USA. Generally, surrogate recoveries of added PAHs prior to soxhlet extraction averaged; d_{10}-phenanthrene 89±17%, d_{10}-pyrene 84±18%, d_{12}-benzo[b]fluoranthene 89±13%, d_{12}-benzo[a]pyrene 87±14%, d_{14}-dibenz[a,h]anthracene 86±12%. A duplicate certified reference material (NIST 1944 New York-New Jersey Waterway) and operational blank was performed with each batch of 10 samples.

Purified sample extracts of SPMD membranes and biological tissues were analysed with the gas chromatograph (GC) using a mass selective detector. The GC was equipped with a HP-5 MS capillary column (30 m long, 0.25 mm ID, and 0.25 μm film thickness) and a split/splitless injection port operating in the splitless mode. Operating conditions for the GC were as follows: constant flow, 1 ml/min; injection port, 290°C; transfer line, 320°C; initial GC oven temperature, 50°C held for 2 min; temperature ramp, 6 °C/min to 300°C, held for 21 min. The total run time was 64.7 min. The mass selective detector was operated in the selected ion monitoring mode for quantifying specific saturated hydrocarbons, polynuclear aromatic hydrocarbons (PAHs), and sulphur heterocyclic constituents.

The concentration of position-specific compounds in the chromatograms was determined by co-injection of reference standards. Reported aliphatic concentrations are summed concentrations of C_{10} to C_{35} *n*-alkanes. Total resolved aromatic concentrations include: naphthalene, acenaphthylene, acenaphthene, fluorene, phenanthrene, anthracene, fluoranthene, pyrene, benzo[*a*]anthracene, chrysene, benzo[*b*]fluoranthene, benzo[*k*]fluoranthene, benzo[*a*]pyrene, indeno[1,2,3-*cd*]pyrene, dibenzo[*a,h*]anthracene, and benzo[*g,h,i*]perylene.

RESULTS AND DISCUSSION

Seabed Metals Concentrations

Barium values for samples of pure barium sulfate averaged 55.1% with a range of 54.5% to 55.6%. This compared favourably with a theoretical concentration of 58.9%, demonstrating the efficiency of the two-step extraction procedure. Results for the drilling mud samples obtained from the Hibernia platform varied from 42.5% to 44.2% Ba. Reproducibility for all metals between replicate sediment samples and reagent blanks was judged very good. At relatively low sediment Ba levels (less than 200 μg/g), the mixed acid digestion (first extraction step) recovered greater than 90% of total Ba. Ba recovery decreased to less than 75% at concentrations greater than 700 μg/g (Fig. 3). The additional alkaline DTPA extraction is therefore

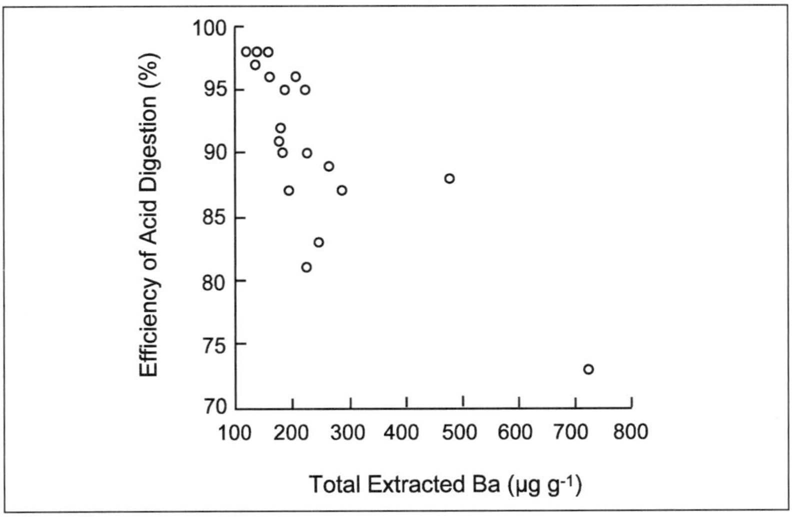

FIGURE 3. Efficiency of barium recovery from seabed samples using only mixed acid digestion (% of total extracted Ba). Total barium was determined using the two step method (mixed acid digestion and alkaline DTPA extraction) described in the text.

believed necessary for quantitative recovery of barite from solid samples.

The determination of other trace elements using mixed acid digestion provided excellent analytical sensitivity, and the minimum detection limits are given in Table 1. Sediment Cu and Pb concentrations were generally at or below the detection limit and are not included in the following analysis. Maximum concentrations of Ba, Cr, Fe, Mn, and Zn were detected at the closest station to the platform (500 m), but only Ba showed a strong negative correlation with distance from the platform (Table 2). Cr, Fe, Mn, Mg, and Zn concentrations were cross-correlated ($r > 0.50$) (Table 2). Maximum Ca, Mg, and Sr concentrations are found in sediments outside 500 m (Table 1). Grain size effects on metals concentration were separated by normalising the concentration to Al. The expected maximum background Ba:Al ratio, calculated from the 95% confidence interval for concentrations in grab samples from the 4000 and 6000 m sites was 0.027. Ba concentrations greater than 160 μg/g are considered to be enriched by drilling operations. All Ba levels measured out to 2500 m from the Hibernia GBS were above this maximum background level. Normalisation to Al indicates that spatial variations in Ca and Sr were within

TABLE 1. Minimum detection limits and location (distance from Hibernia platform) of maximum metal concentrations measured in sediments.

Metal	Minimum Detection Limit (µg/g)	Maximum Concentration (µg/g)	Distance (m)
Al	1	7980	2500
Ba	0.5	724	500
Ca	5	4210	2500
Cr	0.5	7.7	500
Cu	1	<1	-
Fe	5	3200	500
Mg	5	613	1500
Mn	1	81	500
Pb	2	5	2500
Sr	1	67	2500
Zn	1	5	500

TABLE 2. Spearman rank correlations for sediment metal concentrations collected along a transect to the northwest of the Hibernia platform (n = 18).

	Al	Ba	Ca	Cr	Fe	Mg	Mn	Sr	Zn
Ba	0.232								
Ca	0.952	0.238							
Cr	0.267	0.469	0.365						
Fe	0.037	0.202	0.208	0.794					
Mg	0.503	0.367	0.536	0.775	0.633				
Mn	-0.113	0.046	0.055	0.694	0.932	0.486			
Sr	0.887	0.445	0.905	0.439	0.247	0.544	0.141		
Zn	0.341	0.507	0.424	0.452	0.619	0.522	0.413	0.481	
Distance	0.266	-0.720	0.237	-0.472	-0.268	-0.278	-0.217	-0.028	-0.177

background variation, while above background levels of Cr, Fe, Mg, and Mn can be detected out to 1500 m.

Benthic Community Analysis

The most ubiquitous species in both abundance and biomass at Hibernia was the propeller clam, *Cyrtodaria siliqua*, but other principal species included the polychaete, *Chaetozone setosa*, and the sand dollar, *Echinarchnius parma*. These same species were abundant at a site 60 km from Hibernia (Kenchington et al., 2001) at 120-146 m water depth. Both studies used the same VideoGrab sampling equipment and taxonomic expertise.

The macrofauna community was generally homogeneous at each of the six sampling sites, except for two individual grab samples (2500 and 4000 m sites) that displayed a relatively high species richness and biomass (Fig. 4). The number of species per sample ranged from 15 to 75 and averaged 35. Species richness was higher at the site sampled by Kenchington et al. (2001), where an average of 68 species per grab was observed. Olsgard and Gray (1995) showed that total number of species was one of the more sensitive community measures for detecting disturbance in benthic communities from offshore oil and gas operations. Species richness was relatively low at the site closest to the Hibernia platform where Ba concentrations were on average three times higher than

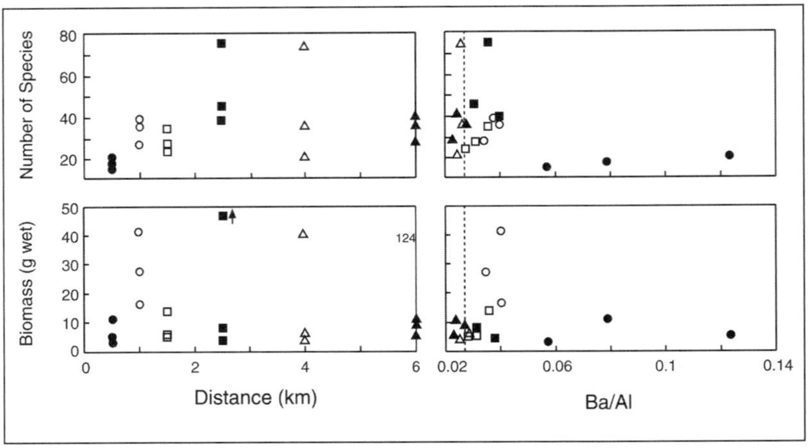

FIGURE 4. Relationships between the total number of benthic species and biomass against distance from the Hibernia platform (left), and against the sediment Ba/Al concentration ratio (right). The maximum background Ba/Al ratio is indicated by the broken line.

background (Fig. 4). One-way analysis of variance was used to test hypotheses that average species richness and biomass at the six sampling sites were equal. No significant difference among the six means was noted for species richness (P-value = 0.126) or biomass (P-value = 0.567) despite the elevated Ba levels (indicated by Ba/Al > 0.027) detected in sediment out to approximately 2500 m from the Hibernia GBS (Fig. 4). It should be noted that the small sample size employed at each site (n = 3) would result in these tests having limited power to detect a significant spatial differences in community parameters.

Caged Bivalves

Bivalve mortalities in cages over the 83-day study period were low, with an average survival of *P. magellanicus, C. islandica,* and *M. edulis* of 87.8, 98.4, and 96.5%, respectively. The results of statistical comparisons (MANOVA) of shell length and tissue mass between initial (June 15, 2000; cage C1) and final (September 9, 2000; reference cage C6) sampling dates are shown in Table 3. For both scallop species, significant differences were observed between the initial and final dates (P-value < 0.001), with the difference in tissue mass attributed to reductions in gonad, digestive gland, and mantle tissue (*P*-value ≤ 0.015). *P. magellanicus* lost an average of 18% of their tissue mass (1.16g) during the study, with 43% of the weight loss resulting from reduction in gonad. *C. islandica* tissue was reduced an average of 17% (1.31 g) with 31% attributed to gonad.

Comparisons between all cages sampled at the end of the experiment showed an overall significant site difference for sea scallops (*P*-value = 0.049) that was related to variations in gonad mass (*P*-value = 0.031). However, *post-hoc* comparison tests did not detect a significant difference between animals from the 6000 m reference cage and from cages located closer to the Hibernia GBS (*P*-value ≥ 0.185). There was no significant difference in Icelandic scallop measurements between cage sites (*P*-value = 0.112), so no *post-hoc* comparisons were performed.

MANOVA results for caged mussels are also summarised in Table 3. Comparison of initial and final samples showed significant differences in tissue dry weight (*P*-value < 0.001), but not in shell length. Tissue mass decreased on average by 32% or 0.22 g during the study. The comparison between all cage sites at the end of the study showed significant differences in both dependant variables (*P*-value ≤ 0.023). However, comparisons between the reference cage (6000 m) and sites closer to the GBS showed no significant differences (*P*-value > 0.05). Sea scallops typically undergo gametogenesis during the summer, with the majority of gonad growth taking place between June and September. This species has been observed to complete gametogenesis at –1°C (MacDonald and Thompson, 1986), indicating that low temperature is unlikely to be the primary factor for the observed gonad resorption at Hibernia. A lack of available food appears to account for reduced gonad production in populations residing in deep-water (Barber et al., 1988). The digestive gland and adductor muscle serve as major nutrient storage organs in scallops (Barber and Blake, 1991) and changes in their relative size provides an indication of the nutritive state of the animals. On average, 40% of initial digestive gland mass was catabolised during the study period

Table 3. Summary table of results from one-way multivariate analysis of variance (MANOVA) on shell height (scallops) and length (mussels), and tissue dry weight (DW) from the caged bivalve study at Hibernia. The cage factor has six levels; one initial (C1) and five final cages (C2 = 500 m, C3 = 1000 m, C4 = 1500 m, C5 = 2500 m, and C6 = 6000 m reference site). The indicated test hypotheses (final = cages C2 to C6) were tested by multivariate *F*-tests based on values computed for the Pillai Trace statistic. *Post-hoc* multiple comparison tests between selected cages and univariate F-tests for each dependent variable of the model were performed only if a significant MANOVA result was detected. **P*-value ≤ 0.05.

	Multivariate Test Statistics				**Univariate *F*-test *P*-values**				
Species and Hypothesis	Pillai Trace	*F*-statistic	*df*	*P*-value	Shell height	Gonad DW	Adductor DW	Digestive DW	Remaining DW
Sea Scallops (*P. magellanicus*)									
C1=C6	0.811	51.619	5, 69	0.000*	0.201	0.000*	0.224	0.000*	0.000*
Final Equal	0.161	1.538	20, 732	0.049*	0.640	0.031*	0.401	0.192	0.120
C2=C6	0.094	1.430	5, 69	0.224	-	-	-	-	-
C3=C6	0.101	1.554	5, 69	0.185	-	-	-	-	-
C4=C6	0.059	0.871	5, 69	0.505	-	-	-	-	-
C5=C6	0.087	1.252	5, 69	0.295	-	-	-	-	-
Icelandic Scallops (*C. islandica*)									
C1=C6	0.806	53.990	5, 65	0.000*	0.392	0.015*	0.157	0.000*	0.000*
Final Equal	0.132	1.404	20, 820	0.112	-	-	-	-	-
Blue Mussels (*M. edulis*)					Shell Length	Tissue DW			
C1=C6	0.439	66.95	2, 171	0.000*	0.969	0.000*			
Final Equal	0.030	2.399	8, 1244	0.014*	0.023*	0.022*			
C2=C6	0.002	0.291	2, 245	0.747	-	-			
C3=C6	0.019	2.392	2, 247	0.095	-	-			
C4=C6	0.023	2.958	2, 248	0.054	-	-			
C5=C6	0.010	1.245	2, 247	0.290	-	-			

while adductor muscle increased by 5%. The scallops may have utilised lipid stored in digestive gland for maintenance metabolism while attempting to increase muscle carbohydrate stores needed to initiate and support gametogeneses.

Icelandic scallop populations are found in waters around Hibernia and should be more suited to the local environmental conditions than sea scallops. However, similarly poor growth trends were observed for both species during the study period. Icelandic scallop gonad and digestive gland tissue dry weight decreased by an average of 26% and 30%, respectively, during the study while adductor muscle increased by 1%. Compared with *P. magellanicus*, relatively little published information is available on the reproductive cycle of *C. islandica*. Spawning was observed to take place in late July to mid August in populations residing in coastal waters in eastern Canada (Arsenault and Himmelman, 1998). The observed decrease in mean gonad mass at Hibernia may therefore be attributed to spawning. However, the relatively small size of gonad at the start of the experiment on June 15, 2000 suggests that spawning occurred prior to cage deployment. Environmental and/or anthropogenic factors appear to have caused the resorption of gamete and digestive gland by the two scallop species.

The positioning of scallops in cages up to 1 m above the seabed, where food supplies can differ substantially from the sediment-water interface, may have been a factor contributing to the observed weight loss. However, this does not preclude the use of bivalve cages for detecting sublethal effects in EEM programs, assuming that the experimental design includes a control site and that mortalities are not excessive. Time-series sampling of wild bivalve populations is preferable for EEM over importing animals in cages, but would require a homogeneous wild population (similar age structure) without natural spatial variations in body condition. The use of a single cohort that has been graded to a narrow size range and placed in cages eliminates natural between-site variability and increases the sensitivity of the method for detecting anthropogenic impacts.

Data collected on contaminant exposure and tissue dose, determined from the hydrocarbon content of SPMDs and bivalve tissue, were used to address the possibility that drilling and production wastes may have contributed to the lack of bivalve growth at all sites, including the reference site. Total resolved *n*-alkane (C_{10} to C_{35}) concentrations in scallop tissue and in SPMDs are shown in Figure 5. There was a general increase in total alkane concentrations adsorbed to SPMDs with distance away from the Hibernia GBS (Fig. 5). Low molecular weight *n*-alkanes ($< C_{23}$) dominated and C_{21} and

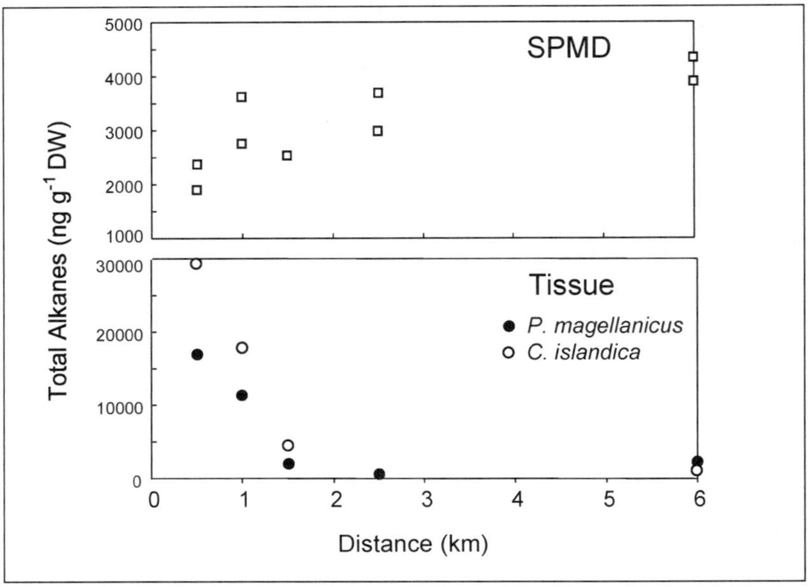

FIGURE 5. Total resolved n-alkane (C10 to C35) concentrations in semi-permeable membrane devices (SPMDs) and in scallop tissue against distance from the Hibernia platform.

C_{11} accounted for an average of 41% and 23% of total alkanes, respectively. The odd:even ratio of the low molecular weight *n*-alkanes averaged 3.8 ± 1.6 suggesting that the origin of the hydrocarbons is from a biogenic marine source (phytoplankton; Zanardi et al., 1999). The pristine to phytane ratios were less than one (averaged 0.84 ± 0.14), which may indicate some oil contamination beside the biogenic inputs. Pristane is present in zooplankton and other higher marine animals while phytane is present in oil and can occur during diagenetic processes (Steinhauer and Boehm, 1992; Wakeham and Carpenter, 1976; Gilbert et al., 1992; Al-Saas and Al-Timari, 1993). Very low levels of PAHs were detected in the SPMD samples and those detected were more water soluble PAHs (<4 log K_{ow}) of low molecular weight.

Scallop tissue *n*-alkane levels decreased with distance away from the GBS for *C. islandica* and *P. magellanicus* (Fig. 5). The primary accumulation site for hydrocarbons was the digestive gland (> 80% of total body burden). Relatively high molecular weight *n*-alkanes (> C_{23}) were predominant in scallops taken from cage sites out to 1.5 km, but were generally not present at more distant sites. No spatial trends were visible in the PAHs and alkylated PAHs, but this is reflective of the low concentrations at the various distances. The major PAHs found in sea and Icelandic scallops tissues were comprised of mostly alkylated 2-3 ring, low molecular weight, and more water-soluble compounds (< 4 logK_{ow}). The same PAHs were found in the SPMD samples. These types of PAHs are preferentially taken up by marine animals through respiration, while more particle bound, high molecular weight PAHs (logK_{ow} >6) would be taken up by filtration. It is clear from our data that the scallops are taking up and accumulating low levels (300–800 ng/g dry weight) of low molecular weight (2-3 ring) PAHs. In some cases, high molecular weight PAHs (> 4 rings) were present in the scallops at low concentrations.

Hydrocarbon-growth relationships indicate growth reductions in *M. edulis* will occur when dose exceeds 0.1–1 μg/g dry tissue weight (Widdows and Donkin, 1989; Widdows et al., 1995). Total hydrocarbon levels in *M. edulis* tissue ranged from 10.8 μg/g at the 500 m site to 4.2 μg/g at the reference site, and body burden was up to three times higher in the two scallop species. As result, hydrocarbon contaminants may have contributed to the observed poor growth rates at Hibernia. The potential effects of contaminant exposure are discussed further below.

Sedimentation and Biodeposition

Drilling waste discharges reported by HMDC during the study were limited to a single period between July 1 and 10 (Fig. 6a) when a total of 455 tonnes of cuttings containing 80 tonnes of synthetic base-oil were discharged. The average 48 hr mean of base-oil on cuttings during this time was 18.6g/100g. Vertical particle flux (mg dry weight/m^2/d) (Fig. 6b) and barium flux (mg Ba/m^2/d) (Fig. 6d), measured using the control traps (no scallops) was relatively high at the 500 and 1000 m sites, compared to the 6000 m site. The organic content

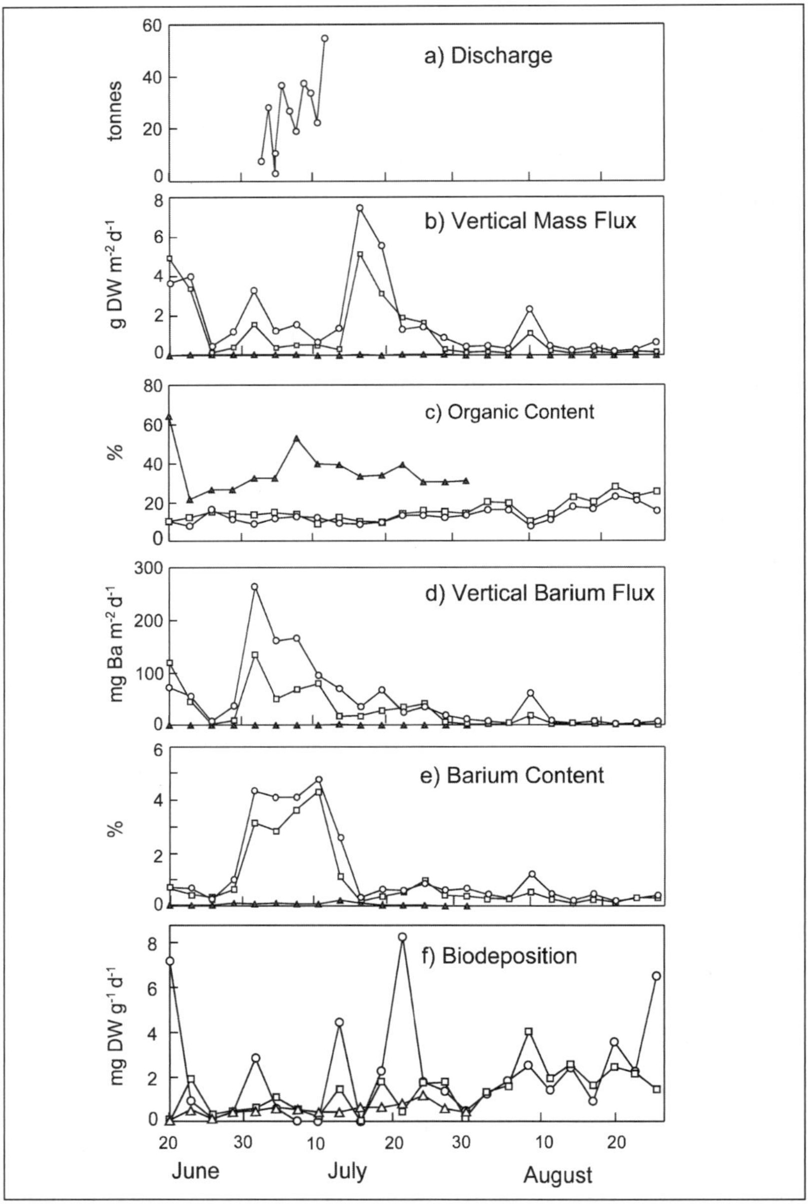

FIGURE 6. Drilling waste discharges from the Hibernia platform (a) and sediment trap measurements at sites located 500 (circles), 1000 (squares), and 6000 m (triangles) from the GBS during 2000. The control traps (no scallops) provided information on sedimentation rates and the composition of settled material (b-e). Traps containing Icelandic scallops (*Chlamys islandica*) were used to measure biodeposition rates (f) (see text for details). The control trap located at 6000 m stopped working on August 1, 2000.

of material collected in traps was lowest at the two sites closest to the GBS (Fig. 6c). Barium flux and content (Ba flux/total flux) increased during the early July waste discharge period at all three sites, with the lowest fluxes observed at the 6000 m site (Fig. 6d and e). Elevated sedimentation rates were observed at the 500 and 1000 m sites during three additional periods (June 20–25, July 20–25, and August 10–12; Fig. 6b). The highest sedimentation rates observed during this study occurred between July 18 and 25, but the barium content of the settled material was relatively low. However, maximum fluxes of Cr, Fe, Mn, Mg, and Zn were observed during this period (data not shown). The lack of an apparent relationship between near-seabed current speed measurements collected during this study (K. Muschenheim, unpublished data) and the sedimentation rate data indicates that the observed sedimentation rate maxima in July did not result from the resuspension of bottom deposits. The more likely explanation is a discharge of low density (low barite) drilling mud.

Hydrocarbon analysis was conducted on several control sediment trap samples, including periods of low (August 28–30) and high particle flux (June 19–21 and July 20–23). The highest hydrocarbon sedimentation rate measured was observed at the 500 m site during the period between July 20–23 (472 μg total hydrocarbons/m^2/d), when barite content was relatively low. Hydrocarbon sedimentation rates at the 1000 m and 6000 m sites during the same period were 145.6 and 0.3 μg/m^2/d, respectively. The high vertical flux of hydrocarbons adjacent to the Hibernia GBS during this period indicates that the material is primarily discharged drilling waste. As there was no clear dominance of odd:even alkane carbon number in the sediment trap sample (odd:even *n*-alkane ratio = 1.14 ± 0.12) and the pristane:phytane ratios were less than one, we conclude that the predominate source of alkanes in the sediment trap samples are from petroleum hydrocarbons (Steinhauer and Boehm, 1992). The aromatic components in the sediments were clearly dominated by alkylated PAHs with 2-3 ring structures as the dominant compounds. Total PAH concentrations were low in sediment trap samples and ranged from 184 to 621 ng/g dry wt.

Time-series of Icelandic scallop biodeposition into sediment traps are shown in Figure 4f. Scallop biodeposition rates were generally similar between all three sites except for several periods of relatively high and low (zero) biodeposition at the 500 m site. Variations in biodeposition rate were not correlated ($r < 0.2$) with drilling waste discharges (Fig. 6a), sedimentation rate (Fig. 6b and d), or the barium content of settled material (Fig. 6e). Suspended drilling waste concentrations as low as 0.5 mg/l have been reported to cause significant reductions in sea scallop clearance rates (Cranford et al., 1999; Armsworthy et al., 2004), a measure that is directly related to biodeposition rate. Assuming that Icelandic scallops respond in a similar manner to drilling waste exposure as sea scallops, it is expected that biodeposition rate and barium levels in settled material would be negatively correlated. However, this impact would only occur if waste concentrations were above the 0.5 mg/l no-effect threshold. Suspended particulate matter (SPM) concentrations measured between June 15 and 22, 2000 from within a 6000 m radius from Hibernia were 0.49 ± 0.15 mg/l (mean ± 2SE) for water collected 10 m above bottom, and 0.89 ± 0.38 mg/l at 0.5 m above bottom (Law et al., 2002). These SPM values are similar to the background levels of 0.91 ± 0.26 mg/l measured at the same sites (10 m above bottom) in July 1995 prior to commencement of production drilling (Milligan et al., 1997). While the presence of drilling waste in the water column can be detected and monitored using the sediment trap approach, the ambient concentration appears to have been lower than the threshold known to affect scallop feeding.

CONCLUSIONS

Offshore oil and gas EEM programs have generally relied on measuring relatively long-term biological responses to contaminant exposure (e.g. annual sampling). This provides only a limited capacity for identifying operational activities primarily responsible for any observed impact (e.g. drilling waste vs. produced water discharges; toxicity vs. physical effects; primary vs. secondary effects of waste remobilization). The bivalve technologies described in this paper, when combined with more traditional EEM approaches, provides information on potential environmental effects over time-scales ranging from days to years. Seabed sampling along a transect leading away from the Hibernia GBS showed that Ba levels were above the maximum background level out to 2500 m away from the Hibernia GBS. Benthic community analysis of these same seabed samples appears to indicate a localized impact adjacent to the GBS (inside 500 m radius; Fig. 4), but we were unable to detect a significant long-term impact due to the large spatial and temporal variability in benthic communities (this paper; Kenchington et al., 2001).

The bivalve cages provide information on near-seabed biological effects occurring over a time scale of a few months. The growth of three bivalve species at Hibernia was monitored between June 17 and September 9, 2000. Although all species lost tissue weight during the study period, there was no indication of growth conditions varying significantly between cage sites (distance from GBS). Reported waste discharges during the study period were limited to just nine days (Fig. 6a), so a major effect on growth was not anticipated. Analysis of SPMDs and caged bivalve tissues show the presence of hydrocarbon contamination near the seabed (Fig. 5). Hydrocarbons retained by SPMDs appeared to have a biogenic source, while those found in bivalve tissues appeared to be anthropogenic and were generally more concentrated in bivalves (particularly in scallops) caged within 1000 m to the GBS.

The major advantage for EEM programs afforded by the use of the HABITRAP Biological Effects Monitoring System is the ability to continuously monitor a sensitive bioindicator while also monitoring contaminant exposure and dose. This permits observation of biological responses to individual discharge events, such as the discharge of cuttings containing synthetic oil-based muds at Hibernia in early July 2000 (Fig. 6a). Analysis of the barium content of trapped material served as a tracer for the exposure of the test organisms (Icelandic scallops) to drilling waste. Exposure to wastes having a low barium content was detected by hydrocarbon analysis of trapped sediment. The similarity of daily biodeposition measurements for scallop cohorts held at distances out to 6 km away from the GBS (Fig. 6f) indicated that the exposure to drilling wastes was below the no-effects threshold previously determined in laboratory experiments (0.5 mg/l; Armsworthy et al., 2004).

The bivalve cages, HABITRAPs, and supporting new methodologies and technologies (subsurface float release and sediment barium extraction method) were shown effective under the demanding environmental conditions present on the Grand Banks of Newfoundland. Incorporation of bivalves as sentinel organisms into offshore oil and gas EEM programs has many advantages (see Introduction) but has at present been limited to the collection of some wild species, and/or the caging of mussels for hydrocarbon contaminant analysis. The described methodologies would be a valuable addition to EEM programs owing to simplicity, low cost, and high sensitivity for detecting and identifying the operational cause of lethal and sublethal impacts.

ACKNOWLEDGEMENTS

We thank the officers and crews of the CCGS *Hudson* and CCGS *Teleost* for their invaluable assistance in carrying out the field component of this project. Glen Morton, Dan Moffitt, and Dwight Reimer provided engineering support in the development, construction and testing of the HABITRAPs and bivalve cages. The two-step sediment barium extraction method was developed by Ross Keen under a contract to the Research and Productivity Council (RPC), Fredericton, NB. Technical support was provided by Victoria Burdett-Coutts, Kirsten Querbach, Murray Scotney, Bob Ryan, Peter Vass, and Kevin MacIssac. Sylvie St.-Pierre conducted the hydrocarbon analysis at the Institute Maurice Lamontagne, Mont-Joli, PQ. Live scallops were collected with the assistance of the captain and crews of scallop draggers operated by Knickle and Sons and Clearwater Fine Foods. Rick McCubben of the Hibernia Management and Development Corporation provided logistical support for this project and provided data on drilling waste discharges. Funding for this project was provided by the Interdepartmental Panel on Energy Research and Development and the Department of Fisheries and Oceans.

REFERENCES

Al-Saas, H.T. and A. Ak Al-Timari. 1993. "Sources of Hydrocarbon and Fatty Acids in Sediments from Hor Al-Hammar March, Shatt Al-Arab, and Northwest Arabian Gulf." *Marine Pollution Bulletin.* 26:559-564.

Armsworthy, S.L., P.J. Cranford, T. King, and K. Lee. 2005. "Chronic Effects of Synthetic Drilling Mud on Sea Scallops (*Placopecten magellanicus*)." In: S.L. Armsworthy, P.J. Cranford, and K. Lee (Eds.), *Proceedings of the Offshore Oil and Gas Environmental Effects Monitoring Workshop: Approaches and Technologies, Bedford Institute of Oceanography, May 26–30, 2003.* Battelle Press (in prep).

Arsenault, D.J. and J.H. Himmelman. 1998. "Spawning of the Iceland Scallop (*Chlamys islandica* Müller, 1776) in the Northern Gulf of St. Lawrence and its Relationship to Temperature and Phytoplankton Abundance." *The Veliger.* 41:180-185.

Barber, B.J., R. Getchell, S. Shumway, and D. Schick. 1988. "Reduced Fecundity in a Deep-water Population of the Giant Scallop *Placopecten magellanicus* in the Gulf of Maine, USA." *Marine Ecology Progress Series.* 42:207-212.

Barber, B.J. and N.J. Blake. 1991. "Reproductive Physiology." In S.E. Shumway (Ed.), *Scallops: Biology, Ecology and Aquaculture,* pp. 377-428. Elsevier, New York.

Barlow, M.J. and P.F. Kingston. 2001. "Observations on the Effect of Barite on Gill Tissues on the Suspension Feeder *Cerastoderma edule* (Linné) and the Deposit Feeder *Macoma balthica* (Linné)." *Marine Pollution Bulletin.* 42:71-76.

Capuzzo, J.M. 1988. "Physiological Effects of a Pollution Gradient—Introduction." *Marine Ecology Progress Series.* 46:111.

Cranford, P.J. 1995. "Relationships between Food Quantity and Quality and Absorption Efficiency in Sea Scallops *Placopecten magellanicus* (Gmelin)." *Journal of Experimental Marine Biology and Ecology.* 189:123-42.

Cranford, P.J. 2003. "Scallops and marine contaminants." In: S. Shumway, and J. Parsons (Eds.), *Scallops: Biology, Ecology and Aquaculture.* In press.

Cranford, P.J. and D.C. Gordon. 1992. "The Influence of Dilute Clay Suspensions on Sea Scallop (*Placopecten magellanicus*) Feeding Activity and Tissue Growth." *Netherlands Journal of Sea Research.* 30:107-120.

Cranford, P.J. and B.T. Hargrave. 1994. "In Situ Time-Series Measurements of Ingestion and Absorption Rates of Suspension-Feeding Bivalves: *Placopecten magellanicus* (Gmelin)." *Limnology and Oceanography*. 39:730-738.

Cranford, P.J., D.C. Gordon Jr., K. Lee, S.L. Armsworthy, and G.-H. Tremblay. 1999. "Chronic Toxicity and Physical Disturbance Effects of Water- and Oil-Based Drilling Fluids and Some Major Constituents on Adult Sea Scallops (*Placopecten magellanicus*)." *Marine Environmental Research.* 48:225-256.

Cranford, P.J., D.C. Gordon Jr., C.G. Hannah, J.W. Loder, T.G. Milligan, D.K. Muschenheim, and Y. Shen. 2003. "Modelling Potential Effects of Petroleum Exploration Drilling on Northeastern Georges Bank Scallop Stocks." *Ecological. Modelling.* 166:19-39.

Daan, R., W.E. Lewis, and M. Mulder. 1990. *Biological Effects of Discharged Oil-Contaminated Drill Cuttings in the North Sea.* Netherlands Institute of Sea Research Report. 1990-5.

Gilbert, M., L. Rivet, A.I. Jawad, and J.C. Bertrand. 1992. "Hydrocarbons Distributions in Low Polluted Surface Sediments from Kuwait, Bahrain and Oman Coastal Zones (before the Gulf War)." *Marine Pollution Bulletin.* 24:622-626.

Gray, J.S., K.R. Clark, R.M. Warwick, and G. Hobbs. 1990. "Detection of Initial Effects of Pollution on Marine Benthos: an Example from the Ekofisk and Eldfisk Oilfields, North Sea." *Marine Ecology Progress Series.* 66:285-299.

Hartley, J.P. 1996. "Environmental Monitoring of Offshore Oil and Gas Drilling Discharges—a Caution on the Use of Barium as a Tracer." *Marine Pollution Bulletin.* 32:727-733.

Kenchington, E.L.R., J. Prena, K.D. Gilkinson, D.C. Gordon, Jr., K. MacIsaac, C. Bourbonnais, P.J. Schwinghaner, T.W. Rowell, D.L. McKeown, and W.P. Vass. 2001. "Effects of Experimental Otter Trawling on the Macrofauna of a Sandy Bottom Ecosystem on the Grand

Banks of Newfoundland." *Canadian Journal of Fisheries and Aquatic Science.* 58:1043-1057.

Law, B.A., G. Wohlgeschaffen, S.E. Cobanli, K. Lee, T.G. Milligan, A.R.J. Stewart, and D.K. Muschenheim. 2002. *Observations of Drilling Wastes at the Hibernia Oil Field Site, 1998–2000.* Canadian Data Report of Fisheries and Aquatic Sciences. 1088:vi+237p.

MacDonald, B.A. and R.J. Thompson. 1986. "Influence of Temperature and Food Availability on the Ecological Energetics of the Giant Scallop *Placopecten magellanicus*. III. Physiological Ecology, the Gametogenic Cycle and Scope for Growth. *Marine Biology.* 93:37-48.

Milligan, T.G., D.K. Muschenheim, K.S. Saunders, and A. Prior. 1997. *Suspended Particulate Material Observations at the Hibernia Oil Field Development Site, July 1995.* Canadian Data Report of Fisheries and Aquatic Science. 1009:ix+110 p.

Neff, J.M. 1987. "Biological Effects of Drilling Fluids, Drill Cuttings and Produced Water." In: D.F. Boesch and N.N. Rabalais (Eds.), *Long-Term Environmental Effects of Offshore Oil and Gas Development,* pp. 469-538. Elsevier Applied Science, London.

Olsgard, F. and J.S. Gray. 1995. "A Comprehensive Analysis of the Effects of Offshore Oil and Gas Exploration and Production on the Benthic Communities of the Norwegian Continental Shelf." *Marine Ecology Progress Series.* 122:277-306.

Reiersen, L.-O., J.S. Gray, K.H. Palmork, and R. Lange. 1989. "Monitoring in the Vicinity of Oil and Gas Platforms; Results from the Norwegian Sector of the North Sea and Recommended Methods for Forthcoming Surveillance. In F.R. Engelhardt, J.P. Ray, and A.H Gilliam (Eds.), *International Conference on Drilling Wastes,* pp. 91-117. Elsevier Applied Science, New York.

Steinhauer, M.S. and P.D. Boehm. 1992. "The Composition and Distribution of Saturated and Aromatic Hydrocarbons in Nearshore Sediments, River Sediments, and Coastal Peat of the Alaskan Beaufort Sea: Implications for Detecting Anthropogenic Hydrocarbon Inputs." *Marine Environmental Research.* 33:223-253.

Wakeham. S.G. and R. Carpenter. 1976. "Aliphatic Hydrocarbons in Sediments of Lake Washington." *Limnology and Oceanography.* 21:711-723.

Widdows, J. and P. Donkin. 1989. "The Application of Combined Tissue Residue and Physiological Measurements of Mussels (*Mytilus edulis*) for the Assessment of Environmental Pollution." *Hydrobiologia.* 188/189:455-461.

Widdows, J., P. Donkin, M.D. Brinsley, S.V. Evans, P.N. Salkeld, A. Franklin, R.J. Law, and M.J. Waldock. 1995. "Scope for Growth and Contaminant Levels in North Sea Mussels *Mytilus edulis,*" *Marine Ecology Progress Series.* 127:131-148.

Wilkinson, L., G. Blank, and C. Gruber. 1996. *Desktop Data Analysis with SYSTAT.* Prentice Hall, New Jersey.

Zanardi, E., M.C. Bícego, L.B. De Miranda, and R.R. Weber. 1999. "Distribution and Origin of Hydrocarbons in Water and Sediments in Sao Sebastiao, SP, Brazil." *Marine Pollution Bulletin.* 38:261-267.

Health Effect Indicators in American Plaice (*Hippoglossoides Platessoides*) from the Terra Nova Development Site, Grand Banks, NL, Canada

Mathieu, A. (amathieu@oceans.nf.net), W. Melvin, B. French, M. Dawe (Oceans Ltd, St. John's, NL, Canada), E.M. DeBlois (Jacques Whitford Environment Ltd, St. John's, NL, Canada), F. Power, and U. Williams (Petro Canada, St. John's, NL, Canada)

ABSTRACT: It is important for fisheries and the oil and gas industry to assure that developments taking place on important fishing grounds are not posing undue risks to fish and shellfish populations. Bioindicator responses can act as an early warning system to help identify the scope of any impacts before they might become significant. Bioindicator studies were carried out on a commercial groundfish, American plaice, at the Terra Nova development site prior to commencing (1997) and during (2000 and 2001) drilling operations. During this period, 6 wells were drilled using a combination of water-based and synthetic-based muds, which were discharged according to regulatory guidelines. No production waters were released during this period, but other minor sources of contaminants included sewage, deck drainage, bilge waters, and those from shipping activity. Fish were collected in the vicinity of the development site (~ 1–3 km) as well as at a reference site (~ 20 km distant) during the 3 annual surveys. A range of bioindicators were investigated in over 250 fish, including fish condition indices, visible skin and organ lesions, histopathological alterations in liver and gill tissues, levels of hepatic mixed-function oxygenase (MFO) enzymes, and differential blood cell counts. Comparable results were obtained between the sites indicating that the health status of American plaice has not been affected by development at the Terra Nova site.

INTRODUCTION

Population level effects in species such as fish would be both highly expensive to investigate and in many instances difficult to detect in the absence of major impacts, especially in the open ocean. Thus, there is increasing emphasis on the use of biochemical, physiological, and pathological indicators of chemical stress, which can provide an appreciation of the degree and severity of any potentially impending fish health problems. These indicators are commonly referred to as early warning or health effect bioindicators (Adams, 1990; Hugget et al., 1992). Assessment of any potential impacts on fish can be of considerable importance for regulators and the oil and fishing industries in regions such as Atlantic Canada, and bioindicators can provide a tool for assessing if effects are occurring, and if so, whether they might be of regulatory or socioeconomic concern. Traditionally, American plaice was the most abundant flatfish in the Northwest Atlantic, and of major commercial importance on the Grand Banks. It is presently under a fisheries moratorium, and during stakeholders discussions held by Petro Canada for their Environment Effects Monitoring (EEM) program, American plaice was identified as the finfish species of primary interest for tainting and fish health studies.

The primary objective of this study was to assess bioindicator responses in American plaice, under natural conditions of exposure around the Terra Nova site. In addition to its commercial importance, American plaice has a long life span, which is beneficial when considering potential for long-term chronic toxicity. It is also a bottom dwelling species in frequent contact with sediments, where contaminant levels can be elevated in sediment fines. American plaice is regarded as a fairly sedentary species in comparison with other commercial groundfish species. Pitt (1969) reported that most fish tagged on the northeast Grand Banks in the general area of the Terra Nova site, were still recovered in a radius less than 48 km even up to 7 years after tagging. It is reasonable to suggest that American plaice are resident for periods of weeks to months in specific areas with early warning indicators of health, such as mixed-function oxygenase (MFO) enzyme induction and selected gill and liver lesions being causally linked to these areas. Given that fish are well known to be attracted to oil and gas platforms (e.g. Love et al., 2000), this could also be happening to some extent on the Grand Banks. However, fish might move in and out of potential "contaminated" zones very close to study areas, such that exposure would be insufficient to result in deleterious effects. Overall, the presence or absence of health effects under natural ecological conditions of exposure is important to assess from a fisheries perspective. It is also important to note in this regard that the objective was to specifically focus on potential health effects in this particular commercial species, not to assess its overall sensitivity as a general monitoring indicator.

The bioindicators studied for the Terra Nova EEM program included fish condition indices, visible lesions on skin and internal organs, histopathology, levels of MFO enzymes, and haematology. Externally visible fish abnormalities have been recommended as bioindicators for biological monitoring programs at the national or international level by the International Council for Exploration of the Sea (ICES, 1999), and have been commonly used in survey work in different European countries, as well as in the USA and Canada. Fish liver histopathology has been adopted as one of the techniques to be used in biomonitoring programs designed to assess the environmental quality of waters in the North Sea area (Stagg, 1998). It has also been used in studies in the UK (Stentiford et al., 2003) as well as on the Pacific (Myers et al., 1994; Stehr et al., 1998) and the northeast Atlantic (Johnson et al., 1993) coasts of the USA. Khan (2000) has also carried out some preliminary studies on the Grand Banks of Newfoundland. MFO induction is currently used by eight of the contracting parties to the Oslo Paris Commission for monitoring biological effects of contaminants in the North Sea area (Stagg, 1998). Studies on general haematological parameters, such as changes in various types of blood cells, were also included since they can provide insight into potential effects on immune functions and resistance to disease (e.g. Weeks et al., 1992).

MATERIALS AND METHODS

Project Area Description

The Terra Nova oil field is situated on the Grand Banks of Newfoundland (Canada), approximately 350 km east-southeast of St. John's (Fig. 1). The oil field is being developed using a floating production storage and offloading facility with a semi-submersible drilling rig. A total of ~ 24 wells are being drilled at ~ 100 m depth through seven subsea templates located in five drill centers (for more details, see DeBlois et al., 2005, this publication). Six wells were drilled in relation to the study period being reported here. The discharges from the Terra Nova development site principally included drill

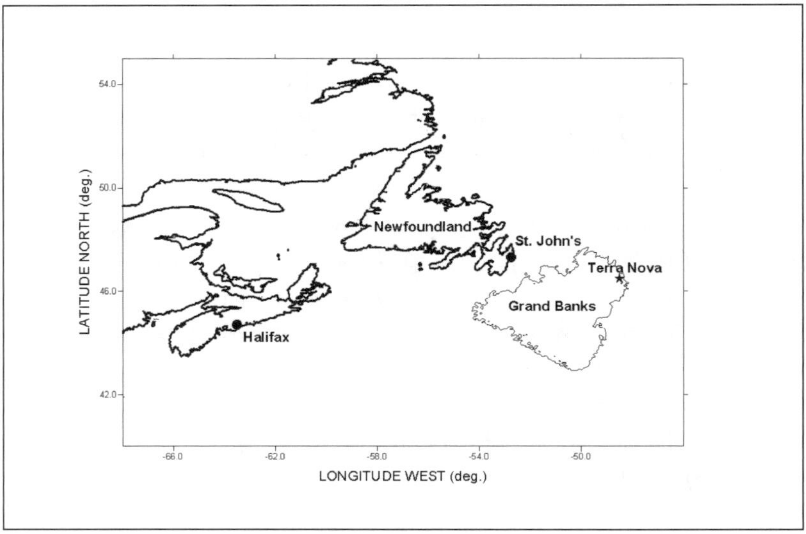

FIGURE 1. Location of Terra Nova oil field.

cuttings, sewage, cooling water, deck drainage, and bilge water, while the burning of fuel on the rig for the generation of electricity might also have resulted in some input of combustion hydrocarbons into the nearby area. Sewage was passed through a secondary treatment prior to discharge, while bilge water and deck drainage were passed through an oil and water separator. The cooling water uses an electric anti-fouling system and was treated to a minimum chlorine residual before discharge.

Both water-based and synthetic-based muds have been used at the site. The composition of the synthetic-based mud (SBM) was approximately 70% base oil (Petro Canada product Pure Drill, previously known as IPAR-3 and IA-35), 29.5% water and 0.5% additives. The base oil is a synthetic isoalkane that is hydroisomerised and hydrogenated. It is composed of aliphatic carbon compounds in the C_{11}–C_{21} range and is free of aromatics. Approximately 1100 tonnes of SBM was discharged at the site from July 1999 to August 2001. Transport and fishing vessels would also be expected to be a source of contaminants for the general area from hydrocarbons derived from fuel combustion and release of bilge waters.

Considering chemical usage at the site, it is noted that any chemicals or chemical mixtures slated for discharge must go through a Chemical Screening and Risk Assessment Process for human health and environmental acceptability before permission is granted for discharge by the Canada Newfoundland Offshore Petroleum Board. Also, although no specific studies have been carried out with American plaice, Payne et al. (2001) obtained negligible toxicity in fish larvae, copepods, and ctenophores acutely exposed to IA-35.

Fish Collection and Necropsy

Adult American plaice were collected by otter trawl from two areas on the Grand Banks, one in the vicinity of the Terra Nova development (~ 1–3 km from the Fisheries Exclusion Zone) and the other, at a reference area (~ 20 km southeast of the rig). More details on collection areas are provided in DeBlois et al. (2004, this publication). Cruises were carried out before drilling activities in mid-November 1997, as well as during drilling activities in early July 2000, and the end of June 2001. We hypothesise that any contaminant conditions of a background nature in a given region in the open sea should be more or less similar, and one reference site was considered to be reasonable and cost effective for this type of study with fish. It is also noted that collection of fish at the reference site still involved trawling over a substantial area which would be expected to provide fairly representative samples.

Upon capture, blood was drawn from a dorsal blood vessel near the tail and blood smears were prepared. Total length and weight were recorded. Observations for gross pathology were carried out under the general framework described by Adams et al. (1993), except no observations were made on thymus and kidney. Fish were dissected and sexed and maturity stage recorded according to procedures used by the Department of Fisheries and Oceans (DFO) in the Newfoundland Region. Otoliths were collected for determination of age.

A 3–4 mm section from the centre portion of each liver (along the longitudinal axis) and the first gill arch were fixed in 10% buffered formalin for histopathological analysis. The remaining liver tissue was frozen on dry ice until returning to port when it was placed in a –65°C freezer for MFO analysis.

Tissue Histopathology

Liver and gill samples fixed in formalin were processed for paraffin embedding by standard histological methods (Lynch et al., 1969), sectioned at 6 μm and stained with hematoxylin and eosin (Luna, 1968). Histological slides were examined by microscopy at 25x, 63x, and 250x magnification. Histological examination was conducted by the same investigator in all years. One slide was examined per fish. Information on the site of capture was not revealed until final diagnoses were made.

Liver lesions previously identified as having a putative chemical etiology (Myers et al., 1987; Johnson et al., 1993) were recorded for each fish as present or absent. The percentage of fish affected by each type of lesion or prevalence of lesion was then calculated. Macrophage aggregation was recorded on a relative scale from 0 to 7 and prevalence was calculated for fish showing a moderate to high aggregation (3 or higher on the scale).

Secondary gill lamellae were counted in 5 randomly selected fields at 250x magnification. Each lamella was then examined for the presence of different conditions including basal and distal hyperplasia, epithelial lifting, clubbing, aneurysm, and fusion. Results for each fish were expressed as the percentage of lamellae presenting the condition in relation to the total number of lamellae counted (approximately 100 to 120 lamellae per fish). The degree of oedema was also recorded on a 0 (absent) to 3 (heavy) relative scale.

Selected slides were also reviewed by Dr. Mark Myers, Chief Pathologist at the Northwest Fisheries Science Center in Seattle, for confirmation of diagnosis.

MFO Enzyme Analysis

Liver samples were thawed within 3 weeks of storage at –65°C and homogenised in 4 volumes of 50 mM Tris buffer, pH 7.5, using 10 passes of a glass Ten Broek hand homogeniser. Homogenates were centrifuged at 9000 g for 15 min at 4°C and the post mitochondrial supernatant (S9 fraction) frozen at –65°C until assayed. Assays were carried out within 3 weeks of storage but essentially no loss of activity has been noted in homogenates stored for up to 9 months.

MFO enzyme activity in the S9 fraction was assessed as 7-ethoxyresorufin O-deethylase (EROD) according to the fluorimetric method of Pohl and Fouts (1980) as modified by Porter et al. (1989). EROD activity was assayed in a final volume of 1 ml containing 50 mM Tris buffer, pH 7.5, 2 μM ethoxyresorufin (Sigma) dissolved in dimethyl sulphoxide, 0.15 mM NADPH and S9 protein. After 15 min incubation at 27°C, the reaction was stopped with 2 ml of methanol (HPLC grade) and samples were centrifuged (3600 g for 5 min) to remove the protein precipitate. The fluorescence of resorufin formed in the supernatant was measured at an excitation wavelength of 550 nm and an emission wavelength of 580 nm using a Perkin-Elmer

LS-5 fluorescence spectrophotometer. Blanks were performed as above with methanol added at the beginning of the incubation. An S9 fraction of fish with known specific activity was used as a positive control with distilled water as a negative control. All the samples were run in duplicate. Protein concentration was determined using the Lowry protein method (Lowry et al., 1951) with bovine serum albumin as standard. The rate of enzyme activity in pmol/min mg protein was obtained from the regression of fluorescence against standard concentrations of resorufin.

Haematology

Blood smears were fixed in methanol and stained with Giemsa for identifying different types of cells (Ellis, 1976). A blood cell differential count was performed on lymphocytes, neutrophils, and thrombocytes and expressed as a percentage of each type of cell on 400 cells counted per slide, using the Exaggerated Battlement Differential counting technique (Lynch et al., 1969).

Statistical Analysis

MFO enzyme activity, biological characteristics, condition indices, and arcsine square root-transformed percentages of blood cells and gill lamellae, were analysed by the Unpaired t-test or the Mann-Whitney Rank Sum test when the groups were not normally distributed. Condition indices were also assessed by analysis of covariance. Degree of oedema was analysed by the Mann-Whitney Rank Sum test and prevalence of gross pathology, liver lesions, and maturity stages by the Fisher exact test. Comparisons between sites having a $P < 0.05$ were considered to be statistically significant.

RESULTS AND DISCUSSION

Biological Characteristics and Condition of Fish

Information on fish biological characteristics is valuable for interpreting results of bioindicator studies. Although a limited number of fish were analysed from a population perspective (e.g. Dutil et al., 1995), such data can also provide a level of information for assessing major effects on animal condition. Whole body weight, length, age, and fish condition (Fulton's condition factor calculated as 100 × whole body weight/ length^3) of male and female fish (expressed as mean and standard deviation) from the reference and development areas during the 3 surveys are summarised in Tables 1 and 2. Site comparisons for each survey were carried out for each sex.

Condition factor, which is generally interpreted as an indicator of energy storage in fish, was slightly lower ($P < 0.05$ by the Unpaired t-test) in male fish from the development area in

TABLE 1. Biological characteristics and condition of male American plaice from the 1997, 2000, and 2001 surveys. All data are expressed as mean and standard deviation; [a]Fulton's condition factor: 100 x whole body weight/length^3; [b]NA= data not available; *Significant difference ($P < 0.05$) by the Unpaired t-test.

	1997		2000		2001	
	Reference Area	Development Area	Reference Area	Development Area	Reference Area	Development Area
Number of Fish	7	19	12	8	18	13
Whole Weight (g)	NA [b]	NA [b]	286 ± 71	264 ± 87	299 ± 95	290 ± 129
Length (cm)	29.7 ± 1.7	28.9 ± 1.0	31.5 ± 3.1	31.5 ± 2.7	32.6 ± 3.3	32.0 ± 4.6
Age (yr)	6.4 ± 1.4	5.9 ± 1.2	7.1 ± 0.5	7.0 ± 0.5	8.2 ± 1.1	8.5 ± 1.7
Fulton's CF [a]	NA [b]	NA [b]	0.897 ± 0.076	0.816 ± 0.077 *	0.837 ± 0.080	0.833 ± 0.038

TABLE 2. Biological characteristics and condition of female American plaice from the 1997, 2000, and 2001 surveys. All data are expressed as mean and standard deviation; [a]Fulton's condition factor: 100 x whole body weight/length^3; [b]NA= data not available; *Significant difference ($P < 0.05$) by the Unpaired t-test.

	1997		2000		2001	
	Reference Area	Development Area	Reference Area	Development Area	Reference Area	Development Area
Number of Fish	18	31	38	36	30	37
Whole Weight (g)	NA [b]	NA [b]	689 ± 290	624 ± 241	657 ± 408	549 ± 247
Length (cm)	35.3 ± 0.9	36.9 ± 1.1	41.7 ± 5.1	40.4 ± 5.1	40.1 ± 6.1	38.1 ± 6.1
Age (yr)	7.8 ± 0.8	7.7 ± 1.8	9.8 ± 1.5	9.8 ± 1.5	10.4 ± 1.4	9.4 ± 1.8 *
Fulton's CF [a]	NA [b]	NA [b]	0.900 ± 0.013	0.888 ± 0.019	0.874 ± 0.091	0.923 ± 0.072 *

2000, and slightly higher ($P < 0.05$ by the Unpaired t-test) in females in 2001. However, there were no statistically significant inter-site differences in condition indices in males or females as assessed by analysis of covariance.

Contaminant mediated oedema in gills (e.g. Grizzle, 1986) or other tissues, especially if accompanied by contaminant linked fatty degeneration and enlargement of organs such as liver (e.g. Sloof et al., 1983) could possibly result in small changes in condition indices in fish. However, small changes in condition indices could be also due to differences in feeding (e.g. Adams et al., 1982; Dutil et al., 1995) or a small gain or loss of energy reserves during sexual maturation (Craig, 1977; Goede and Barton, 1990).

Maturity stages were also recorded in the 2000 and 2001 surveys (Tables 3 and 4). In 2001, there was indication of differences between sites in maturity stages in males and females, although the differences were not statistically different (Fisher exact test).

TABLE 3. Maturity stages of male American plaice from the 2000 and 2001 surveys. Maturity stages were defined according to procedures used by DFO and expressed as percentages.

	Site	Number of Fish	Immature (%)	Maturing to Spawn (%)	Partly spent (%)	Spent (%)
Survey 2000	Reference Area	12	0.0	0.0	66.7	33.3
	Development Area	8	0.0	12.5	50	37.5
	P-Value		1.000	0.400	0.648	1.000
Survey 2001	Reference Area	18	6.0	22	55	17
	Development Area	13	0.0	54	46	0.0
	P-Value		1.000	0.128	0.726	0.245

TABLE 4. Maturity stages of female American plaice from the 2000 and 2001 surveys. Maturity stages were defined according to procedures used by DFO and expressed as percentages.

	Site	Number of Fish	Immature (%)	Maturing to Spawn (%)	Partly spent (%)	Spent (%)
Survey 2000	Reference Area	37	13.5	8.1	0.0	78.4
	Development Area	34	2.9	8.9	0	88.2
	P-Value		0.201	1.000	1.000	0.349
Survey 2001	Reference Area	30	3.3	13.3	0.0	83.4
	Development Area	37	10.8	10.8	10.8	67.6
	P-Value		0.370	1.000	0.122	0.168

Gross Pathology

Other than two fish from the reference site in 2001, one with lymphocystis (Templeman, 1965), and the other with a growth on the fins, there were no other visible lesions on the skin, gills, or internal organs of fish (liver, digestive tract, musculature, spleen) from either the development or reference sites in any survey (Table 5). Observations were carried out on a total 267 fish.

TABLE 5. Prevalence (% of fish affected) of visible abnormalities in American plaice from the 1997, 2000, and 2001 surveys.

	1997		2000		2001	
	Reference Area	Development Area	Reference Area	Development Area	Reference Area	Development Area
Number of fish	25	50	50	44	48	50
Lesions on skin and fins	0	0	0	0	4.2	0
Lesions on internal organs	0	0	0	0	0	0

Tissue Histopathology

Liver tumors or other lesions typically observed at chemically contaminated sites elsewhere (e.g. Myers et al., 1987; Hinton et al., 1992; Stehr et al., 1998) were not observed or appeared at very low prevalence in both areas (Table 6). No cases of eosinophilic foci, basophilic foci, clear cell foci, carcinoma, cholangioma, cholangiofibrosis, or hydropic vacuolation were observed in any fish from any survey. Frequencies of megalocytic hepatosis, nuclear pleomorphism and macrophage aggregation were very low and no significant differences (Fisher exact test) were observed between the reference and development areas in any survey.

Gill as well as liver histopathology was assessed since gill epithelia are quite sensitive to certain classes of chemicals including relatively low levels of hydrocarbons (Solangi and Overstreet, 1982; Kiceniuk and Khan, 1987). All fish from both the development and reference areas displayed some type of hyperplasia (distal, tip, or basal) of secondary lamellae indicating general lamellar thickening putatively of a background nature. No cases of epithelial lifting were observed in any fish in any survey (Table 7).

TABLE 6. Prevalence (% of fish affected) of hepatic lesions in American plaice from the 1997, 2000, and 2001 surveys. [a]Moderate to high aggregation (3 or higher rating on a 1–7 relative scale).

	1997		2000		2001	
	Reference Area	Development Area	Reference Area	Development Area	Reference Area	Development Area
Number of Fish Examined	25	50	50	44	48	50
Nuclear Pleomorphism	0	2.0	0	2.3	0	0
Megalocytic Hepatosis	0	0	0	2.3	0	0
Cell Foci Alterations	0	0	0	0	0	0
Carcinoma	0	0	0	0	0	0
Cholangioma	0	0	0	0	0	0
Macrophage Aggregation [a]	0	0	4.0	2.3	0	0
Hydropic Vacuolation	0	0	0	0	0	0

TABLE 7. Percentage of gill lamellae displaying pathological conditions in American plaice from the 1997, 2000 and 2001 surveys. [a]Mean percentage of lamellae presenting the condition (in relation to the total number of lamellae counted); [b]Mean of rating on a relative 0–3 scale; *Significantly different ($P < 0.05$) after the Mann-Whitney Rank Sum test analysis.

	1997		2000		2001	
Conditions	Reference Area	Development Area	Reference Area	Development Area	Reference Area	Development Area
Number of Fish Examined	25	50	48	43	47	50
Epithelial Lifting[a]	0	0	0	0	0	0
Aneurysm [a]	< 0.1	< 0.1	0.3	< 0.1	< 0.1	< 0.1
Fusion [a]	0.1	0.3	0.6	0.5	0.5	0.0
Oedema [b]	0.7	0.7	1.4	1.4	1.3	1.0 *

Histopathological effects have been observed in fish and shellfish in association with some oil developments in the Gulf of Mexico (Gallaway et al., 1981; Grizzle, 1986; Wilson-Ormond et al., 1994), but equivocal results have also been obtained (Menzie, 1982). However, relatively high levels of lesions were often noted in fish from reference sites, and such pathological effects may be background in nature and linked to general water quality conditions in the Gulf. This would be expected to confound resolution of any specific rig-related effects. It is of interest to note, in this regard, that histopathological effects were essentially non existent at both the Terra Nova and reference sites on the Grand Banks.

MFO Enzyme Activity

There were no significant differences in MFO enzyme activity, assessed as EROD, in either male or female fish from the reference and development areas during the 1997, 2000, and 2001 surveys (Figs. 2 a–f). Several studies have reported on the sensitivity of fish to induction by selected aromatic compounds with induction persisting for a few weeks to months (e.g. George et al., 1995) in the case of some episodic exposures to petroleum hydrocarbons.

A capacity for induction of MFO enzymes has been observed in a variety of fish at late stages of gonad maturation including during spawning (Collier et al., 1986; Payne et al., 1987; Porter et al., 1989). However, since reproductive state can influence enzyme activity (e.g. Hansson et al, 1982; Mathieu et al., 1991; Whyte et al., 2000) and probably result in some loss of sensitivity for resolving contaminant mediated differences, a comparison of enzyme activity was also carried out on spent female fish alone in 2001 (data not shown). Enzyme activity was slightly elevated in fish from the development site but the differences were not statistically significant.

There may be some loss of sensitivity to enzyme induction since fish were collected towards the end of their spawning cycle. However, collection of fish outside Petro-Canada's normal EEM cruise, when a suitable trawler was available, would have had major cost considerations. The different MFO "baseline" levels observed in different years also indicate that caution is warranted when carrying out comparisons between reference and experimental sites other than on an annual basis. Such MFO variability may be natural but a changing background of low levels of chemical inducers, e.g. from general ship traffic, cannot be ruled out. The point to note here in relation to the development site is that no differences were noted in MFO enzyme levels between development and reference site during 3 annual surveys.

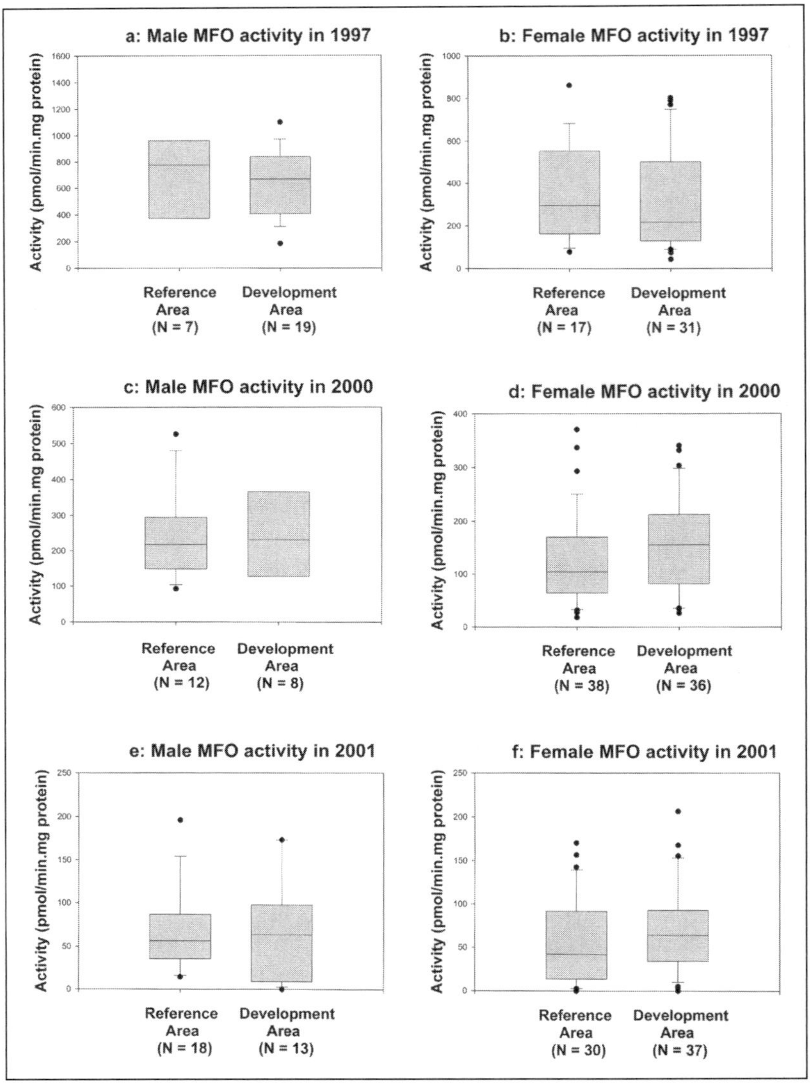

FIGURE 2. Hepatic MFO activity in male and female American plaice (**a-f**). Data plotted are median, 10th, 25th, 75th and 90th percentiles as vertical boxes with error bars. Data points beyond the 5th and 95th percentiles are also displayed.

MFO induction has been observed in a number of fish species around production platforms in the North Sea compared to nearby reference sites. A 2–3 fold induction was noted in cod, haddock, and whiting (Davies et al., 1984), species that are generally considered to be much more migratory than flatfish. MFO induction in this case was measured as benzo(a)pyrene hydroxylase, which is often induced to a lesser degree than EROD. Stagg et al. (1995) documented considerably higher levels of induction in a flatfish (dab) (4 fold in males and 10–20 fold in females) around a group of oil platforms in the northern North Sea. Elevated levels of MFO enzymes have also been observed in fish larvae collected around the same group of platforms (Stagg and McIntosh, 1996). The induction observed in both adult fish and larvae in the North Sea studies was likely due to discharge of diesel-based muds and/or production waters which have not been discharged at the Terra Nova site. MFO induction in fish has also recently been used as a monitoring tool at an oil development site in the Caspian Sea (Moore et al., 2003).

MFO induction was not observed in fish collected around 3 platforms in the Gulf of Mexico (McDonald et al., 1996). However, it is noted that the reference fish as well as the fish collected near the platforms often contained quite high levels of PAH metabolites in their bile, suggesting extensive background contamination by hydrocarbons at the reference sites. This would be expected to confound any platform associated effects. Nevertheless, in perspective, it could be suggested that any enzyme induction potential associated with the platforms at various sites in the Gulf of Mexico may be not unlike the induction potential associated with waters of the Gulf in general. It is understood in this regard that the Gulf of Mexico has traditionally received relatively high levels of hydrocarbon contamination from a variety of sources including large numbers of developing oil and gas wells, runoff from the Mississippi River, ship discharges, and urban wastewaters.

Haematology

With respect to potential for indicating potential immunological effects and disease status or susceptibility, there were no significant differences in the percentage of white blood cells (lymphocytes and neutrophils) or thrombocytes in fish from the development and reference Areas in 2001 (Table 8).

There were also no apparent qualitative differences in morphology or staining characteristics of red blood cells. No blood smears were available for 1997 and 2000.

TABLE 8. Blood cell differential counts in American plaice from the 2001 survey. *Mean percentage of each type of cell derived from 10 fish (400 cells per fish) per site.

	Reference Area	Development Area
Lymphocytes *	77.9	78.4
Neutrophils *	3.7	3.7
Thrombocytes *	18.4	17.9

CONCLUSIONS

Petro Canada has included a fish health component in their EEM program, and a range of health effect indicators have been studied in American plaice, a designated species of commercial importance at their development site. Results were found to be similar in fish taken from the development and reference areas during 3 annual surveys. Notable in this regard was the virtual absence of visible abnormalities and histopathological lesions in liver and gill tissues of approximately 265 fish analysed. These findings support the hypothesis that the health status of American plaice has not been affected by the development at the Terra Nova site. Since a number of wells were drilled at the site, the results are also of general interest with respect to sites where discharges would be much less, such as at single exploratory wells.

ACKNOWLEDGMENTS

We thank Dr. Myers, Chief Pathologist at the Northwest Fisheries Science Center in Seattle, for confirmation of diagnosis of pathological lesions.

REFERENCES

Adams, S.M., A.B. Brown, and R.W. Goede. 1993. "A Quantitative Health Assessment Index for Rapid Evaluation of Fish Condition in the Field." *Transactions of the American Fisheries Society*. 122:63-73.

Adams, S.M. (Ed.). 1990. *Biological Indicators of Stress in Fish*. American Fisheries Symposium 8, Bethesda, Maryland.

Adams, S.M., R.B. McLean, and A. Parotta. 1982. "Energy Partitioning in Largemouth Bass under Conditions of Seasonally Fluctuating Prey Availability." *Transactions of the American Fisheries Society*. 111:549-558.

Collier, T.K., J.E Stein, R.J. Wallace, and U. Varanasi. 1986. "Xenobiotic Metabolizing Enzymes in Spawning English Sole Exposed to Organic-Solvent Extracts of Marine Sediments from Contaminated and Reference Areas." *Comparative Pharmacology and Toxicology*. 84:291-298.

Craig, J.F. 1977. "The Body Composition of Adult Perch, *Perca fluviatilis*, in Windermere, with Reference to Seasonal Changes and Reproduction." *Journal of Animal Ecology*. 46:617-632.

Davies, J.M., J.M. Addy, R.A. Blackman, J.R. Blanchards, J.E. Ferbrache, D.C. Moore, H.J. Sommerville, A. Whitehead, and T. Wilkinson. 1984. "Environmental Effects of the Use of Oil-Based Drilling Muds in the North Sea." *Marine Pollution Bulletin*. 15:363-370.

DeBlois, E.M., C. Leeder, K.C. Penney, M. Murdoch, M.D Paine, F. Power, and U. Williams. 2005. "Terra Nova Environmental Effects Monitoring Program: From Environmental Impact Statement Onward." In S.L. Armsworthy, P.J. Cranford, and K. Lee (Eds.), *Proceedings of the Offshore Oil and Gas Environmental Effects Monitoring Workshop: Approaches and Technologies, Bedford Institute of Oceanography, Dartmouth, Nova Scotia, May 26–30, 2003*, Battelle Press. This volume.

Dutil, J.D., Y. Lambert, G.A. Chouinard, and A. Frechet. 1995. *Fish Condition: What Should We Measure in Cod (Gadus morhua)?* Department of Fisheries and Oceans (DFO) Atlantic Fisheries Research Document 95/11.

Ellis, A.E. 1976. "Leucocytes and Related Cells in the Plaice *Pleuronectes platessa*." *Journal of Fish Biology*. 8:143-156.

Gallaway, B.J., L.R. Martin, R.L. Howard, G.S. Boland, and G.D. Dennis. 1981. "Effects on Artificial Reef and Demersal Fish and Macrocrustacean Communities." In *Environmental Effects of Offshore Oil Production: The Buccaneer Gas and Oilfield Study*, pp. 237-299. Middleditch, BS.

George, S.G., J. Wright, and J. Conroy. 1995. "Temporal Studies of the Impact of the Braer Oil Spill on Inshore Feral Fish from Shetland, Scotland." *Archives of Environmental Contamination and Toxicology*. 29(4):530-534.

Goede, R.W. and B.A. Barton. 1990. "Organismic Indices and an Autopsy-Based Assessment as Indicators of Health and Condition of Fish." In S.M. Adams (Ed.), *Biological Indicators of Stress in Fish*, pp. 93-108. American Fisheries Symposium 8, Bethesda, Maryland.

Grizzle, J.M. 1986. "Lesions in Fishes Captured near Drilling Platforms in the Gulf of Mexico." *Marine Environmental Research*. 18:267-276.

Hansson, T., L. Forlin, J. Rafter, and J.A. Gustafsson. 1982. "Regulation of Hepatic Steroid and Xenobiotic Metabolism in Fish." In E. Hietanen, M. Laitinen and O. Hanninen (Eds.), *Cytochrome P-450 Biochemistry, Biophysics and Environmental Implications*. Elsevier, Amsterdam.

Hinton, D.E., P.C. Baumann, G.R. Gardner, W.E. Hawkins, J.D. Hendricks, R.A. Murchelano, and M.S. Okihiro. 1992. "Histopathologic Biomarkers." In R.J. Hugget, R.A. Kimerle, P.M. Mehrle, and H.L. Bergman (Eds.), *Biomarkers: Biochemical, Physiological and Histological Markers of Anthropogenic Stress*, pp. 155-209. Society of Environmental Toxicology and Chemistry (SETAC) Special Publication Series, Lewis Publishers, Chelsea, MI.

Huggett, R.J., R.A. Kimerle, P.M. Mehrle, and H.L. Bergman (Eds.). 1992. *Biomarkers: Biochemical, Physiological, and Histological Markers of Anthropogenic Stress*. Society of Environmental Toxicology and Chemistry (SETAC) Special Publication Series, Lewis Publishers, Chelsea, MI.

ICES (International Council for the Exploration of the Sea). 1999. Report of the ICES Advisory Committee on the Marine Environment, Copenhagen, Denmark, 31 May–5 June 1999. *ICES Cooperative Research Report # 239*.

Johnson, L.L., C.M. Stehr, O.P. Olson, M.S. Myers, S.M. Pierce, C.A. Wigren, B.B. McCain, and U. Varanasi. 1993. "Chemical Contaminants and Hepatic Lesions in Winter Flounder *(Pleuronectes ameri-*

canus) from the Northeast Coast of the United States." *Environmental Science and Technology*. 27(13):2759-2771.

Khan, R.A. 2000. "Comparison of Tissue Lesions in Four Species of Benthic Fish Sampled in 1972–1973 and 1997–1998 on the Grand Banks off Newfoundland." *Bulletin of Environmental Contamination and Toxicology*. 65:78-83.

Kiceniuk, J.W. and R.A. Khan. 1987. "Effect of Petroleum Hydrocarbons on Atlantic Cod, *Gadus morhua*, Following Chronic Exposure." *Canadian Journal of Zoology*. 65:490-494.

Love, M.S., J.E. Caselle, and L. Snook. 2000. "Fish Assemblage around Seven Oil Platforms in the Santa Barbara Channel Area." *Fisheries Bulletin*. 98:96-117.

Lowry, O.H., N.J. Rosebrough, A.L. Fan, and R.J. Randall. 1951. "Protein Measurement with the Folin Phenol Reagent." *Journal of Biological Chemistry*. 193:265-275.

Luna, L.G. 1968. *Manual of Histological Staining Methods of the Armed Forces Institute of Pathology*. McGraw-Hill, New York.

Lynch, M., S. Raphael, L. Mellor, P. Spare, and M. Inwood. 1969. *Medical Laboratory Technology and Clinical Pathology*. W. B. Saunders Company.

Mathieu, A., P. Lemaire, S. Carriere, P. Drai, J. Giudicelli, and M. Lafaurie. 1991. "Seasonal and Sex Linked Variations in Hepatic and Extra Hepatic Biotransformation Activities in Striped Mullet *Mullus barbatus*." *Ectoxicology and Environmental Safety*. 22:45-57.

McDonald, S.J., K.L. Willett, J.T. Thomsen, K.B. Beatty, K. Connor, T.R. Narasimhan, C.M. Erickson, and S.H. Safe. 1996. "Sublethal Detoxification Responses to Contaminant Exposure Associated with Offshore Production Platforms." *Canadian Journal of Fisheries and Aquatic Sciences*. 53:2606-2617.

Menzie, C.A. 1982. "The Environmental Implications of Offshore Oil and Gas Activities." *Environment and Science Technology*. 16(8):454-472

Moore, M.J., I.V. Mitrofanov, S.S. Valentini, V.V. Volkov, A.V. Kurbskiy, E.N. Zhimbey, L.B. Eglinton, and J.J. Stegeman. 2003. "Cytochrome P4501A Expression, Chemical Contaminants and Histopathology in Roach, Goby and Sturgeon and Chemical Contaminants in Sediments from the Caspian Sea, Lake Balkhash and the Ily River Delta, Kazakhstan." *Marine Pollution Bulletin*. 46:107-119.

Myers, M.S., C.M. Stehr, O.P. Olson, L.L. Johnson, B.B. McCain, S.L. Chan, and U. Varanasi. 1994. "Relationships between Toxicopathic Hepatic Lesions and Exposure to Chemical Contaminants in English Sole (*Pleuronectes vetulus*), Starry Flounder (*Platichthys stellatus*), and White Croaker (*Genyonemus lineatus*) from Selected Marine Sites on the Pacific Coast, USA." *Environmental Health Perspectives*. 10:200-215.

Myers, M.S., L.D. Rhodes, and B.B. McCain. 1987. "Pathologic Anatomy and Patterns of Occurrence of Hepatic Neoplasms, Putative Preneoplastic Lesions, and Other Idiopathic Hepatic Conditions in English Sole (*Parophrys vetulus*) from Puget Sound, Washington." *Journal of National Cancer Institute*. 78(2):333-363.

Payne, J.F., L. Fancey, C. Andrews, J. Meade, F. Power, K. Lee, G. Veinott, and A. Cook. 2001. "Laboratory Exposures of Invertebrate and Vertebrate Species to Concentrations of IA-35 (Petro-Canada) Drill Mud Fluid, Production Water, and Hibernia Drill Mud Cuttings. *Canadian Manuscript Report of Fisheries and Aquatic Sciences*, No *2560*:1-27.

Payne, J.F., L. Fancey, A. Rahimtula, and E. Porter. 1987. "Review and Perspective on the Use of Mixed-Function Oxygenase Enzymes in Biological Monitoring." *Comparative Biochemistry and Physiology*. 86C(2):233-245.

Pitt, T.K. 1969. "Migrations of American Plaice on the Grand Bank and in St. Mary's Bay, 1954, 1959, and 1961." *Journal of Fisheries Research Board of Canada*. 26(5): 1301-1319.

Pohl, R.J. and J.R. Fouts. 1980. "A Rapid Method for Assaying the Metabolism Of 7-Ethoxyresorufin by Microsomal Subcellular Fractions." *Analytical Biochem*istry. 107:150-155.

Porter, E., J.F. Payne, J.W. Kiceniuk, L. Fancey, and W. Melvin. 1989. "Assessment of the Potential for Mixed-Function Oxygenase Enzyme Induction in the Extrahepatic Tissues of Cunners during Reproduction." *Marine Environmental Research*. 28:117-121.

Sloof, W., C.F. Van Kreijl, and A.J. Baars. 1983. "Relative Liver Weights and Xenobiotic-Metabolising Enzymes of Fish from Polluted Surface Waters in the Netherlands." *Aquatic Toxi*cology. 4:1-14.

Solangi, M.A. and R.M. Overstreet. 1982. "Histopathological Changes in Two Estuarine Fishes, *Menidia beryllina* (Cope) and *Trinectes maculatus* (Bloch And Schneider), Exposed to Crude Oil and its Water-Soluble Fractions." *Journal of Fish Diseases*. 5:13-35.

Stagg, R.M. 1998. "The Development of an International Program for Monitoring the Biological Effects of Contaminants in the OSPAR Convention Area." *Marine Environmental Research*. 46(1-5):307-313.

Stagg, R.M. and A. McIntosh. 1996. "Hydrocarbon Concentrations in the Northern North Sea and Effects on Fish Larvae." *Science of the Total Environment*. 186:189-201.

Stagg, R.M., A. McIntosh, and P. Mackie. 1995. "Elevation of Hepatic Monooxygenase Activity in the Dab (*Limanda limanda L.*) in Relation to Environmental Contamination with Petroleum Hydrocarbons in the Northern North Sea." *Aquatic Toxicology*. 33:245-264.

Stehr, C.M., L.U. Johnson, and M.S. Myers. 1998. "Hydropic Vacuolation in the Liver of Three Species of Fish from the U.S. West Coast: Lesion Description and Risk Assessment Associated with Contaminant Exposure." *Diseases of Aquatic Organisms*. 32:119-135.

Stentiford, G.D., M. Longshaw, B.P. Lyons, G. Jones, M. Green, and S.W. Feist. 2003. "Histopathological Biomarkers in Estuarine Fish Species for the Assessment of Biological Effects of Contaminants." *Marine Environmental Research*. 55:137-159.

Templeman, W. 1965. "Lymphocystis Disease in American Plaice of the Eastern Grand Bank." *Journal of Fisheries Research Board of Canada*. 22:1345-1356.

Weeks, B.A., D.P. Anderson, A.P. DuFour, A. Fairbrother, A.J. Goven, G.P. Lahvis, and G. Peters. 1992. "Immunological Biomarkers to Assess Environmental Stress." In R.J. Huggett, R.A. Kimerle, P.M. Mehrle and H.L. Bergman (Eds.), *Biomarkers: Biochemical, Physiological, and Histological Markers of Anthropogenic Stress*, pp. 211-233. Society of Environmental Toxicology and Chemistry (SETAC) Special Publication Series, Lewis Publishers, Chelsea, MI.

Whyte, J.J., R.E. Jung, C.J. Schmitt, and D.E. Tillitt. 2000. "Ethoxyresorufin-O-deethylase (EROD) Activity in Fish as a Biomarker of Chemical Exposure." *Critical Review in Toxicology*. 30(4):347-570.

Wilson-Ormond, E.A., M.S. Ellis, and E.N. Powel. 1994. "The Effect of Proximity to Gas Producing Platforms on Size, Stage of Reproductive Development and Health in Shrimp and Crabs."

Overview of Potential Impacts from Produced Water Discharges in Atlantic Canada

Lee, K. (LeeK@dfo-mpo.gc.ca), K. Azetsu-Scott, S.E. Cobanli, J. Dalziel, S. Niven, G. Wohlgeschaffen, and P. Yeats (Department of Fisheries and Oceans, Dartmouth, NS, Canada)

ABSTRACT: With the development of Canada's offshore oil and gas reserves, large volumes of produced water may be discharged into the ocean. This is of concern to regulators and the environmental community, since these waters typically contain high concentrations of potential contaminants. Furthermore, emerging evidence from other offshore oil fields in the world have suggested that production water discharges may impact the biota (invertebrates, fish, and larvae) at greater distances from operational platforms than originally envisaged. Preliminary laboratory studies with produced water from hydrocarbon reserves off the continental shelf of Atlantic Canada have provided evidence that produced water contaminants discharged into the water column may be transported to the benthic environment and the surface microlayer. This overview includes a description of studies initiated to evaluate the ecological significance of production water discharges in the Atlantic Region of Canada to ensure protection of our marine habitat and the sustainability of our fisheries resources.

SOURCES AND VOLUME OF PRODUCED WATER DISCHARGES

The aqueous fraction extracted along with oil and gas from geological formations is defined as produced water. It consists of formation water naturally found in the geological structure, injection water introduced into the formation to maintain pressure and increase production efficiency, and other chemical additives used in routine operations (treatment chemicals such as hydrate inhibitors, emulsion breakers, flocculants, coagulants, defoaming agents, scale and corrosion inhibitors, bactericides, etc).

The discharge of produced water is of growing concern as it is continuously generated through all stages of development, whereas other wastes (e.g. drilling muds and fluids) are generated only at specific stages. Indeed based on volume, produced water has been classified as the largest contributor to the offshore waste stream (Krause, 1995), constituting approximately 80% of the wastes and residuals produced from natural gas production (McCormack et al., 2001). In most offshore operations, produced water is treated to reduce the dispersed oil content to regulatory limits and subsequently discharged to the ocean (Neff et al., 1987).

Produced water discharge in Atlantic Canada is currently limited as the offshore oil and gas resources are in the early stages of development; however, its volume is anticipated to increase as new oil and gas fields are developed in the future. In addition, the volume of produced water usually increases with the age of a well (Henderson et al., 1999). In nearly depleted fields, production may be 98% produced water and 2% fossil fuel (Stephenson, 1992; Shaw et al., 1999). With increased environmental concern regarding greater discharge volumes, regulatory agencies and members of the environmental community are now considering a call for zero discharge of produced water into the marine environment.

The only potential alternative to discharge of produced water is underground injection. However, this practice offshore is dependent upon a number of factors, including the access to a suitable disposal formation and chemical characteristics that result in the formation of precipitates that may plug the receiving formation. Re-injection feasibility therefore must be evaluated on a case-by-case basis for each field. Furthermore the net environmental benefit of re-injection must be considered. On the basis of energy requirements, it has been estimated that 2.6–4.3 grams of CO_2 emissions are produced for each litre of produced water re-injected into a subsurface well (Shaw et al., 1999). Additional knowledge on the environmental impact of produced water associated with the development of eastern Canada's frontier oil and gas reserves is needed to identify acceptable disposal limits.

CONSTITUENTS OF CONCERN

The formation-water component of production water is, in effect, brine which derives its salinity from the major ions found in seawater (GESAMP, 1993). However, depending on the nature of the formation from which it is withdrawn and operational procedures, it also contains a number of metal and organic constituents of environmental interest including heavy metals, hydrolysis metals, organic compounds including petroleum hydrocarbons, nutrients, radionuclides, and treating chemicals (Røe, 1999; Neff, 2002). Its chemical composition varies over a wide range and depends mainly on attributes of the reservoir's geology. Thus, the composition of produced water may also change through the production lifetime of the reservoir as different sectors of the formation become exploited during production (CAPP, 2000; Cranford et al., 2001). In terms of discharge capacity, it is generally reported that the total amount of water produced from gas fields is much smaller than from oil production fields. While many gas fields discharge less than 10 m^3 of produced water per day, most oil fields discharge hundreds or even thousands of m^3 of produced water per day (IAOGP, 2002).

In terms of inorganic constituents, the heavy metals most commonly associated with produced water include barium, cadmium, chromium, copper, lead, mercury, nickel, silver, strontium, and zinc (Table 1). While many are known to be highly toxic and their concentrations can be very high (100–10,000 times seawater concentrations), they are assumed to be of minor environmental concern due to rapid dilution following discharge. There are numerous case studies describing rapid dilution of produced water components in the near field. Somerville, et al. (1987) modelled the dilution of produced water discharged from the Brent platform in the North Sea. For a 10,000 m^3/day discharge rate, they estimated a 100-fold dilution at 50 m and a 2,800-fold dilution at 1,000 m from the platform. Furuholt (1996) modelled a 2,000 m^3/day discharge and found a 1,000:1 dilution approximately 50 m downstream of the discharge point. For a 20,000 m^3/day discharge he found a 1,000:1 dilution approximately 500 m from the discharge point.

The hydrolysis metals, predominantly Fe, Mn, and Al, are a group that readily form precipitates in seawater. The high concentrations of these metals in produced water can generate inorganic metal oxide precipitates at 20–50 mg/l concentra-

TABLE 1. Concentration of metals (μg/l) in a representative produced water sample from the Scotian Shelf of Atlantic Canada (Copan field) and in ambient seawater. N.D. is indicative of concentrations below the detection limit value in brackets. (Sable Offshore Oil and Gas Inc., 1995; Yeats, pers. comm., 2004).

Metal	Produced Water	Seawater
Al	N.D. (80) - 2100	1 - 10
Cd	N.D. (2) - 6	0.004 - 0.04
Cr	N.D. (80)	0.2 - 2.00
Co	N.D. (2) - 2	0.10 - 0
Cu	N.D. (8) - 4200	0.18 - 0.31
Fe	12000 - 28000	0.25 - 1.10
Pb	N.D. (8) - 45	0.005 - 0.05
Mn	1300 - 2300	0.24 - 0.45
Ni	N.D. (20) - 420	0.15 - 0.40
Zn	90 - 26000	0.15 - 0.06

tions. Both inorganic and organic constituents of concern within produced water may be removed from the water column by adsorption on suspended solids. Thus, in addition to naturally occurring suspended particulate material, precipitates formed from hydrolysis metals while not in themselves considered to be toxic, may provide an important vector in sequestering of both dissolved and particulate contaminants to the benthic environment (by sedimentation) and the surface microlayer (by interaction with buoyant oil droplets).

The salinity of produced water may range from a few parts per thousand (ppt) up to 300 ppt (Rabalias and Boesch, 1987). Produced water samples from the Hibernia platform off the coast of Atlantic Canada have been found to have an average salinity of 195 ppt (Parker and Ayers, 2001). Increased salinity can have major implications on the survival of existing organisms, development of young larvae, and on the dispersion of the discharge through the water column (Payne et al., 2001 and Querbach et al., 2005).

The organic components in produced water occur in a dispersed (primarily oil droplets) or dissolved (soluble) form. While oil-water separators are used to remove discrete oil droplets from the water phase prior to discharge, the equipment is not 100% efficient. The soluble components include aliphatic hydrocarbons, phenols, carboxylic acids, and low molecular weight hydrocarbons (Stephenson, 1992). Soluble oil is not easily removed from the process stream. As a result, petroleum hydrocarbon concentrations in produced water are approximately 1,000 times ambient. Hydrocarbons in produced waters have been shown to impact the species diversity of benthic communities in areas of shallow water discharge (Neff, 1987). In terms of regulatory guidelines within Canada for new installations, the maximum permissible oil-in-water limits for produced water is a 30-day average of 30 mg/l and a 24-hour average of 60 mg/l for new installations (National Energy Board et al., 2002). There is also a requirement for chemical analysis and twice-yearly reporting of produced water composition and for annual reporting of its aquatic toxicity. This information will help support effects monitoring programs and aid in determining whether produced water constituents other than oil should be regulated, for example, the organic acids that have been found to account for ~ 94% of the organic constituents in produced waters discharged into the North Sea (Garland and Aquitaine, 1998). To date, the focus on organic constituents has been on the hydrocarbons since they are more persistent and toxic.

In terms of contaminant hydrocarbon effects, produced water discharges are unlikely to have large-scale environmental impacts. The National Research Council (1985) estimates that worldwide petroleum hydrocarbon input to the oceans from produced water represents less than 0.4% of the total amount of petroleum hydrocarbons entering the world's oceans from all sources. Nevertheless, there remains concern that petroleum hydrocarbons and other potentially toxic components within produced water discharge streams may cause localised impacts.

Nutrient concentrations can be very high in produced water. Produced water ammonia concentrations as high as 40 mg/l have been measured at the Copan site on the Scotian Shelf of Atlantic Canada. While such levels may be associated with detrimental effects on biota, toxic effects are expected to be mitigated by natural dilution processes in the open ocean environment. Nevertheless, these compounds may have a major ecological impact as they may represent a significant nutrient source for the receiving waters. Indeed, Rivkin et al. (2000) suggested that changes in ecosystem structure could result from nutrients derived from produced water discharge. Using an ecosystem model to simulate the effect of produced water nutrients in an ecosystem typical of the Grand Banks area off Newfoundland, they predicted a change in the ecosystem structure that favoured small-celled phytoplankton and bacteria to the detriment of meso-zooplankton, on which pelagic larval and juvenile fish feed. Theoretically, blooms of phytoplankton, bacterioplankton, and zooplankton may produce high concentrations of colloidal organic material (flocs) that sequester contaminants from the water column for rapid transport to the bottom.

Elevated concentrations of Naturally Occurring Radioactive Materials (NORMs) are frequently found in produced water. Radium-226 (^{226}Ra) and radium-228 (^{228}Ra) resulting from the decay of uranium and thorium associated with specific clay and parent rocks of hydrocarbon reserves usually account for the bulk of the radionuclides in produced water (Snavely, 1989; Neff, 2002). In addition to human health considerations, radium is of interest because of its high solubility, potential for co-precipitation reactions in produced water, and its tendency to bioaccumulate in food chains. Concentrations of ^{226}Ra and ^{228}Ra within produced water samples recovered on the Scotian Shelf have been found to be enriched by factor of 20,000 (Table 2). While certainly above background levels, these concentrations are not considered to be an environmental health issue to personnel and the marine species surrounding development sites (Garland and Aquitaine, 1998). The presence of n-alkanes in sediments has previously been used as a tracer of produced water chemicals in sediments (Neff et al., 1992). There is potential for the use of NORM profiling to delineate the impact zone from produced water discharges since they are highly source-specific and less influenced by biodegradation processes relative to traditional organic tracers.

In addition to the chemicals that occur naturally in produced water, the oil field operator may add other chemicals during the production process to treat or prevent operational problems. Examples of production chemicals that are used include scale and corrosion inhibitors, biocides, and emulsion breakers. Some of these chemicals remain in the produced water discharged to the ocean and may account for some of the

TABLE 2. Concentration of Natural Occurring Radioactive Materials (NORMs) in a produced water sample recovered from the offshore platform at the Copan site on the Scotian Shelf.

Isotope	Background Levels (mBq/l)	Measured Results (Bq/l)	Enrichment Factor
^{226}Ra	1.0 – 2.0	11.6 ± 2.1	10,000
^{228}Ra	0.5 – 1.0	15.5 ± 2.4	20,000

observed differences in produced water toxicity between sites (Brendehaug et al., 1992).

BIOLOGICAL EFFECTS

Bioassays have been conducted with produced water samples with a number of reference test organisms including indigenous species over a range of trophic levels. Toxicity tests to determine 24-h and 48-h LC_{50} values for production waters recovered from an offshore platform on the Scotian Shelf have been conducted with brine shrimp nauplii (*Artemia franciscana*) and marine copepods (*Calanus finmarchicus*). The results showed very low acute toxicity with 100% survival at concentrations up to 5% produced water (Payne et al., 2001). To further evaluate the impact of produced water on early life stages of marine organisms—which are considered more sensitive to toxicants—Querbach et al. (2005) conducted 96-h exposure tests on haddock (*Melanogrammus aeglefinus*), lobster (*Homarus americanus*), and sea scallop (*Placopecten magellanicus*). Over an experimental gradient of produced water concentrations (0, 0.1, 1, 10, and 25%), the onset of significant reductions in survival for yolk-sac haddock larvae, stage-I lobster larvae, and scallop veligers were identified at the 10% or greater level. These results concur with previous studies in other regions that reported acute toxicity values for whole produced water to marine organisms with LC/EC_{50} values ranging from 1 to 50% (Brendehaug et al., 1992; CAPP, 2000; Holdway, 2002).

Based on the concentrations of toxic chemicals in most produced waters and predicted dispersion rates for sites of concern, it is envisioned that there would be only limited potential for acute toxicity beyond the immediate vicinity of rig sites within Atlantic Canada. This is attributed to 1) sensitivity of the biotests—primarily regulatory acute toxicity assays, 2) the rapid dispersion of the process stream, and 3) the limited volumes released at this stage of our offshore operations.

While there are numerous case studies describing rapid dilution of produced water components in the near field (e.g. Somerville et al., 1987; Furuholt, 1996; Berry and Wells, 2005), a level of environmental concern remains as the result of previous field studies. For example, Raimondi and Schmitt (1992) demonstrated that produced water plumes discharged into high-energy open coast environments adversely affected planktonic larvae.

To fully assess the potential impact of produced water discharges, the chronic impacts associated with long-term exposures must be quantified. Continual long-term chronic exposure may cause sub-lethal changes in organisms including decreased community and genetic diversity, lower reproductive success, decreased growth and fecundity, respiratory problems, behavioural and physiological disorders, decreased developmental success, and endocrine disruption (Holdway, 2002).

Evidence to support the need for detailed chronic effect studies include the observation of chronic toxic responses in the clam, *Donax faba*, exposed to produced water concentrations as low as 0.08 ppm (Din and Abu, 1992), and detrimental effects of low produced water concentrations on the reproductive success and development of early life stages of sea urchins (Krause et al., 1992; Krause, 1994). Bioaccumulation studies using oysters (*Crassostrea virginica*) showed that polyaromatic hydrocarbons (PAHs), trace metals, and radium associated with the produced water stream were taken up and accumulated in close proximity to discharges, and also at distances up to 1,000 m (Neff et al., 1992). In chronic studies with early life stages of sea scallops (*Placopecten magellanicus*), Querbach et al. (2005) noted that concentrations of 10% produced water significantly reduced veliger survival and size. In community, population, and ecosystem effect studies, Osenberg et al. (1992) and Rabalais et al. (1992), reported differences in benthic infaunal distributions with distance from the release point of produced water.

In fish, Brown et al. (1998) stated that while acute effects on juvenile turbot larvae were probably limited to the immediate vicinity of the point source of produced water discharge pipe (< 100 m), chronic effects were likely to extend further because produced water dilutions of 10,000 to 100,000 were observed to inflict gill damage. Evidence of detrimental effects on fish exposed to sediments contaminated by produced water includes the induction of ethoxyresorufin O-deethylase (EROD) activity enzymes involved in PAH metabolism (Stagg and McIntosh, 1996). Other effects include gill hyperplasia, reduced phagocytic activity of macrophages and pancreatic necrosis (Hinkle-Conn et al., 1998).

It has been hypothesized that feeding in areas of PAH contaminated sediment would result in increased susceptibility to disease from suppressed immune function, reduced growth, and delayed sexual maturity. However, potential effects on the fisheries are not only limited to PAHs. Alkylphenols detected in produced water from various oil production platforms in the Norwegian sector of the North Sea are now of environmental concern due to their hydrophobic properties, slow biodegradation rates and possible links with estrogenic effects identified in laboratory studies (Tollefsen et al., 1998; Thomas et al., 2004). The full ecological significance of "endocrine disruptors" has not been established, because standard field assessment protocols are still being developed (Lye, 2000). Nevertheless, proposals to control the discharge of these chemicals into the marine environment are being made by the Oslo and Paris Commission (OSPAR) and the Helsinki Commission (HELCOM).

Planktonic species form the base of the marine food chain, constituting 50% of the total marine biomass, and are an essential food supply for all other members (Brakstad et al., 1996). Furthermore, bacteria are also responsible for primary metabolic processes in the ocean including nutrient regeneration and the biodegradation and biotransformation of contaminants. Therefore, the potential decrease in population sizes and the potential for accumulation of toxicants, resulting in transfer to other organisms, is a major threat to the health of the marine ecosystem as a whole. Concerns over the impact of produced water on survival, growth rates, and community structure of planktonic organisms has been identified (Gamble et al., 1987; Somerville et al., 1987; Burns and Codi, 1999; Querbach et al., 2005).

Within Canada, there is much interest in determining the influence of produced water on primary production. Due to elevated ammonia concentrations in produced water discharge (in addition to elevated concentrations of P, Si, Mn, and Fe), changes in the phytoplankton species composition and abundance may occur around offshore platforms. There is a potential for shifts in species succession associated with these blooms that may lead to an increase in desirable benign species or undesirable toxigenic species that could impact the fisheries (Rivkin et al., 2000). Any changes induced from produced water exposure not only affect the given population, but also influence the functioning of species higher in the food web, which flags an 'ecologically significant' impact.

In the open ocean, there is typically a tight coupling between bacterial growth rate and bacterial grazing, resulting in nearly constant amounts of bacterial biomass over time within a biotope. It is hypothesized that produced water may uncouple this link by either increasing bacterial growth because of the possible stimulation of indigenous oil-degrading bacteria, or by reducing grazer efficiency as a toxic response. As bacteria respond very quickly to their environmental conditions a microbial biotest was devised to monitor metabolic responses at sea around the Hibernia platform. Radiotracer studies based on the quantification of ^{3}H-thymidine uptake rates into DNA were conducted to quantify changes in the rate of bacteria-specific growth on exposure to produced water. At the time of the field study, no significant differences were observed in water column depth profiles in transect lines from 0.5 to 20 km away from the Hibernia gravity-based structure (Fig. 1). This finding is consistent with that of Hodgins (1993) who concluded that the zone of impact for acute toxicity around the Hibernia platform (> 1% produced water 96-hr LC_{50}) is confined to the plume and extends out from the discharge port to a maximum distance of 50 m.

SIGNIFICANCE OF CHEMICAL KINETICS

It is suggested that chemical kinetics (natural processes that occur as chemical equilibrium is re-established following

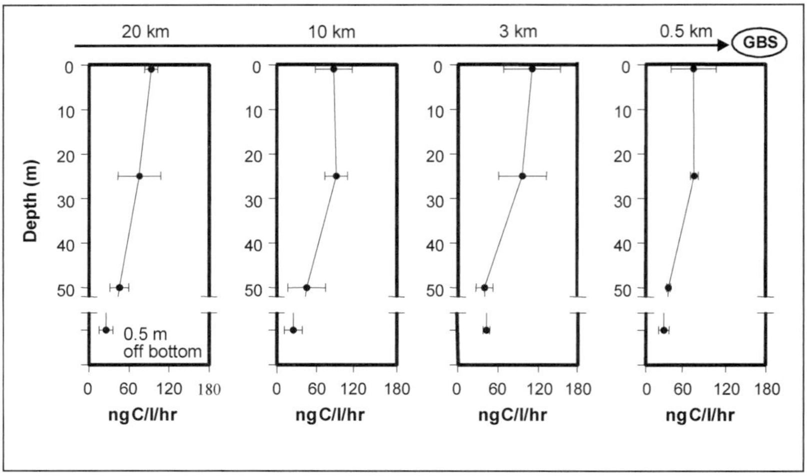

FIGURE 1. Depth profiles of bacterial productivity (surface to 0.5 m off bottom) over a transect line extending from 0.5 to 20 km from the Hibernia gravity base structure (GBS) off the coast of Atlantic Canada.

produced water discharge) is a principal factor controlling the magnitude of produced water toxicity following its discharge into the ocean. For risk analysis, a survey of samples from active east coast production wells has been undertaken to characterise the chemical composition and reactivity of various produced waters following their discharge into the ocean. Laboratory studies to date have demonstrated that certain metals within produced water (predominantly Fe, Mn, and Al) hydrolyse on release into aerobic seawater to form insoluble precipitates. These precipitates flocculate to form aggregates (Fig. 2) that sequester or scavenge other contaminant hydrocarbons and metals (Fig. 3) to facilitate the rapid transport of toxic produced water constituents to the benthic environment where they could impact benthic organisms. Thus, contaminants may be effectively transported out of the water column and concentrated at higher levels within the sediments in the vicinity of the platform. These precipitates have also been shown to attach to buoyant oil droplets (Fig. 4) following their discharge into the sea. As a result, elevated concentrations of co-precipitating toxic metals and residual hydrocarbons may occur within the surface micro-layer. Considering the

FIGURE 2. Formation of hydrolysis metal precipitates from produced water following injection into aerobic seawater. Inset images illustrate the formation of larger particulates from the aggregation of colloids over time.

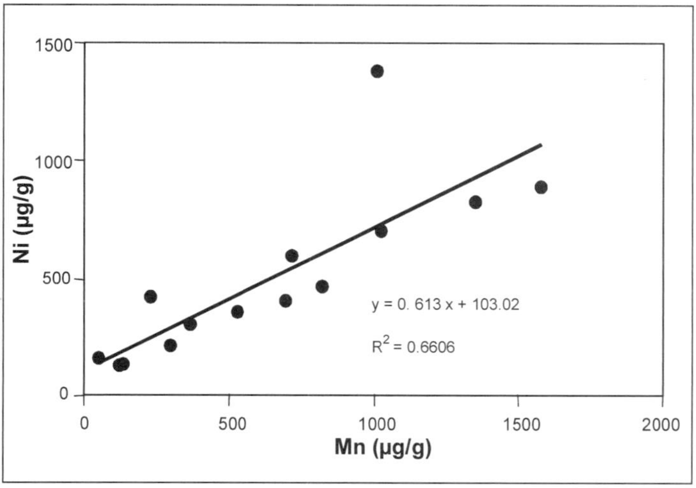

FIGURE 3. Co-precipitation of metals (e.g. Ni and Mn) on produced water precipitates.

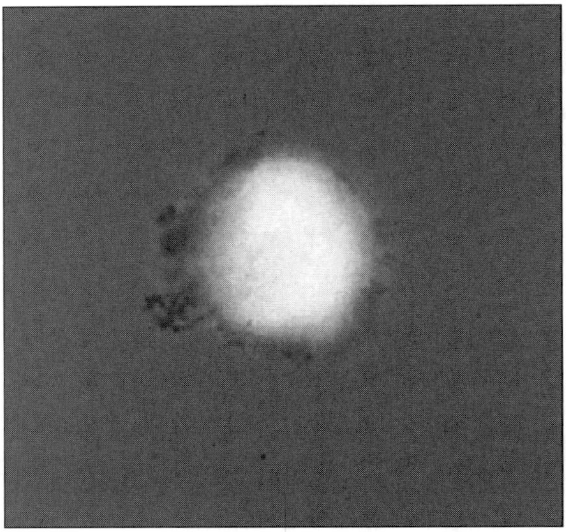

FIGURE 4. Photomicrograph showing the scavenging of particles by oil droplets.

ecological importance of the surface micro-layer to ocean productivity, nutrient recycling and provision of habitat, a specific chemical/toxicological study is warranted to quantify the significance of this contaminant transport mechanism.

Due to the complexity of the potential transport pathways, simple dispersion models cannot accurately predict the fate and transport of produced water contaminants. In terms of biological effects, there is a need to understand the multitude of processes, as illustrated in Figure 5, governing the fate and transport of specific contaminant fractions of concern.

ENVIRONMENTAL EFFECTS MONITORING

Based on the economic reality that the sites of commercial fisheries must coexist with offshore industry, a comprehensive study is needed to quantify the environmental impact of produced water discharges. While it is generally accepted that acute toxicological effects associated with production water may be reduced to acceptable regulatory limits by dilution within a short distance from the point of release (Neff, 1987; Reed and Johnsen, 1995), there is ongoing concern over the long-term effects generated by constant low-level exposure

FIGURE 5. Illustration of processes that influence the transport and fate of the contaminant fractions within produced water discharges.

(Gillbrand et al., 1995). As the composition and concentration of toxic compounds within produced water can vary between sites and within sites over time (Table 3), on-site EEM research programs are needed to provide an early warning of any undesirable change resulting from offshore development operations.

TABLE 3. EC_{25} values from the Microtox biotest (Effective produced water concentration (%) producing 25% inhibition of bioluminescence in the marine bacterium *Vibrio fisheri*) for produced water samples recovered from the same offshore platform over a 7-month period.

Sample Date	Salinity (ppt)	EC_{25} (%)
05-10-98	89	47.2
19-10-98	88	1.8
01-02-99	90	58.1
08-02-99	90	67.8
26-04-99	88	21.8
13-05-99	88	75.1

The degree of produced water toxicity is not just dependent on its composition, bioavailability, and the species exposed. Physical and chemical mechanisms that influence produced water toxicant exposure include dilution, volatilization, chemical reactions, adsorption on suspended solids, and biodegradation (Stephenson, et al., 1994; Berry and Wells, 2005). Variability in regards to contaminant exposure is a major factor of concern to be addressed in the assessment of produced water toxicity in the natural environment. This is due to variability in the composition and discharge rates of the produced water stream and changes in exposure due to both active and passive movement of pelagic organisms (Cranford and Lee, 2004). Furthermore, structural design characteristics of the platform and operational procedures, such as timing of discharges and the position of the disposal out-fall pipe also influence exposure.

The significance of these physico-chemical reactions on toxicity over time are clearly illustrated in the results of laboratory toxicity studies using the Microtox bioassay, which is based on the reaction of the bioluminescent bacterium, *Vibrio fischeri*, to contaminants present in a test sample. *V. fischeri* is a naturally occurring, nonpathogenic marine gram negative species which gives off light as a byproduct of its cellular respiration. Changes in metabolism, resulting from exposure to the test sample, cause changes in light output of the test organism that are measured. Reduction of the light is generally proportional to the biological impact of the sample. The endpoint for the test is the concentration that effects a percentage reduction in light output, in this case the EC_{25} (25% reduction). The Microtox system was used to monitor time-series changes in seawater following the addition of produced water (Fig. 6).

NUMERICAL MODELS FOR RISK ASSESSMENT

Several models have been developed to simulate the dispersion and the toxic effects of produced water on the marine environment (Brandsma and Smith, 1996; Murray-Smith et al., 1996; Røe et al., 1999; Rye et al., 1996; Schobben et al., 1996; Riddle et al., 2001; Sabeur et al., 2002; Neff, 2002). Most studies

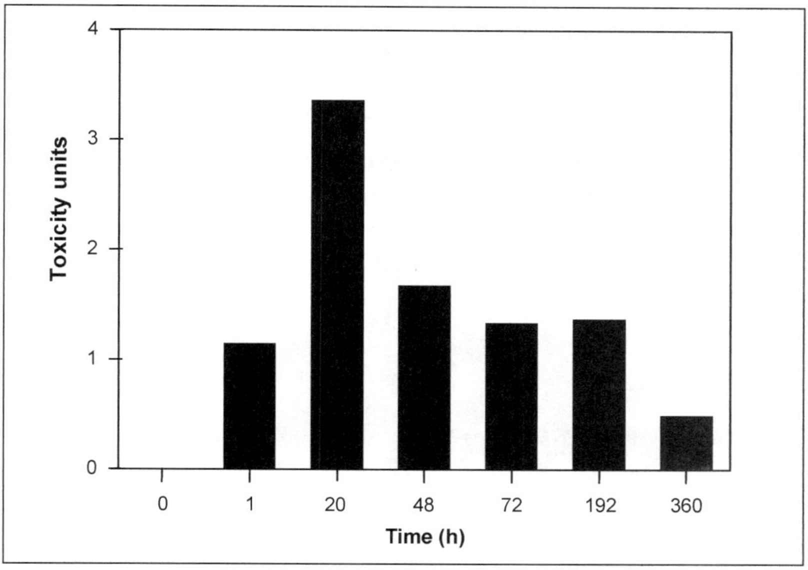

FIGURE 6. Relative time-series alteration in the toxicity of produced water following its release into seawater (monitored with the Microtox biotest). Toxicity increased to a maximum value at 20-hours as a result of both chemical and physical processes that influenced contaminant bioavailability as chemical equilibrium was established following produced water addition.

have focused on the potential impacts of the petroleum hydrocarbon components. However, recent studies have focused on the potential effects of the inorganic nutrients (e.g. ammonia and phosphorus) and dissolved organic carbon fractions (e.g. volatile fatty acids) that may alter planktonic community structure (Rivkin et al., 2000; Khelifa et al., 2003). Environmental changes at this trophic level may alter both ecosystem structure and function.

The numerical models have provided insight into potential cumulative impacts associated with the anticipated future expansion of Canada's offshore oil and gas industry. For example, simulations conducted by Khelifa et al. (2003) for a one year period showed negligible (less than about 2%) impact on the production of the planktonic community when a small discharge of produced water was diluted over a large area (e.g. 1 m^3/day over 0.1 km^2), even at the maximal concentrations of ammonia and volatile fatty acids reported in the literature. However, when the discharge increases by a factor of 10 or 100, the impact on biomass became significant. Annual productivity for the bacterioplankton increased by ten-fold. Alterations in bacterial, phytoplankton, and zooplankton production with depth could be resolved. This information can be used to effectively alter diffuser design and placement to minimise the potential impact of produced water on the marine environment.

CONCLUSIONS

For ecological relevance, we must take a holistic approach to assess the biological impacts of produced water discharges. Due to inherent levels of variability in natural environments, it is extremely difficult to predict the potential long-term impacts of the offshore oil and gas industry. While chemical methods can be used to identify the presence of contaminant gradients in the field, verification of chronic long-term biological effects may be much more difficult as a consequence of delayed toxicity mechanisms such as endocrine disruption or the slow increase in contaminants above background levels. Such factors can only be accurately assessed by long-term studies of chronic impacts on whole populations, communities, and even ecosystems. The effects of chronic toxicity may only become evident after monitoring several life stages or generations. It is important to acknowledge the consequence of long-term effects from offshore installations seeing that they may have a life-span of 15–20 years.

To determine the potential cumulative effects of produced water discharges in the future, mixed populations of indigenous organisms should be exposed to naturally variable effluent concentrations at the levels expected when expansion of production operations are well underway. Biological data from in situ mesocosm (Lee et al., 1985; Gamble et al., 1987) or cage (Serigstad et al., 2002) studies can not only provide new insights into the spatial and temporal distribution of ecological impacts, but when combined with specific physical and chemical data on the fate and transport of specific compounds of concern within the produced water stream, may delineate the mechanism of toxic responses in the field.

REFERENCES

Berry, J.A. and P.G. Wells. 2005. "Environmental Modelling of Produced Water Dispersion with Implications for Environmental Effects Monitoring Design." In: S.L. Armsworthy, P.J. Cranford, and K. Lee (Eds.), *Offshore Oil and Gas Environmental Effects Monitoring: Approaches and Technologies.* Battelle Press (this volume).

Brandsma, M.G. and J.P. Smith. 1996. "Dispersion Modeling Perspectives on the Environment Fate of Produced Water Discharges". In: M. Reed and S. Johnsen (Eds.), *Produced Water 2*, pp. 215-224. Plenum Press, New York.

Brakstad, O.G., Olsen, A.J., Nordtug, T.K., Frost, Aunaas, T. and S. Johnsen. 1996. *Uptake and Degradation of Discharged Produced Water Components in Marine Microorganisms.* SPE Report. 35910:219-229.

Brendehaug, J., S. Johnson, K.H. Bryne, A.L. Gjose, T.H. Eide, and E. Aamot. 1992. "Toxicity Testing and Chemical Characterization of Produced Water a Preliminary Study." In: J.P. Ray and R.F. Engelhardt (Eds.), *Produced Water: Technological/Environmental Issues and Solutions*, pp. 245-256., Plenum Press, New York.

Brown, J.A., S.M. Stephens, and R.M. Stagg. 1998. "Oil Produced Water: Chronic Impacts on Juvenile Turbot." In: *Fish Response to Toxic Chemicals. Proceedings of the International Congress on the Biology of Fish, July 27–30*, 1998, pp.121-127. Baltimore, MD.

Burns, K.A. and S. Codi. 1999. "Non-volatile Hydrocarbon Chemistry Studies Around a Production Platform on Australia's Northwest Shelf." *Estuarine, Coastal and Shelf Science.* 49(6):853-876.

CAPP. 2000. *Produced Water Waste Management.* Publication by the Canadian Association of Petroleum Producers (CAPP). November 2000. 59 p.

Cranford, P.J. and K. Lee. 2005. "A Perspective on Offshore Oil and Gas Environmental Effects Monitoring." In: S.L. Armsworthy, P.J. Cranford, and K. Lee (Eds.), *Offshore Oil and Gas Environmental Effects Monitoring: Approaches and Technologies.* Battelle Press (this volume).

Cranford, P.J., K. Lee, J.W. Loder, T.G. Milligan, D.K. Muschenheim, and J. Payne. 2001. "Scientific Considerations and Research Results

Relevant to the Review of the 1996 Offshore Waste Treatment Guidelines." *Canadian Technical Report on Fisheries and Aquatic Sciences.* 2364:vi + 25 p.

Din, Z.B. and A.B. Abu. 1992. "Sublethal Effects of Produced Water from Crude Oil Terminals on the Clam *Donax faba*." In: J.P. Ray and F.R. Engelhardt (Eds.), *Produced Water (Technological/Environmental Issues and Solutions)*, pp. 445-454. Plenum Press, New York.

Furuholt, E. 1996. "Environmental Effects of Discharge and Reinjection of Produced Water." In: M. Reed and S. Johnsen (Eds.), *Produced Water 2. Environmental Issues and Mitigation Technologies*, pp. 275-288., Plenum Press, New York.

Gamble, J.C., J.M. Davies S.J. Hay, and F.K. Dow. 1987. "Mesocosm Experiments on the Effects of Produced Water Discharges from Offshore Oil Platforms in the Northern North Sea." *Sarsia.* 72:383-386.

Garland, E.M. and E. Aquitaine. 1998. *Produced Water in the North Sea: a Threat for the Environment or a Threat for the Industry.* SPE Report. 46706:1-7.

GESAMP. 1993. "Impact of Oil and Related Chemicals and Wastes on the Marine Environment." IMO/FAO/UNSECO/WMO/WHO/IAEA/UN/UNEP—Joint Group of Experts on the Scientific Aspects of Marine Pollution. *Rep. Stud. GESAMP.* 50:180pp.

Gillbrand, R. Stagg, W.R. Turrell, and G. Furnes. 1995. "Observations and Monitoring of the Results of Low-level Hydrocarbon Discharges in the North Sea." *International Council for the Exploration of the Sea.* Report CM 1995/C: 11, pp. 1-7.

Henderson, S.B., S.W. Grigson, P. Johnson, and B.D. Roddie. 1999. "Potential Impact of Production Chemicals on Toxicity of Produced Water Discharges in North Sea Oil Platforms." *Marine Pollution Bulletin.* 38(12):1141-1151.

Hinkle-Conn, C., J.W. Fleeger, J.C. Gregg, and K.R. Caman. 1998. "Effects of Sediment-bound Polycyclic Aromatic Hydrocarbons on Feeding Behavior in Juvenile Spot (*Leiostomus xanthurus*)." *Journal of Experimental Marine Biology and Ecology.* 227:113-132.

Hodgins, D.O. 1993. *Hibernia Effluent Fate and Effects Modeling.* Report prepared for Hibernia Management and Development Company Ltd. by Seaconsult Marine Research Ltd. November 1993 (Revised January 1994).

Holdway, D.A. 2002. "The Acute and Chronic Effects of Wastes Associated with Offshore Oil and Gas Production on Temperate and Tropical Marine Ecological Processes." *Marine Pollution Bulletin*. 44:185-203.

IAOGP (International Association of Oil & Gas Producers). 2002. "Aromatics in Produced Water: Occurrence, Fate and Efforts, and Treatments." *OGP Publications*. Report No 1.20/324, p. 24.

Khelifa, A., M. Pahlow, A. Vezina, K. Lee, and C. Hannah. 2003. "Numerical Investigation of Impact of Nutrient Inputs from Produced Water on the Marine Planktonic Community." In: *Proceedings of the 26th Arctic and marine Oilspill Program (AMOP) Technical Seminar, Victoria, BC., June 10–12, 2003*. pp. 323-334.

Krause, P.R. 1994. "Effects of an Oil Production Effluent on Gametogenesis and Gamete Performance in the Purple Sea Urchin (*Strongylocentrotus purpuratus* Stimpson)." *Environmental Technology and Chemistry*. 13:153-1161.

Krause, P.R. 1995. "Spatial and Temporal Variability in Receiving Water Toxicity near an Oil Effluent Discharge Site." *Archives of Environmental Contamination and Toxicology*. 29:523-529.

Krause, P.R., C.W. Osenberg, and R.J. Schmitt. 1992. "Effects of Produced Water on Early Life Stages of a Sea Urchin: Stage-Specific Responses and Delayed Expression." In: J.P. Ray and F.R. Engelhardt (Eds.), Produced Water: Technical/Environmental Issues and Solutions. *Environmental Science Research*. 46: 431-444.

Lee, K., C.S. Wong, W.J. Cretney, F.A. Whitney, T.R. Parsons, C.M. Lalli, and J. Wu. 1985. "Microbial Response to Crude Oil and Corexit 9527: SEAFLUXES Enclosure Study." *Microbial Ecology*. 11:337-351.

Lye, C.M. 2000. "Impact of Oestrogenic Substances from Oil Production at Sea." *Toxicological Letters*. 113:265-272.

McCormack, P., P. Jones, M.J. Hetheridge, and S.J. Rowland. 2001. "Analysis of Oilfield Produced Water and Production Chemicals by Electrospray Ionization Multistage Mass Spectrometry [ESI-MSn]." *Water Research*. 35(15):3567-3578.

Murray-Smith, R., D. Gore, S.A. Flynn, I. Vance, and R. Stagg. 1996. "Development and Appraisal of a Particle Tracking Model for the Dispersion of Produced Water Discharge from an Oil Production Platform in the North Sea." In: M. Reed and S. Johnsen (Eds.), *Produced Water 2*, pp. 225-246. Plenum Press, New York.

National Energy Board (NEB), Canada–Newfoundland Offshore Petroleum Board, and Canada–Nova Scotia Offshore Petroleum Board. 2002. *Offshore Waste Treatment Guidelines*. 21 pp.

National Research Council. 1985. *Oil in the Sea*. National Academy Press, Washington, DC.

Neff, J.M. 2002. *Bioaccumulation in Marine Organisms: Effect of Contaminants from Oil Well Produced Water*. Elsevier Science Publishers, Amsterdam. 452 pp.

Neff, J.M. 1987. "Biological Effects of Drilling Fluids, Drill Cuttings and Produced Waters." In: D.F. Boesch and N.N. Rabalais (Eds.), *Long-term Environmental Effects of Offshore Oil and Gas Development*. pp. 469-538. Elsevier Applied Science Publishers Ltd., London.

Neff, J.M., N.N. Rabalais, and D.F. Boesch. 1987. "Offshore Oil and Gas Development Activities Potentially Causing Long-term Environmental Effects." In: D.F. Boesch and N.N. Rabalais (Eds.), *Long-term Environmental Effects of Offshore Oil and Gas Development*. pp. 149-174. Elsevier Applied Science Publishers Ltd., London.

Neff, J.M., T.C. Sauer Jr., and N. Maciolek. 1992. "Composition, Fate and Effects of Produced Water Discharges to Nearshore Marine Waters." In: J.P. Ray and R.F. Engelhardt (Eds.), *Produced Water: Technological/Environmental Issues and Solutions*, pp. 371-385. Plenum Press, New York.

Osenberg, C.W., R.J. Schmitt, S.J. Holbrook, and D. Canestro. 1992. "Spatial Scale of Ecological Effects Associated with an Open Coast Discharge of Produced Water." In: J.P. Ray and R.F. Engelhardt (Eds.), *Produced Water: Technological/Environmental Issues and Solutions*, pp. 387-402. Plenum Press, New York.

Parker, M. and R.C. Ayers. 2001. *Produced Water Waste Management*. Technical Report 2001-0030. Canadian Association of Petroleum Producers, Halifax, NS.

Payne, J., L. Fancey, C. Andrews, J. Meade, F. Power, K. Lee, G. Veinott, and A. Cook. 2001. "Laboratory Exposures of Invertebrate and Vertebrate Species to Concentrations of IA-35 (Petro-Canada) Drill Mud Fluid, Production Water and Hibernia Drill Mud Cuttings." *Canadian Manuscript Report on Fisheries and Aquatic Sciences*. No. 2560: iv + 27.p.

Querbach, K., G. Maillet, P. Cranford, C. Taggart, K. Lee, and J. Grant. 2005. "Potential Effects of Produced Water Discharges on the Early

Life Stages of Three Resource Species." In: S.L. Armsworthy, P.J. Cranford and K. Lee (Eds.), *Offshore Oil and Gas Environmental Effects Monitoring: Approaches and Technologies*. Battelle Press (this volume).

Rabalais, N.N. and D.F. Boesch. 1987. "Offshore Oil and Gas Development Activities Potentially Causing Long-term Environmental Effects." In: D.F. Boesch and N.N. Rablais (Eds.), *Long-term Environmental Effects of Offshore Oil and Gas Development*. Elsevier Applied Science, London, 708 pp.

Rabalais, N.N., B.A. McKee, D.J. Reed, and J.C. Mearns. 1992. "Fate and Effects of Produced Water Discharges in Coastal Louisiana, Gulf of Mexico, USA." In: J.P. Ray and R.F. Engelhardt (Eds.), *Produced Water: Technological/Environmental Issues and Solutions*, pp. 355-369. Plenum Press, New York.

Raimondi, P.T. and R.J. Schmitt. 1992. "Effects of Produced Water on the Settlement of Larvae: Field Test Using Red Abalone." In: J.P. Ray and R.F. Engelhardt (Eds.), *Produced Water: Technological/Environmental Issues and Solutions*, pp. 415-430. Plenum Press, New York.

Riddle, A.M., E.M. Beling, and R.J. Murray-Smith. 2001. "Modelling the Uncertainties in Predicting Produced Water Concentrations in the North Sea." *Environmental Modelling and Software*. 16(7):659-668.

Roe Utvik, T.I. 1999. "Chemical Characterization of Produced Water from Four Offshore Oil Production Platforms in the North Sea." *Chemosphere*. 39:2593-2606.

Roe Utvik, T.I., G.S. Durell, and S. Johnsen. 1999. "Determining Produced Water Originating Polycyclic Aromatic Hydrocarbons in North Sea Waters: Comparison of Sampling Techniques." *Marine Pollution Bulletin*. 11:977-989.

Rye, H., M. Reed, A. Melbye, and S. Johnson. 1996. "An Intercomparison between Field Measurements and Three Different Calculation Models for Estimates of Dilution Factors." In: M. Reed and S. Johnsen (Eds.), *Produced Water 2*, pp. 257-269, Plenum Press, New York.

Schobben, H.P.M., E.A. Vik, G.G. Hutjes, C.C. Karman, and D.G. Ofjord. 1996. "An Introduction to the CHARM Model." In: M. Reed and S. Johnsen (Eds.), *Produced Water 2*, pp. 295-301. Plenum Press, New York.

Serigstad, J., E. Thain, K. Hylland, G. Becker, J. Klungsoyr, T. Lang, A. McIntosh, K.V. Thomas, T.I.R. Utvik, D. Vethaak, and W. Wosniok.

2002. "The Use of In Situ Deployment of Live Organisms and Passive Samplers to Monitor Contaminants in Pelagic Ecosystems." In: *ICES (International Council for the Exploration of the Sea) Theme Session on Biological Effects of Contaminants in Marine Pelagic Ecosystems.* ICES CM: 2002/X:06.

Shaw, D.G., J.W. Farrington, M.S. Connor, B.W. Trippm, and J.R. Schubel. 1999. *Potential Environmental Consequences of Petroleum Exploration and Development on Grand Banks.* New England Aquarium Aquatic Forum Series Report 99-3, Boston. 64 pp.

Snavely, E.S. 1989. *Radionuclides in Produced Water.* American Petroleum Institute, Washington, D.C.

Somerville, H.J., D. Bennett, J.N. Davenport, M.S. Holt, A. Lynes, A. Mahieu, B. McCourt, J.G. Parker, R.R. Stephenson, R.J. Watkinson, and T.G. Wilkenson. 1987. "Environmental Effects of Produced Water from North Sea Oil Operations." *Marine Pollution Bulletin.* 18:549-558.

Stagg, R.M. and A. McIntosh. 1996. "Hydrocarbon Concentration in the Northern North Sea and Effects on Fish Larvae." *Science of the Total Environment.* 186:189-201.

Stephenson, M.T. 1992. "A Survey of Produced Water Studies." In: J.P. Ray and R.F. Engelhardt (Eds.), *Produced Water: Technological/Environmental Issues and Solutions*, p. 46. Plenum Press, New York.

Stephenson, M.T., R.C. Ayers, L.J. Bickford, D.D. Caudle, J.T. Cline, G. Cranmer, Duff, E. Garland, T.A. Herenius, R.P.W.M. Jacobs, C. Inglesfield, G. Norris, J.D. Petersen, and A.D. Read. 1994. *North Sea Produced Water: Fate and Effects in the Marine Environment.* Report No. 2. 62/204. E&P Forum, London, England. 48 pp.

Tollefsen, K.E., K. Ingebrigtsen, A.J. Olsen, K.E. Zachariassen, and S. Johnsen. 1998. "Acute Toxicity and Toxicokinetics of 4-heptylphenol in Juvenile Atlantic Cod (*Gadus morhua* L.)." *Environmental Toxicology and Chemistry.* 17:740-746.

Thomas, K.V., J. Balaam, M.R. Hurst, and J.E. Thain. 2004. "Identification of In Vitro Estrogen and Androgen Receptor Agonist in North Sea Offshore Produced Water Discharges." *Environmental Toxicological Chemistry.* 23:1156-1163.

Potential Effects of Produced Water Discharges on the Early Life Stages of Three Resource Species

Querbach, K. (querbachk@mar.dfo-mpo.gc.ca) (Fisheries and Oceans Canada, Dartmouth, NS, Canada), G. Maillet (Fisheries and Oceans Canada, St. John's, NL, Canada), P.J. Cranford (Fisheries and Oceans Canada, Dartmouth, NS, Canada), C. Taggart (Dalhousie University, Halifax, NS, Canada), K. Lee (Fisheries and Oceans Canada, Dartmouth, NS, Canada), and J. Grant (Dalhousie University, Halifax, NS, Canada)

ABSTRACT: Acute and chronic effects of exposure to produced water (PW) from offshore drilling were quantified for the early life stages of haddock (*Melanogrammus aeglefinus*), lobster (*Homarus americanus*), and sea scallop (*Placopecten magellanicus*) in terms of survival, growth, and fertilization success. During 96-h exposures over a range from 0–25% PW, yolk-sac haddock larvae, fed stage-I lobster larvae, and scallop veligers each displayed significant reductions in survival at concentrations greater than 10%. The average size of scallop veligers was significantly reduced after exposure to 10 and 25% PW. Scallop fertilization success was significantly reduced at all concentrations ≥ 1%. During 18 d chronic exposures to over a range of concentrations between 0 and 10% PW, significant reductions in scallop veliger survival and size were observed at the highest treatment level. Chronic exposure of the diatom, *Thalassiosira pseudonana* to 10% PW resulted in a significant reduction in physiological condition, though there was no effect on chlorophyll-*a* concentration.

INTRODUCTION

The potential environmental effects of exploration, production, and transport related to offshore hydrocarbon development highlight concerns related to operational wastes and their disposal. Many regions of the world, including eastern Canada, are currently experiencing a rapid increase in the exploration and development of oil and gas reserves on the continental shelves. Recent international studies in regions of extensive oil and gas drilling activity have documented evidence of environmental impacts at relatively large scales (Cameron and Berg, 1992; Olsgard and Gray, 1995; Stagg and McIntosh, 1996; Raimondi et al., 1997; Patin, 1999) that may be correlated to the discharge of drilling fluids and production waters.

The largest volume of waste produced during oil and gas production is produced water (PW). Over the life of a producing field, PW can exceed the volume of hydrocarbons extracted by an order of magnitude (Stephenson, 1992). PW is connate water naturally occurring in the oil and/or gas reservoir and contains a number of metal and organic constituents depending on the geological formation and field operations, and can also contain various amounts of injection waters (GESAMP, 1993). The discharge of PW has been linked to a significant fraction of the contaminant petroleum hydrocarbons found in ocean environments (Syversten, 1996). Other primary contaminants of concern within PW discharge streams include: hydrolysis metals, heavy metals, radionuclides, and treatment chemicals (added after removal from reservoir). Although, in some instances, PW can be re-injected into the well for enhanced hydrocarbon recovery (or disposal; e.g. USA), the most frequently used practice is direct discharge into the marine environment (Rabalais et al., 1992; Stephens et al., 2000).

The discharge of produced water has been considered by many to be environmentally inconsequential based on the assumption that it is rapidly dispersed (diluted) at sea (Flynn et al., 1996; Neff and Sauer, 1996; Murray-Smith et al., 1996; Stagg and McIntosh, 1996; Neff, 2002). However, there is increasing concern of potential impacts on marine organisms. Benthic organisms have generally been the focus of environmental effects monitoring programs, as the bulk of solid wastes discharged from drilling platforms sediment rapidly and flocculation processes can concentrate the contaminant fraction and accelerate its transport to the benthic environment (Muschenheim et al., 1995). Nonetheless, fisheries biologists are now reconsidering the potential effects of PW on fish following impacts reported in the North Sea (Stagg and McIntosh, 1996; Brown et al., 1999; Stephens et al., 2000) and indirect effects observed on fish larvae (cod and herring) resulting from direct impacts on phytoplankton and zooplankton (Gamble et al., 1987).

Acute toxicity tests, generally based on a 96-h LC_{50} (lethal concentration to 50% of the test population) protocol, have characterised PW as slightly toxic (LC_{50} = 0.1% PW) to practically non-toxic (LC_{50} = 1% PW) (Neff, 1987). Such studies are generally conducted with test organisms that are readily maintained in the laboratory such as mysids, sea urchins, amphipods, and sheepshead minnows. However, the acute toxicity of PW has been shown to be highly species-specific (Neff, 1987; Schiff et al., 1992) and there is limited empirical evidence of the biological effects of operational drilling wastes on the commercial finfish and shellfish species residing in the production area. In addition to multi-species considerations, a comprehensive evaluation of PW discharges on fisheries demands assessment across the range of developmental stages typical of most marine organisms. Toxicological studies on the early developmental stages are particularly warranted as these are generally more sensitive to toxicants than adults (e.g. Linden, 1975; Lonning, 1977; Solbakken et al., 1984; Britt et al., 1987; Neff, 1987) and many are unable to relocate. Consequently, we initiated this preliminary study to directly assess the toxicity of PW (at concentrations expected very near the discharge point) to the early life stages of haddock (*Melanogrammus aeglefinus*), lobster (*Homarus americanus*), and sea scallop (*Placopecten magellanicus*). These species were chosen because: 1) they exhibit typical developmental stages for most marine fish, crustaceans, and molluscs; 2) each is of significant economic value throughout their ranges; and 3) they inhabit the range of offshore hydrocarbon exploration and development along continental margins of the north Atlantic.

Inherent sensitivity of some of the early-life stages of the species studied resulted in mortality of laboratory control populations that exceed established "classical" lab-based criteria for assessing bioassay results for regulatory compliance. It is not possible at this time to implement typical toxicity-based criteria (i.e. standard "regulatory" biotests) for assessing PW impacts on these species early-life stages, as these organisms' biology does not permit it. Natural mortality of the early life stages of these species both in nature and in the laboratory is extremely high. Nevertheless, considering their economic importance, and the capacity of suitably controlled experimental conditions to detect significant biological effects, this factor should not preclude the use of important resource species and sensitive life stages for assessing the potential for anthropogenic impacts.

MATERIALS AND METHODS

Test Organisms

Fertilized eggs from naturally spawning haddock (*Melanogrammus aeglefinus*) were obtained at an age of 24–48 h from a captive breeding stock maintained by the National Research Council facility at Sandy Cove, Nova Scotia. Eggs and larvae were incubated at 8–9°C in static, 0.2 μm filtered natural seawater in a 100 l tank with overhead fluorescent lighting that provided a surface irradiance of 25 μmol photons/m^2/s on a 12 h light:12 h dark photoperiod. The insulated incubation tank was circular, with a black interior, and was covered with clear acrylic. Aeration was provided with compressed air filtered to 0.2 μm and passed through activated charcoal. Non-viable (negatively buoyant and opaque) eggs and dead larvae were removed from the bottom of the tank every 1–3 days. First-feeding larvae were offered cultured algae (*Nanochloris* spp.) and a marine rotifer (*Brachionus plicatilis*) *ad libitum*.

Female adult lobsters (*Homarus americanus*) carrying fertilized eggs were collected from Georges Bank by the commercial fishery and transferred to a continuous-flow, 1500 l tank maintained at ambient seawater temperatures (5–12°C). Adult lobsters were fed *ad libitum* a dry pellet formulation (Maine Lobster Technology Inc., Hancock, Maine).

Adult sea scallops (*Placopecten magellanicus*) were collected from Georges Bank by a commercial dragger and from a mariculture facility in Chester Basin, Nova Scotia. Scallops were transferred to a continuous-flow seawater tank maintained at 13°C with a daily addition of the cultured algae *T. Isochrysis galbana* (*T-iso*) for feed. Spawning was induced by thermal shock (4°C increase in temperature). The marine diatom *Thalassiosira pseudonana* was grown at 20°C and at an average light intensity of 35 μmol photons/m^{-2} s. The light source was provided by a bank of 40 W cool-white fluorescent fixtures on a 16 h light:8 h dark photoperiod.

Produced Water

For the study of chemical kinetics following the discharge of produced water, a number of 2 l amber glass bottles were completely filled with samples of PW from the operational discharge line of the CoPan offshore production site (PanCanadian Petroleum Ltd.) located on the Scotian Shelf near Sable Island on May 11 and September 1, 1998. The samples were transported to the laboratory by helicopter within hours of recovery. Salinity of these fresh samples which had no obvious color or turbidity was 83 ppt and 90 ppt. Sample bottles destined for the toxicity studies were maintained at 8–13°C under nitrogen gas until used in the experiments. The concentration of total GC-detectable petroleum hydrocarbons (TPH) was determined on dichloromethane extracted samples with a Hewlett-Packard (HP) 5890 gas chromatograph equipped with a flame-ionization detector (FID). The concentration of total GC-resolved hydrocarbons in the samples was 10.7 and 9.8 ppm, respectively. Nutrient concentrations for the respective samples were: 25.9 and 27.0 mg/L NH_4; 0.019 and 0.015 mg/l PO_4 and non-detectable concentrations of NO_2 and NO_3.

Acute Toxicity Evaluations

Acute toxicity exposure conditions are summarised in Table 1. All glassware was soaked in 10% HCl for 24 h, rinsed with Nanopure water and dried at 60°C prior to use. Produced water concentrations were prepared using glass volumetric pipettes, which dispensed the PW into vials containing 0.2 μm filtered seawater. Although a concentration as high as 25% PW is not likely to be sustained at any distance away from the release point of the discharge pipe, this high concentration was included to provide insight into threshold values for these animals. It should further be noted that production waters are often highly saline and that alone may contribute to any toxicity at the high concentrations. Test organisms were added after a 48–72 h incubation period at experimental temperatures. Lobster larvae were in the first stage as described by Aiken and Waddy (1989). The 'fed lobster' larvae treatments were supplied with a ground dry lobster feed as noted above for adults. The addition of PW, in its reduced state, to aerobic test waters resulted in the precipitation of hydrolysis metals (flocculation) prior to exposure to test organisms.

Haddock eggs were individually placed in the test vials using soft aluminum forceps and larvae were added using wide-bore Pasteur pipettes. Lobster larvae were transferred individually to vials using soft-forceps. Scallop embryos, sperm and eggs were added using an Eppendorf repeater pipette and additional triplicate samples were added to vials containing buffered formaldehyde for determination of initial numbers. All vials were randomly placed in a rotator set at 3 rpm for the duration of the experiment to maintain the suspension of PW particulates. Acute toxicity tests were of 96-h duration, except for the scallop gamete fertilization test which was terminated after 48-h as the embryos degraded quickly after death. Oxygen content (Strathkalvin oxygen meter fitted with a Radiometer electrode), pH, temperature, and salinity (Orion

TABLE 1. Summary of experimental conditions for acute toxicity tests with produced water.

Condition	**Species** *Placopecten magellanicus* (sea scallop)	*Homarus americanus* (lobster)	*Melanogrammus aeglefinus* (haddock)
Test type	Static	Static	Static
Dilution water	Filtered sea water	Filtered sea water	Filtered sea water
Age of organism	1 - 2 h gametes	7 - 10 d larvae	1 - 4 d embryo (ESE)
	5 - 7 d veliger	14 - 17 d larvae	8 -12 d embryo (LSE)
			3 - 7 d larvae (YSL)
			13 - 17 d larvae (OSL)
Test duration	gametes - 48 hours	96 hours	96 hours
	veligers - 96 hours		
Temperature	13°C	8°C	8°C
Photoperiod	12 h light:12 h dark	12 h light:12 h dark	12 h light:12 h dark
Light Irradiance (μmol photons/m^2/s)	25	25	25
Dilution (%)	0, 0.1, 1, 10, and 25	0, 0.1, 1, 10, and 25	0, 0.1, 1, 10, and 25
No. of Replicates	10	10	10
Organisms/Replicate	150 - 200	3	10
Vessel Volume	24 ml	24 ml	24 ml
Water Volume	24 ml	24 ml	20 - 24 ml
Feeding	*T. Isocrysis galbana*	7-10 d fed 14-17 d not fed	algal cells and rotifers
Aeration	none	none	none
Rotations per minute	3	3	3
Effects Measured	Gametes - Fertilization	Death	Death
	Veligers - Death, Size		
Other Measurements	Oxygen	Oxygen	Oxygen
	pH	pH	pH
	Salinity	Salinity	Salinity

conductivity meter) were monitored in vials periodically during the exposure period.

Following incubation, the vials were removed from the rotator and the numbers of live and dead individuals in each replicate were determined. Mortality was determined by direct observation using a Wild dissecting microscope. Live and dead animals were assessed using the following criteria: clarity and buoyancy for haddock eggs; heart activity for haddock and lobster larvae; and motility of flagella, cilia, and/or internal organs for sea scallop veligers. In addition to addressing the lethal effects of each PW treatment on scallop veligers, the effect on growth was determined by measuring veliger size distribution. The size distribution of initial and final samples of scallop larvae from each treatment were determined using a Coulter Multisizer II Particle Analyzer equipped with a calibrated 400 μm aperture tube. Size was measured as Equivalent Spherical Diameter (ESD).

Fertilization success of sea scallop eggs during the 48-h incubation of eggs and sperm to each PW concentration was assessed by determining the ratio of the average number of live trochophore larvae in five replicate 0.5 ml sub-samples to the estimated number of fertilized eggs (using the preserved samples) added to each vial.

Chronic Toxicity Evaluations

Chronic toxicity test conditions for each species are summarised in Table 2. Glassware was treated as detailed above for acute toxicity evaluations. Produced water concentrations were prepared with filtered sea water (0.2 μm) using glass volumetric pipettes to dispense the PW into each bottle. These concentrations were chosen to represent likely dilutions of PW directly after release into the marine environment including concentrations that might be found very close to the platform (e.g. 10%). Chambers containing PW were allowed to equilibrate for 48–72 h at experimental temperatures before the addition of organisms. Typically high mortality in the haddock and lobster control populations during the acute tests prevented an evaluation of the chronic effects of PW. Sea scallops proved easier to maintain and approximately 13,000, 5–7 d old larvae were added to each bottle by pipette at the beginning of an 18 d exposure period along with the *T-iso* food source. The bottles were placed on a rotator at 3 rpm for the 18 d exposure.

At ~ 2-d intervals, 20 ml sub-samples were collected from each bottle for determination of chlorophyll-*a* concentration (see below), veliger mortality (assessed as above) and preservation in formaldehyde. On days 5, 8, 12, 15, and 18, sub-samples (80 ml) were collected to determine veliger size distribution (as above). After sub-samples were removed from exposure bottles, similar volume (to that removed) of seawater containing

TABLE 2. Summary of experimental conditions for chronic toxicity tests with produced water.

	Species	
Condition	***Placopecten magellanicus*** **(sea scallop)**	***Thalassiosira pseudonana*** **(algae, diatom)**
Test type:	Static	Static
Dilution water:	Filtered sea water	Filtered sea water
Age of organism:	5 - 7 d veliger	-
Test duration:	18 d	10 d
Temperature:	13°C	8°C
Photoperiod:	12 h light:12 h dark	12 h light:12 h dark
Light Irradiance (μmol photons m^{-2} s^{-1}):	25	35
Dilution (%):	0, 0.01, 0.1, 1, and 10	0, 0.01, 0.1, 1, and 10
No. of Replicates:	2	2
Organisms/Replicate:	18,000	50 μg/l Chl-*a*
Vessel Volume:	1000 ml	1000 ml
Water Volume:	1000 ml (small air space)	1000 ml (small air space)
Feeding:	*T. Isochrysis galbana* at 2 d intervals	enriched f/2 media
Aeration:	none	none
Rotations per minute:	3	3
Effects Measured:	Death, Growth	Chl-*a*, Condition (CFC)
Other Measurements:	Oxygen	Oxygen
	pH	pH
	Salinity	Salinity
	Chl-*a*	

the correct compensating PW concentration and fresh *T-iso* food particles was added.

To assess the impact of PW on larval food sources, a 10-d chronic exposure experiment was conducted using the marine diatom, *Thalassiosira pseudonana*. A diatom inoculum in an enriched f/2 seawater media (Guillard, 1975) was added to each bottle containing the desired PW concentration (0, 0.01, 0.1, and 10%). The bottles were rotated during exposure as above. On every second day of the exposure, duplicate sub-samples were collected (without replacement) for determination of chlorophyll-*a* concentration (μg/l), and cellular fluorescence capacity (CFC) using a Turner Designs fluorometer (model 10-001). Samples for chlorophyll-*a* analysis were filtered (Whatman GF/F) and extracted in 10 ml of 90% acetone for 24 h at 10°C before measurement. Cellular fluorescence capacity is a physiological index of relative photosynthetic efficiency, and was estimated according to Vincent (1980).

Calculations and Statistical Analysis

The lethal concentration for 50% of the initial population (LC_{50}) was computed using the Toxstats program (created by Charles Stephan and modified by Gary Westlake of the Ontario Ministry of Environment and Energy in 1997). LC_{50} values were estimated using the probit method only when the survival in the control populations averaged 90% or above.

All data were tested for normality using the Kolmogorov-Smirnov test in Graphpad Prism® (San Diego, CA) and were subsequently normalised using the arcsine square root transformation. Analysis of variance (ANOVA) was used to assess exposure time and/or treatment (PW concentration) effect on the response variables (survivorship, growth, and fertilization success) from acute and chronic exposure experiments. Where significant treatment effects were observed, multiple comparison tests (Bonferroni corrected) were used to determine which treatments differed based on the null hypothesis that mean response values are equal among treatments at the $\alpha = 0.05$ level of significance. Statistical power ($1 - \beta$) was determined for all proportional data to validate the ANOVA results, where β is the probability of committing a Type II error. Power was calculated using Sample Power Ver. 1.00 (SPSS Inc., Chicago, Ill.) and power values > 0.8 were assumed sufficient to guard against Type II errors.

Survival as described by Ouellet et al. (1992) was used to compare the percent mortality of scallop veliger larvae between treatments during the chronic exposure. Kruskal-Wallace distribution comparisons were conducted to verify significant differences between size distributions of scallop veligers and for CFC data from the *T. pseudonana* chronic exposure.

Particle size data collected during the chronic exposure study for the highest concentrations included both PW particles and scallop veliger larvae. Measurements of veliger ESD obtained using the Multisizer were confirmed by direct observation and image analysis. The control, 1%, and 10% PW concentrations were used for this comparison. Due to the difficulty of measuring veliger width, images of width and area were collected (sub-sample = 30 veligers) using a dissecting microscope and Zipshot digital camera system. These morphometric variables were estimated using SigmaScan Pro and the relationship between area and width was determined using linear regression analysis ($r^2 = 0.69$). The regression parameters were used to estimate the width of all other veligers on the basis of the area estimate. ESD was estimated using an oblate spheroid approximation; this shape best represented the scallop veliger. Finally, a sample of PW was analysed using the Multisizer and 200 and 50 μm apertures to further aid in deciphering veligers from PW particles.

RESULTS

Acute Toxicity Evaluations–Haddock

All life-stages of haddock suffered a decrease in survival at the 25% PW concentration (Fig. 1a). Yolk-sac larvae were particularly sensitive with a decreased survival at the 10% PW level. Survival of control haddock (Fig. 1a) was high (> 80%) for late stage embryos (LSE) and yolk-sac larvae (YSL), but relatively low (~50%) for the early-stage embryos (ESE) and older feeding-stage larvae (OSL). LC_{50} values were not calculated for most life-stages owing to < 90% survival in the controls, although it was estimated at 21.8% PW for the LSE. Significant effects ($P < 0.05$, ANOVA) of PW on survival of LSE, YSL, and OSL were apparent (Table 3). The ESE appears to be relatively insensitive as no significant effect on mortality was observed over the concentration range. We caution the interpretation of this latter result as the test-power was 0.6. The LSE and OSL showed significant reductions in mean survival at 25% PW and YSL showed a significant reduction in survival at the 10% concentration (Table 3).

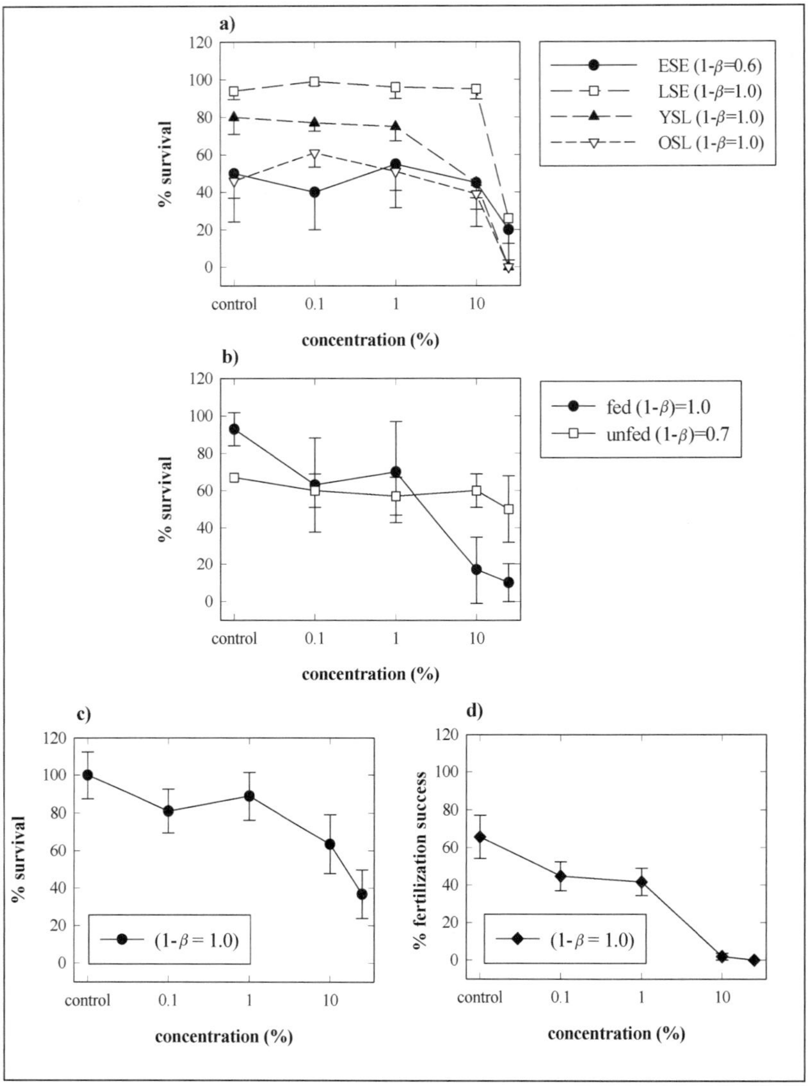

FIGURE 1. The percent survival of haddock **(a)**, lobster **(b)** and scallop **(c)** early life stages (abbreviations in legends are defined in the text) after acute exposure (96-h) to produced water (PW). **(d)** Percent fertilization success of sea scallop eggs during acute exposure (48-h) of gametes to PW. Error bars include –2 standard error (SE) for **(a)** and ± 2 SE for **(b)**, **(c)**, and **(d)**.

TABLE 3. Results of multiple comparison tests after ANOVA of percent survival data for the different life-stages of haddock exposed to different concentrations of produced water in 96-h acute toxicity tests. Treatments that differed significantly are indicated in each column. Treatment a = Control, b = 0.1%, c = 1%, d = 10%, and e = 25%. * = $P < 0.05$, ** = $P < 0.01$, and *** = $P < 0.001$. (e.g. for the YSL in 1% produced water; survival in 1% produced water differed significantly from both 10 and 25% treatments with a P-value of < 0.001).

Stage	Produced Water Concentration				
	0%	0.1%	1%	10%	25%
ESE	-	-	-	-	-
LSE	e***	e***	e***	e***	a,b,c,d***
YSL	d, e***	d, e***	d, e***	a,b,c,e***	a,b,c,d***
OSL	e***	e***	e***	e***	a,b,c,d***

Acute Toxicity Evaluations—Lobster

We were unable to determine any significant change in survival of unfed lobster larvae with increasing PW concentration (Fig 1b), partially as a result of low statistical power ($1 - \beta = 0.7$). However, a significant ($P < 0.0001$, ANOVA) reduction in survival was apparent in all PW treatments for fed larvae (Fig. 1b) where the LC_{50} value was 0.9% PW. Multiple comparison tests showed that lobsters in the 10 and 25% PW treatments experienced significantly greater mortality than the control population. As a result of test enclosure size and slow mixing (i.e. rotation rate), cannibalism among lobster larvae was evident. Although mortality due to cannibalism may be constant across treatments, we cannot attribute observations of reduced lobster survival in PW treatments to the exposure alone.

Acute Toxicity Evaluations—Sea Scallop

Survival of veliger-stage scallops was high (> 80%) for PW concentrations below 1% but declined at higher levels (Fig. 1c) and LC_{50} was estimated at 21% PW. Significant differences in survival over the range of PW concentrations ($P < 0.0001$, ANOVA) were determined and the multiple comparison tests indicated that the 10 and 25% PW treatments significantly reduced survival ($P < 0.01$).

The fertilization success (percent of initial unfertilized eggs developing to the trocophore stage) of scallop eggs was less than 70% in all treatments and decreased with increasing PW concentration (Fig. 1d). There was a highly significant effect of the PW treatments ($P < 0.0001$) and multiple comparison tests showed all PW concentrations ≥ 1% resulted in reduced fertilization relative to the control ($P < 0.01$). Few eggs were successfully fertilized at PW concentration > 10% (Fig. 1d).

The mean size (ESD μm) of early veliger larvae after acute exposure to PW showed a general decrease with increasing concentration (Fig. 2a). Veligers in the 0, 0.1, and 1% PW treatments increased in mean size over the 96-h exposure period, while the 10 and 25% PW treatments resulted in a decrease (Fig. 2a). There was a highly significant treatment effect on mean size ($P < 0.0001$) and multiple comparisons revealed that this resulted from significantly smaller veligers in the 10 and 25% PW treatments. Skewness calculations of these distributions revealed that PW concentrations > 10% resulted in fewer larger individuals than observed in the control ($S_{control} = 1.01$, $S_{10\%} = 2.31$ and $S_{25\%} = 2.54$).

Chronic Toxicity Evaluations—Sea Scallop

It is apparent from Figure 2b that veliger size at age after exposure (a surrogate for growth) was similar in all treatments except for 10% PW, where no significant positive growth was observed (large overlap in standard error). The exposure to 10% PW was terminated on day 15 due to the low number of individuals remaining. Image analysis confirmed that the particles for which we calculated mean values were scallops veligers and PW particles ranged in size from 0.9 to 30 μm (<< the initial size of the veligers). Much of the increase in mean size in all treatments occurred within the first 10 days of exposure. By day 12 there was a significant treatment effect on mean size ($P < 0.004$) due to the impact of the 10% PW concentration. The relatively high chlorophyll-*a* concentration in the 10% treatment is indicative of reduced feeding rate by veligers (Fig. 2c).

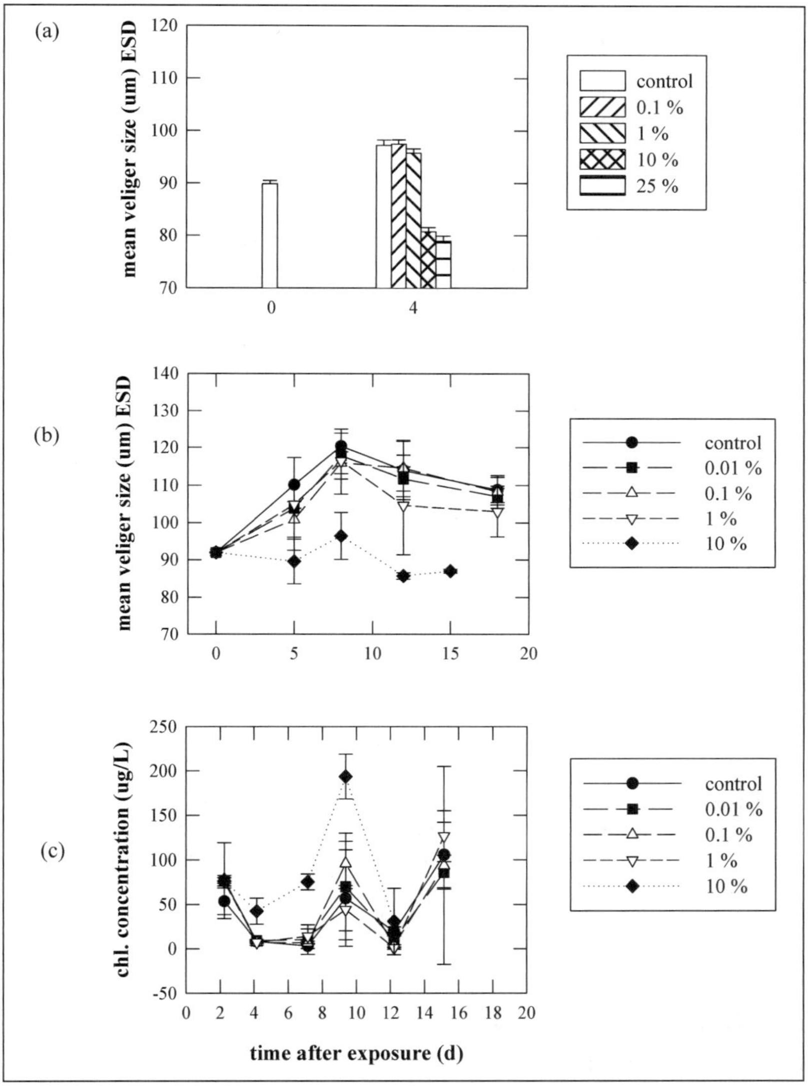

FIGURE 2. Mean size as equivalent spherical diameter (ESD, μm) of sea scallop veligers **(a)** before and after acute exposure (96 h), and **(b)** during chronic exposure (18 d) to the different produced water concentrations. **(c)** Chlorophyll-a concentration (μg/l) of *T. I. galbana* during the chronic exposure. The error bars include ± 2 SE.

With the exception of the 10% PW treatment, scallop survival was generally high during the test (Fig. 3). Mortality in the 10% treatment was extreme, with > 50% of the initial population dead by day 10.

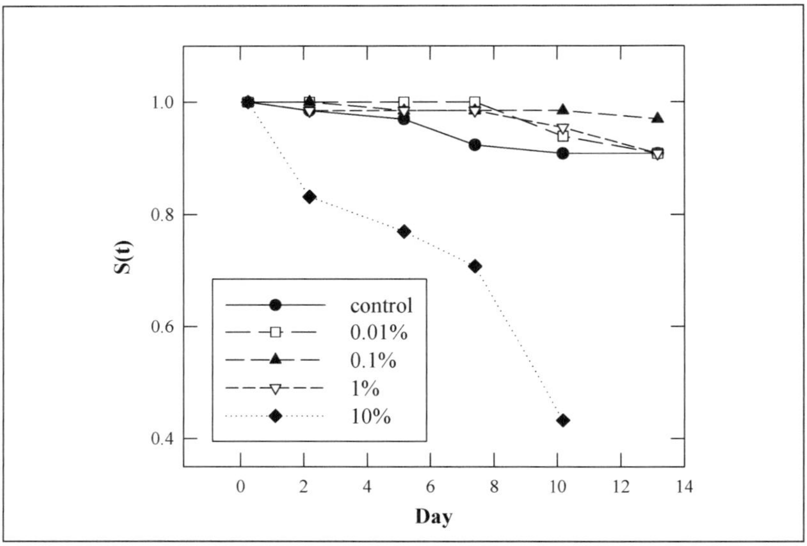

FIGURE 3. Survival of sea scallop veligers during 18 d chronic exposures to 0, 0.01, 0.1, 1, and 10% produced water. S(t) is the cumulative probability of surviving beyond a given time. The data have been normalised between 0 and 1.

Chronic Toxicity Evaluations—Marine Diatom

An exponential increase in algal biomass occurred in all treatments during the test (Fig. 4a). There was no indication of reduced biomass in the highest PW concentrations at the end of the test (large overlap in standard error), and there was even a slight, but not significant increase. However, cellular fluorescence capacity decreased in all treatments during the test (Fig. 4b), with reduction being greatest at the 10% PW concentration.

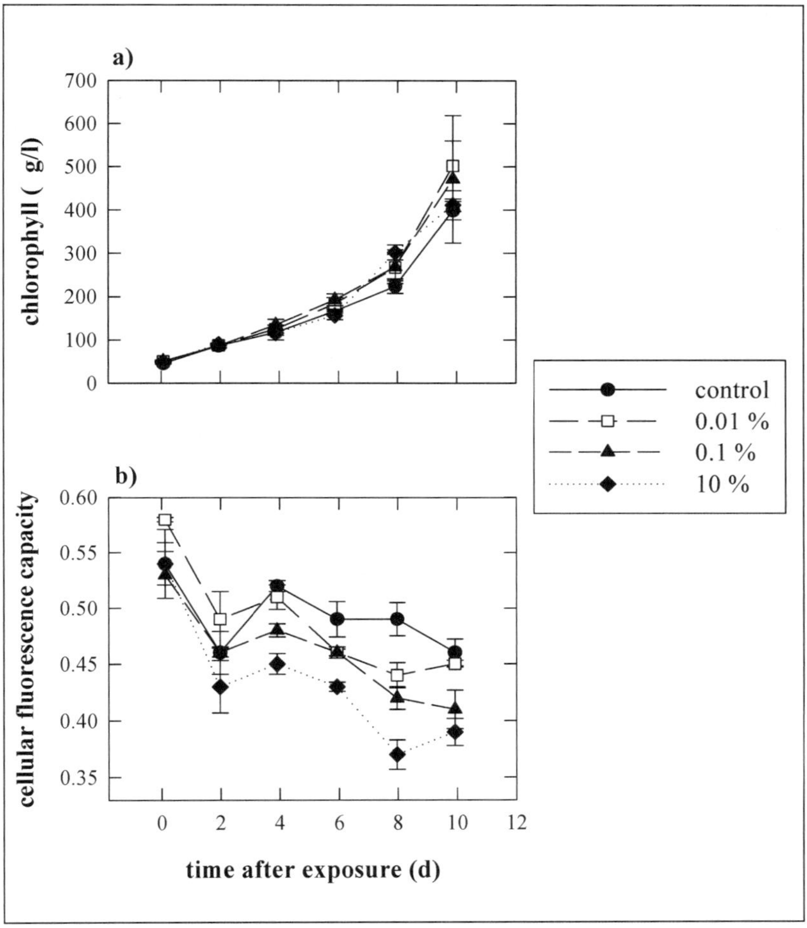

FIGURE 4. (a) The chlorophyll-a concentration (μg/l), and **(b)** cellular fluorescence capacity (CFC) of T. pseudonana during chronic 18 d exposures to each produced water concentration. Error bars include ± 2 SE. A concentration of 1% was not used in this exposure due to a limited number of cultures.

DISCUSSION

Acute Toxicity Evaluations—Haddock

Acute toxicity test results have demonstrated detrimental effects on the early life stages of haddock, sea scallop, and lobster. The most critical developmental stage for survival tested in haddock was the yolk-sac larvae, which may be particularly sensitive, as it included the transition from embryo to larva (i.e. hatching occurred in the 96-h exposure period). Britt et al. (1987) found that the larval stages of capelin (*Mallotus villosus*) and lumpsucker (*Cyclopterus lumpus*) were more sensitive to hydrocarbon exposure than the embryonic stages. Further, the yolk-sac larvae stage of turbot was the most sensitive stage when exposed to the water-soluble fraction of crude oil (Stephens et al., 1997). It was suggested that this could be a result of higher surface area to volume ratio of larvae, in addition to the higher lipid content in the yolk where contaminants can accrue. Rapid growth and development of larval stages can also cause them to be more susceptible to effects (Baldwin et al., 1992; Patin, 1999). Lastly, fish larvae do not appear to have the capacity to metabolise contaminants as is evident in adults.

The lower sensitivity of early-embryonic stage haddock to PW, relative to the older embryos and yolk-sac larvae, may be attributed to the presence of the egg chorion layer that may act as an effective barrier to toxic substances. However, in our study, the relatively large variation in results and low statistical power ($1 - \beta = 0.6$) in survival data among replicates for early-stage embryos limited our ability to detect significant differences among these treatments. The high survival of feeding stage larvae (OSL), relative to the control, in the 10% PW concentration indicated that the acute toxicity of PW is reduced, but not eliminated, as haddock develop. Stephens et al. (2000) found that 50-day old juvenile turbot experienced very little mortality at PW levels between 0.001 and 1% over six weeks, which might suggest a lack of sensitivity in older animals.

Acute Toxicity Evaluations—Lobster

The acute toxicity of PW to first-stage lobster larvae varied considerably among tests, depending on whether they were starved or fed. Significant reductions in the survival of larvae in the 10 and 25% PW treatments were observed only with fed animals. The 96-h LC_{50} value of 0.9% (9,000 ppm) PW is consistent with a toxic effect at relatively low concentrations (< 10,000 ppm; Neff, 1987). The elevated mortality in tests where food was provided may have resulted from higher metabolic rates and/or ingestion of PW contaminants adsorbed to food particles. Due to high quantities of food particles, oxygen deprivation was another consideration, but oxygen measurements showed normal levels. Interpretation of the mortality data for the lobster larvae tests is compromised by the cannibalistic nature of these animals. Mortality from cannibalism may be assumed to be constant among the different PW treatments, including the control. This is apparent during the unfed exposure, as mortality was consistent over the range of PW concentrations. If cannibalism was affected by PW concentration, a graph similar to Figure 2b would have appeared. Although the mortality appears consistent, it still prevents us from attributing observed mortality solely to PW exposure.

Although only a few species have been used to test the acute toxicity of drilling wastes, lobster larvae have previously been identified as being one of the more sensitive test organisms (Neff, 1987). Lobster larvae have also shown sensitivity to petroleum hydrocarbons, and Capuzzo et al. (1984) found that the development and metamorphosis were altered after exposure to South Louisiana crude oil. It would appear from the relatively high, acute toxicity observed for the fed lobster larvae (LC_{50} of 0.9% PW), that this species and life stage is sensitive to toxic constituents in PW.

Acute Toxicity Evaluations—Sea Scallop

The LC_{50} value of 21% for scallop veligers is substantially higher than that found for some other larval benthic species and represents an unrealistically high value for field considerations. Somerville et al. (1987) found an LC_{50} value of 5% PW for embryos and larvae of the oyster, *Crassostrea gigas*. In addition to the reduced survival of sea scallop veligers during acute exposure, the mean size of veligers decreased at > 10% PW. Examination of individual size distributions indicates that this decrease was not the result of negative growth, but was caused by selective mortality of larger individuals. It may be that the effects of PW exposure are only manifested once the larvae grow to a certain size or that it impacts the most metabolically active organisms (i.e. the rapid growers).

The scallop fertilization test resulted in significant reduction in fertilization success and the median effect concentration (EC_{50}) was estimated to be between 0.1 and 1% PW. Toxicity testing with sea urchins, *Strongylocentrotus purpuratus*, based on the fertilization success of eggs after sperm was exposed to PW for 60 minutes, resulted in a similar toxic response with EC_{50} values between 0.74% to 1.2% for different samples of PW (Schiff et al., 1992). Krause et al. (1992) conducted tests on the acute effects of PW on *S. purpuratus*, and found that 10 minutes of exposure to 0.0001% (1 ppm) significantly reduced fertilization success, and ranked the sensitivity of early life stages as: sperm > eggs >> zygotes. An important observation from that study was that effects resulting from sperm exposure to PW can be delayed and expressed as abnormalities in a later life stage.

The acute lethal toxicity of PW to marine organisms has been studied to some extent (e.g. Brendehaug et al., 1992; Moffitt et al., 1992; Terrens and Tait, 1994; Stromgren et al., 1995; Bakke et al., 1996), and available data do indicate a wide range of toxicity depending on the species tested and the source of the PW (see Neff, 2002; Table 20 for a complete list of acute, chronic, and sublethal toxicity to both freshwater and marine species). The sensitivity to PW of the species tested in this experiment fall into the lower LC_{50} range noted in the literature. However, Payne et al. (2001) reported a low acute toxicity potential of production waters from a CoPan site to both invertebrate and vertebrate species. Due to the high variation in the effect of PW observed among species, one must take a cautionary approach when extrapolating from well-maintained laboratory species to organisms of interest (e.g. resource species).

Chronic Toxicity Evaluations—Sea Scallop

Chronic lethal and sub-lethal toxicity tests with sea scallop veligers showed significant effects on survival at 10% PW, a concentration similar to that for an acute toxicity response. It appears that prolonged exposure to relatively low (< 10%)

concentrations of PW does not increase mortality in scallop veligers. This suggests that the toxic threshold for PW is in the range of 1 to 10%. Veligers in all treatments except 10% PW showed similar changes in population size and similar trends in feeding (presumably similar feeding behaviour) as the control. The apparent cessation of growth after day 8 may indicate the presence of a "bottle effect" that constrained larval growth in all experimental treatments.

Chronic Toxicity Evaluations—Marine Diatom

Observed changes in the chlorophyll-*a* concentration and cellular fluorescence capacity (CFC) of the marine diatom *T. pseudonana* during the exposure were similar for all treatments except the 10% concentration, which produced a significant reduction in CFC, compared to the control. The chlorophyll-*a* data did not show any significant impact on phytoplankton biomass, and possibly a slight increase as compared to the control. A model produced by Rivkin et al. (2000) suggest that an increase in primary and secondary production could result in response to PW, as the normally limiting nutrient NH_4 and organic carbon are available in higher quantities. Indeed elevated concentrations of nutrients (25.9 and 27.0 mg/l NH_4) were found in the produced water samples used in this toxicity study. However, the CFC in our study indicates reduced photosynthetic efficiency of algal cells during exposure to 10% PW, which would result in lower productivity. The reduction in CFC during all treatments is likely related to sub-saturating light irradiance for this species. Although the photosynthetic efficiency is lessened, perhaps the algae are still growing and potentially at a higher rate than the control due to the availability of NH_4 and carbon. Produced water toxicity studies on algae growth are limited; however, Brendehaug et al. (1992) and Strømgren et al. (1995) examined growth inhibition in *Skeletonema costatum* exposed to PW from different sources and reported a range of EC_{50} values from 2.1% to 50% PW.

Extrapolation of Laboratory Results

The effects of PW on marine organisms will depend on actual exposure concentrations and duration at the hydrocarbon (oil/gas) production site. Much of the total dilution of the PW discharge can occur during its initial descent through the water column as turbulence mixes the jet/plume with the surrounding seawater (Neff, 2002). During the initial descent phase, PW can be diluted to 1–5% in the immediate vicinity of the discharge (Brandsma et al., 1992). Advection, mixing, and diffusion continue to disperse the waste plume, which can be diluted to 0.1% within 0.5 to 4 km of the discharge site (Somerville et al., 1987; Strømgren et al., 1995). Any larvae entrained in the descent plume may be exposed to high waste concentrations for only a short period and it would be related to the buoyancy of the organism relative to the plume. However, contaminants are known to accumulate in the surface microlayer and readily adsorb to particulate matter that can concentrate in convergence zones at tidal fronts and in the benthic boundary layer. In addition, stratification of water columns in spring and summer months may effectively sustain PW plumes at depth (Washburn et al., 1999). Embryos and larvae may become concentrated in these zones increasing their exposure to PW contaminants (Holdway, 2002). Raimondi and Schmitt (1992) have shown that planktonic larvae could be adversely impacted by PW plumes in a high-energy environment; therefore, more research is required on the actual fate of PW constituents before definitive conclusions on their impact can be made.

Our analyses are based on limited acute chronic toxicity data and do not address the physical/chemical interaction of PW with seawater following discharge. Changes in chemical speciation of contaminants can occur through processes such as flocculation that can alter transport kinetics and bioavailability. The short-term tests reported here present first approximations of immediate biological responses to PW and only begin to provide the insight needed on the effects of continuous, long-term discharges on fishery resources. Validation of reported observations that brief exposure of early life stages of sea urchins to low PW concentrations (0.0001%) may result in a developmental response at a later life stage (Krause et al., 1992), was not considered in this study.

Krause (1994) demonstrated that urchins held in cages down current from a PW outfall suffered a decrease in gamete performance (egg fertilization success and sperm attack rate) that was directly related to the distance from the outfall up to a distance of 80 m. The potential of PW from the Scotian Shelf (same PW as used in the current study) to impact sea urchin fertilization has also been confirmed (K. Lee, unpublished data). Significant effects on performance were detected to a distance of 80 m from the outfall. The Raimondi and Schmitt (1992) field experiments on abalone, *Haliotis rufescens*, larvae indicated the viability of larvae (based on survivorship, settlement ability, and metamorphosis) was reduced at all distances up to 100 m from the PW outfall. Osenberg et al. (1992) studied the growth and survivorship of mussels (*Mytilus edulis* and *M. californianus*) held in cages in the water column at different distances from a PW outfall and they detected biological effects up to a distance of at least 100 m from the discharge. All of these field studies detected biological impacts at distances from the outfall that were similar to the zone of impact predicted using waste dilution models and acute toxicity data (e.g. LC_{50} and EC_{50} concentrations of PW predicted within 500 m from discharge).

Field studies have focused primarily on PW effects on the benthos, and relatively little information is available on possible effects on the life history of pelagic organisms (Neff, 1987). Sub-lethal evaluations on turbot suggest that detrimental impacts of produced waters on young fish will most likely be restricted to concentrations less than 1000x dilution (Stephens et al., 2000).

The residual effects on organisms resulting from acute exposure to drilling wastes is a concern and the potential to resolve this issue with available information is quite limited (Boesch et al., 1987). Factors controlling stock recruitment are also poorly known, limiting our ability to predict the consequences of varying larval mortality to fisheries production. This does not suggest that the effects on fisheries are absent, only that our present data and knowledge base is insufficient to adequately address the environmental significance of PW impacts in a rigorous manner.

CONCLUSIONS

The results of this study indicate some sensitivity of the developmental stages of haddock, lobster, and sea scallop to produced water. LC_{50} thresholds for the different haddock, lobster, and sea scallop life-stages tested in the present study were comparable to a range of toxicity observed for other marine organisms. Fed, stage I lobster larvae were the most sensitive with an observed LC_{50} of 0.9%. Feeding stage haddock larvae and scallop veligers were the least sensitive with LC_{50} values of 22 and 21%, respectively. Chronic exposure studies with the sea scallop and algae showed that the diatom *T. pseudonana* was relatively insensitive to exposures of produced water. In contrast, a decrease in mean size of scallop veligers was observed at 10% produced water. It was concluded that while the toxicity of produced water tends to be species and life-stage specific, at elevated concentrations, it can be potentially harmful to the fertilization success, survival, and growth of the developmental stages of important resource species. This study has shown that resource species can be used effectively in toxicological evaluations even though the relatively high natural mortality of these organisms typically exceeds established lab-based criteria for assessing biotest results.

ACKNOWLEDGEMENTS

This study was sponsored by the Georges Bank Review Panel through a joint project agreement between Fisheries and Oceans Canada and the Department of Oceanography at Dalhousie University. Thanks are extended to Dave Knickle of Adams and Knickle Ltd. for providing adult sea scallops from Georges Bank; Mike Dadswell of the Great Maritime Scallop Trading Co. Ltd. who provided live scallops; Captain Richard Aitkinson and the crew of the F.V. Ryan Atlantic on behalf of Continental Seafoods Ltd. for berried female lobsters; and to Stewart Johnson and Brian Blanchard of the National Research Council (NRC) who provided the fertilized haddock eggs. We thank Steve McKenna of NRC for advice and assistance with spawning of scallops.

REFERENCES

Aiken, D.E. and S.L. Waddy. 1989. "Culture of the American Lobster, *Homarus americanus*". In A.D. Boghem (Ed.), *Cold Water Aquaculture in Atlantic Canada*, pp. 79-122. University of Moncton, New Brunswick, Canada.

Bakke, S., E.A. Vik, H. Gruner, and T.A. Hjelmas. 1996. Produced Water Reinjection (PWRI). Experiences from the Ula Field. In M. Reed and S. Johnsen (Eds.), *Environmental Science Research 52: Produced Water 2. Environmental Issues and Mitigation Technologies*, pp 447-458. Plenum Press, New York, New York.

Baldwin, J.D., M.C. Pillai, and G.N. Cherr. 1992. "Response of Embryos of the Sea Urchin *Strongylocentrotus purpuratus* to Aqueous Petroleum Waste Includes the Expression of High Molecular Weight Glyocoprotein." *Marine Biology*. 114:21-30.

Boesch, D.F., J.N. Butler, D.A. Cacchione, J.R. Geraci, J.M. Neff, J.P. Ray, et al. 1987. "An Assessment of the Long-term Environmental Effects of U.S. Offshore Oil and Gas Development Activities: Future Needs." In D.F. Boesch and N.N. Rabalais (Eds.), *Long-term Environmental Effects of Offshore Oil and Gas Development*, pp. 1-54. Elsevier Applied Science, London, England.

Brandsma, M.G., J.P. Smith, J.E. O-Reilly, R.C. Jr. Ayers, and A.L. Holmquist. 1992. "Modeling Offshore Discharges of Produced Water." In J.P. Ray and F.R. Englehardt (Eds.), *Produced Water: Technological/Environmental Issues*, pp. 59-72. Plenum Press, New York, New York.

Brendehaug, J., S. Johnse, K.H. Bryne, A.L. Gjøse, T.H. Eide, and E. Aamot. 1992. "Toxicity Testing and Chemical Characterization of Produced Water—A Preliminary Study." In J.P. Ray and F.R. Englehardt (Eds.), *Produced Water: Technological/Environmental Issues*, pp. 245-256. Plenum Press, New York, New York.

Britt, I., F. Peterson, and E. Kjorsvik. 1987. "Acute Toxicity Tests of the Effects of Oils and Dispersants on Marine Fish Embryos and Larvae—A Review". *Sarsa*. 72:411-413.

Brown, J.A., S.M. Stephens, and R.M. Stagg. 1999. "Oil Produced Water: Chronic Impacts on Juvenile Turbot." In C. Kennedy and D.

MacKinlay (Eds.), *Fish Response to Toxic Environments*, pp 121-126. International Congress on the Biology of Fish, Baltimore, MD.

Cameron, P. and J. Berg. 1992. "Morphological and Chromosomal Aberrations during Embryonic Development in Dab (*Limanda limanda*)." *Marine Ecology Progress Series*. 91:163-169.

Capuzzo, J.M., B.A. Lancaster, and G.C. Saski. 1984. "The Effects of Petroleum Hydrocarbons on Lipid Metabolism and Energetics of Larval Development and Metamorphosis in the American Lobster (*Homarus americanus* Milne Edwards)." *Marine Environmental Research*. 14:201-228.

Flynn, S.A., E.J. Butler, and L. Vance. 1996. "Produced Water Composition, Toxicity, and Fate. A Review of Recent BP North Sea Studies." In M. Reed and S. Johnsen (Eds.), *Environmental Science Research 52: Produced Water 2. Environmental Issues and Mitigation Technologies*, pp 69-80. Plenum Press, New York.

Gamble, J.G., J.M. Davies, S.J. Hay, and F.K. Dows. 1987. "Mesocosm Experiments on the Effect of Produced Water Discharges from Oil Platforms in the Northern North Sea." *Sarsia*. 72:383-386.

GESAMP. 1993. *Impact of Oil and Related Chemicals and Wastes on the Marine Environment*. IMO/FAO/UNSECO/WMO/WHO/IAEA/UN/UNEP—Joint Group of Experts on the Scientific Aspects of Marine Pollution. Rep. Stud. GESAMP, 50.

Guillard, R.L. 1975. "Culture of Phytoplankton for Feeding Marine Invertebrates." In W.L. Smith and M.H. Chanley (Eds.), *Culture of Marine Invertebrate Animals*, pp. 29-60. Plenum Press, New York, New York.

Holdway, D.A. 2002. "The Acute and Chronic Effects of Wastes Associated with Offshore Oil and Gas Production on Temperate and Tropical Marine Ecological Processes." *Marine Pollution Bulletin*. 44:185-203.

Krause, P.R., C.W. Osenberg, and R.J. Schmitt. 1992 "Effects of Produced Water on Early Life Stages of a Sea Urchin: Stage-specific Responses and Delayed Expression." In J.P. Ray and F.R. Englehardt (Eds.), *Produced Water: Technological/Environmental Issues*, pp. 431-444. Plenum Press, New York, New York.

Krause, P.R. 1994 "Effects of Oil Production Effluent on Gametogenesis and Gamete Performance in the Purple Sea Urchin (*Strongylocentrotus purpuratus Stimpson*)." *Environmental Toxicology and Chemistry*. 13:1153-1161.

Linden, O. 1975. "Acute Effects of Oil and Oil/Dispersant Mixing on Larvae of Baltic Herring." *Ambio*. 4:130-133.

Lonning, S. 1977. "The Effects of Ekofisk Oil and Oil Products on Marine Fish Larvae." *Astarte*. 10:37-47.

Moffitt, C.M., M.R. Rhea, P.B. Dorn, J.F. Hall, J.M. Bruney, and S.H. Evans. 1992. "Short-term Chronic Toxicity of Produced Water and Its Variability as a Function of Sample Time and Discharge Rate." In J.P. Ray and F.R. Englehardt (Eds.), *Produced Water: Technological/Environmental Issues*, pp. 235-244. Plenum Press, New York, New York.

Muschenheim, D.K., T.G. Milligan, and D.C. Jr. Gordon. 1995. *New Technology and Suggested Methodologies for Monitoring Particulate Wastes Discharged from Offshore Oil and Gas Drilling Platforms and their Effects on the Benthic Boundary Layer Environment*. Canadian Technical Report of Fisheries and Aquatic Sciences. 2049, x + 55p.

Murray-Smith, R., D. Gore, S.A. Flynn, L. Vance, and R. Stagg. 1996. "Development and Appraisal of a Particle Tracking Model for the Dispersion of Produced Water Discharged from an Oil Production Platform in the North Sea." In M. Reed and S. Johnsen (Eds.), *Environmental Science Research 52: Produced Water 2. Environmental Issues and Mitigation Technologies*, pp 225-245. Plenum Press, New York, New York.

Neff, J.M. 1987. "Biological Effects of Drilling Fluids, Drill Cuttings and Produced Water." In D.F. Boesch and N.N. Rabalais (Eds.), *Long-term Environmental Effects of Offshore Oil and Gas Development*, pp. 469-538. Elsevier Applied Science, London, England.

Neff, J.M. and T.C. Sauer Jr., 1996. "An Ecological Risk Assessment for Polycyclic Aromatic Hydrocarbons in Produced Water Discharges to the Western Gulf of Mexico." In M. Reed and S. Johnsen (Eds.), *Environmental Science Research 52: Produced Water 2. Environmental Issues and Mitigation Technologies.* Plenum Press, New York, New York.

Neff, J.M. 2002. "*Bioaccumulation in Marine Organisms*", 452 pp. Elsevier Science Ltd, Osford, UK.

Olsgard, F. and J.S. Gray. 1995. "A Comprehensive Analysis of the Effects of Offshore Oil and Gas Exploration and Production on the Benthic Communities of the Norwegian Continental Shelf." *Marine Ecology Progress Series*. 122:277-306.

Osenberg, C.W., R.J. Schmitt, S.H. Holbrook, and D. Canestro. 1992. "Spatial Scale of Ecological Effects Associated with an Open Coast Discharge of Produced Water." In J.P. Ray and F.R. Englehardt (Eds.),

Produced Water: Technological/Environmental Issues, pp. 387-402. Plenum Press, New York, New York.

Ouellett, P., C.T. Taggart, and K.T. Frank. 1992. "Lipid Condition and Survival in Shrimp (*Pandalus borealis*) Larvae." *Canadian Journal of Fisheries and Aquatic Science*. 49:368-378.

Patin, S. 1999. *Environmental Impact of the Oil and Gas Industry*. Ecomonitor Publishing, New York. 425 pp.

Payne, J., L. Fancey, C. Andrews, J. Meade, F. Power, K. Lee, G. Veinott, and A. Cook. 2001. *Laboratory Exposures of Invertebrates and Vertebrate Species to Concentrations of 1A-35 (Petro-Canada) Drill Mud Fluid, Production Water and Hibernia Drill Mud Cuttings*. Canadian Manuscript Report of Fisheries and Aquatic Sciences. No 2560. 27pp.

Rabalais, N.N., B.A. McKee, D.J. Reed, and J.C. Means. 1992. "Fate and Effects of Produced Water Discharges in Coastal Louisiana, Gulf of Mexico, USA." In J.P. Ray and F.R. Englehardt (Eds.), *Produced Water: Technological/Environmental Issues,* pp. 355-369. Plenum Press, New York, New York.

Raimondi, P.T. and R.J. Schmitt. 1992. "Effects of Produced Water on Settlement of Larvae: Field Studies using Red Abalone." In J.P. Ray and F.R. Englehardt (Eds.), *Produced Water: Technological/Environmental Issues,* pp. 415-430. Plenum Press, New York, New York.

Raimondi, P.T., A.M. Barnett, and P.R. Krause. 1997. "The Effect of Drilling Muds on Marine Invertebrate Larvae and Adults." *Environmental Toxicology and Chemistry*. 16:1218-1228.

Rivkin, R.B., R. Tian, M.R. Anderson, J.F. Payne, and D. Deibel. 2000. *Ecosystem Level Effects of Offshore Platform Discharges – Identification, Assessment and Modeling*. Canadian Technical Report of Fisheries and Aquatic Sciences. 2331.

Schiff, K.C., D.J. Reish, J.W. Anderson, and S.M. Bay. 1992. "A Comparative Evaluation of Produced Water Toxicity." In J.P. Ray and F.R. Englehardt (Eds.), *Produced Water: Technological/Environmental Issues,* pp. 199-208. Plenum Press, New York, New York.

Solbakken, J.E., S. Tilseth, and K. Palmork. 1984. "Uptake and Elimination of Aromatic Hydrocarbons and Chlorinated Biphenyl in Eggs and Larvae of Cod (*Gadus morhua* L.)." *Marine Ecology Progress Series*. 16: 297-301.

Somerville, H.J., D. Bennett, J.N. Davenport, M.S. Holt, A. Lynes, A. Mathieu, et al. 1987. "Environmental Effect of Produced Water from North Sea Oil Operations." *Marine Pollution Bulletin*. 18:549-558.

Stagg, R.M. and A. McIntosh. 1996. "Hydrocarbon Concentration in the Northern North Sea and Effects on Fish Larvae." *The Science of the Total Environment*. 186:189-201.

Stephens, S.M., A.Y.A. Alkindi, C.P. Waring, and J.A. Brown. 1997. "Corticosteroid and Thyroid Responses of Larval and Juvenile Turbot Exposed to the Water-soluble Fraction of Crude Oil." *Journal of Fish Biology*. 50:953-964.

Stephens, S.M., S.C. Franklin, R.M. Stagg, and J.A. Brown. 2000. "Sub-Lethal Effects of Exposure of Juvenile Turbot to Oil Produced Water." *Marine Pollution Bulletin*. 40:928-937.

Stephenson, M.T. 1992. "A Survey of Produced Water Studies." In J.P. Ray and F.R. Englehardt (Eds.), *Produced Water: Technological/Environmental Issues*, pp. 1-11. Plenum Press, New York, New York.

Strømgren, T., S.E. Sørstrøm, L. Schou, I. Kaarstad, T. Aunaas, O.G. Brakstad, et al. 1995. "Acute Toxic Effects of Produced Water In Relation To Chemical Composition and Dispersion." *Marine Environmental Research*. 40:147-169.

Syversten, E.E. 1996. "Regulation of Produced Water on the Norwegian Continental Shelf". In M. Reed and S. Johnsen (Eds.), *Environmental Science Research* 52: Produced Water 2. *Environmental Issues and Mitigation Technologies*, pp 7-11. Plenum Press, New York, New York.

Terrens, G.W. and R.D. Tait. 1994. "Effects on the Marine Environment of Produced Formation Water Discharges from Offshore Development in Bass Strait, Australia." *The APEA Journal*. 730-740.

Vincent, W.F. 1980. "Mechanisms of Rapid Photosynthetic Adaptation in Natural Phytoplankton Communities. II. Changes in Photochemical Capacity as Measured by DCMU-induced Chlorophyll Fluorescence." *Journal of Phycology*. 16:568-577.

Washburn, L., S. Stone, and S. MacIntyre. 1999. "Dispersion of Produced Water in a Coastal Environment and Its Biological Implications." *Continental Shelf Research*. 19:57-78.

Marine Coastal Monitoring: Designing an Effective Offshore Oil and Gas Environmental Effects Monitoring Program

Green, R.H. (rgreen@uwo.ca) (University of Western Ontario, London, ON Canada and Environment and Natural Resources Institute, University of Alaska Anchorage, Anchorage, AK USA)

ABSTRACT: Marine coastal monitoring design strategies vary among three broad categories: major spills (oil spills), chronic point-source pollution (outfalls and oil/gas platforms), and multiple inputs (from rivers). In all cases, the study design has both a temporal and a spatial component. Environmental effects monitoring of offshore oil and gas exploration and exploitation could fall into either of the first two categories—the first if a catastrophic event like a blowout was a concern, and the second where long-term chronic pollution, such as by drilling muds or leakage of hydrocarbons, was the primary concern.

With the first category, major spills, the temporal and spatial components interact strongly; the spatial extent and the nature of the impact change rapidly over a short time. "Before-spill" data are extremely desirable, but rarely available. Usually impacted and unimpacted sites are in different areas leading to a confounding of impact effects with natural spatial pattern.

With chronic point source pollution a "bullseye" design usually captures the spatial component (directions and distances), while an equally-spaced series of times, often of indefinite duration, captures the temporal component. Spatial and temporal components interact slightly, such as with seasonal variation in current direction. Examples of this scenario are found in the Gulf of Mexico, offshore northwestern Australia, and Atlantic Canada.

Cost-effective implementation of monitoring chronic point source pollution for offshore oil and gas environmental effects monitoring (EEM) requires difficult choices within the bullseye design concept. The goals of this paper are to discuss design choices that must be made, and to outline the characteristics of a good design.

INTRODUCTION

Marine pollution generally impacts shorelines, whether it originates there or comes ashore from elsewhere. The open sea is also impacted, such as with plastic debris trapping or choking turtles and birds. However, most impacts are in coastal waters, on shorelines, or in estuaries. Study design strategies vary among three broad categories:

- Major spills (oil spills)
- Chronic point-source pollution (outfalls and oil/gas platforms)
- Multiple inputs (from rivers)

This paper was derived from a keynote presentation titled, "Marine Coastal Monitoring Designs: Examples for Major Spills, Chronic Point Sources, and River Inputs", given at the *Asia-Pacific Conference on Marine Science and Technology* in Kuala Lumpur, Malaysia in May of 2002, and emphasised the multiple inputs category as exemplified by the Straits of Malacca. Both that keynote presentation and this paper briefly consider the major spills category, however, the main emphasis here, is on chronic point source pollution, as exemplified by oil and gas platforms.

Cost-effective implementation of monitoring chronic point source pollution for offshore oil and gas EEM requires choices within a scheme that is typically based on the bullseye design concept (see Point Sources section below). Design concepts that should be considered include:

- How far away from the platform or platforms should the greatest distance be?
- How should the distances be scaled—arithmetic or geometric?
- How many directions should there be, and must/should the angles between them be equal? Is there a directional current pattern, either generally or seasonally?
- If there is more than one platform do their potential effects overlap?
- Are multiple platforms close enough that direction from one platform has the same meaning as that direction from another platform?
- Must/should there be replicate sampling within each distance x direction combination for each platform? If not, then where does the error variance in statistical analysis come from, or should we forget hypothesis-testing and just describe effects?
- Can multivariate statistical methods be used effectively and unambiguously for describing effects e.g. at the level of the biological community?
- Should statistical analysis be based on distance as a category variable (such as "rings"), or distance as a continuous variable e.g. in km? What are the attributes of a good design for offshore oil and gas EEM?
- Is time a fixed or a random effect in EEM designs?
- Are there some untried design possibilities? For example, sampling in a completely randomized design within the

spatial component but with allocation proportional to predicted distance and direction of effects.

The goal of this paper is to discuss design choices that must be made, and to outline the characteristics of a good design. The focus is on design. The scope of this paper does not extend to sampling methodology, choices of variables, or statistical analysis methods—except where these topics relate to design questions. Of course, there is usually synoptic sampling of biological and environmental variables (the latter includes both anthropogenic and natural variation) within the chosen study design. For the concepts in this paper, I assume a certain level of knowledge of the basic principles of sampling design and statistical design (e.g. ANOVA), of the sort that is in basic university biostatistics courses and textbooks for such courses. However, I do cite good sources for such information.

GENERAL DESIGN PRINCIPLES

There are general principles of design that apply to all three monitoring design categories. Good references on environmental study design include Elliott (1977), Green (1979), and Keough and Mapstone (1995). A more advanced and statistical analysis-oriented text is Underwood (1997). The "bible" on sampling design is still Cochran (1963; see also Cochran, 1983). Good basic statistic texts with emphasis on sampling and experimental design are Snedecor and Cochran (1967), Mead (1988), Zar (1996), and Dytham (1999). Reviews and opinions of where environmental impact and monitoring study designs should "go" are found in Stewart-Oaten (1996a; 1996b), Green and Montagna (1996), and Osenberg and Schmitt (1996). Clarke and Warwick (1994) provide an overview of environmental study design and analysis as implemented in their PRIMER package (www.primer-e.com), which is easy to learn and emphasizes both descriptive and hypothesis-testing multivariate analyses. Robust randomization tests are used. Workshops are offered at many locations.

Particular environmental study designs of general importance are "Before-After-Control-Impact" (BACI) (Green 1978; 1979; Green and Smith, 1997; Underwood, 1991; 1992; 1993; 1994;

1997), repeated measures (Green, 1993a), and the Sediment Quality Triad (Chapman et al., 1987; 1991; Green at al., 1993; Green and Montagna, 1996; Chapman et al., 1997). BACI is the optimal design (Green, 1979) if control sites and before-impact times are available. Repeated measures designs have a long history in studies on humans in the fields of medicine, pharmacology, and psychology, and are appropriate where the same sites are sampled repeatedly over time. The Sediment Quality Triad involves synoptic samples for biological, contaminant, and behavioural variables, allowing the relationships among these types of variables to be described.

Distinguishing between natural and impact-related variation in space and time is discussed by Green (1987), Osenberg et al. (1996), Thrush et al. (1996), and by Underwood in his papers (1991; 1992; 1993; 1994) and his book (1997). Choosing the appropriate scale in space and time for sampling in environmental studies is crucial, and this is discussed in Green (1984) and Keough and Black (1996). Livingston (1987) discusses the scale of replicate samples to ensure that they are statistically independent. Criteria for choosing response variables are discussed in Green (1979; 1987) and Jones and Kaly (1996). "Lessons learned" from previous studies are reviewed in Peterson (1993) and Ritchie (1995).

Hypothesis testing in environmental studies and related subjects—power, replication, pseudoreplication, and valid error for tests are covered in Green (1984), Hurlbert (1984), Mapstone (1996), Stewart-Oaten (1996a; 1996b), and Underwood (1991; 1992; 1993; 1994; 1997). The necessary number of replicate samples and other topics related to the power of a study to detect impact effects are found in Elliott (1977), Cuff and Coleman (1979), Clarke and Green (1988), Green (1989; 1994), Hyland et al. (1994), and Osenberg et al. (1996). Combining observational and manipulative approaches is often valuable. The latter contribute a more classically scientific component to the study, and can be a field experiment (e.g. Hinch and Green, 1989), or an experiment using mesocosms (e.g. Gray et al., 1988).

The study design will always have a spatial and a temporal component if there is sampling at more than one time and at more than one location. One station sampled over time would have no spatial component, and a spatial array sampled once would have no temporal component. Neither would be appropriate to offshore oil and gas EEM, given that monitoring is always over time, and evidence of impact is from change in spatial pattern. Examples will be given that illustrate the importance of spatial and temporal components of EEM study designs.

Confounding is a problem in all observational studies, and usually there is no study design that can entirely remove it. It occurs when levels of design components that are intended to represent impact, effects, or degree of impact, coincide with natural differences or gradients. An apparent impact effect may be due to natural variation. Confounding may occur with a spatial component (impacted sites may differ from unimpacted sites on natural environmental variables), or with a temporal component (the climate may change around the same time that the impact begins).

MAJOR SPILLS

Accidental, unforeseeable spills often occur in coastal waters, for example, oil spills. In March 1978, the *Amoco Cadiz* ran aground in a storm and spilled 1,619,000 barrels of crude oil that polluted more than 300 km of Brittany (France) coastline. In March 1989, the tanker *Exxon Valdez* ran aground leaving the port of Valdez and spilled 11.2 million US gallons of crude oil into Prince William Sound, Alaska, and the oil came ashore along approximately 750 km of shoreline. In January 1993, the tanker *Braer* ran aground south of the Shetlands and broke up, spilling 85,000 tons (620,000 barrels) of light crude oil. Environmental damage by the *Braer* spill was minimal because it was light crude and because high winds (ca. 160 km) and rough seas caused most of the oil to evaporate or sink, and not much went onshore (see Ritchie, 1997). In February 1996, the tanker *Sea Empress* struck a rock in Milford Haven harbour, Wales, and spilled 70,000 tons of light crude oil into the Irish Sea.

Again, winds and tides at the time, and light rather than heavy crude, reduced the potential environmental damage. Spills can also occur onshore and impact coastal waters, as happened with a ruptured oil storage tank on the Caribbean shore of Panama (Jackson et al., 1989).

When a spill occurs, the spatial extent and nature of the impact typically change rapidly over a short time. This leads to a strong interaction between the spatial and temporal components in the design. For example, with a heavy crude oil in cold calm water a "mousse" forms from the surface oil slick, becomes dense and heavy, and sinks or clings to the shore (e.g. the *Exxon Valdez* spill, initially), and becomes tar-like over time. With lighter crude in rough water, the oil is driven below the surface, where some sinks, but wind and wave energy drive much of it into the air (e.g. the *Braer* spill). But whatever the type of oil, weather, or water temperature, the situation changes rapidly after the spill occurs. The impact/monitoring study design must capture these changes.

Before-spill data are highly desirable because they permit a BACI design if unimpacted but similar sites can be found, which is usually possible. Since spills are accidents and therefore unexpected, we usually have to be innovative to obtain before-spill data. In the *Exxon Valdez* spill and the Panama spill, there were some data from previous studies that had been done for other purposes, which might be called opportunistic "before" data. Such data are valuable because they provide a pre-spill contrast with post-spill conditions. The study design should incorporate them somehow. Another approach is to use paleoecological techniques. For example, in the Panama spill studies, annual bands in sections of coral heads were used to derive growth indices for 10 years before the spill, to be compared to post-spill years (Jackson et al., 1989).

As previously indicated, confounding is often a problem in that the spatial component of an oil spill study is confounded with natural spatial variation. In the Panama study, the impacted shoreline was quite different from the closest unoiled shoreline. The *Exxon Valdez* spill primarily impacted western Prince William Sound and the adjacent Gulf of

Alaska, which is biogeographically different from the relatively unoiled eastern Prince William Sound (Bousfield and McAllister, 1962)—the obvious candidate area for control sites. One solution is to use a paired design in which impact and control sites are near each other and similar to each other except for the oiled-unoiled contrast.

There is an early review of the various *Exxon Valdez* oil spill studies by Paine et al. (1996). Peterson et al., (2001) provided an extensive and detailed critique of the relationship between the study designs and the conclusions reached. This paper was sharply criticized by Gilfillan et al. (2002), and a response to that criticism is in Peterson et al., (2002). A recent important review paper is provided by Peterson et al., (2003).

POINT SOURCES

Oil and gas drilling platforms are the classic point-source polluter (actual or potential) in coastal waters. A recent important study is the Gulf of Mexico Offshore Operations Monitoring Experiment (GOOMEX), published as a series of eight papers in issue 11 of Volume 53 of the *Canadian Journal of Fisheries and Aquatic Sciences* (1996). The study design is the traditional "bullseye" design (Fig. 1) with radial transects going out from the platform (five in GOOMEX), which are sampled at distances on an exponential scale. In GOOMEX, there were three platforms, each with its bullseye design, and four cruises over two years. There is no sample replication at sites (see discussion of this issue in the next section). The analysis design was based on error terms which were random effects (i.e. the "among radii" term and all interactions with it) (Table 1). There is usually a weak interaction between spatial and temporal components. For example in GOOMEX, seasonal variation in current direction would be described by interactions of the temporal factor "cruise" with the spatial factors "platform", "distance", and "radii". See Kennicutt et al. (1996a) for the sampling design and the rationale for the statistical analysis design based on it. Kennicutt et al. (1996b) described the contaminant distribution in relation to this design and multivariate techniques that were used to model it. Green and Montagna (1996) applied innovative analysis and interpretation approaches, and recommended design approaches for future similar studies. Peterson et al. (1996) summarised and interpreted all the GOOMEX results, compared them with other studies, and made recommendations.

In GOOMEX spatial confounding with natural environmental factors complicated interpretation (Table 2). The three platforms were at different distances from shore with farther offshore = deeper = finer sediments = more easterly = historically dirtier. The results showed greater biological impact along the same gradient, but how can one say what the explanation is when potential explanatory variables are confounded? In the end, the among-platform variation in demonstrable biological impacts, while real, was simply described.

How relevant is GOOMEX to the Grand Banks (GB) and Scotian Shelf (SS) EEM situation? The environments certainly differ. In the Gulf of Mexico (GOM), currents are weaker and seasonally variable. On the GB and SS the current is strong and consistent in direction. Storms frequently disturb the bottom, resulting in coarse unstable sediment with little meiofauna (which was a useful community for EEM in GOOMEX). The distribution of detectable effects around platforms regarding distance and direction would probably be different.

In other offshore oil and gas monitoring studies, temporal confounding has been a problem. In the CAlifornia offshore Monitoring Program (CAMP) Phase 3 studies, the beginning of drilling happened to coincide with the beginning of a several year dry period and thus was confounded with annual discharge from the coastal drainage.

Chapman et al. (1991) described the application of the Sediment Quality Triad to an oil platform study. The group associated with J.S. Gray has published numerous papers related to assessment of oil and gas industry impacts on the Norwegian continental shelf. Gray et al. (1999) summarises much of this work.

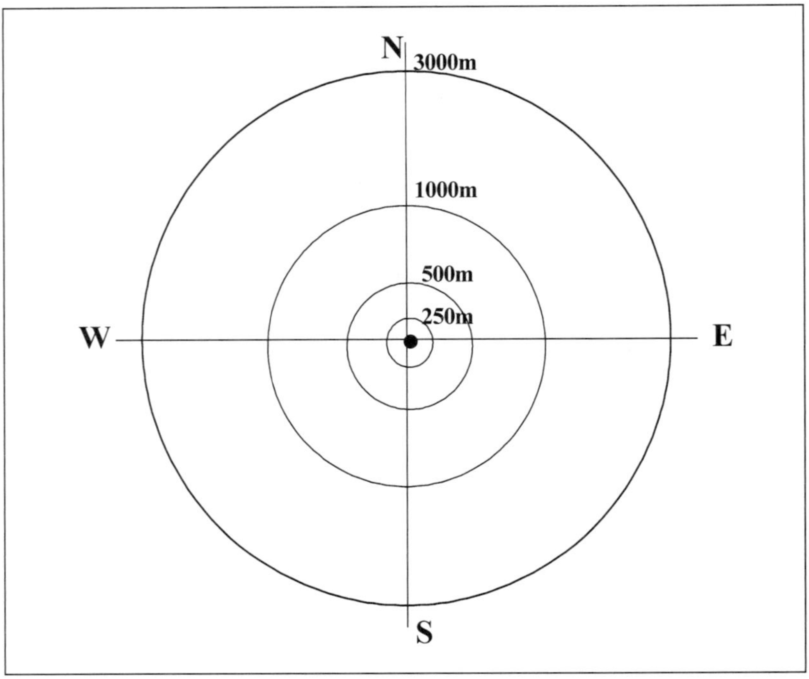

FIGURE 1. Example of a bullseye design for offshore oil and gas EEM.

TABLE 1. The GOOMEX study design and hypothesis tests.

Gulf of Mexico Oil/gas Platforms: 3 Platforms, 4 Times, 5 Distances, 5 Radii
SAS Model Statement: Y = P T R(P) T*R D T*P P*D D*R D*T P*D*T D*T*R
R is random (*R*); P, T, and D are fixed (F).

Row	Source	Degrees of Freedom		F or R	Test vs. Row
1	P	(3 - 1)	= 2	F	2
2	R(P)	(5 - 1)3	= 12	*R*	(7 + 8) - 11
3	D	5 - 1	= 4	F	7
4	T	4 - 1	= 3	F	8
5	P*D	(3 - 1)(5 - 1)	= 8	F	7
6	P*T	(3 - 1)(4 - 1)	= 6	F	8
7	D*R(P)	(5 - 1)(5 - 1)3	= 48	*R*	11
8	T*R(P)	(4 - 1)(5 - 1)3	= 36	*R*	11
9	D*T	(5 - 1)(4 - 1)	= 12	F	11
10	P*D*T	(3 - 1)(5 - 1)(4 - 1)	= 24	F	11
11	D*T*R(P)	(5 - 1)(4 - 1)(5 - 1)3	=144	*R*	12
12	Rep. error	3*4*5*5*(1 - 1)	= 0		
	Total	3*4*5*5*1 - 1	=299		

Comments on table:
For one platform rows with P (rows 1, 5, 6, 10) drop out.
This example is with no replication, so D*T*R(P) (row 11) cannot be tested.
Anything with R in it is a random effect and can be used as an error term.
Highest level interactions must be tested first (i.e. rows 10 and 11, then 5 - 9, then 1 - 4).

TABLE 2. Confounding of among-platform component of GOOMEX design.

Platform	Depth and Distance From Shore *	Longitude	Historical Contaminant Discharges	Biological Impacts
MAI - 686	shallow/near	westernmost	least	least
MU - A85	mid/mid	mid	mid	mid
HI - A389	deep/far	easternmost	most	most

* includes related environmental information including sediment, temperature, current, etc.

Cohen (2002) describes offshore oil and gas monitoring studies on the Australian North West Shelf, where the Harriet Field platforms differ in kind, and overlap in their effects. "The three Harriet platforms provide a method of estimating the separate effects of discharged drilling wastes and produced formation water on benthic communities surrounding long-established offshore structures, because all three platforms are located in close proximity and in a similar environment, yet only one platform discharges produced formation water into the marine environment." Control sites were pristine areas away from platforms, which imply confounding of effects of platform structure (organic "rain" from hard substrate fouling organisms, modification of current, etc.) with effects of discharges in the design, if platforms are contrasted with control sites. Montagna et al. (2004) has shown that such platform structure effects can be important.

DESIGN CHOICES FOR OFFSHORE OIL AND GAS EEM

Cost-effective implementation of monitoring chronic point source pollution for offshore oil and gas EEM requires choices within a scheme based on the bullseye design concept.

How far away from the platform or platforms should the greatest distance be? For given cost and effort, a very large distance including "just in case there is an unexpected far-out effect", will mean a less precise description of effect magnitude as a function of distance closer in. Another concern is that beyond a certain distance, it becomes more difficult to stay within a comparable environment (e.g. depth, substrate, current pattern, biogeographic zone). The consensus, including at the Offshore Oil and Gas EEM Workshop (Armsworthy et al., 2004), is that biological effects are not detected beyond 500 m, and that the limit is usually 200 or 250 m. The type of drilling mud may influence this; with the old oil-based muds having the longer range effects. One report of effects on communities beyond 1000 m was in this category, and was based on possible community response using multivariate descriptors. For routine offshore oil and gas EEM, I would suggest an "outer ring" at 1000 m distance. If there is a blowout or spill, then of course effects may go further, but this is not usually what EEM is for. Such an event does not really require detection. It is possible that newer methods to detect sublethal effects may find those effects at a greater distance from platforms, and in that case, this question will have to be re-evaluated.

Can multivariate (MV) statistical methods be used effectively and unambiguously for describing effects, for example, at the level of the biological community? MV statistical analysis is often more powerful than univariate statistics when there are correlated response variables. However, the results require careful interpretation, and it is sometimes easy to see what you are looking for. Detection of effects using descriptive MV techniques should be convincing to you and also to your most skeptical colleague. There are formal tests of hypotheses, but statistical significance is easily obtained with MV tests. Something is usually going on in the real world that would affect at least one of your response variables. Clearly relating statistical significance to impact is more difficult. In my experience, good graphical display is more convincing than formal significance per se.

Should the control sites be at distances beyond expected effects from platforms, at imaginary platforms (pristine sites), or both? Certainly the former is necessary. As for the latter, Cohen (2002) sampled bullseyes at imaginary platforms established at pristine locations, and this sort of sampling has been done elsewhere. The problem is that in any contrast between platforms that are discharging (whether produced water or drilling wastes) and control sites of this kind, platform structure effects will be indistinguishable from discharge effects. More studies are needed using two kinds of controls in addition to active production platforms: platforms or platform-like structures at sites where there hasn't been any production, and pristine sites without any structure or past activity. Montagna et al. (2004) has shown the way on this, and also has shown how important platform structure effects can be. It may be that much of what we have been attributing to discharge effects is really platform structure effects.

How should the distances be scaled—arithmetic or geometric? If the implied model is an exponential decay of effect magnitude with distance, which seems reasonable, then distances ("rings") should be on a geometric scale (e.g. 50, 100, 200, etc.).

How many directions should there be, and must/should the angles between them be equal? There should be at least four directions, preferably five, and equally spaced if possible. Sometimes it isn't possible. For example, one of the GOOMEX platforms (HI-A389) was close to the edge of the continental shelf, and to have sites at similar depths, the radii had to be concentrated in directions along or away from the drop-off. The number of radii will be influenced by current strength and direction. If there is no current or variable current, then more radii are needed to "map" the effects. A constant current speed and direction requires fewer radii, perhaps as few as two (180° apart and aligned with the current).

Is there a directional current pattern, and is it constant or does it vary seasonally? As previously noted, this will influence the number and orientation of radii needed in the bullseye design to capture the spatial (distance from platform) x temporal (e.g. among times of year) interaction that is implied. If there are multiple platforms then possible overlap of effects at some times of year should be considered for the design and statistical analysis based on it.

If there is more than one platform, do their potential effects overlap? If the answer is yes, then teasing apart effects of different platforms is most easily done if platforms differ in their discharges, as in Cohen's (2002) case. Also, a more sophisticated approach to statistical design and analysis, such as he used, will be required.

Are the platforms close enough that direction from one platform has the same meaning as that direction from another platform? If the answer is no (as in the case of GOOMEX), then directions of radii do not have comparable meaning among platforms. For example, a radius running 270° from one platform is not commensurate with one running 270° from another platform and cannot be interpreted similarly. If they are close enough for directions to have the same meaning, then the statistical design should reflect this, and sources of error for hypothesis testing will have to be different (see below).

Should there be replicate sampling within each distance x direction combination for each platform? There is no need to take replicate boxcores (or other types of samples) at each site (distance x radius combination) for statistical design reasons. Use of variation among replicate boxcores within sites as error for testing hypotheses related to impact would be improper. It is "pseudoreplication" error (sensu Hurlbert, 1984), and would only be a proper error term for testing differences between sites, such as two sites on the same ring. This seems unlikely to be of interest. The only argument for taking two or more samples at each site might be to increase coverage (area sampled) at each site to estimate site means more precisely. However, Cuff and Coleman (1979) noted that in general, for a given number of samples, the best design to detect and describe spatial pattern is a design that allocates one sample to each site so as to maximise the number of sites. I found this difficult to accept at the time, but I now believe it to be correct.

If there is no within-site replicate sampling, then where does the error variance for hypothesis testing come from? As noted above, variation among replicates within sites is not appropriate for testing impact-related (platform, distance, and radius related) hypotheses. If directions of radii are not commensurate among platforms (see above), then radii can be considered a random effect in the design, nested within platform. Thus among radii and all interactions with radius (e.g. with distance) are also random effects, and they can be used to test impact-related hypotheses, as appropriate. This was the case in GOOMEX. As can be seen in Table 1, there are ample error degrees of freedom for testing the various impact-related hypotheses. If directions of radii are commensurate among platforms then replication is needed, but at a higher level in the design than within-site. One could have replicate platforms (within type of platform). With fixed-effects within replicates (e.g. distances within replicate platforms), it would be a split-plot design (see any of the basic statistic texts emphasising sampling/experimental design that are cited in the General Design Principles section above). Alternatively, the entire "bullseyes at platforms" design can re replicated, as in Montagna et al. (2004).

Should statistical analysis be based on distance as a category variable ("rings"), or distance as a continuous variable? The former leads to an ANOVA with model degrees of freedom (*df*) equal to the number of rings less one. No assumption is made about the relationship between impact and distance from platform. The latter leads to a regression analysis. The regression model would generally cost only 1 *df*, but the relationship is assumed to be linear (probably after transformation). It comes down to this: If you want to assume nothing about the relationship between impact and distance from platform then use rings and an ANOVA to test whether or not they differ. It is possible to do both; to do an ANOVA among rings, and then if significance is obtained, to "model" the trend using regression.

Attributes of a Good Design

A good offshore oil and gas EEM design would include:

- 2 (preferably at least 3) or more replicate platforms within each type of platform, the replicate platforms being within the same environmental conditions;
- alternatively, the entire study can be replicated (see above), either at the same time or at subsequent times if conditions are thought to be comparable;
- 2 (preferably at least 3) or more imaginary platforms in pristine areas, within the same environmental conditions;

- 2 (preferably at least 3) or more installed, but never used platforms (or structures simulating platforms), within the same environmental conditions;
- all platforms far enough away from other platforms that their expected effects will not overlap;
- 4–6 radii from each platform, with comparable directions for replicate platforms of a platform type;
- distances geometrically scaled with furthest distance beyond expected effects (probably 50, 100, 200 or 250, 500 and 1000 m). If uncertain about distance beyond expected effects, have one at ≥ 3000 m, but within a comparable environment; and
- repeated sampling over time to capture seasonal and annual variation, unless there are good reasons to not expect any.

With such a design, all impact-related hypotheses of interest can be tested: Do platform types differ? Is there any impact, even close to the platform? If there is then how does it decrease with distance, and does that pattern vary with direction from platform? Are platform effects (in general and for particular platform types) greater than the effects of platform structure alone would explain? Do any of these spatial patterns differ among times? And so on. As usual, with complex ANOVA designs, the highest level interactions would be tested first. For example, differences among rings (distances from platform) would be tested only after finding all interactions involving distance to be nonsignificant. The reason is logical, not technical. If the distance x radius interaction is significant, then the test of distance alone has no meaning—one would have to say that the relationship with distance depends on which direction (radius) you are talking about. The same applies to all interactions involving distance, radius, platform, type of platform and time, with the highest level interactions tested first (potentially 5-way!).

Is Time a Fixed or a Random Effect in EEM Designs?

An issue that has been debated (mostly informally e.g. at conferences and workshops) is whether "time" would be a fixed or a random effect in the ANOVA design. There are three traditional criteria for deciding: (1) Are the "levels (e.g. times) randomly sampled from a larger population or otherwise representative of it? (2) Are you interested in the levels for their own sake, or only in what you can infer about the larger population? and (3) Could you have chosen other levels "in-between" the ones you did, or are the ones you picked an arbitrary selection? In practical applications these criteria rarely agree. My own feeling, and that of many others, is that choosing to treat time as a random effect has the higher threshold to surmount.

Most scientists agree that with criterion (1) it is whether the sample is representative that matters, not whether it was randomly selected. Random sampling is a way to obtain a representative sample, and not otherwise of importance in itself. Criterion (2) is important, but the logic isn't always applicable. For example, in a dose-response study with different drug levels—a classic fixed effects design—you are often not interested in the doses used in the experiment but rather in the dose-response relationship (i.e. whether there is one and describing it if there is). Criterion (3) suggests that when the levels chosen are arbitrary, then a random effect is implied, but when the predictor variable is continuous (as in the drug dose-response study), the levels chosen are almost always arbitrary.

What about time in an EEM study? The three criteria would suggest the variable of time is fixed for (1), random for (2), and random for (3). With respect to criterion (1), a typical EEM study is over a series of *n* consecutive years, which is short in relation to any long sequence of *N* years that it falls within, but a long time period has much environmental pattern in it (Green, 1984), and a short consecutive sequence of years cannot be representative of it. The short sequence of years may be representative of a period of time not much longer than the length of the study, but then the ratio *n/N* will not be near zero as it is supposed to be according to sampling theory. In summary, times in an EEM study are never randomly selected and are rarely representative of any larger population of times that is conceptually useful. Related to this is the practical reality that offshore oil and gas EEM studies rarely continue for more than 2 or 3 years, which would result in only 1 or 2 degrees of "among years" degrees of freedom. Estimation of a variance on so few *df* would be very poor. I once heard the semi-humorous comment that if there are ≤ 6 times, it's a fixed effect because there isn't anything else useful that it can be. So criterion (1) implies a fixed effect. Regarding criterion (2), we are usually interested in what is true in general, i.e. we do want to extrapolate the results beyond the period of the EEM study. This implies a random effect. Regarding criterion (3), the years when an EEM study is done is something probably determined more by budget and political process than any importance of the years themselves. The study could have been done over the same number of years starting a year or two earlier or a year or two later. This implies a random effect.

The effect of time in an EEM study is not a trivial issue because it has important practical consequences for the statistical design especially with respect to hypothesis testing. On the one hand, time as a random effect would generate many interactions (any interactions involving time), which would also be random effects having potential use as error terms for testing hypotheses, much as described above for radius when it is considered a random effect. On the other hand, procedures for testing random effects (as opposed to using them as error to test something else) are less straightforward, as are interpretations of results. Add to that, the practical reality that EEM studies rarely have enough times to estimate an among-times variance very well (see above comments). Finally, add in the reality that significance of a random effect is generally interpreted as saying something about the larger conceptual population, and here there really isn't one that the study years can represent. To conclude, time in an offshore oil and gas EEM study should almost always be treated as a fixed effect. There are opinions on both sides. Most, that I have seen or heard, agree with this conclusion. Two opinions to the contrary, being in favour of time being a random effect, are held by A.J. Underwood and A. Stewart-Oaten (pers. comm.).

Is the Bullseye Design (Rings and Radii) the Best One?

Should the predictors, distance and direction (radius), be fixed and discontinuous (category) or random (probabilistic) and continuous? The bullseye design based on the fixed effects distance and radius has been around so long that one is tempted to look for alternatives. One would be to sample in a completely randomised design within the spatial component, with sample allocation proportional to predicted distance and direction of effects. A distribution of n sites (out of a much larger population of N locations that potentially could have been chosen) could easily be generated based on these criteria. The attractions are simplicity and random allocation of sites. Since there would be no category variables, a regression analysis approach would be appropriate. Distance, possibly transformed, would be a straightforward predictor. Direction (radius) would be trickier because it has a circular distribution. A cosine function of direction might work if there is one dominant current direction, but not if there are more, like seasonal variation. Zar (1996) has a nice chapter dealing with circular distributions which arise fairly often, for example, water depth over the tidal cycle, birth date, time of day, and direction of movement of released animals or plant seeds. Thus, it is not a problem unique to offshore oil and gas platform EEM studies. An adequate model which "costs" only 3 model *df* might be:

$$\mathrm{Y} = b_0 + b_1\mathrm{D} + b_2\cos(\mathrm{R}+a) + b_3\mathrm{D}^*\cos(\mathrm{R}+a)$$

The parameter a, which sets the direction of maximum distance effect would have to be based on dominant current direction, or estimated from the data (iteratively since it is inside the cosine function). If it is the latter, then a is estimated from the data so it costs 1 *df*; if it is the former, then the value of a is a priori so it doesn't.

Against this completely randomised approach is the theoretical argument that predictor variables, which are fixed effects, are expected in a Model I analysis, be it ANOVA or regression. Non-standard Model II estimation and testing procedures are required when both response and predictor variables are random effects. Furthermore, there is also the practical argument that ships' captains probably prefer to sample along a fixed transect (e.g. radii) rather than go to randomly selected locations.

REFERENCES

Armsworthy, S.L., A. Muecke, and P.J. Cranford. 2004. *Workshop on the Offshore Oil and Gas Environmental Effects Monitoring, Bedford Institute of Oceanography, Dartmouth, Nova Scotia, May 26–30, 2003*, Environmental Science Research Fund Report. In press.

Bousfield, E.L., and D.E. McAllister. 1962. "Station List of the National Museum Marine Biological Expedition to South Eastern Alaska and Prince William Sound." *National Museum of Canada Bulletin No. 183, Contributions to Zoology*, 1960–1961, pp. 75-103.

Chapman, P. M., B. Anderson, R. S. Carr, V. Engle, R. H. Green, J. Hameedi, M. Harmon, P. S. Haverland, J. Hyland, C. G. Ingersoll, E. R. Long, J. Rodgers,Jr., M. H. Salazar, P. K. Sibley, P. J. Smith, R. C. Swartz, B. Thompson, and H. Windom. 1997. "General Guidelines for Using the Sediment Quality Triad." *Marine Pollution Bulletin.* 34:368-372.

Chapman, P.M., E.A. Power, R.N. Dexter, and H.B. Andersen. 1991. "Evaluation of Effects Associated with an Oil Platform, Using the Sediment Quality Triad." *Environmental Toxicology and Chemistry.* 10:407-424.

Chapman, P.M., R.N. Dexter, and E.R. Long. 1987. "Synoptic Measures of Sediment Contamination, Toxicity and Infaunal Community Structure (The Sediment Quality Triad)." *Marine Ecology Progress Series.* 37:75-96.

Clarke, K. R., and R. H. Green. 1988. "Statistical Design and Analysis for a 'Biological Effects' Study." *Marine Ecology Progress Series.* 46:213-226.

Clarke, K.R. and R.M. Warwick. 1994. *Change in Marine Communities: An Approach to Statistical Analysis and Interpretation.* Natural Environment Research Council, Plymouth, UK.

Cochran, W.G. 1963. *Sampling Techniques.* 2nd edition. Wiley, New York.

Cochran, W.G. 1983. *Planning and Analysis of Observational Studies.* Wiley, New York.

Cohen, B.F. 2002. *Decoupling the Confounded Effects of Multiple Discharges around Established Offshore Oil and Gas Production Platforms: A Study of the Benthic Fauna of the Harriet Field, North West Shelf, Australia.* Ph.D. Thesis, University of Western Australia. 152p.

Cuff, W., and N. Coleman. 1979. "Optimal Survey Design: Lessons from a Stratified Random Sample of Macrobenthos." *Canadian Journal of Fisheries and Aquatic Science.* 36:351-361.

Dytham, C. 1999. *Choosing and Using Statistics: A Biologist's Guide.* Blackwell, London.

Elliott, J.M. 1977. "Some Methods for the Statistical Analysis of Samples of Benthic Invertebrates." *Science Publication, No. 25, Freshwater Biological Association UK*, Ferry House, UK.

Gilfillan, E.S., E.J. Harner, and D.S. Page. 2002. "Comment on Peterson et al Sampling Design Begets Conclusions". *Marine Ecology Progress Series.* 231:303-308.

Gray, J.S., M. Aschan, M.R. Carr, K.R. Clarke, R.H. Green, T.H. Pearson, R. Rosenberg, and R.M. Warwick. 1988. "Analysis of Community Attributes of the Benthic Macrofauna Oo Frierfjord/Langesundfjord and in a Mesocosm Experiment." *Marine Ecology Progress Series.* 46:151-165.

Gray, J.S., T. Bakke, H.J. Beck, and I. Nilssen. 1999. "Managing the Environmental Effects of the Norwegian Oil and Gas Industry: From Conflict to Consensus." *Marine Pollution Bulletin.* 38: 525-530.

Green, R.H. 1978. "Optimal Impact Study Design and Analysis." In K. L. Dickson, J. Cairns, Jr., and R. J. Livingston (Eds.), *Biological Data in Water Quality Assessment*, Philadelphia: American Society for Testing and Materials STP 652, pp. 3-28, Philadelphia.

Green, R.H. 1979. *Sampling Design and Statistical Methods for Environmental Biologists*. Wiley, New York.

Green, R.H. 1984. "Statistical and Nonstatistical Considerations for Environmental Monitoring Studies." *Environmental Monitoring and Assessment.* 4:293-301.

Green, R.H. 1987. "Statistical and Mathematical Aspects: Distinction between Natural and Induced Variation." In V.B. Vouk, G.C. Butler, D.V. Upton, D.V. Parke, and S.C. Asher, (Eds.), *Methods for Assessing the Effects of Mixtures of Chemicals*, pp. 335-354. Wiley, Chichester, U.K.

Green, R.H. 1989. "Power Analysis and Practical Strategies for Environmental Monitoring." *Environmental Research.* 50:195-205.

Green, R.H. 1993a. "Application of Repeated Measures Designs in Environmental Impact and Monitoring Studies." *Austrailian Journal of Ecology.* 18:81-98.

Green, R.H. 1994. "Aspects of Power Analysis in Environmental Monitoring." In D.J. Fletcher and B.F.J. Manly, (Eds.), *Statistics in Ecology and Environmental Monitoring*, pp. 173-182, Otago Conference Series, Otago, New Zealand.

Green, R.H. and P.A. Montagna. 1996. "Implications for Monitoring: Study Designs and Interpretation of Results." *Canadian Journal of Fisheries and Aquatic Science.* 53:2629-2636.

Green, R.H., and S.R. Smith. 1997. "Sample Program Design and Environmental Impact Assessment on Coral Reefs." In J.B.C. Jackson, (Ed.), *Proceedings of the International Coral Reefs Symposium*, Smithsonian Tropical Research Institute, Panama.

Green, R.H., J.M. Boyd, and J.S. Macdonald. 1993. "Relating Sets of Variables in Environmental Studies: the Sediment Quality Triad as a Paradigm." *Environmetrics.* 4:439-457.

Hinch, S. G. and R. H. Green. 1989. "The Effects of Source and Destination on Growth and Metal Uptake in Freshwater Clams Reciprocally Transplanted Among South-Central Ontario Lakes." *Canadian Journal of Zoology.* 67: 855-863.

Hurlbert, S.H. 1984. "Pseudoreplication and the Design of Ecological Field Experiments." *Ecological Monographs.* 54:187-211.

Hyland, J., D. Hardin, M. Steinhauer, D. Coats, R. Green, and J. Neff. 1994. "Environmental Impact of Offshore Oil Development on the Outer Continental Shelf and Slope off Point Arguello, California." *Marine Environmental Research.* 37:195-229.

Jackson, J.B.C., J.D. Cubit, B.D. Keller, V. Batista, K. Burns, H.M. Caffey, R.L. Caldwell, S.D. Garrity, C.D. Getter, C. Gonzalez, H.M. Guzman, K.W. Kauffman, A.H. Knap, S.C. Levings, M.J. Marshall, R. Steger, R.C. Thompson, and E. West. 1989. "Ecological Effects of a Major Oil Spill on Panamanian Coastal Marine Communities." *Science.* 243:37-44.

Jones, G.P. and U.L. Kaly. 1996. "Criteria for Selecting Marine Organisms in Biomonitoring Studies." In R.J. Schmitt and C.W. Osenberg, (Eds.), *Detecting Ecological Impacts: Concepts and Applications in Coastal Habitats*, pp. 29-48. Academic Press, New York.

Kennicutt, M.C. Jr., R.H. Green, P.A. Montagna, and P.F. Roscigno. 1996a. "Gulf of Mexico Offshore Operations Monitoring Experiment (GOOMEX), Phase 1:Sublethal Responses to Contaminant Exposure

—Introduction and Review." *Canadian Journal of Fisheries and Aquatic Science.* 53:2540-2553.

Kennicutt, M.C. Jr., P.N. Boothe, T.L. Wade, S.T. Sweet, R. Rezak, F.J. Kelly, J.M. Brooks, B.J. Presley, and D.A. Wiesenburg. 1996b. "Geochemical Patterns in Sediments near Offshore Production Platforms." *Canadian Journal of Fisheries and Aquatic Science.* 53:2554-2566.

Keough, M.J. and B.D. Mapstone. 1995. *Protocols for Designing Marine Ecological Monitoring Programs Associated with BEK Mills.* Technical Report No. 11. CSIRO, Canberra.

Keough, M.J., and K.P. Black. 1996. "Predicting the Scale of Marine Impacts: Understanding Planktonic Links between Populations." In R.J. Schmitt and C.W. Osenberg, (Eds.), *Detecting Ecological Impacts: Concepts and Applications in Coastal Habitats*, pp. 199-234. Academic Press, New York.

Livingston, R.J. 1987. "Field Sampling in Estuaries: The Relationship of Scale to Variability." *Estuaries.* 10:194-207.

Mapstone, B.D. 1996. "Scalable Decision Criteria for Environmental Impact Assessment: Effect Size, Type I and Type II Errors." In R.J. Schmitt and C.W. Osenberg (Eds.), *Detecting Ecological Impacts: Concepts and Applications in Coastal Habitats*, pp. 67-80. Academic Press, New York.

Mead, R. 1988. *Statistical Principles for Practical Applications.* Cambridge University Press, Cambridge.

Montagna, P.A, S.C. Jarvis, and M.C. Kennicutt II. 2004. "Distinguishing between Artificial Reefs and Platform Effects on Benthos." (abstract only). *Workshop on the Offshore Oil and Gas Environmental Effects Monitoring, Bedford Institute of Oceanography, Dartmouth, Nova Scotia, May 26–30, 2003*, Environmental Science Research Fund Report. In press.

Osenberg, C.W. and R.J. Schmitt. 1996. "Detecting Ecological Impacts Caused by Human Activities." In R.J. Schmitt and C.W. Osenberg, (Eds.), *Detecting Ecological Impacts: Concepts and Applications in Coastal Habitats*, pp. 3-16. Academic Press, New York.

Osenberg, C.W., R.J. Schmitt, S.J. Holbrook, K.E. Abu-Saba, and A.R. Flegal. 1996. "Detection of Environmental Impacts: Natural Variability, Effect Size, and Power Analysis." In R.J. Schmitt and C.W. Osenberg, (Eds.), *Detecting Ecological Impacts: Concepts and Applications in Coastal Habitats*, pp. 83-108. Academic Press, New York.

Paine, R.T., J.L. Ruesink, A. Sun, E.L. Soulanille, M.J. Wonham, C.D. Harley, D.R. Brumbaugh, and D.L. Secord. 1996. "Trouble on Oiled Waters: Lessons from the Exxon Valdez Oil Spill." *Annual Review of Ecology and Systematics.* 27:197-235.

Peterson, C.H. 1993. "Improvement of Environmental Impact Analysis by Application of Principles Derived from Manipulative Ecology: Lessons from Coastal Marine Case Studies." *Austrailian Journal of Ecology.* 18:21-52.

Peterson, C.H., L.L. McDonald, R.H. Green, and W.P. Erickson. 2001. "Sampling Design Begets Conclusions: The Statistical Basis for Detection of Injury to and Recovery of Shoreline Communities after the 'Exxon Valdez' Oil Spill." *Marine Ecology Progress Series.* 210:255-283.

Peterson, C.H., L.L. McDonald, R.H. Green, and W.P. Erickson. 2002. "The Joint Consequences of Multiple Components of Statistical Sampling Designs." *Marine Ecology Progress Series.* 231:309-314.

Peterson, C.H., M.C. Kennicutt Jr., R.H. Green, P.A. Montagna, D.E. Harper, E.N. Powell, and P.F. Roscigno. 1996. "Ecological Consequences of Environmental Perturbations Associated with Offshore Hydrocarbon Production: A Perspective on Long-Term Exposures in the Gulf of Mexico." *Canadian Journal of Fisheries and Aquatic Science.* 53:2637-2654.

Peterson, C.H., S.D. Rice, J.W. Short, D. Esler, J.L. Bodkin, B.E. Ballachey, and D.B. Irons. 2003. "Long-term Ecosystem Response to the Exxon Valdez Oil Spill." *Science.* 302:2082-2086.

Ritchie, W. 1995. "Maritime Oil Spills—Environmental Lessons and Experiences with Special Reference to Low-Risk Coastlines." *Journal Coastal Conservation.* 1:63-76.

Ritchie, W. 1997. "The *Braer* Oilspill: Environmental Circumstances and the Remit of the Ecological Steering Group on the Oilspill in Shetland." In J.M. Davies and G. Topping, (Eds.), *The Impact of an Oil Spill in Turbulent Waters: The Braer*, pp. 1-12. HMSO, Edinburgh.

Snedecor, G.W., and W.G. Cochran. 1967. *Statistical Methods.* 6th edition. Iowa State University Press, Ames, Iowa.

Stewart-Oaten, A. 1996a. "Goals in Environmental Monitoring." In R.J. Schmitt and C.W. Osenberg, (Eds.), *Detecting Ecological Impacts: Concepts and Applications in Coastal Habitats*, pp. 17-27. Academic Press, New York.

Stewart-Oaten, A. 1996b. "Problems in the Analysis of Environmental Monitoring Data." In R.J. Schmitt and C.W. Osenberg (Eds.), *Detecting Ecological Impacts: Concepts and Applications in Coastal Habitats*, pp. 109-131. Academic Press, New York.

Thrush, S.F., R.D. Pridmore, and J.E. Hewitt. 1996. "Impacts on Soft-sediment Macrofauna: the Effects of Spatial Variation on Temporal Trends." In R.J. Schmitt and C.W. Osenberg (Eds.), *Detecting Ecological Impacts: Concepts and Applications in Coastal Habitats*, pp. 49-66. Academic Press, New York.

Underwood, A.J. 1991. "Beyond BACI: Experimental Designs for Detecting Human Environmental Impacts on Temporal Variations in Natural Populations." *Austrailian Journal of Marine and Freshwater Research.* 42:569-587.

Underwood, A.J. 1992. "Beyond BACI: the Detection of Environmental Impacts on Populations in the Real, but Variable, World." *Journal of Experimental Marine Biology Ecology.* 161:145-178.

Underwood, A.J. 1993. "The Mechanics of Spatially Replicated Sampling Programmes to Detect Environmental Impacts in a Variable World." *Austrailian Journal of Ecology.* 18:99-116.

Underwood, A.J. 1994. "On Beyond BACI: Sampling Designs that Might Reliably Detect Environmental Disturbances." *Ecological Applications.* 4:3-15.

Underwood, A.J. 1997. *Experiments in Ecology: Their Logical Design and Interpretation Using Analysis of Variance.* Cambridge, U.K.

Zar, J.H. 1996. *Biostatistical Analysis.* Prentice-Hall, Upper Saddle River, New Jersey.

Numerical Modelling of Spreading of Oil on Ice

Bjorndalen, N., S. Mustafiz, A. Basu, M.R. Islam (Rafiqul.Islam@dal.ca) (Dalhousie University, Halifax, NS, Canada), and K. Lee (Bedford Institute of Oceanography, Fisheries and Oceans Canada, Dartmouth, NS, Canada)

ABSTRACT: The development of environmentally friendly remediation protocols is a major goal for the oil spill response community. This is especially true for the removal of oil from ice-infested waters where frigid temperatures and ice impede clean-up procedures and many remediation techniques become ineffectual. To understand the factors controlling the dynamics of a spill under these environmental conditions and its subsequent propagation, a numerical model was developed and validated with laboratory data. This study examined the significance of various parameters that influenced the fate and transport of the oil in the environment including the viscosity and density of the oil and temperature.

INTRODUCTION

Oil spills on land or water can range from minor incidents to great environmental disasters. The role that ice-infested waters play on oil spill interactions is dynamic indeed. There have been many occurrences of oil spills in ice-infested waters including the Othello spill (Metge and Telford, 1979), a spill of 100 tons of fuel oil at Deception Bay, NWT (Ramaseir et al., 1973), and a 600 ton spill of fuel oil in Sweden in 1972 (Jerbo, 1973). As well, spills in Buzzards Bay (Deslaurier, 1979) and Cabot Strait (Vandermeulen and Buckley, 1985) have been studied extensively, but the observations made in relation to the movement of the oil were conducted weeks after the spills. Therefore, the immediate impact of the spill was not determined.

The density of the oil spilled plays an important role in the final location of the oil with respect to ice. Since the density of oil can be either less than or greater than the density of ice, it is possible that the oil will either float on or under the ice (Milne and Herlinveaux, 1973). It has been observed that oil can be swept under the ice, splash onto the ice, and absorb into the ice (Buist and Bjerkelund, 1986). Furthermore, under freezing conditions, it was detected that the oil can exist as particles in packed ice (Wilson and Mackay, 1986). Visually entrained drops seemed to be created by wave action. The oil is often swept beneath the underside of ice, which is highly porous in nature. From there they tend to diffuse into the solid ice mass. During the thawing season, the oil that has diffused will surface (Milne and Herlinveaux, 1973). Wilson and Mackay (1986) studied the effect of ice particle size on the amount of oil entrapped in ice. It was determined that there is a maximum amount of oil that can be diffused for any given oil density. This maximum occurs when the ice particle size is between 5 and 8 mm. It was stated that at greater particle sizes the void spaces are greater, and therefore the oil is able to flow and thus it does not become entrapped.

This paper outlines a numerical model developed to understand the nature of oil spreading on ice. The numerical model was verified using experimental results collected in the laboratory and including a study on the development of a spill remediation procedure for use in ice-infested waters (Mustafiz et al., 2003). By utilising the numerical model, many conclusions can be drawn as to the effect of different parameters on the spreading system.

EXPERIMENTAL SET-UP AND PROCEDURE

A square tray (12 × 12 inch or 30.48 × 30.48 cm) was separated into 4 sections and filled with 500 ml of seawater. The seawater was obtained from the Atlantic Ocean at Crystal Crescent Beach, Nova Scotia. The tray with seawater was placed in a –25°C freezer. The tray was levelled to obtain a horizontal ice surface and was left overnight for freezing. Once the ice was formed, the tray was levelled again to ensure that the spreading of the oil spill would not be effected by a vertical gradient. Depending on the experiment performed, the ice tray was either left in the freezer during the experiment (–25°C experiment) or embedded in snow (4°C experiment). A Canon Optura 100 MC Digital Video Camcorder was placed on a tripod at a fixed height above the ice tray and the experiment was recorded throughout. A predetermined amount of Santa Barbara Crude Oil of 500 cp viscosity and 831.5 kg/m^3 density was dropped on the center of the ice surface. For the trials without sorbent material, the initial time was taken to be after the last drop of oil had been spilled. The spill was filmed for the entire 3-minute duration from which still images were extracted at 0, 15, 30, 45, 60, 120, and 180 second intervals. The area of the spill was obtained by evaluating the pictures with the Computer Image Analyzer.

NUMERICAL SIMULATION

In-depth work has been accomplished on the modelling of oil spills (Brebbia, 2001), but most of this work is limited to spills on water. For oil spills on ice, the work is more limited. The equation utilised in the numerical simulator was developed from the theories of Fay and Hoult (1971). Fay and Hoult examined oil spills on water and based their work on the assumption that there are four forces acting on an oil spill; gravity, interfacial tension, inertia, and friction. They defined gravity force as the creation of an unbalanced pressure distribution between the pool of oil and the surrounding water that results in horizontal spreading. The net surface tension is defined as the sum of the interfacial tension at the oil-air and oil-water surfaces and the inertial force is the force that retards oil-spread and is proportional to the mass of the oil and horizontal acceleration (Fay and Hoult, 1971). The frictional force comes from the surface underneath the slick and is dragged along below the expanding oil slick. In fact, the progression of an oil spill can be divided into three phases based on the four forces, where certain forces act for and against the spread at differing intervals. Figure 1(a) illustrates the three phases. Initially the gravity-inertial phase is dominant, which is followed by the gravity-viscous phase and the viscous-surface phase. Glaeser and Vance (1971) utilised the two initial phase equations developed by Fay and Hoult to model oil spills in the Arctic and found some positive results, especially for the gravity-viscous phase. Chen et al. (1974) based their work of spreading of oil spills on ice on the work of Fay and Hoult. They modelled the gravity-viscous phase of a spill for ice applications. McMinn (1972) stated that gravity and inertia forces dominate the spread of oil on ice and that it is independent of the properties of the oil as well as the temperature. It was also stated that the ratio of inertia to viscous forces depends on the slick thickness, and in turn, on the time frame

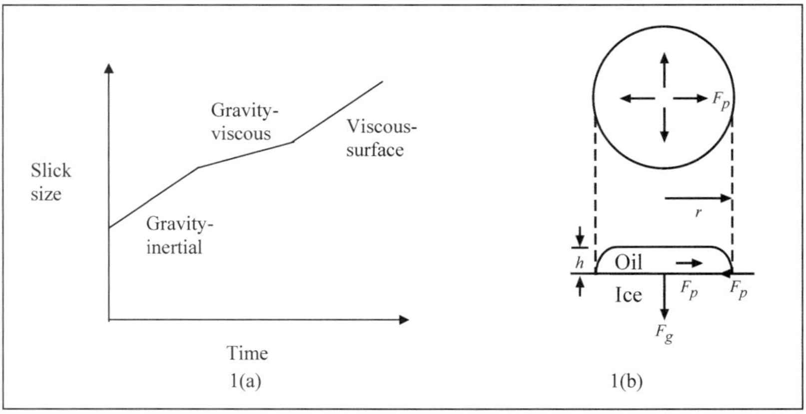

FIGURE 1. (a) Simplified diagram of oil slick as a function of time and **(b)** forces acting on an oil slick, gravity-viscous spreading phase (from Chen et al., 1974).

of the slick. For short times, the inertia forces dominate, while over longer periods, the viscous forces dominate. Then it can also be stated that for thinner slicks, viscous forces dominate. Figure 1(b) illustrates the forces acting on oil slick under gravity-viscous spreading phase. Since the experimental work presented in this paper is based a very small quantity of oil, it can be assumed that the spill thinned very quickly and that the viscous forces dominated. The model developed by Glaeser and Vance (1971) was based on assumptions and the elimination of differential terms. Therefore, numerical modelling through a differential equation has been developed in this paper to increase the accuracy of the modelling process. The numerical model developed was based on the following equation:

$$F_D = F_P - F_V \tag{1}$$

where F_D is the driving force, F_P is the force due to hydrostatic pressure and F_V is the viscous force. F_P is related to gravitational force (F_g) by:

$$F_P = kF_g \tag{2}$$

where k is the proportionality constant which is less than or equal to 1. The gravitational force can be expressed as:

$$F_g = \rho ghA \tag{3}$$

where ρ is the density of oil, g is the gravitational force, h is the height of spill and A is the area of spill. The force due to viscosity is defined as:

$$F_v = \tau A \tag{4}$$

where τ is the shear stress on the oil. τ can further be defined by:

$$\tau = \mu \frac{du}{dh} \tag{5}$$

where μ is the viscosity of the oil and u is the velocity of the spread. u can further be defined as dr/dt where r is the radius of the spill and t is the duration of time after spill. Therefore driving force can be defined as:

$$K_D = k\rho ghA - \mu A \frac{d^2r/dt^2}{dh/dt} \tag{6}$$

Since $h = V/\pi r^2$ where V is the volume of the spill, $dh/dt = (-2V/\pi r^3)\, dr/dt$. Equation (6) can then be rewritten as:

$$\frac{d^2r}{dt^2} + \frac{2V^2k\rho g - 2VF_D}{\pi^2\mu r^5}\frac{dr}{dt} = 0 \qquad (7)$$

Finally, Equation (7) can be reduced to:

$$\frac{d^2r}{dt^2} + \frac{S(Vk\rho g - F_D)}{r^5}\frac{dr}{dt} = 0 \qquad (8)$$

where $S = 2V/\pi^2\mu$. This problem was discretized using explicit method and the initial and boundary conditions were as follows:

$$r(0) = r_{initial}$$

$$\frac{dr}{dt} = \frac{r_{initial} - r_{final}}{elapsed} \qquad (9)$$

Velocity of the spill was determined based on an average velocity of the experimental work. The initial radius was assumed to be the radius measured immediately following the last drop of the spill for a particular run.

RESULTS AND DISCUSSION

Table 1 shows the variables used to develop the numerical results. It should be noted that an average radius was estimated for the initial radius of the oil spill. Since it is very difficult to form an oil spill of a uniform radius this value had to be assumed and was estimated from various trial runs. Figure 2 is a graphical representation of the results for the –25°C run. As well, the 4°C experimental run is presented. It is evident from the figure that the numerical model predicts the experimental run more accurately in the later stages of the experiment. This

TABLE 1. Parametric values used in numerical model.

Average Velocity	5.82×10^{-7} m/s
Initial Radius (r_i)	0.035 m
Volume of Spill (V)	6.0×10^{-6} m
Density of Oil (ρ)	831.5 kg/m^3
Time Step	15s
Viscosity (μ)	0.5 Pa-s
Driving Force (F_D)	5.0×10^{-8}
Proportionality Constant (k)	5.0×10^{-8}

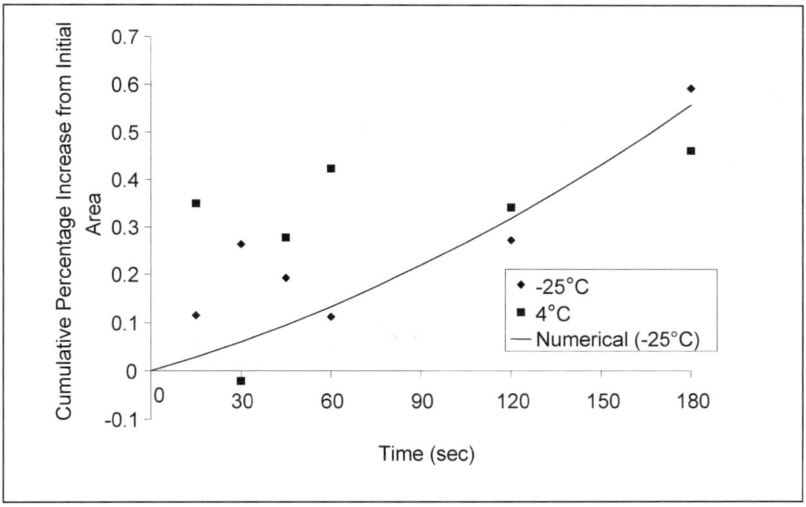

FIGURE 2. Comparison of numerical and experimental results. Experimental work from Mustafiz et al. (2003).

may be due to inertial forces still affecting the experiment in the early stages. Since the initial time begins immediately following the addition of the last drop of oil, the forces due to that can still be in effect. This is also noticeable in the 4°C run in that there is a greater increase in propagation during the initial phases of the experiment. The experimental work also shows that there is a decrease in cumulative percentage area during the first minute of the experiment for both of the runs. This is illogical since there should not be a decrease in propagation without an addition of sorbent. Given that variation of light can affect the digitization of the image, there may have been some loss of information during the image evaluation stage. This is true for all data reported, but since the total cumulative area increase was so small in this case (less than 0.60%), the increase in area due to propagation could not overcome the image processing error. All of the points on the graph are greater than the initial area values determined with one or two exceptions, and this assures the authors that slight propagation was occurring. Comparing the experimental data, one can see that the 4°C run had a great increase in area or propagation for most of the data points. This result is supported by Schrader (1993) who determined that as temperature increases propagation should increase. After 2 minutes of the experiment, the –25°C run surpassed the 4°C one. This can be explained by the fact that as the mild (4°C) one propagates, its gravitational force decreased much faster than that of cold (–25°C) one. Moreover, an additional liquid-liquid surface appeared as a result of melt of the seawater. It is to be noted that seawater above –2.1°C (melting point of seawater) was observed during the experiment. The surface tension created by this film of water did not promote the propagation of the spill, but rather inhibited it. A fairly flat slope that the 4°C run underwent after 1 minute sustains this fact. The role of surface tension in the spreading of oil is also discussed by Sayed and Løset (1993).

Figure 3 illustrates the variation of numerical results with a change in oil viscosity. A large range of viscosities were used to better illustrate the effect of viscosity on radius. The viscosity force is the retarding force in the differential equation. It is evident from the figure that as the viscosity of the oil increases, the ability to spread decreases. However, in plotting this graph, only the viscosity value was varied, although a change in viscosity should have affected the initial spill-radius for a given volume. It is to be noted that the propagation rate has a positive increasing slope at a lower viscosity.

Figure 4 shows how a change in oil density affects the numerical results. A wide range of values were intentionally applied to illustrate the effect of density. As the density of the oil

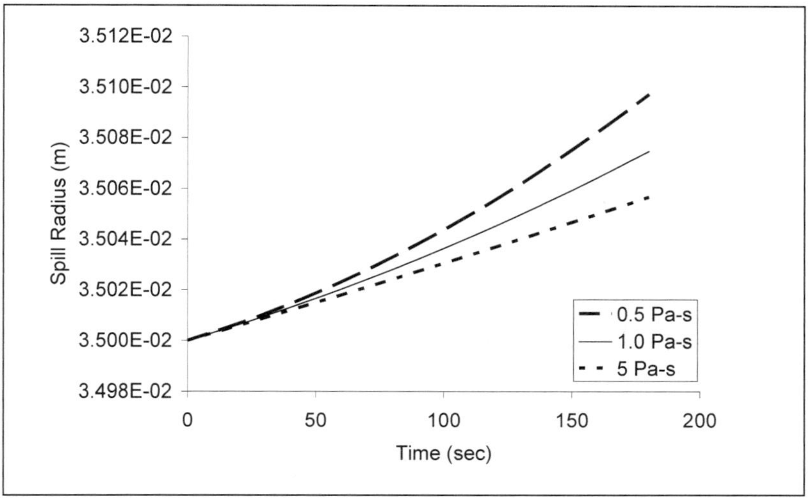

FIGURE 3. Effect of oil viscosity on spill radius.

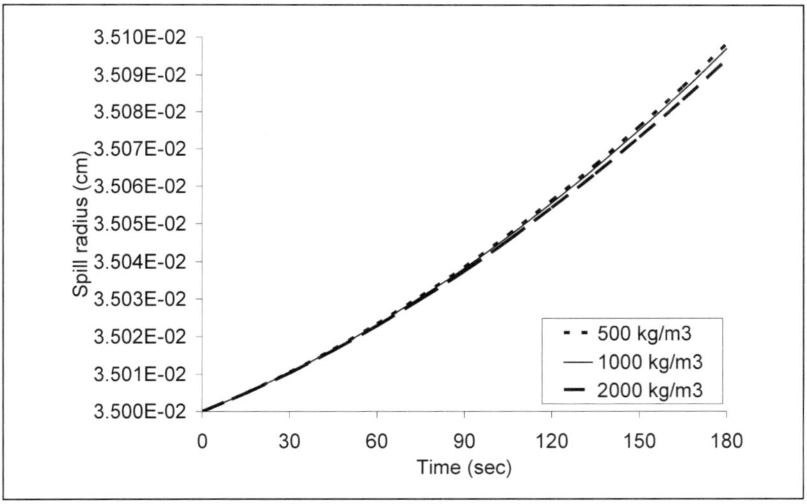

FIGURE 4. Effect of oil density on spill radius.

increases, the spreading rate decreases. In fact, it can be stated that density has virtually no effect on the spreading of the oil since even at extremely high densities (2000 kg/m^3), the increase in radius is similar to that of the 500 kg/m^3. This change in spreading is not as drastic as that shown in Figure 3. Although, Figure 4 illustrates a slight change due to density, one must note that with a change in the density of oil, there is a change in the viscosity of oil. For the generation of Figure 4, only the density of oil was varied and the viscosity was fixed. Even though Figure 4 illustrates that density does not seem to have a great affect on spreading indirectly it will through viscosity.

The effect that driving force has on propagation is illustrated in Figure 5. With an increase in the driving force, there is an increase in the propagation rate. The driving force is the difference between the gravitational and viscous forces in the system, and its amount directly dictates the spreading. It can be correlated that for a greater driving force, either the viscosity of the system is very low, or the gravitational force is very high.

The effect of the volume of crude oil spilled is given in Figure 6 when volumes were varied from 6 ml to 60 l. The height of the spill was assumed to be constant and the initial radius and

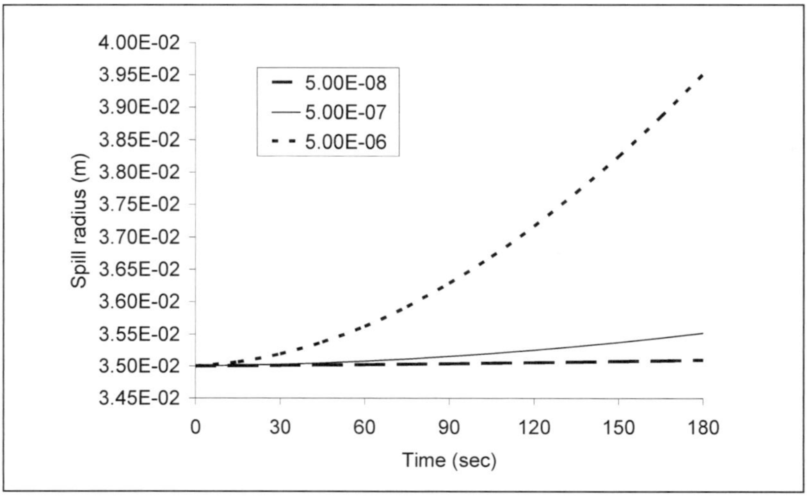

FIGURE 5. Effect of driving force on spill radius.

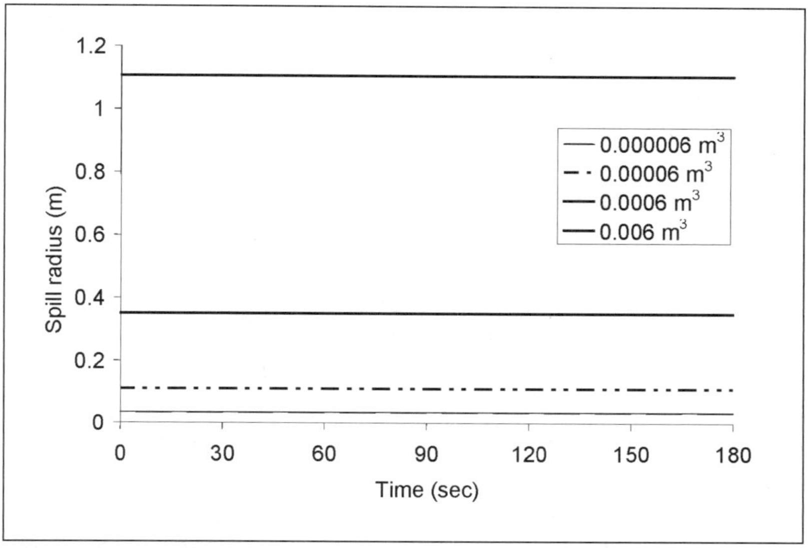

FIGURE 6. Effect of volume on spill radius.

propagation was determined for differing volumes. It is important to note that the model is predicting the gravity-viscous phase at the initial time the last drop of oil has been completely spilled, therefore, the initial radius of the spill should change with the volume spilled. The difference in successive radii at the selected time frame is very small, and this is why the graph shows horizontal lines for all runs. As well, the range of radii is large for the volume selected. One can compare the 0.000006 m^3 results to the results of Figure 2. Figure 3 shows the increase in the radius whereas Figure 6 does not. In Figure 6, 0.006 m^3 or 60 l shows that at initial time in the gravity-viscous phase the spill radius is over 1 m.

CONCLUSIONS AND RECOMMENDATIONS

An experimental and numerical study with a model based on gravitational force and viscosity on the spreading of oil on ice was completed. The model was found to be in reasonable agreement with the experimental data collected in the laboratory. The agreement between experimental and mathematical model was particularly good in the later phase of the experiment. Inertial forces may have affected the initial phase of the experiment. The equations used for the gravity-viscous phase do not cover these forces. It was found that viscosity, density, and driving force are the factors in determining the propagation rate. Although the model's results are somewhat limited by an assumption that the initial radius of the oil spill is uniformly round, conclusions can still be drawn regarding the factors controlling the extent of spreading. Numerical runs were conducted for volumes ranging from 6 ml to 60 l, and show that the initial radius of the spill will be between 3.5 cm and 1.1 m when entering into the gravity-viscous phase. An extension of this work is the addition of a term for diffusion to account for spills at temperature close to freezing. As well, it is proposed that the model can be expanded to predict the influence of sorbent additions by incorporating a supplementary sorbent term in the initial equation.

ACKNOWLEDGEMENTS

Funding of the research was possible through several grants from the federal government and petroleum industries.

NOMENCLATURE

A = area of spill (m^2)
F_D = driving force (N)
F_{DF} = force due to diffusion (N)
F_P = force due to hydrostatic pressure (N)
F_s = force due to sorbent (N)
F_V = viscous force (N)
g = gravitational force (m/s^2)
h = height of spill (m)
k = proportionality constant
m = mass of spill (g)
r = radius of the spill (m)
t = duration of time after spill (sec)
V = volume of spill (m^3)

ρ = density of oil (kg/m^3)
μ = viscosity of oil (Pa-s)
τ = shear stress on oil (N/m^2)

REFERENCES

Brebbia, C.A. 2001. *Oil Spill Modelling and Processes.* WIT Press, Boston.

Buist, I.A. and I. Bjerkelund. 1986. "Oil Packed in Ice: Preliminary Results of Three Experimental Spills." *9th Arctic marine Oilspill Program Technical Seminar*. Edmonton AB, June 10–12, 379-397.

Chen, E.C., C.K. Overall, and C.R. Phillips. 1974. "Spreading of Crude Oil on an Ice Surface." *Canadian Journal of Chemical Engineering*. 52:71-74.

Deslaurier, P.C. 1979. "Observation of Oil Behavior in Ice Floes and the 1977 Ethel H. Spill." *Workshop on Oil, Ice and Gas*. Toronto, ON.

Fay, J.A. and D.P. Hoult. 1971. *Physical Processes in the Spread of Oil on a Water Surface.* USCG Report AD-726 281, Washington, D.C.

Glaeser, LTJG J.L. and LCDR G.P. Vance. 1971. *A Study of the Behavior of Oil Spills in* the Arctic. USCG, Washington D.C.

Jerbo, A. 1973. "Two Types of Oil Spills in Swedish Inland Waters." Oil Spill Conference. Washington, D.C.

McMinn, T.J. 1972. *Crude Oil Behavior on Arctic Winter Ice, Project 734108*. USCG Washington D.C. 75p.

Metge, M. and A.S. Telford. 1979. "Oil in Moving Pack Ice Study", Conf. Ports and Oceans in Arctic Conditions. Oslo, Norway.

Milne, A.R. and R.H. Herlinveaux. 1973. "Crude Oil in Cold Water." In R.J. Childerhose (Ed.), *Beaufort Sea Project*, Minister of Supply and Services Canada, Ottawa, ON.

Mustafiz, S. 2002. "A Novel Method for Heavy Metal Removal from Aqueous Streams." M.A.Sc. Thesis, Dalhousie University.

Mustafiz, S., N. Bjorndalen, A Basu, K. Lee, and M.R. Islam. 2003. "A New Technique for the Remediation of Oil Spills from Ice-infested Waters." Canadian Society for Civil Engineers 31st Annual Conference. Moncton, NB, June 4–7, 2003.

Ramseier, P.O., G.S. Gontcheff, and L. Colby. 1973. *Oil Spill at Deception Bay, Hudson Strait, Canada*. Centre for Inland Water Scientific Series No. 29, Environment Canada Ottawa, ON.

Sayed, M. and Løset, S. 1993 "Laboratory Experiments of Oil Spreading in Brash Ice." International Journal of Offshore and Polar Engineering. 3(4):306-312.

Schrader, E.L. 1993. "A Practical Composition of Organic, Synthetic and Inorganic Sorbents." Clean Gulf 93 and the American Chemical Society Conference: Emerging Technology in Hazardous Waste Management.

Vandermeulen, J.H. and D.E. Buckley, (Eds.) 1985. *The Kurdistan Oil Spill of March 16–17, 1979; Activities and Observation of the Bedford Institute of Oceanography Response Team.* Canadian Technical Report of Hydrography and Ocean Sciences No. 35, Bedford, NS.

Wilson, D.G. and D. Mackay. 1986. "The Behavior of Oil in Freezing Situations." *9th Arctic Marine Oilspill Program Technical Seminar.* Edmonton, AB.

PART 3

Applications and Regional Experience

Offshore Oil and Gas Environmental Effects Monitoring Investigations Conducted by the U.S. Minerals Management Service

Ahlfeld, T.E. (Thomas.ahlfeld@mms.gov) (Minerals Management Service Herndon, VA, USA)

ABSTRACT: The Minerals Management Service (MMS), an agency of the United States Department of the Interior, is responsible for leasing, permitting, and regulating oil and natural gas activities in the U.S. offshore waters within its Exclusive Economic Zone. An element of the MMS mission is "to manage the mineral resources of the Outer Continental Shelf (OCS) in an environmentally sound and safe manner". To assure that offshore development occurs under these guidelines, MMS inspects all offshore facilities, reviews plans of exploration and development, performs environmental analyses, and funds environmental and engineering research related to OCS oil and gas activities. A significant portion of the MMS research budget has been directed to marine environmental monitoring projects. The monitoring projects discussed here include investigations of the environmental effects of oil and gas exploration activities, long-term offshore production, oil spills, and potential impacts to habitats and biological communities of special concern such as the Flower

Garden Banks National Marine Sanctuary. Provided here is an overview of what has been learned through the MMS environmental effects monitoring investigations since the initiation of the program for accelerated development of U.S. offshore oil and gas resources 30 years ago.

THE MINERALS MANAGEMENT SERVICE AND THE ENVIRONMENTAL STUDIES PROGRAM

The MMS was established in 1982, primarily through the combination of components from two other bureaus within the Department of the Interior; the Bureau of Land Management, and the U.S. Geological Survey. The mission of the MMS related to this discussion is to manage the mineral resources on the Outer Continental Shelf (OCS) in an environmentally sound and safe manner A second component of the MMS mission is to collect, verify, and distribute mineral revenues from Federal and Indian lands, both offshore and onshore, in a timely manner. The term OCS includes those areas of the U.S. continental shelves beyond jurisdiction of the coastal states. This generally is the seabed beyond three miles from the coastline except in the Gulf of Mexico off the coasts of Florida and Texas where State jurisdiction extends to three marine leagues or about 8.76 nautical miles. The OCS extends across the continental slope and rise and abyssal plain out to the 200 nautical mile extent of the U.S. Exclusive Economic Zone.

The offshore area managed by the MMS is vast, encompassing approximately 1.76 billion acres. Currently there are about 8,000 active oil and gas leases and 4,000 offshore oil and gas structures operating on the OCS. This offshore activity currently results in the production of nearly 25 percent of the U.S. domestic natural gas and 30 percent of the domestic crude oil.

In the Gulf of Mexico, where by far the greatest amount of offshore activity is occurring now and has occurred in the past, emphasis in recent years has been on developing reserves in deep water areas. Production is now taking place in water depths greater than 2,300 m, and exploratory wells in the Gulf of Mexico have been drilled at depths of over 3,000 m. In the Pacific, there has been no new leasing in nearly 20 years, and the current moratorium prohibiting any new development extends through the year 2012. There are, however, 23 active producing platforms in Federal waters off southern California, and decommissioning of these offshore facilities will be a major issue for the MMS. In the Alaskan Beaufort Sea, the first production of Federal offshore petroleum has recently begun from BP's Northstar site, and another prospect to the east, Liberty, is very promising for development in the future.

Within the offshore oil and gas program, the Environmental Studies Program was established in 1973, originally as part of the Bureau of Land Management and now as part of the Minerals Management Service, to provide environmental and socioeconomic information to decision makers. The information provided by the Environmental Studies Program is used to predict, assess, and manage impacts from offshore oil and gas and marine mineral exploration, development, and production activities on human, marine, and coastal environments. Since 1973, the Environmental Studies Program has sponsored research and data collection projects totaling about $750 million, with approximately $125 million spent on environmental monitoring studies. Level funding of about $17 million/year is anticipated for the near term.

LAWS AFFECTING THE OFFSHORE OIL AND GAS PROGRAM

The OCS Lands Act Amendments of 1978 established a mandate for conducting environmental studies and monitoring. With respect to managing the resources of the Outer Continental Shelf, the Secretary of the Interior is charged with monitoring the human, marine, and coastal environments. According to this requirement, which the Secretary has delegated to the Minerals Management Service, monitoring will provide time series information that can be compared with previously collected data to identify significant changes in the quality or productivity of an area, to establish trends, and to design experiments to identify the causes of observed changes.

There are many other laws that directly affect offshore oil and gas development activities, and some of these laws have additional monitoring requirements that may be carried out by MMS, the offshore operators, or other Federal Agencies. An example is the Federal Water Pollution Control Act (Clean Water Act) which regulates the discharges from oil and gas activities into marine waters through National Pollution Discharge Elimination System (NPDES) permits established by the U.S. Environmental Protection Agency. Others include the Clean Air Act, Endangered Species Act, Marine Mammal Protection Act, Fishery Conservation and Management Act, National Historic Preservation Act, Coastal Zone Management Act, and the National Environmental Policy Act.

COMPLIANCE AND ENVIRONMENTAL EFFECTS RESEARCH MONITORING

Two different aspects of "monitoring", compliance, and environmental effects research, represent significant areas of activity for MMS. Compliance monitoring is important to assuring environmentally sound operations. In general terms, this type of monitoring provides a process to assure the offshore industry's conformity with regulations, permit requirements, and mitigation measures. An example is the sampling and analysis of produced water and drilling fluid discharges to verify that offshore operations are in compliance with Environmental Protection Agency regulations (NPDES permits). The focus here, however, is on the effects research aspect of monitoring rather than compliance monitoring. Effects monitoring investigations have been conducted by MMS to assess natural variability, establish baseline conditions, and identify and quantify changes that may be due to offshore oil and gas activities.

In some cases, MMS research has resulted in the development of new lease management requirements. The MMS sponsored two major studies designed to characterize the structure and functions of chemosynthetic communities associated with naturally occurring hydrocarbon seeps in the Gulf of Mexico (MacDonald et al., 1996; MacDonald, 2002). These studies provided information used to develop a Notice to Lessees and

Operators, or NTL, with requirements designed to protect these specialized communities living in areas of oil and gas development in water depths of 400 m and greater. This NTL may require the operator to monitor for potential impacts caused by proposed activities.

MONITORING EFFECTS OF OFFSHORE OIL AND GAS EXPLORATION ACTIVITIES

Early in the program when offshore oil and gas leasing was accelerated in several new or frontier areas, the MMS conducted a number of monitoring projects to investigate environmental impacts associated with offshore oil and gas exploratory activities. The more significant of these projects were conducted off the U.S. Atlantic coast where no Federal oil or gas has been produced and moratoria restrict any consideration of future activity until 2012.

At two exploratory drilling sites on Georges Bank, 12 cruises were conducted between July 1981 and June 1984. The program included intensive quantitative sampling of soft-bottom benthic communities near, upcurrent, and downcurrent of drilling rigs. Study methods included analysis of bottom photographs for epifauna and microtopography, dredge and trawl collections, analysis of chemical contaminants in sediments and benthic fauna, and total organic carbon and sediment grain-size analyses. Sampling station locations were established to determine both near-rig and regional environmental impacts. Twenty-nine of a total 46 stations were located in a tight radial array around an exploratory drilling rig positioned in a water depth of 80 m. A second group of three stations was located in proximity to a second exploratory drilling site positioned in a water depth of 140 m to determine near-field impacts. These stations were located from 200 to 2,000 m up and down current from the drilling rig. Remaining stations were positioned over a broad expanse of the bank and within areas of potential drilling mud deposition, including stations at the head of Lydonia and Oceanographer canyons. Although eight wells had been drilled in the study area, no biological impacts attributable to drilling activities could be detected at any of the stations (Maciolek-Blake et al., 1985).

Concentrations of hydrocarbons in Georges Bank sediments were related to grain size, and the highest concentrations were associated with fine-grained sediments. Polycyclic aromatic hydrocarbon concentrations in the study area were consistent with previously recorded levels at both regional and site-specific stations. No evidence of long-term accumulation of petroleum hydrocarbons in near-rig sediments was apparent during the monitoring program, and no uptake or accumulation of drilling fluid constituents was detected in analyzed faunal tissues. Trace metal and hydrocarbon levels were consistently low and comparable to previously recorded background levels (Payne et al., 1985).

Of 12 trace elements measured in bottom sediments, only barium, a good indicator of the deposition of drilling muds due to its high density and insolubility, increased in concentration on Georges Bank during exploratory drilling from July 1981 to September 1982. The greatest post-drilling barium concentration reached 172 parts per million in sediments, which was 5.9 times greater than pre-drilling levels. No changes in the concentrations of the 11 other metals were observed. It was estimated that 25 percent of the discharged barium was present in the sediments within 6 km of drilling activities, and barium levels decreased rapidly during the first post-drilling year. Evidence of barium transport was found 115 km west and 35 km east of drilling activities. Sediment trap samples had significantly higher concentrations of barium in post-drilling samples than in pre-drilling samples. The findings also showed that sediments were periodically resuspended to at least 25 m above the seafloor (Bothner et al., 1985).

In the Middle Atlantic OCS Planning Area off the coasts of New Jersey and Delaware, an array of 14 benthic sampling stations, located at water depths of 1,500 and 2,100 m, was placed at varying distances from two exploratory drilling sites on the continental slope and rise. Stations were sampled six times over a 21 month period from March 1984 to November 1985. Temporal and spatial variations in epifaunal and infaunal communities were studied through a variety of sampling methods, including camera transects, crab traps, bottom trawls, and box corers. Infaunal recolonization rates were measured from trays deployed for 6 or 12 month periods. Sediment textures and chemistry were examined from box core samples. Trace metals and hydrocarbons were examined in two common echinoderms, the brittle star, *Ophiomusium lymani*, and the sea urchin, *Echinus affinis*.

In one downslope area, decreases in sea pen densities occurred subsequent to drilling. It was unclear, however, what factors, whether natural or man-made, caused the observed changes (Maciolek et al., 1987). There was no evidence of drilling related increases in metals or hydrocarbons within the tissues of brittle stars and sea urchins analyzed. Physical impacts from drilling activities were minimal, although plastic pipe casings were seen on the seafloor. Barium levels only increased at one station following drilling. This result was surprising, since amounts of mud discharged at another drilling site were ten times higher. Despite predictions that slight (1–2 millimeters) accumulations of drilling muds or cutting solids may produce dramatic reductions in infaunal diversity, there was no evidence for drilling related changes in the fauna at either site (Maciolek et al., 1987).

In the late 1970s, two exploratory rigs in the northwestern Gulf of Mexico were monitored with a radial sampling pattern of stations at distances of 100, 500, 1,000, and 2,000 m from the drill site. Results from these projects were similar to the previously reported exploration effects monitoring, with only near-field and short-term effects observed for benthic communities and detectable levels of contaminant concentrations (University of Texas Marine Science Institute, 1977).

MONITORING EFFECTS OF OFFSHORE OIL AND GAS DEVELOPMENT AND PRODUCTION

The MMS also has monitored the effects of development and production activities in the Gulf of Mexico, the Pacific Ocean off southern California, and in the Beaufort Sea off the coast of Alaska.

The Gulf of Mexico Offshore Operations Monitoring Experiment (GOOMEX) was a 3 year, $4.5 million project completed in 1995. Five platforms were evaluated as candidate study sites, and three were selected for long-term study. It was determined that study sites must be outside the perpetual influence of the Mississippi River and must possess a long-term production history of at least 10 years where measurable impacts would be expected. These conditions resulted in the selection of three platforms in the northwestern Gulf of Mexico off the coast of Texas. The study design employed a radial pattern of stations at 30 to 50, 100, 200, 500, and 3,000 m distances from the platforms. This approach supported a dose-response model to test the hypothesis that biological, chemical, and biochemical variations were due to platform-derived contaminants. Study components included contaminant analysis (trace metals and hydrocarbons) in sediments, pore waters, and biological tissues. Analysis of benthic meiofaunal, macroinfaunal, and epifaunal assemblages; assessment of community health based on life-history and reproduction studies; and the induction of detoxification responses were other major study elements.

Results of GOOMEX demonstrated that increased levels of contaminants from the discharge of drill muds and cuttings were restricted to sediments within hundreds of meters of the platforms. Concentrations of polycyclic aromatic hydrocarbons in sediments were well below levels known to be associated with toxic biological effects (<4,000 parts per billion), and no significant bioaccumulation of hydrocarbons was observed in megafaunal invertebrates or fish collected near platforms (Kennicutt, 1995).

When exposed to chemical contaminants, many organisms have systems that eliminate the toxicants. In this study, no increase in the detoxification responses of tested organisms was observed when comparing near versus far field suggesting that little or no increased exposure was experienced by benthic organisms around platforms (Kennicutt, 1995).

The California Monitoring Program (CAMP) was a 12-year, $15 million project completed in 1995. The first phase of this project was a 3-year characterization of the Santa Maria Basin and Western Santa Barbara Channel. Benthic habitats, biological assemblages, and hydrocarbon and trace metal concentrations in sediments were studied at 107 soft-bottom sampling stations located along 16 transects. An additional 15 hard-bottom stations were investigated using manned submersible dives (Lissner et al., 1986). Production drilling in the study area began in 1986, and Phase 2 of CAMP was initiated to detect and measure potential changes in the marine environment in the vicinity of these activities. In this phase of the study, sampling design consisted of a series of regional stations and two arrays of site-specific stations near existing and planned oil production platforms (Battelle, 1991). The final phase of this project was designed to develop a better understanding of the environmental fate and effects of chronic, low-level inputs of drilling wastes and their potential for cumulative impacts to benthic communities. New tasks included in situ larval settling experiments and laboratory toxicity studies testing drilling muds (Lissner, 1995).

Of the trace metal contaminants studied, only barium showed increased levels in both bottom sediments and suspended sediment traps. After drilling stopped, levels of barium in the sediment decreased, but after 1.5 years had not returned to background concentrations. In sediment traps, barium concentrations did return to background levels 1 year after drilling (Lissner, 1995).

Hydrocarbon concentrations in sediments were occasionally elevated, but these changes were not consistent with drilling activities. Examination of the total petroleum hydrocarbon and polycyclic aromatic hydrocarbon distributions indicated that natural petroleum seeps common in the study area were likely responsible for randomly elevated hydrocarbon concentrations. Elevated levels of petroleum hydrocarbons observed at some stations in 1991 were attributed to sources other than offshore production, including a land-based spill (Lissner, 1995).

No drilling-related effects were evident from in situ natural larval settlement experiments. Overall, larval settlement rates were very low and there was high spatial and temporal variability. Laboratory toxicity tests indicated that drilling muds did result in mortality to brown cup corals but did not negatively affect the fertilization mechanism and early development stages of red abalone (Lissner, 1995).

In the Alaska Beaufort Sea, MMS conducted a five-year monitoring project, Arctic Nearshore Impact Monitoring in the Development Area or ANIMIDA. A multi-year continuation of ANIMIDA is planned to extend this monitoring project.

The ANIMIDA study area is centered in the Beaufort Sea around the Northstar Production Island, the Liberty Prospect, and the Boulder Patch. The Boulder Patch is an area of large rocks populated by kelp and a diverse invertebrate assemblage.

A number of potential offshore petroleum development disturbances were considered in designing ANIMIDA. These include:

- changes to the physical environment related to circulation and sedimentation from construction activities including gravel island, causeways, and pipeline trenching;
- effects on biota of changes in sediment inputs;
- inputs of contaminants to the water, primarily metals and hydrocarbons, but also particles;
- increased bioavailability of chemical contaminants to fish and their prey;
- changes in local subsistence whaling or fishing; and
- acoustical changes and possible effects on bowhead whale migration.

Some of the significant preliminary results of ANIMIDA to date include:

- Pre-Northstar and Liberty baseline data were established.
- Post-Northstar construction levels of hydrocarbons and metals are within the historical range for the region and reflect natural background from regional sources. There are no indications of petroleum exploration and production-related contamination.

- The contaminant concentrations in sediments are generally well below the range that would be expected to result in adverse biological effects.
- Geochronology data support previous studies indicating that the nearshore Beaufort Sea is a net-erosional area.
- Rivers are a major source of sediment to the study area with more than 50 percent of the input occurring over a period of two to three weeks associated with the spring thaw.
- River-borne sediment is carried well offshore both over and under the ice.
- Ambient concentrations of suspended sediment in the study area during open-water are wind dependent, a product of resuspension.
- Concentrations of dissolved metals in Beaufort Sea coastal waters were found to be well below values identified for acute and chronic State and Federal guidelines.
- Dissolved and particulate levels of several metals in the coastal Beaufort Sea were found to follow predictable distribution coefficients, thereby allowing modeling of dissolved and particulate behavior.

MONITORING SPECIFIC HABITATS, ORGANISMS, AND OIL SPILLS

Another aspect of marine environmental effects monitoring conducted by the MMS involves specific habitats, organisms, and oil spills. These include the following examples:

- monitoring of the east and west Flower Garden Banks in the Gulf of Mexico;
- monitoring of California rocky intertidal communities;
- monitoring bowhead whale migration in the Alaska Beaufort Sea; and
- long-term assessment of a major oil spill in Panama.

The Flower Garden Banks National Marine Sanctuary is located about 110 nautical miles southeast of Galveston, Texas, in an area of significant offshore petroleum development. One platform is located within the sanctuary boundary and has been producing natural gas and condensate for many years prior to sanctuary designation. The MMS sponsored several years of research to characterize these banks and their living coral reef communities before initiating the monitoring program. The banks are characterized by a very high percentage of living coral cover—about 50 percent compared to less than 10 percent for many of the Caribbean and Florida coral reefs. The monitoring project is now a cooperative program with co-sponsorship from MMS and the National Oceanic and Atmospheric Administration, the agency responsible for marine sanctuary management. Monitoring concerns include potential impacts on the banks from fishing, recreational diving, and the development of oil and gas resources.

Photography, growth spikes, and sclerochronolgy are used to measure living coral cover as well as accretionary and lateral growth rates. Some coral bleaching has been observed since 1997 but with little permanent damage. All growth measures applied indicate that corals on the Flower Garden Banks are growing and generally healthy (Gittings et al., 1992; Continental Shelf Associates, Inc, 1996; Dokken et al., 1999; Dokken et al., 2003).

In the Pacific Region, the MMS Intertidal Team (MINT) conducts an in-house study project which has monitored rocky intertidal communities adjacent to Federal offshore operations off the California coast since 1991. The Multi-Agency Rocky Intertidal Network or MARINe was established with MMS leadership in 1996 to coordinate and standardize intertidal monitoring in California. There are now 23 local, State, and Federal agencies and private organizations monitoring 70 intertidal sites. This program provides information needed to determine impacts to shoreline resources from possible oil spills. Additional information is now available at the MARINe website, www.MARINe.gov.

The Bowhead Whale Aerial Survey Program is another in-house monitoring project conducted by the MMS. Since 1987, the MMS Alaska Region has monitored the annual fall migration of bowhead whales for timing and migration corridors in the Beaufort Sea.

Migration patterns are monitored upstream, in the vicinity of, and downstream from offshore seismic exploration activities (Treacy, 2002). Real-time data are used to implement seasonal restrictions and limitations on geological and geophysical exploration.

A major on-shore crude oil spill of between 60,000 and 100,000 barrels occurred at Bahia las Minas, Panama, in 1986 near coral reef, seagrass, and mangrove communities. The MMS, together with the Smithsonian Institution's Tropical Research Institute, responded to this opportunity to study the impacts of spilled oil on sensitive marine communities similar to those off the coast of southern Florida. A three year study was conducted to document changes in the distribution and abundance of marine organisms in the spill area and to understand the ecological processes causing changes.

The oil spill had widespread lethal and sublethal effects on prominent organisms in all intertidal and subtidal environments examined. About 7 percent, or 64 hectares, of the mangrove forest in the spill area was killed (Keller and Jackson, 1993).

Cover, size, and diversity of live corals decreased on two oiled reefs but increased on unoiled reefs, and sclerochronology indicated a reduction in growth of corals during the year of the spill. Macroalgae and sessile invertebrates of fringe reef flat communities had heavy mortality initially but showed nearly complete recovery within a year (Keller and Jackson, 1993).

CONCLUSIONS

Monitoring projects conducted by MMS have provided a wealth of information concerning the effects of offshore oil and gas exploration, development, and production on the marine environment. They also provide models for the refinement of the design of future environmental monitoring efforts.

Based on the results of MMS sponsored and other monitoring projects, a review of the environmental effects of offshore oil and gas development concluded that discharges from exploratory rigs in the offshore environment would not result in significant long-term effects, except possibly in areas with rare, slowly recovering communities (Boesch et al., 1987). Following recommendations of Boesch et al. (1987) and Carney (1987), the focus of more recent offshore oil and gas environmental monitoring has been on the long-term effects of development and production operational discharges on benthic communities as discussed above for the CAMP and GOOMEX Programs.

Looking toward the future, additional deepwater monitoring in the Gulf of Mexico is anticipated. At least two sites with major aggregations, or "thickets", of the coral, *Lophelia sp.*, have been reported in the northern Gulf of Mexico. Studies to date have shown the deepwater corals and chemosynthetic communities to be long-lived and slow growing. Early indications of any impacts associated with offshore drilling activities are important to the protection of these organisms. To provide necessary baseline information for the determination of potential impacts, MMS initiated a study during 2003 designed to describe the environmental conditions that result in the development of high density communities such as the *Lophelia* thickets.

Future deepwater monitoring projects in the Gulf of Mexico also may be initiated by MMS in response to the development of gas hydrates as an energy resource. There has been an increased interest in the recovery of gas hydrates from the OCS as a source of natural gas. To be prepared as the recovery of gas hydrates becomes economically viable, MMS will need to collect environmental information to be used in evaluating the potential environmental impacts of the recovery of this energy resource. The potential impacts to the environment will depend on the methodology used to recover the natural gas and the proximity of the resource extraction activity to biological communities. It is already known that gas hydrates in the Gulf of Mexico exist below the seafloor and, in some cases, are actually exposed at the seafloor. The gas hydrates at the surface support chemosynthetic communities, which MMS presently protects. However, removal of hydrates from the subsurface could affect larger areas of the primarily soft mud seafloor. The biological communities that live in these areas could be affected potentially resulting in disruption of the ecological balance of deep sea communities. Also, because of the importance of geochemical processes in the sediment and physical processes at the sediment water interface to deep sea marine life, these processes may be studied since they could be perturbed as a result of the hydrate recovery process.

REFERENCES

Battelle. 1991. *California OCS Phase II Monitoring Program.* Final Report for the U.S. Dept. of the Interior, Minerals Management Service, Pacific OCS Region, Camarillo, CA.

Boesch, D.F, J.N. Butler, D.A. Cacchione, J.R. Geraci, J.M. Neff, J.P. Ray, and J.M Teal. 1987. "An Assessment of the Long-Term Environmental Effects of U.S. Offshore Oil and Gas Development Activities: Future Research Needs." In D.F. Boesch and N.N. Rabalais (Eds.), *Long-Term Environmental Effects of Offshore Oil and Gas Development,* pp. 1-54. Elsevier Applied Science, New York, NY.

Bothner, M.H., R.R. Rendigs, E.Y. Campbell, M.W. Doughten, C.M. Parmenter, C.H. O'Dell, G.P. DiLisio, R.G. Johnson, J.R. Gillison, and N. Rait. 1985. *The Georges Bank Monitoring Program: Analysis of Trace Metals in Bottom Sediments during the Third Year of Monitoring.* Final Report Prepared by the U.S. Geological Survey for the U.S. Dept. of the Interior, Minerals Management Service, Atlantic OCS Region, Vienna, VA.

Carney, R.S. 1987. "A Review of Study Designs for the Detection of Long-Term Environmental Effects of Offshore Petroleum Activities". In D.F. Boesch and N.N. Rabalais (Eds.), *Long-Term Environmental Effects of Offshore Oil and Gas Development,* pp. 651-696. Elsevier Applied Science, New York, NY.

Continental Shelf Associates, Inc. 1996. *Long-Term Monitoring at the East and West Flower Garden Banks.* Final Report for the U.S. Dept. of the Interior, Minerals Management Service, Gulf of Mexico OCS Region, New Orleans, LA. OCS Study MMS 96-0046.

Dokken, Q.R., I.R. MacDonald, J.W. Tunnell, Jr., T. Wade, K. Withers, S.J. Dilworth, T.W. Bates, C.R. Beaver, and C.M. Rigaud. 2003. *Long-Term Monitoring at the East and*

West Flower Garden Banks, 1998–2001: Final Report. U.S. Dept. of the Interior, Minerals Management Service, Gulf of Mexico OCS Region, New Orleans, LA. OCS Study MMS 2003-031.

Dokken, Q.R., I.R. MacDonald, J.W. Tunnell, Jr., C.R. Beaver, G.S. Boland, and D.K. Hagman. 1999. *Long-term Monitoring of the East and West Flower Garden Banks, 1996–1997.* U.S. Dept. of the Interior, Minerals Management Service, Gulf of Mexico OCS Region, New Orleans, LA. OCS Study MMS 99-0005.

Gittings, S.R., G.S. Boland, K.J.P. Deslarzes, D.K. Hagman, and B.S. Holland. 1992. *Long-Term Monitoring at the East and West Flower Garden Banks.* U.S. Dept. of the Interior, Minerals Management Service, Gulf of Mexico OCS Region, New Orleans, LA. OCS Study MMS 92-0006

Keller, BD. and J.B.C. Jackson (Eds.). 1993. *Long-Term Assessment of the Oil Spill at Bahia las Minas, Panama: Synthesis Report (3 Vols.).* Prepared by the Smithsonian Tropical Research Institute for the U.S. Dept. of the Interior, Minerals Management Service, Gulf of Mexico OCS Region, New Orleans, LA. OCS Study MMS 93-0047-49.

Kennicutt, M.C., II, (Ed.). 1995. *Gulf of Mexico Offshore Operations Monitoring Experiment, Phase I: Sublethal Responses to Contaminant Exposure.* Final Report Prepared by the Geochemical and Environmental Research Group, Texas A&M University for the U.S. Dept. of the Interior, Minerals Management Service, Gulf of Mexico OCS Region, New Orleans, LA. OCS Study MMS 95-0045.

Lissner, A., C. Phillips, D. Cadien, R. Smith, B. Bernstein, R. Cimberg, T. Kauwling, and W. Anikouchine. 1986. *Assessment of Long-Term Changes in Biological Communities of the Santa Maria Basin and Western Santa Barbara Channel—Phase I.* Final Report Prepared by Science Applications International Corporation for the U.S. Dept. of the Interior, Minerals Management Service, Pacific OCS Region, Camarillo, CA.

Lissner, A. (Ed.). 1995. *Monitoring Assessment of Long-Term Changes in Biological Communities in the Santa Maria Basin: Phase III.* Final Report Prepared by Science Applications International Corporation for the U.S. Dept. of the Interior, Minerals Management Service, Pacific OCS Region, Camarillo, CA.

MacDonald, I.R., (Ed.). 2002. *Stability and Change in Gulf of Mexico Chemosynthetic Communities. Volume II: Technical Report.* Prepared by the Geochemical and Environmental Research Group, Texas A&M University for the U.S. Dept. of the Interior, Minerals Management

Service, Gulf of Mexico OCS Region, New Orleans, LA. OCS Study MMS 2002-036.

MacDonald, I.R., W.W. Schroeder, and J.M. Brooks. 1996. *Chemosynthetic Ecosystems Study.* Final Report Prepared by Texas A&M University Geochemical and Environmental Research Group for the U.S. Dept. of the Interior, Minerals Management Service, Gulf of Mexico OCS Region, New Orleans, LA.

Maciolek, N., J.F. Grassle, B. Hecker, P.D. Boehm, B. Brown, B. Dade, W.G. Steinhauer, E. Baptiste, R.E. Ruff, and R. Petrecca. 1987. *Study of Biological Processes on the U.S. Mid-Atlantic Slope and Rise.* Final Report Prepared by Battelle Ocean Sciences, Woods Hole Oceanographic Institution, and Lamont-Doherty Geological Observatory of Columbia University for the U.S. Dept. of the Interior, Minerals Management Service, Atlantic OCS Region, Vienna, VA.

Maciolek-Blake, N, J.F. Grassle, and J.M. Neff (Eds.). 1985. *Georges Bank Benthic Infauna Monitoring Program: Final Report for the Third Year of Sampling.* Prepared by Battelle New England Marine Research Laboratory and Woods Hole Oceanographic Institution for the U.S. Dept. of the Interior, Minerals Management Service, Atlantic OCS Region, Vienna, VA.

Payne, J.R., J.L. Lambach, G.H. Farmer, C.R. Phillips, M.K. Beckel, J.G. Sutton, and R.R. Sims. 1985. *Georges Bank Monitoring Program: Analysis of Hydrocarbons in Bottom Sediments and Analysis of Hydrocarbons and Trace Metals in Benthic Fauna during the Third Year of Monitoring.* Final Report Prepared by Science Applications International Corporation for the U.S. Dept. of the Interior, Minerals Management Service, Atlantic OCS Region, Vienna, VA.

Treacy, S.D. 2002. *Aerial Surveys of Endangered Whales in the Beaufort Sea, Fall 2001.* In-House Study Report Prepared for U.S. Dept. of the Interior, Minerals Management Service, Alaska OCS Region, Anchorage, AK. OCS Study MMS 2002-061.

University of Texas Marine Science Institute. 1977. *Environmental Studies, South Texas Outer Continental Shelf, Final Report Rig Monitoring Program, 1976*. Final Report to the U.S. Dept. of the Interior, Bureau of Land Management, Gulf of Mexico OCS Office, New Orleans, LA.

Harmonised Monitoring of Offshore Drilling Waste Effects in Norway

Bakke, T. (torgeir.bakke@niva.no) (Norwegian Institute for Water Research, Oslo, Norway) and I. Nilssen (Statoil Research Centre, Trondheim, Norway)

ABSTRACT: The fate and effects of drilling waste discharged to the sea from Norwegian offshore installations have been monitored annually according to essentially the same standard procedures (defined below) since 1988. The results are reported to the Norwegian Pollution Control Authority and assessed for compliance and scientific quality by an independent Scientific Expert Group. The physical, chemical, and biological characterisation of surface sediments covers about 50 fields and single installations. Since 1996, the petroleum fields have been grouped in six latitudinal regions, monitored in a 3-year cycle. Each year, 350–400, sediment stations are surveyed according to regulatory guidelines. The prime elements of success for the programme have been: 1) agreed procedural standardisation at all steps; 2) documentation and formalisation of quality assurance requirements; 3) mandatory set of univariate and multivariate techniques for analysis and interpretation of physical/chemical and biological data; 4) generation of long time series facilitating temporal trend analysis; 5) close interaction between operators, authorities, and the scientific community in conducting and evaluating the monitoring; and 6) strong coordination among the operators in executing the surveys. Main future challenges are to agree on a common objective interpretation of subtle changes in bottom faunal assemblages, and to strengthen the taxonomic expertise regarding bottom fauna.

BACKGROUND

The first permission to explore for oil on the Norwegian continental shelf was granted to Phillips in 1965, and in late December 1969, oil was found at the Ekofisk field in the southern North Sea. In 1971, the first oil was produced from a Norwegian field, and several new discoveries were made in the years after. From 1973, the Norwegian Pollution Control Authority (SFT) required that all the companies given licenses on the continental shelf should submit annual reports on the environmental conditions in their impact areas. The requirement was one of the conditions for discharge permits, and guidance was given on the minimum scope and content of the environmental surveys to be performed by the operators.

Until the mid 1980s, these survey reports were produced and submitted annually to SFT but they were not to any extent reviewed or further assessed. In 1985, an independent scientific expert group, with two marine chemists and two marine biologists, was established by SFT. The terms of reference were to review all monitoring reports annually for compliance, scientific quality, and reliability, and to produce an overview of the environmental status at Norwegian offshore fields. These reviews quickly demonstrated the need for better harmonisation of the monitoring methods to be used, and in 1987, SFT and the offshore operators jointly hosted a workshop to agree on common strategy and methodology for the offshore baseline and monitoring surveys. The outcome of this workshop was developed into a guideline document that was put into force in Norway in 1988. In the same year, the Paris Commission adopted the guidelines for use in the convention waters (PARCOM, 1989). Later, the guidelines were turned into a SFT Manual (SFT, 1990), which became mandatory for monitoring surveys around petroleum installations in Norwegian waters since 1991.

After an effective prohibition on discharge of cuttings contaminated with oil-based mud in 1993, the environmental management focus shifted gradually to the increasing discharges of oil and chemicals with produced water. Concurrently, a denser pattern of installations at several fields and decentralised drilling discharges called for modification of the standardised sampling design. In 1997 and 1999, the Norwegian guidelines were revised to reflect these changes in the pollution situation (SFT, 1997; 1999). The revisions were done in close co-operation with the Norwegian Oil Operators Association (OLF) and the scientific community. The continental shelf was divided into 11 sea bottom regions by latitudes (Table 1). Each region with petroleum activity (at present 6 regions) was to be monitored as an entity every third year, and the monitoring should comprise both field specific stations and regional stations. For the former, the core of the previous sampling design was retained with an emphasis to take into account the discharge pattern of each individual field. Also approximately 10 regional stations should be located in each

TABLE 1. Names and borders of sediment monitoring regions at the Norwegian continental shelf.

Region	Name	Borders
	North Sea	
I	Ekofisk region	56 – 58°N
II	Sleipner region	58 – 60°N
III	Oseberg region	60 – 61°N
IV	Tampen region	61 – 62°N
	Norwegian Sea	
V	Møre region	62 – 64°N
VI	Halten region	64 – 66°N
VII	Nordland region	66 – 68°N
	Barents Sea	
VIII	Troms region	68 – 70°N
IX	Finnmark region	70 – 72°N
X	Barents south region	72 – 75°N
XI	Barents north region	North of 75°N

region to reflect the general background conditions. Only minor changes were made in the sampling and analytical procedures. In addition, guidelines for environmental monitoring of the pelagic ecosystem were made to cover potential impacts of produced water.

In 2001, the OSPAR Commission adopted a slightly simplified version of the new guidelines for use within the Commission waters (OSPAR, 2001). From January 2002 the guidelines were incorporated into the HSE Regulations for the Norwegian offshore petroleum activities (SFT, 2002).

PURPOSE AND PRINCIPLES

The purpose of the monitoring program is to provide data 1) to assess if regulations enforced upon the oil industry are effective, and 2) for environmental quality status reporting. The basic principles underlying the sediment monitoring guidelines are:

- Synoptic physical, chemical, and biological characterisation of the same sediments stations.
- Surveys performed at the same season each year.
- The development of standard operating procedures (SOPs), uniform across all fields and regions, in all phases of a survey from planning of sampling design to final report format.
- Requirement of formalised quality assurance procedures and documentations by the performing consultants.

To a large extent the procedural specifications have been the same since the implementation in 1988, but certain modifications have been introduced over time on the basis of experience and due to the development of improved methodology. The main changes have been:

- Reduced demand on analysis of PAH and specific oil related hydrocarbons (only in samples where total hydrocarbons exceed limits of significant contamination (LSC)) and of mercury (only at selected stations).
- Introduction of new analytical procedures to cover synthetic-based drilling fluids.

- Freedom to use alternative extraction procedures for analysis of hydrocarbons.
- Standard procedure for calculating the LSC.

Sampling Design

The monitoring is based on quantitative point sampling of the sediment by use of grab (0.1 m^2 long-arms vanVeen grab) or boxcorer. Station layout for both baseline and monitoring surveys of a field is specified. The most common layout for field specific monitoring is a transect design (Fig. 1). A grid design may be used in baseline surveys if the installation position is unknown. The design may cover both depositional and dispersive bottoms.

The standard program calls for 3 replicates for chemical and 5 for bottom fauna analysis from each station. At reference stations, 5 chemical and 10 fauna samples are taken. The

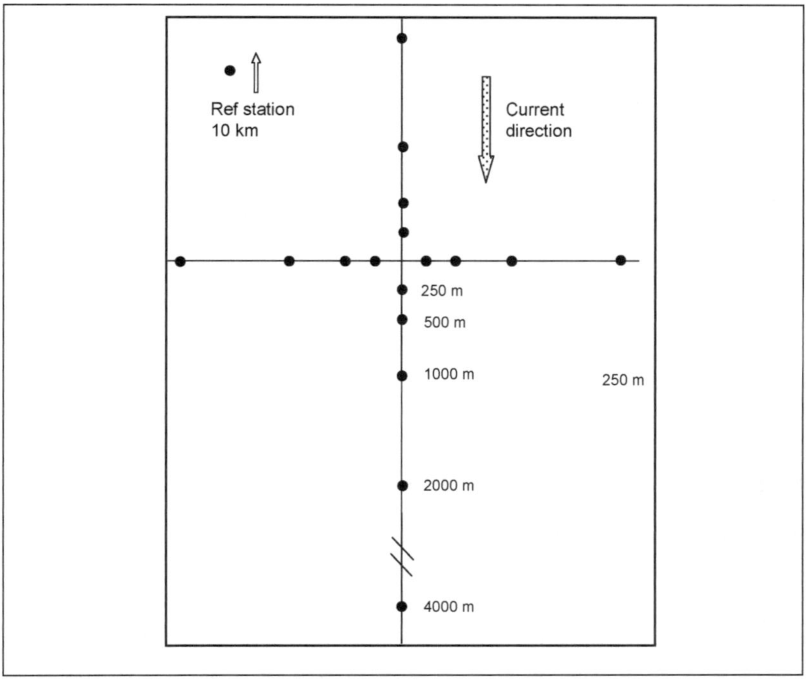

FIGURE 1. Standard station transect design for field specific monitoring.

chemical analyses cover the upper 0–1 cm of the sediment, grain size distribution the upper 0–5 cm, and the macrofauna analysis the > 1 mm fraction of the whole sample. The vertical distribution of chemical parameters is assessed at selected stations. Current measurements are not required, but the operators are encouraged to place such data, if available, at the disposition of the consultants.

Survey Timing

The surveys are conducted after 1 May, to avoid harsh weather conditions that reduce sample quality, and before 15 June to avoid intensive settlement of transitory juveniles in the bottom fauna.

Analytical Program

The SFT guidelines specify the principle of methods and the detection limits (when applicable) for all analytical parameters (Table 2).

TABLE 2. Suite of analytical parameters required in the SFT guidelines. [1)]Sum of naphtalene, phenanthrene, dibenzothiophene and their alkyl homologues, [2)]Only at stations where total hydrocarbon concentration (THC) exceeds LSC, [3)]Only for those drilling fluids reportedly being used.

Physical parameters	Chemical parameters	Biological parameters
Sediment surface appearance	Organic content	Macrofauna species number
Presence of conspicuous fauna	Total hydrocarbons	Species list
Colour	NPD aromatics [1) 2)]	Abundance of each species
Odour	Polyaromatic hydrocarbons [2)]	
Grain size distribution	Alkyl decalins [2)]	
% silt and clay	Synthetic-based fluids [3)]	
	Ba, Cd, Cr, Cu, Pb and Zn	
	Hg, Al, Li [2)]	

Calculations and Statistical Data Treatment

The guidelines specify a minimum requirement for data handling and analysis, but encourage extensions of this. For each chemical parameter a minimum requirement is to calculate mean concentration at each station in comparison to statistical limits of significant contamination (LSC) calculated from reference and regional stations. The LSC for a specific parameter is calculated as:

$$LSC = mean + t \times st.dev. \times \sqrt{1 + 1/N}$$

from the mean concentration and standard deviation of the parameter across the N reference/regional stations used in the calculation, and the students t-value (two way test, $P < 0.05$; df = N - 1).

A suite of macrofauna community index values are required: species richness, total abundance, diversity (Shannon and Weaver, 1963; Hurlbert, 1971), evenness (Pielou, 1966), dominance, table of 10 top dominant species, and the log-normal distribution of species among individuals (Gray et al., 1988). In addition, analysis of similarities in fauna structure among the stations is required by use of the Bray-Curtis similarity index (Bray and Curtis, 1957) in coupling with multivariate analysis by hierarchic agglomerative clustering (Lance and Williams, 1967) and non-parametric multidimensional scaling (MDS) (Kruskal and Wish, 1978).

The area of significant chemical contamination at each field is estimated from an assumed elliptical distribution model where the axes are defined by the distance from the centre to the outermost station that exceeds the LSC for the particular parameter along each transect. Areas of faunal effects are estimated in a similar way from the distribution of stations defined as impacted based on the uni- and multivariate community analysis.

Report Specifications, Data Storage

The guidelines specify which reports shall be produced (summary, main, and appendix reports) and their size and content. Also OLF requires that all data from the surveys are stored in a common database.

Quality Assurance (QA)

QA requirements are specified where relevant. In addition, the tender process for selecting consultant companies involves both a strict pre-qualification on basis of QA protocols and in-house capacities, and a requirement that the consultants are officially accredited for the tasks to be done according to the standard NS-EN ISO/IEC 17024.

EXAMPLE OF RESULT

Approximately 250 field-specific and regional surveys have been completed on the Norwegian shelf since the harmonised environmental monitoring started, covering a total of about 50 fields and single installations (Fig. 2).

The 2001 survey around the Veslefrikk field at a water depth of about 170 m (Mannvik et al., 2002) may be used as an example of the outcome of a typical field-specific survey. Before 1994, 600 tonnes of oil on cuttings had been discharged from Veslefrikk. From 1994 to 2000, annual discharges continued, but only of water-based cuttings. The 2001 survey covered a total of 12 stations. Sampling was conducted on May 4, 2001 and resulted in 38 grab samples for physical-chemical analyses and 65 samples for benthic fauna.

The levels of total hydrocarbons (THC) for the last 3 surveys (Fig. 3) show a gradual and strong increase from 1995 to 2001

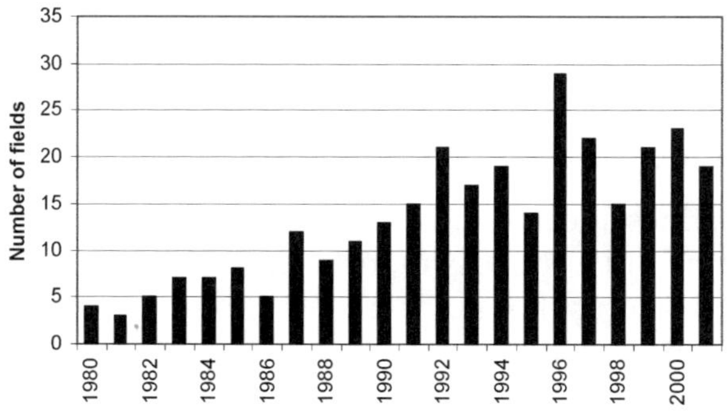

FIGURE 2. Number of Norwegian fields/single installations monitored each year during 1980–2001.

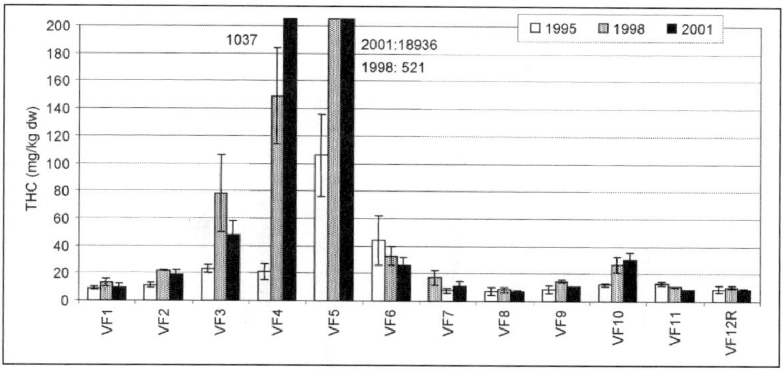

FIGURE 3. Levels of total hydrocarbons in sediment at the Veslefrikk field in 1993, 1998, and 2001 (redrawn from Mannvik et al., 2002).

at the two innermost stations in the main current direction (VF5 at 250 m and VF4 at 500 m), and reasonably stable and far lower levels along the other transects. The barium distribution (not shown) gives the same impression. Since the increase happened during a period without oily cuttings discharges, the cause is thought to be resuspension of older cuttings deposits at the centre of the field. There are, however, no local bottom current data available to support this assumption. In spite of the strong THC increase at the innermost stations, the area extent of significant contamination did not increase from 1995 to 2001 (Fig. 4).

Several features of the benthic fauna structure reflected these near field changes. Total abundance was highest and number of species, hence also diversity, was lowest at the two stations with high THC (Fig. 5). Dominance expressed as distribution of species in geometric abundance classes was highest at the same two stations, and four typical species indicating eutrophic bottoms constituted about 90–95% of the individuals: the polychaetes, *Chaetozone setosa, Raricirrus beryli,* and *Cirratulus cirratus,* and the bivalve, *Thyasira sarsi.* These stations were clearly separated from the others in the multivariate analysis of faunal similarity.

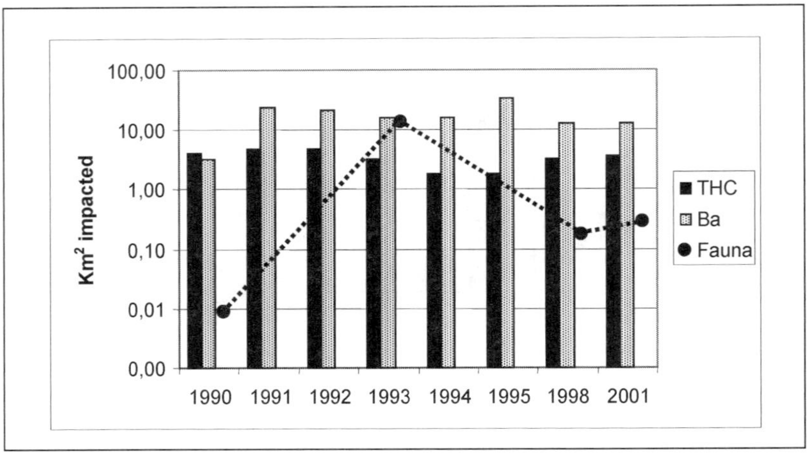

FIGURE 4. Seabed areas around the Veslefrikk field contaminated with THC and barium and showing effects on fauna during 1990–2001.

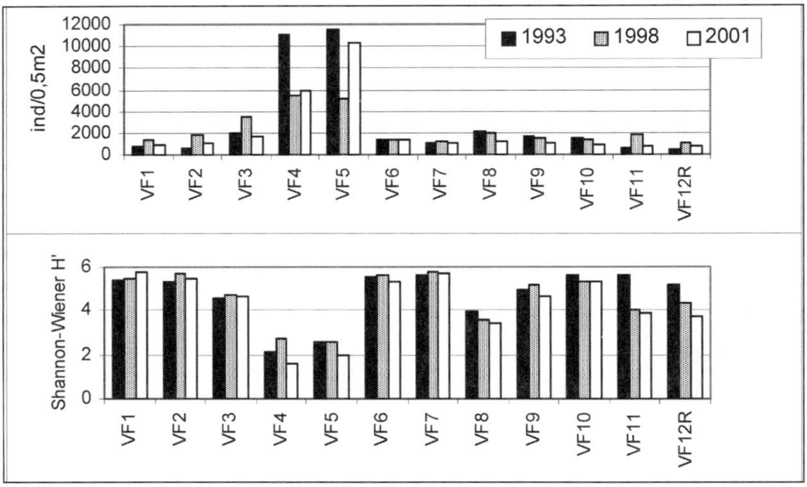

FIGURE 5. Macrofauna community attributes at the Veslefrikk field 1993–2001(from Mannvik et al., 2002). Upper: total faunal abundance, lower: Shannon-Wiener diversity (redrawn from Mannvik et al., 2002).

Veslefrikk is one of 11 petroleum fields included in the Region III surveys covering a total of 160 field related and 18 background stations. The depth range within the region is from 93 to 358 m, and grain size distribution, background values of THC and barium, and natural community structure are distinctly different in the shallow and deep sub-regions (Mannvik et al., 2002).

The Veslefrikk example illustrates the typical nature of results derived from the harmonised surveys. Valuable time series on the environmental conditions have been generated around all Norwegian fields of any age, since the same approach has been followed with only minor modifications since 1988. Changes in levels and spatial distribution of organic drilling fluids (hydrocarbons, esters, ethers, olefins) relative to drilling history seem fairly consistent, as do the responses of the fauna to the concentrations of hydrocarbons and synthetic drilling fluids in the sediments. All the data from the regional surveys and earlier data, back to 1990, have been fed into a common database administered by the oil companies. The database, which is available for external use too, has been used, for example, to generate an overall status report for the Norwegian sector of the North Sea (Carrol et al., 2000). Simple indicator values of areas with contaminated sea bottom and affected fauna are calculated annually both for individual fields, regions, and the whole shelf to assess the changes over time (Fig. 6). SFT uses these indicators in the national reports of the environmental impact of the offshore petroleum activities, and in the reports to OSPAR.

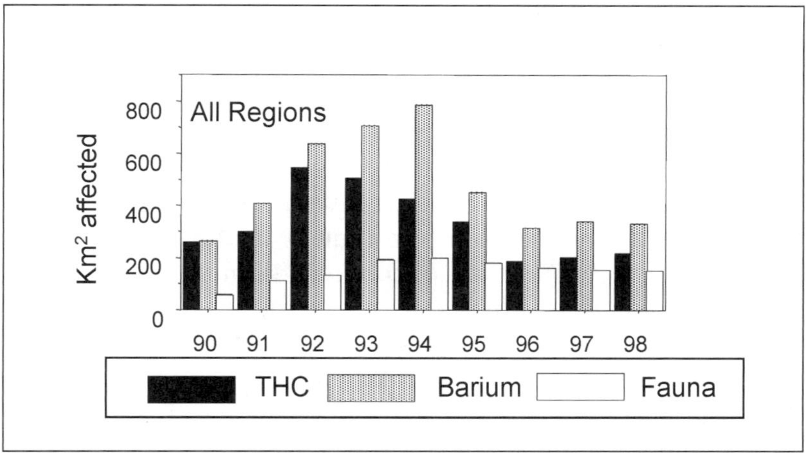

FIGURE 6. Norwegian oil and gas fields contaminated with THC and barium and showing effects on the fauna during 1990–1998 (redrawn from Carrol et al., 2000).

LESSONS LEARNED AND FUTURE CHALLENGES

The most important success factor for the offshore environmental monitoring has been the strict standardisation enforced upon the operators and their consultants through the very detailed SFT guidelines. Since these guidelines were developed and revised as a joint effort among authorities, operators, consultants, and the Scientific Expert Group; there is a strong feeling of consensus and ownership among all parties that supports the use of the SOPs. This is further augmented by an open, 1–2 day annual workshop with the same participants, arranged jointly by SFT and the operators. Here, previous surveys are evaluated, plans for future surveys discussed, and ideas to rationalise and improve the guidelines are brought forward for open discussion. After a few years of trial and error, compliance to the guidelines by the consultants became high and is still improving. This has facilitated comparison of results both over time, and within and between regions.

A disadvantage of standardisation is that new and improved procedures are difficult to incorporate. As it is important to be able to utilise new analytical principles and instrumentation to advance our knowledge, a recent change in the guidelines allows for procedural alternatives, as long as their accuracy and precision can be documented to be satisfactory (e.g. by use of common certified reference material).

Certain aspects of the procedures are also not easy to standardise. The outcome of the fauna analyses relies heavily on the taxonomic skill among the experts identifying the animals, and differences in fauna have, in several cases, been assigned to differences in taxonomic resolution among consultants. One discrepancy is the use of different nomenclature, which has been partially overcome by a comprehensive species list annexed to the guidelines (SFT, 1999), and through the introduction of the common database. This is also a minor problem during the statistical analysis of fauna data within one survey. More serious is the variable level of taxonomic discrimination from one consultant to another, which affects the species-station abundance matrix, and hence all subsequent univariate and multivariate analyses. The guidelines recommend that taxonomic resolution shall be to the species level, but does not require that this is done. This has modest consequences when assessing station differences within one and the same survey. The problem arises when fauna interpretation across surveys is required and when one wants to identify which species are influencing the station groupings most. A typical feature seen when running multivariate similarity analysis of all stations at one impacted field over several years is a stronger between-year fauna difference than between impacted and unimpacted groups of stations. This indicates that some of the differences in fauna over time are differences in taxonomic discrimination from one year to another. Another indication of taxonomic insufficiency is the fact that none of the surveys since 1985 have resulted in the discovery of new, undescribed species, or even new species for the North Sea region.

Therefore a certain degree of pooling of individuals into higher taxonomic levels is necessary prior to data analysis across different surveys. Identification to higher taxonomic level appears to be acceptable when the purpose is to discriminate groups of stations with similar community structure (Dauvin et al., 2003; Gomez Gesteira et al., 2003). Still, the interpretation problems encountered ought to be a stimulus to constantly improve the taxonomic qualifications among the consultants involved.

Another aspect of uncertainty is the interpretation of the similarities and differences in community structure among stations, especially on basis of the multivariate analyses. Both cluster and multidimensional scaling (MDS) analysis present gradients in faunal similarity among stations, and there are no common criteria for when a group of stations is truly a group separated from other groups. Normally, the interpretation seeks support from the univariate community characteristics, dominance patterns, indicator species, and from environmental characteristics such as grain size conditions and levels of hydrocarbons. This is the main reason why some of the less sensitive univariate characteristics such as diversity and evenness indices are still required by the guidelines. Interpretation of the cluster and MDS patterns is still too subjective, and frequently the consultants and the Scientific Expert Group have disagreed in their interpretation as to whether a station is impacted or not.

In an effort to improve the common ground for interpretation of the fauna data, an international workshop was held in Oslo in 2001 to compare the performance of several alternative multivariate techniques on 6 years of data from the Norne field, at the north border of Region VI. The tasks were to identify more objectively any impacted sites, to model the species-environment relationships, and to distinguish natural differences between years from those caused by impact (Legendre and Hooper, 2001). The outcome of the workshop has not yet resulted in revision of the guidelines, but the consultants apply some of the techniques on a voluntarily basis, such as canonical correspondence analysis (CCA) to assess relationships between community structure and the physical and chemical environmental variables.

The environmental variables identified for monitoring under the SFT guidelines are substances being discharged from the offshore operations, and/or natural variables that are believed to alter the structure of macrofaunal communities. Carrol et al. (2000) used CCA to analyse the contribution of these variables in explaining the total variance of the fauna for the whole Norwegian shelf during 1996–1998. The results showed that the natural and anthropogenic variables monitored explained in sum only about 18% of the total variance. Hence, although we tend to believe that the monitoring programme covers the most important environmental and anthropogenic sediment variables structuring the offshore benthic ecosystem, more that 80% of the variability in the fauna is due to factors that we have little or no specific information about. Some of these may not even be linked to the benthos, but may affect pelagic recruits to the benthos. It should be a challenge to scientists, authorities, and operators to jointly identify the most important of these factors and include them in the EEM monitoring.

ACKNOWLEDGEMENTS

This document is based on a large number of monitoring survey reports produced by numerous consulting companies on contract from the offshore operators. The ones referred directly to are cited in the references, the others are hereby thanked collectively for their contribution. The authors also thank two anonymous referees for constructive comments to the original manuscript. This paper was prepared under contract no 4002128 from the Norwegian Pollution Control Authority (SFT) to the Norwegian Institute for Water Research.

REFERENCES

Bray, J.R. and J.T. Curtis. 1957. "An Ordination of the Upland Forest Communities of Southern Wisconsin". *Ecological Monographs*. 27:325-349.

Carrol M., T. Pearson, E. Dragsund, and K. Lie Gabrielsen. 2000. *Environmental Status of the North Sea Based on the Petroleum Offshore Regional Monitoring Programme, 1996-1998*. Akvaplan.niva report no APN-411.1777.01.

Dauvin, J.C., J.L. Gomez Gesteira, and M. Salvande Fraga. 2003. "Taxonomic Sufficiency: An Overview of its Use in the Monitoring of Sublittoral Benthic Communities after Oil Spills". *Marine Pollution Bulletin*. 46:552-555.

Gomez Gesteira J.L., J.C. Dauvin, M. Salvande Fraga. 2003. "Taxonomic Level for Assessing Oil Spill Effects on Soft Bottom Sublittoral Communities". *Marine Pollution Bulletin*. 46:562-572.

Gray, J.S., M. Aschan, M.R. Carr, K.R. Clarke, R.H. Green, T.H. Pearson, R. Rosenberg, and R.M. Warwick. 1988. "Analyses of Community Attributes of the Benthic Macrofauna of Frierfjord/Langesundsfjord and in a Mesocosm Experiment". *Marine Ecology Progress Series*. 46:151-165.

Hurlbert, S.H. 1971. "The Nonconcept of Species Diversity: A Critique and Alternative Parametres". *Ecology*. 52:577-586.

Kruskal J.B. and M. Wish. 1978. *Multidimensional Scaling*. Sage Publications, Beverly Hills, California.

Lance, G.N. and W.T. Williams. 1967. "A General Theory of Classificatory Sorting Strategies. I. Hierarchical Systems". *Comput. Journal.* 9:373-380.

Legendre, P. and E. Hooper. 2001. *Analysis of the Norne Oil Field Survey Data.* Report presented to Det Norske Veritas (DNV), Oslo, Norway, 20 February 2001.

Mannvik, H.P., A. Pettersen, K. Lie Gabrielsen, and F. Mikkola. 2002. *Environmental Survey in Region III 2001* (in Norwegian with English summary). Akvaplan.niva report no APN-411.2203.

OSPAR. 2001. OSPAR Guidelines for Monitoring the Environmental Impact of Offshore Oil and Gas Activities. OSPAR Commission Agreement 01-10.

PARCOM. 1989. *Guidelines for Monitoring Methods to Be Used in The Vicinity of Platforms in the North Sea.* Paris Commission Publication, Chameleon Press London.

Pielou, E.C. 1966. The Measurement of Diversity in Different Types of Biological Collections. *Journal of Theoretical Biology.* 13:131-144.

SFT. 1990. *Manual for Monitoring Surveys around Petroleum Installations in Norwegian Marine Areas* (in Norwegian). SFT (Norwegian Pollution Control Authority) Report 90:01. TA-nr 699/1990.

SFT. 1997. *Guidelines for Environmental Monitoring of the Petroleum Activity on the Norwegian Shelf* (in Norwegian). SFT (Norwegian Pollution Control Authority) Report 97:01. TA-nr 1424/1997.

SFT. 1999. *Environmental Monitoring of the Petroleum Activity on the Norwegian Shelf* (in Norwegian). SFT (Norwegian Pollution Control Authority) Report 99:01. TA-nr 1641/1999.

SFT. 2002. Appendix 1 to the Information Duty Regulations Requirements for Reporting from Offshore Petroleum Activities. http://www.sft.no/arbeidsomr/petroleum

Shannon, C.E. and W. Weaver. 1963. *The Mathematical Basis for Communication.* Univ. Illinois Press Urbana. 117 pp.

The Challenge of Deep Water Oilfield Environmental Monitoring West of Britain

Hartley, J.P. (jph@hartleyanderson.com) Hartley Anderson Ltd., Blackstone, Dudwick, Ellon Aberdeenshire UK

ABSTRACT: Hydrocarbon exploration and production continues to develop in the deep water area to the west of Britain. This area remains poorly understood, and the environmental monitoring approaches adopted for the North Sea have only limited application. This is because of water depth (to > 2400 m), complexity of the hydrographic regime particularly to the north of the Wyville Thomson Ridge, varied topography, and surficial sediments containing much rocky material deposited and sculpted during the last glaciation and the subsequent redistribution of finer particles. In addition, the fauna of the area is poorly known both in terms of identity and in basic ecology. The design and conduct of meaningful environmental monitoring in such areas presents a number of challenges including what, how, and where to monitor. These challenges are compounded by legislative and company controls on discharges to the sea. The approaches taken to date include standard shallow water strategies, nested series of geophysical surveys linked to remote sampling and/or photographic survey, and deployment of benthic landers. It is clear that innovative thinking and collaborative efforts are key to providing the understanding of deep water areas needed for effective biodiversity protection and environmental management.

INTRODUCTION

The area to the north and west of Britain (subsequently referred to as the Atlantic Margin) is subject to a range of industrial activities including hydrocarbon exploration and production. This paper provides summaries of the regional environment, existing scale of hydrocarbon development, legislative and other controls on activities and discharges, and monitoring of the North Sea oilfields. These summaries provide the context for a consideration of approaches to deep water effects monitoring in the Atlantic Margin area. Deep waters refer to the area off the continental shelf below 200 m.

EARLY SCIENTIFIC EXPLORATION

The Atlantic Margin and in particular the area around the Shetland Isles has a long history of scientific study. The first British specimens of the cold water coral, *Lophelia pertusa* (which in the late 1990's became a cause célèbre with some environmental campaigning groups), were collected in 1832 by fishermen in this area (Wilson, 1979), and the Shetland Isles were the subject of a series of eight dredging and collecting cruises (see Jeffreys et al., 1869). These cruises were aimed at collecting for subsequent description the wide variety of species present, and it is clear from the list of taxa obtained that, in qualitative terms, this early work provided a comprehensive picture of the benthic fauna. The work of Jeffreys et al. also provided insights into the relative abundance of species and biogeography of the fauna since their collecting expeditions were not restricted to the Shetland Isles, but ranged between the Mediterranean and the Arctic.

The next major advance in understanding came from observations and seabed samples collected from the Atlantic Margin by HMS *Lightning* in 1868, and more successfully from HMS *Porcupine* in her third cruise of 1869. These cruises were for scientific exploration of the area and included bathymetric, oceanographic, and biological investigations. Benthic sediment was collected with a variety of dredges, sometimes with "hempen tangles" attached to snag surface living animals, particularly echinoderms. The benthic sampling was focussed on the deeper waters, although some material was collected on the continental shelf. The cruises resulted in a series of papers (e.g. Carpenter et al., 1870) and a classic marine biological text, *The Depths of the Sea* (Wyville Thomson, 1874), which established the major water masses of the area and the vertical zonation of many species in relation to temperature. In 1880, the *Knight Errant* expedition carried out soundings, dredging, and trawling in the Faroe-Shetland Channel and discovered the Wyville Thomson Ridge, the existence of which had been predicted on the basis of theoretical considerations of water temperatures (Tizard and Murray, 1882). HMS *Triton* made further observations in the region in 1882 to extend the survey work of the *Knight Errant*. The *Michael Sars* expedition of 1910 covered the North Atlantic and included parallel sampling transects across the Faroe-Shetland Channel and also resulted in a classic and enduring text, *The Depths of the Ocean* (Murray and Hjort, 1912).

It is evident that perspectives on the level of understanding of the seabed fauna can alter over time so that in 1912, Murray and Hjort stated that "British investigators have made the plateau round the Shetland islands to a depth of about 200 metres one of the most familiar". Although by 1995 Kingston et al. concluded that "The offshore benthic fauna around the Shetland Islands is not particularly well known...". These divergent perspectives are believed to be a reflection of the shift over the last century from qualitative to quantitative sampling in soft bottom benthic ecology. In terms of the occurrence and distribution of the common larger species, the early dredging work provided insights which still apply (see Murray, 1886). However, some changes can be inferred to have occurred; Wyville Thomson (1874) noted that in the warm Atlantic water area the coral, *Lophelia*, "forms stony copses covering the bottom for many miles". Recent extensive sonar mapping, sampling, and photography carried out during the Atlantic Frontier Environmental Network (AFEN), UK Department of Trade and Industry (DTI) and other surveys of the area have failed to find substantial colonies of the species on the shelf and shelf break to the west of Shetland (see Roberts et al., 2003).

THE ATLANTIC MARGIN ENVIRONMENT

The environment of the Atlantic Margin is described in a number of recent of conference proceedings including Mauchline (1986), Collins and Sternberg (2001), and publications of survey results such as Pollock et al. (2000) and AFEN (2000; 2001).

The Atlantic Margin is dominated by a number of significant topographical features reflecting the past geological and glacial history (Fig. 1). To the north, the gently sloping North Sea Fan reaches depths of 2400 m in the north, rising to 900 m in the south, where it joins the Faroe-Shetland Channel. The southern end of the channel meets the perpendicularly aligned Faroe Bank Channel and dividing this from the Rockall Trough, the Wyville Thomson Ridge (WTR) rises to within 400 m of the surface. The Rockall Trough contains a number of seamounts and banks and deepens to over 4500 m in the Porcupine abyssal plain.

The seafloor of the area ranges from exposed bedrock through rocky glacial moraines, current scoured gravels and sands to mud (Stoker et al., 1993). The majority of sediments of the area contain numerous cobbles and boulders derived from glacial deposition and ice rafting. There is evidence from bedforms of strong water currents over much of the Faroe-Shetland Channel including currents > 1 m/s at 1000 m depth (Wynn et al., 2002). The entire shelf break and upper slope to around 500 m has been sculpted by icebergs and ice sheets during the last glaciation when sea level was substantially lower than today (Belderson et al., 1973).

The main water masses to the west of Shetland are warm North Atlantic water flowing northeast and sub-zero Faroe-

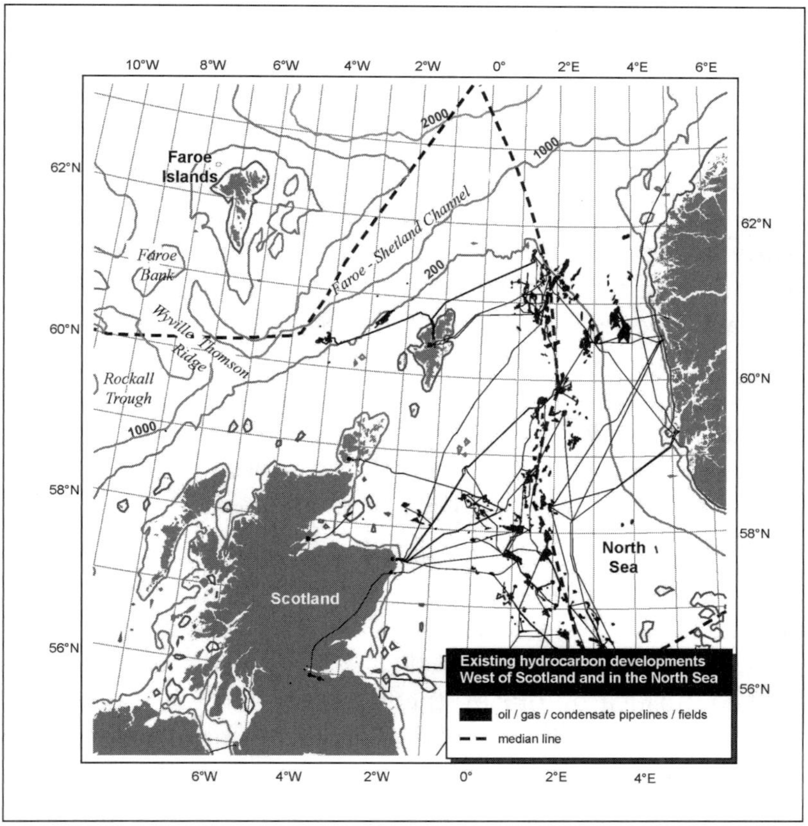

FIGURE 1. Main features of the Atlantic Margin area to the west of Britain and the adjacent central and northern North Sea.

Shetland Channel bottom water flowing southwest (Turrell et al., 1999). The number of water masses and their movements is complex particularly to the north of the WTR, influenced by mesoscale eddies and internal waves breaking in mid depths on the continental slope. To the south of the WTR the North Atlantic water is underlain by Labrador Sea water (Holliday et al., 2000) although below the seasonal thermocline, water temperatures tend to progressively decline with depth.

Zooplankton is dominated by the copepod, *Calanus finmarchicus,* which represents an important food source for the young of many fish species, and is important in the recruitment of fish stocks of the area. The Faroe-Shetland Channel is an important over-wintering area for *Calanus finmarchicus* (Heath and Jónasdóttir, 1999), which are transported into the North Sea in spring by inflows of Atlantic water.

Seabed communities in the area are characteristic of the interface of several biogeographic zones with water temperature, bathymetry, and sediment type identified as the primary environmental influences on distribution patterns in both community and species composition (AFEN, 2000; Bett, 2001—who includes the results of multivariate analyses of the infaunal dataset). A substantial proportion of benthic species are new (e.g. between 20 and 40% of peracarid crustaceans being undescribed) (AFEN, 2001), which has implications for benthic monitoring and drawing data from different sources into a synthesis.

The Darwin Mounds on the southern flank of the WTR appear to be unique geological and biological features, which were first discovered by the 1998 AFEN survey, and a second area found nearby during the 1999 DTI survey. The mound tails, visible to side scan sonar, appear to have no physical expression, but are inhabited by dense populations of xenophyophores (AFEN, 2000). The central mound appears to consist of blocky rubble with the cold water coral, *Lophelia pertusa*, usually present. *Lophelia* and xenophyophores are widely distributed elsewhere in the region, but the ecological significance of the mounds is unclear and the area has been proposed as a Natura 2000 offshore conservation site.

The Atlantic Margin shelf and slope are heavily fished with trawls for demersal species including cod, haddock, and monkfish. The cold water of the deep Faroe-Shetland and Faroe Bank Channels supports a sparse and distinct fish fauna, of little commercial value, and with very few of the Atlantic deep water species of the upper slopes of the Faroe-Shetland Channel being present. The main commercial pelagic species found in the area are mackerel, Norway pout, and blue whiting, with herring and greater silver smelt also being present (Gordon, 2001).

Twenty-three species of seabirds breed in the area of which 21 are present in internationally important numbers (Pollock et al., 2000). Major breeding colonies are on the Faroe, Shetland, and Orkney Islands and on Scotland and Ireland. Offshore surveys show that seabird distribution is closely correlated to water depth, with more birds found over shallower continental shelves than the deeper oceanic waters. Birds present in the deeper slope and oceanic waters comprise mainly pelagic species such as fulmar, gannet, kittiwake, and storm petrel. The deep water areas are probably too distant from colonies for breeding individuals of most species to visit during spring and early summer. Seabirds tend to move through these areas in late summer and autumn in passage to wintering grounds, or in spring while en route to breeding colonies.

The Atlantic Margin is considered important for marine mammals; 12 species occur regularly over the area, with another 11 occasionally sighted at sea (Pollock et al., 2000). Cetaceans inthe area can be broadly distinguished into several groups, on the basis of distribution and feeding:

- baleen whales (blue, fin, sei, and humpback) are mainly recorded in deep water
- white-beaked and Risso's dolphins are concentrated in shelf waters and minke whales are summer visitors to shelf areas
- sperm and beaked whales are concentrated on the eastern flank of Rockall Trough, the southern flank of Wyville Thomson Ridge, and along the eastern side of the Faroe-Shetland Channel
- Atlantic white-sided and common dolphins are widely distributed in deep water
- killer whale, bottlenose dolphin and harbour porpoise are all widely distributed over deep and shelf waters.

Grey and harbour seals are common in shelf depths and hooded seals regularly feed in the deep water of the Faroe-Shetland Channel and to the north.

Contaminant concentrations in sediments and sea water in the area are low and generally at, or close to background levels (OSPAR, 2000; AFEN, 2000; ERT, 2000).

OIL AND GAS EXPLORATION AND DEVELOPMENTS

Hydrocarbon exploration and production licensing the UK Continental Shelf (UKCS) is divided into quadrants of 1° of latitude by 1° of longitude (except where the coastline, or a boundary line intervenes). Each quadrant is further partitioned into 30 blocks each of 10 x 12 minutes. The average block size is about 250 km^2 (roughly 100 square miles) and blocks on the Atlantic Margin were first offered for licensing in 1965. Since then, some 186 exploration and appraisal wells have been drilled along the UK Atlantic Margin, including 10 exploration wells drilled west of Hebrides (Fig. 2).

The Clair oilfield was discovered in July 1977 with reserves estimated at 1.75 billion barrels of oil. However, because of the nature of the reservoir and the crude oil, Clair is currently the largest undeveloped oilfield in the UK, although development plans have recently been given regulatory approval.

Interest in the hydrocarbon potential of the Atlantic Margin surged following the discovery of the Foinaven and Schiehallion oilfields in October 1992 and 1993, respectively. Both

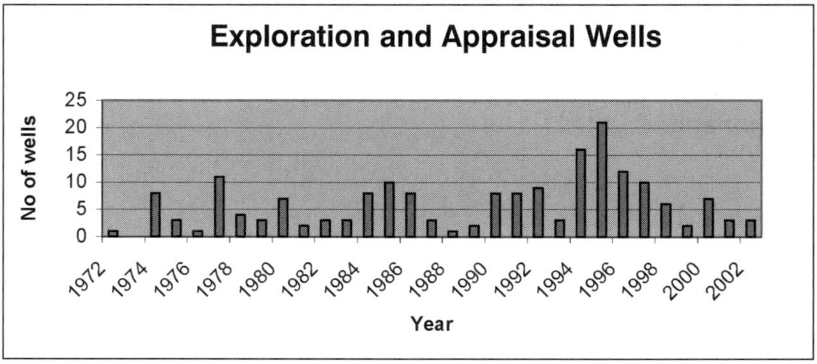

FIGURE 2. Number of exploration and appraisal wells drilled per year on the UK Atlantic Margin (data courtesy of Department of Trade and Industry).

Foinaven and Schiehallion (with a subsea tie back development of the Loyal field) have been developed using floating production, storage, and offloading facilities (FPSO) and are in production. Since then, exploration success has been limited to a number of small discoveries and drilling activity has declined (Fig. 2).

Following the discovery of Foinaven and Schiehallion, a collaborative group, the Atlantic Frontier Environmental Network (AFEN), was established in recognition of the limited applicability of the North Sea experience for environmental management of oil and gas activities in deeper waters. The AFEN group consisted of all oil companies with licensed blocks on the Atlantic Margin, the main regulator the Department of Trade and Industry, and their statutory conservation and fisheries advisers. The AFEN group commissioned a series of studies and two major seabed mapping and sampling surveys, which covered all the then licensed acreage (AFEN, 2000; 2001). A proportion of the work funded by AFEN has now been taken up by the DTI following its decision in 1999 to undertake Strategic Environmental Assessments prior to offering new acreage for licensing (see www.offshore-sea.org.uk).

OVERVIEW OF LEGISLATION APPLYING TO UK OFFSHORE OIL AND GAS

In recent years, there has been a substantial increase in the number of regulations applying to UK offshore oil and gas activities. This has coincided with increased oversight, both internally driven within companies through the application of Environmental Management Systems, and externally by DTI inspections and audits.

The Petroleum Act 1998 provides the basis for granting licences to explore for and produce oil and gas. Under the terms of a Production Licence, licence holders require the authorisation of the Secretary of State before activities commence.

Approval for development programmes and consent for wells, extended well tests, incremental projects, and production consents are contingent on complying with the requirements of the Offshore Petroleum Production and Pipe-lines (Assessment of Environmental Effects) Regulations 1999. Environmental Statement (ES) is mandatory for certain projects, including new developments with expected production > 500 tonnes of oil/day or 500,000 m^3 of gas/day and new pipelines > 40 km in length and 800 mm in diameter. A number of projects may not need an ES to be prepared if a preliminary assessment demonstrates to the satisfaction of the Secretary of State that the project is unlikely to cause a significant adverse environmental impact. The Offshore Petroleum Activities (Conservation of Habitats) Regulations 2001 implement European Directives for the protection of habitats and species in relation to oil and gas activities carried out on the UKCS. Consent is required for a seismic survey (including low power site investigation surveys) in fulfilment of these Regulations. Consent application includes risk assessment and proposed mitigation measures.

A permit is required in advance for the use of chemicals offshore including drilling, well workover, production, and pipeline chemicals (Offshore Chemicals Regulations, 2002). Permit application includes mandatory risk assessment and any variation in use from permit specifications must have prior approval. Chemical use and discharge must be reported at the end of the activity. Chemicals are ranked by hazard, based on a PEC:PNEC (Predicted Effect Concentration : Predicted No Effect Concentration) approach. Chemicals on the OSPAR prescribed lists (e.g. those with endocrine disruptor properties or high concentrations of heavy metals) and those with high toxicity and bioaccumulation potential or with poor biodegradability are subject to an active programme targeting substitution with less hazardous alternatives.

No organic phase drilling fluids may be used without prior authorisation and discharge of cuttings to sea with a concentration > 1% by weight of oil on dry cuttings is prohibited (OSPAR Decision 2000/3 on the Use of Organic-Phase Drilling Fluids (OPF) and the Discharge of OPF-Contaminated Cuttings). OPF contaminated cuttings from exploration wells are either reinjected to deep rock strata or shipped to shore for treatment and disposal at licensed sites.

Discharge of oil or oily mixtures to sea from any offshore installation or pipeline is prohibited (Prevention of Oil Pollution Act 1971), with exemptions to allow lawful discharge of treated produced water, sand, and other operational discharges. The current standard for produced water discharges is maximum monthly average of 40 mg/kg oil-in-water. OSPAR Recommendation 2001/1 for the Management of Produced Water from Offshore Installations provides for a reduction in the discharge of oil in produced water by 15% over a five year period and a lowering of the discharge concentration from each installation to 30 mg/l over the same period. The recommendation also includes a presumption against the discharge to sea of produced water from new developments; existing (and proposed) developments on the Atlantic Margin dispose of around 98% of produced water by reinjection into deep rock strata.

The Merchant Shipping (Oil Pollution Preparedness, Response and Co-operation Convention) Regulations 1998 require all offshore installations (including rigs) and oil handling facilities (e.g. pipelines) to have an approved oil spill contingency plan.

It should be noted that similar controls (primarily derived from European Union and OSPAR initiatives) apply to oil industry activities across the NE Atlantic.

APPLICABILITY OF NORTH SEA FIELD MONITORING APPROACH

The monitoring of the environmental effects of offshore developments in the North Sea began in 1973 with a survey of the seabed and water column around the Ekofisk field (Dicks, 1976), with the first benthic survey in the UK carried out around the Forties field in 1975. Since then, there have been many hundreds of such surveys carried out by the field operators for internal management reasons or as required by legislation. In the UK, an inventory and database of these seabed monitoring surveys has been compiled, (UKOOA, 2000) and further analyses of the data focused principally on potential changes at "background" sites beyond the known zone of gross contamination in the vicinity of platforms and wells (Harries et al., 2001). A similar exercise has also been undertaken in Norway (Carroll et al., 2000). Most monitoring of offshore hydrocarbon developments in the North Sea has focused on the acute effects of drilling discharges in the vicinity of platforms. In the early years of the development in the North Sea, this monitoring provided valuable information on the intensity and extent of impact on the benthic environment, and lead to increasingly stringent discharge controls, particularly of the prime cause of effects: the discharge of oil contaminated drill cuttings. However, for at least the last ten years such an approach has provided little new information over that which could have been predicted from the earlier studies (Harries et al., 2001). That said, the time series for many fields from Norwegian statutory monitoring using generally consistent methods provides valuable confirmation of predictions of sediment hydrocarbon degradation and benthic faunal recovery following cessation of OBM discharges. However, past and current UK and Norwegian monitoring practices have left a number of important gaps in understanding of effects, contaminant transfer, and other processes especially with regard to oily drill cuttings piles (Hartley et al., 2003), since the defined sampling station locations do not include these most severely affected areas.

Around central and northern North Sea installations where multiple wells were drilled with oil-based mud and the cuttings discharged, the faunal succession observed has typically followed the zones outlined by Davies et al. (1984). The zones were:

- an impoverished and highly modified benthic community very close to a platform
- a transition zone where benthic faunal effects could be detected

- a zone where chemical contamination could be detected but benthic faunal effects were not evident
- normal seabed

In contrast, the field ecological effects of the discharge of water-based muds (WBM) and cuttings have been subtle, if detectable at all, even in the comparatively shallow waters of the southern North Sea (e.g. Daan and Mulder, 1996). Counter to the general picture of limited effects of WBM discharges, Cranford and Gordon (1992) reported low tolerance of dilute bentonite clay suspensions in scallops (*Placopecten magellanicus*). Cranford et al. (1999) found that used WBM and its major constituents, bentonite and barite caused effects on the growth, reproductive success, and survival of *Placopecten*, which were attributed to chronic toxicity and physical disturbance. It may be that *Placopecten* is especially sensitive to drill muds (or fine sediments in general), or that in the field WBM discharges rapidly disperse to below effective concentrations.

There has been limited focus on water column monitoring, largely due to the variability in the system and difficulty in detecting a discharge plume or biological or chemical effects. Studies of produced water toxicity and dispersion (e.g. OLF, 1998) concluded that the necessary dilution to achieve a No Effect Concentration (NEC) would be reached at 10 to 100 m and certainly less than 500 m from the discharge point.

MONITORING APPROACHES USED ON THE ATLANTIC MARGIN

The majority of early wells drilled on the Atlantic Margin were in shelf depths and were not subject to benthic monitoring (see UKOOA, 2000). Monitoring of deeper water operations began in the mid 1990's and typically adopted the North Sea sampling strategy and methods. As a typical example, Figure 3 illustrates the arrangement of sampling stations around the Vrackie exploration well in about 300 m with stations extending to 5000 m from the wellhead. The survey report indicates that the seabed was stony and that the transect was repositioned to improve sampling success rate (ERT, 1999). The focus

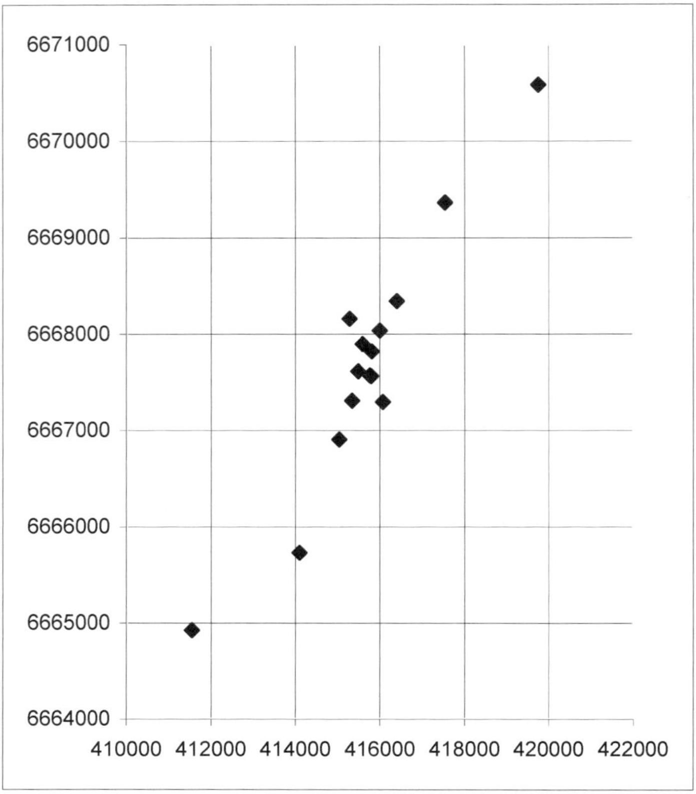

FIGURE 3. Vrackie well survey sampling station arrangement (from ERT, 1999).

of the survey was chemical and physical analyses, and at each station duplicate samples were taken for hydrocarbons and single samples for heavy/trace metals and granulometry. Single biological samples were taken per station and at some, sediment from several grab attempts were combined to provide sufficient sample volume (at least 5 litres). Clearly such samples can only provide a semi quantitative description of the fauna of the area and not a basis for monitoring ecological effects over time.

The AFEN 1996 survey achieved excellent coverage in the UK areas of the West Shetland shelf, slope, and Faroe-Shetland Channel, licensed for oil and gas exploration in the 16th and earlier licensing rounds, while the AFEN 1998 survey covered areas licensed in the 17th UKCS round, in the northern Faroe-Shetland Channel and south of the WTR (see AFEN, 2000; 2001). The survey design selected was a combination of seabed mapping using sidescan sonar imaging followed by seabed sampling (at a total of over 300 individual stations) using a range of equipment to suit the sediment types identified. Notable was the use of a comparatively large sample area multicorer ("megacorer" with 12 x 10 cm diameter cores) that provided high quality samples in muddier sediments. These two principal approaches, of mapping and sampling were supplemented by seabed photography to help 'ground truth' the sonar image data and provide a broader view than that derived from the samples alone.

One of the main objectives of the AFEN surveys (and subsequent DTI surveys using essentially the same strategy) was to provide a regional context into which site or field specific surveys could be placed. However, the scale of the area covered meant that within a given area the number of data points was low. To illustrate this, the 1998 AFEN survey included the Suilven oilfield area (in about 800 m of water, downslope from the Schiehallion field) although there were no physical samples nearby and only two Wide Angle Seabed Photography (WASP) system deployments, one at 560 m and the other at 1100 m, both well outside the field area.

For the baseline environmental investigations of the Suilven field, an alternative survey strategy was adopted taking account of AFEN and other survey results, the water depth, current strength and direction, the anticipated stony nature of the seabed, and the predicted nature and scale of discharges and physical disturbance. The objectives of the Suilven field investigations were to provide an initial site characterisation to identify the range of habitats and megafauna present so that potentially sensitive features could be factored into field development plans; the survey was also aimed at informing decisions about suitable approaches to post development monitoring. During a Suilven field development site and pipeline route survey of September 2002 (Fig. 4), side scan

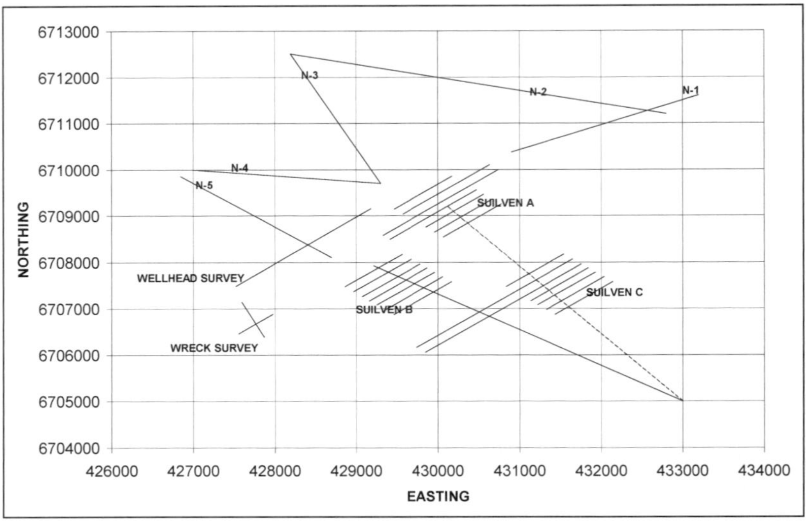

FIGURE 4. Survey coverage in the Suilven field area. Swathe and side scan sonar lines are indicated and Suilven A, B, and C represent potential development locations.

sonar and swathe bathymetry results were assessed on board, and specific habitats and features targeted for further evaluation by still photographs and video. Photographs and video were taken from a Stealth 3003 Remotely Operated Vehicle (ROV) with a Kongsberg Simrad OE14-108 underwater digital stills camera. Samples of specific fauna were collected by ROV manipulator to assist in identification of photographed and videoed animals. Specimens were brought to the surface in a sample box and preserved with 10% buffered formalin or 70% methanol as appropriate.

The seabed sediments of the Suilven field are mixed muddy sands with substantial quantities of coarser material including gravel, pebbles, cobbles, and boulders. The nature of the Suilven fauna is Arctic and distinct from that of the Schiehallion and Foinaven fields, which is attributed to the different water masses and temperature regimes of the two areas.

Side scan sonar indicated the presence of numerous seabed mounds and depressions in the Suilven area and a selection were investigated since these could be sensitive features and/or provide potential photographic monitoring sites. The mounds were up to 3m high and consisted of a pile of (presumably) ice rafted cobbles and boulders extensively colonised by epifauna. Seabed depressions in all inspected cases proved to have a single (or exceptionally, two) large boulder (with abundant large epifauna) at the centre surrounded by a scour moat.

The area of a suspended wellhead (204/14-1) was examined. The area, visibly influenced by surface hole cuttings, extended about 20 m from the wellhead where the fauna was limited to sea anemones and tubes of worms, which presumably were able to burrow back to the surface. In addition, there were numerous Arctic skate (*Raja hyperborea*) in the area. Similar concentrations of Arctic skate were noted in the finer sediment infill of seabed scars believed to be the result of previous rig anchoring.

Previous AFEN survey results had indicated a well-developed to extensively-developed sponge and encrusting fauna in mid slope depths. This epifauna was present in the immediate vicinity of the Schiehallion field installations but was not seen in the other areas of the pipeline route surveyed, although there was clear evidence of extensive disturbance of the seabed including the overturning of large boulders and burial of epifauna presumably by trawling (see Gage et al., in press).

Identification of fauna was difficult; although the presence of new and previously unreported species in the area was not necessarily a reflection of the rarity of the animals, but rather the limited extent of scientific investigations of the fauna in situ in the region.

The workclass ROV encountered no difficulty operating in the current strengths found in the Suilven field area and although numerous pelagic crustaceans were attracted to the ROV lights these did not affect observations of sessile (or apparently of fish or other motile) megafauna.

CONSIDERATIONS FOR FUTURE MONITORING

The rocky nature of the seabed in the Suilven field (as in much of the Atlantic Margin) is clear from the photos and video, and there is evidence from rig anchor scars that apparently soft sediments also contain numerous pebbles, cobbles, and boulders. Such a seabed presents challenges to standard environmental monitoring methods reliant on remote sampling. Grab and core samplers will frequently either fail to penetrate the sediment or to obtain an adequate sample on account of the number of pebbles, cobbles, and boulders present. A high failure rate when sampling in deeper waters can significantly increase the duration and hence costs of the survey. In addition, such samples, as are collected, are self-selecting to the patches of softer sediments, which may not be representative. Similarly, epifauna would be underrepresented in physical samples, emphasising the importance of photographs and video in assessing the biodiversity of the area.

The presence of scattered large boulders colonised by a range of large sessile species over the field area, of which the positions are precisely known, presents the opportunity for an alternative monitoring approach. The construction, operation, and inspection of the field facilities will necessitate the routine use of deep water ROVs.

A credible, cost effective alternative to remote sampling for monitoring the relative health of the seabed of the area is the establishment of fixed photographic quadrats on large boulders. These would be periodically revisited and photographed by ROV, with the photos quantitatively analysed onshore. Such techniques are well established in shallow waters and have been successfully used on rock faces and boulders with photos taken by divers (e.g. Terlizzi et al., 2002). The spread of trawl fishing into deeper waters and rougher grounds has consequent local (and ecosystem) effects, which pose potential difficulties for the use of photographic monitoring since the selected sites may be overturned or scraped of larger epifauna by fishing gear (see Roberts et al., 2000; Koslow et al., 2001; Gage et al., in press). The routine availability of ROV in deep water fields would allow sediment and/or faunal samples to be periodically taken (e.g. by corer or manipulator) to test predictions of contaminant concentrations and biological effects (including by biomarker analysis). Similarly, ROVs can be used to deploy and recover sensors with data loggers to provide long term information on environmental conditions and contaminant concentrations.

A more fundamental point relates to what should be monitored. In NE Atlantic waters, through OSPAR, European Union, and national regulation, stringent controls are now in place on offshore oil industry activities, emissions, and discharges. Such discharges as are permitted, are unlikely to result in detectable ecological effects or at least effects detectable using "traditional" benthic monitoring methods such as those designed to detect gross effects that have been formalised into OSPAR guidelines. The use of arrays of sentinel sensors or biomarkers are possible alternatives but again in view of the low concentrations permitted in discharges, the establishment of cause and effect relationships with the operation being monitored (and also their ecological significance) will be difficult. Additional challenges to the use of biomarkers for deep water Atlantic Margin oilfield monitoring include the generally poor understanding of basic biology of the species (e.g. what is the extent of spawning migrations of the fish species present?) and the presence of numerous shipwrecks (from two World Wars as well as peacetime casualties), which are likely to be sources of chronic oil and other contaminant release. However, this may be more of a perceived than a real issue as Aas et al. (2003) found not detectable or low DNA adduct levels in 11 fish species from "pristine" Atlantic Arctic and sub-Arctic areas. In the absence of gross effects, a major challenge to meaningful monitoring (in the sense of providing a feedback loop informing company or regulator mitigative action) is the identification of the cause(s) of any subtle effects detected. Such subtle effects on the benthos may result for example from the reef effect of installation presence (Montagna and Harper, 1996), fishing activity (Koslow et al., 2000), seabed subsidence (Hartley, 1996), or regional change (Buchanan and Moore, 1986).

This is emphatically not a call for stopping all benthic studies and surveillance of oilfield operations, rather an argument against the routine application of prescriptive approaches framed in a past era. At the planning stages of a well or a development, surveys involving geophysical methods, sampling, and photography can play a vital role in mapping the range of habitats present and identification any sensitive features so that they can be avoided, or other measures put in place to mitigate potential effects.

Operational monitoring of hydrocarbon developments in the UK has been subject to a general move to regional surveys and away from regularly repeated surveys focussed on single offshore installations. For example, the Fladen Ground is a large muddy depression in the central North Sea which could be expected to act as a sink for contaminants from the extensive oil and gas exploration and production in the area. Recent evidence from a wide area survey (Walsham et al., 2002) suggests that earlier reports of a rise in "background" concentrations of hydrocarbons are likely to reflect a temporary phenomenon attributable (primarily) to extensive discharges of oil contaminated cuttings at the time, and that since cessation of such discharges, a substantial decline in hydrocarbon concentrations has occurred. These reductions in sediment hydrocarbon concentrations occurred over the period of significant increases in produced water (and associated oil) discharges and underscore the importance of regional context for interpretation of monitoring results.

Numerous questions remain about the environment within which oilfield operations take place (particularly in deep water), and there are some sources of potential effect that have yet to be fully assessed. The issue of underwater noise from seismic, drilling and production operations and effects on marine mammals (see Richardson et al., 1995) and other groups is a notable gap in understanding that would benefit from further investigation and potentially routine monitoring. This is particularly the case in the Atlantic Margin where a wide range of marine mammal species are present, some in high densities and some populations still recovering from past whaling. As a measure of shifting perceptions, underwater noise was not considered in the environmental assessment of the Schiehallion field development (BP, 1997) and presumably was not a design criterion for the FPSO vessel. Hydrophone monitoring for cetacean vocalisations in the vicinity of the field found an inverse relationship between fin whale detection probability and noise level, although it was not clear if this reflected a change in vocal behaviour or masking (Swift et al., 2002). In the environmental assessment for the Clair field development (BP, 2001) the consideration of underwater noise ran to some 24 pages and included a commitment to undertake cetacean monitoring during pipeline installation and pile driving operations.

Decisions regarding deepwater oilfield monitoring are made against a shifting background of ecological understanding, improving environmental science techniques and understanding of effects, and changes to control legislation, industry technology, and societal expectations. Past monitoring priorities and methods may not be a guide to the future.

ACKNOWLEDGEMENTS

My thanks to Dave Bingham and Edward Salter of BP for the opportunity to participate in the Suilven survey and to use the results. The ideas developed here have benefited from discussions with members of the AFEN seabed survey subgroup, and Brian Bett, John Gage, Sue Hartley, and Stuart Anderson. Thanks also to Shelley Armsworthy and two anonymous referees whose comments have improved the paper.

REFERENCES

Aas, E., B. Liewenborg, B.E. Grøsvik, L. Camus, G. Jonsson, J.F. Børseth, and L. Balk. 2003. "DNA Adduct Levels in Fish from Pristine Areas Are Not Detectable or Low When Analysed Using the Nuclease PI Version of the 32p-postlabelling Technique." *Biomarkers.* 8:445-460.

AFEN. 2000. *Atlantic Margin Environmental Surveys of the Seafloor 1996 & 1998*. CD Atlantic Frontier Environmental Network and UK Offshore Operators Association.

AFEN. 2001. *The UK Atlantic Margin Environment: Towards a Better Understanding*. Atlantic Frontier Environmental Network ISBN 0 9541648 0 6, 78pp.

Belderson, R.H., N.H. Kenyon, and J.B. Wilson. 1973. "Iceberg Plough Marks in the Northeast Atlantic." *Palaeogeography, Palaeoclimatology, Palaeoecology*. 13:215-224.

Bett, B.J. 2001. "UK Atlantic Margin Environmental Survey: Introduction and Overview of Bathyal Benthic Ecology." *Continental Shelf Research*. 21:917-956.

BP. 1997. *Schiehallion Development and the Environment*. BP Exploration, December 1997.

BP. 2001. *Clair Phase 1 Development Environmental Statement for Offshore Activity*. BP Exploration, August 2001.

Buchanan, J.B. and J.J. Moore. 1986. "A Broad Review of Variability and Persistence in the Northumberland Benthic Fauna 1971–85." *Journal of the Marine Biological Association of the United Kingdom*. 66:641-657.

Carpenter, W.L., J.G. Jeffreys, and C. Wyville Thomson. 1870. "Preliminary Report of the Scientific Exploration of the Deep Sea in H.M. Surveying-vessel '*Porcupine*', during the Summer of 1869." *Proceedings of the Royal Society of London*. 18:397-492 + appendices.

Carroll, M., T. Pearson, E. Dragsund, and K.L. Gabrielsen. 2000. *Environmental Status of the Norwegian Offshore Sector Based on the Petroleum Regional Monitoring Programme 1996–1998*. Akvaplan-niva Report No. 411-1777.03 prepared for the Norwegian Oil Industry Association (OLF) 38pp.

Collins, M. and R.W. Sternberg (Eds.). 2001. "The Marine Environment of the North East Atlantic Margin." *Continental Shelf Research*. 21:Nos. 8-10.

Cranford P.J. and D.C. Gordon Jr. 1992. "The Influence of Dilute Clay Suspensions on Sea Scallop (*Placopecten magellanicus*) Feeding Activity and Tissue Growth." *Netherlands Journal of Sea Research*. 30:107-120.

Cranford P.J., D.C. Gordon Jr., K. Lee, S.L. Armsworthy, and G.H. Tremblay. 1999. "Chronic Toxicity and Physical Disturbance Effects of Water- and Oil-based Drilling Fluids and Some Major Constituents on Adult Sea Scallops (*Placopecten magellanicus*)." *Marine Environmental Research*. 48:225-256.

Daan, R. and M. Mulder. 1996. "On the Short Term and Long Term Impact of Drilling Activities in the Dutch Sector of the North Sea." *ICES Journal of Marine Science*. 53:1036-1044.

Davies, J.M., J.M. Addy, R.A.A Blackman, J.R. Blanchard, J.E. Ferbrache, D.C. Moore, H. Somerville, A. Whitehead, and T. Wilkinson. 1984. "Environmental Effects of the Use of Oil-based Drilling Muds in the North Sea." *Marine Pollution Bulletin.* 15:363-370.

Dicks, B.M. 1976. "Offshore Biological Monitoring." In: JM Baker (Ed.), *Marine Ecology and Oil Pollution.* Applied Science Publishers, Barking Essex pp. 325-440.

ERT. 1999. *BP Vrackie (Block 204/28-B) environmental baseline survey.* Final report. Environment and Resource Technology Ltd. Report No P98/014 to BP Exploration.

ERT. 2000. *DTI "White Zone" Environmental Sampling Programme.* Final data report. Environment and Resource Technology Ltd. report R00/126.

Gage, J.D., J.M. Roberts, J.P. Hartley, and J.D. Humphery. (In press). "Potential Impacts of Deep-Sea Trawling on the Benthic Ecosystem along the Northern European Continental Margin". *American Fisheries Society.*

Gordon, J.D.M. 2001. "Deep-water Fisheries at the Atlantic Frontier." *Continental Shelf Research.* 21:987-1004.

Harries, D., P.F. Kingston and C.G. Moore. 2001. *An Analysis of U.K. Offshore Oil and Gas Environmental Surveys 1975–95.* Report of study carried out by Heriot Watt University for UKOOA. 165pp plus appendices.

Hartley, J.P. 1996. "Environmental Monitoring of Offshore Oil and Gas Drilling Discharges—A Caution on the Use of Barium as a Tracer." *Marine Pollution Bulletin.* 32:727-733.

Hartley, J.P., R.J. Trueman, S.J. Anderson, J.M. Neff, P.R. Dando, and K. Fucik. 2003. *Drill Cuttings Initiative: Food Chain Effects Literature Review.* Report to UK Offshore Operators Association 118pp + appendix.

Heath, M.R. and S.H. Jónasdóttir. 1999. "Distribution and Abundance of Overwintering *Calanus finmarchicus* in the Faroe Shetland Channel." *Fisheries Oceanography.* 8(suppl 1):40-60.

Holliday, N.P., R.T. Pollard, J.F. Read, and H. Leach. 2000. "Water Mass Properties and Fluxes in the Rockall Trough; 1975–1998." *Deep-Sea Research.* 47:1303-1332.

Jeffreys, J.G., A.M. Norman, W.C. McIntosh, and E. Waller. 1869. "Last Report on Dredging Among the Shetland Isles." *Report of the British Association for the Advancement of Science for 1868* pp. 231-345.

Kingston, P.F., I.M.T. Dixon, S. Hamilton and D.C. Moore. 1995. "The impact of the *Braer* Oil Spill on the Macrobenthic Infauna of the Sediments off the Shetland Isles." *Marine Pollution Bulletin.* 30:445-459.

Koslow, J.A., G.W. Boehlert, J.D.M. Gordon, R.L. Haedrich, P. Lorance, and N. Parin. 2000. "Continental Slope and Deep-sea Fisheries: Implications for a Fragile Ecosystem". *ICES Journal of Marine Science.* 57:548-557.

Koslow, J.A., K. Gowlett-Holmes, J.K. Lowry, T. O'Hara, G.C.B. Poore, and A. Williams. 2001. "Seamount Benthic Macrofauna off Southern Tasmania: Community Structure and Impacts of Trawling". *Marine Ecology Progress Series.* 213:111-125.

Mauchline, J. (Ed.). 1986. "The Oceanography of the Rockall Channel." *Proceedings of the Royal Society of Edinburgh.* 88B:1-356.

Montagna, P.A. and D.E. Harper. 1996. "Benthic Infaunal Long-term Response to Offshore Production Platforms in the Gulf of Mexico." *Canadian Journal of Fisheries and Aquatic Science.* 53:2567-2588.

Murray, J. 1886. "The Physical and Biological Conditions of the Seas and Estuaries about North Britain." *Proceedings of the Philosophical Society of Glasgow.* 17:306-333.

Murray, J. and J. Hjort. 1912. *The Depths of the Ocean.* London, Macmillan & Co., 821pp.

OLF. 1998. *Produced Water Discharges to the North Sea: Fate and Effects in the Water Column.* Summary Report. Oljeindustriens Landsforening. 39pp.

OSPAR. 2000. *Quality Status Report 2000, Region II—Greater North Sea.* OSPAR Commission, London, 136 +xiii pp.

Pollock, C.M., R. Mavor, C.R. Weir, A. Reid, R.W. White, M.L. Tasker, A. Webb, and J.B. Reid. 2000. *The Distribution of Seabirds and Marine Mammals in the Atlantic Frontier, North and West of Scotland.* Joint Nature Conservation Committee. 92pp.

Richardson, W.J., C.R. Greene Jr., C.I. Malme, and D.H. Thomson. 1995. *Marine Mammals and Noise.* Academic Press, 576pp.

Roberts, J.M., D. Long, J.B. Wilson, P.B. Mortensen, and J.D. Gage. 2003. "The Cold-Water Coral *Lophelia pertusa* (Scleractinia) and Enigmatic Seabed Mounds along the North-East Atlantic Margin: Are They Related?" *Marine Pollution Bulletin.* 46:7-20.

Roberts, J.M., S.M. Harvey, P.A. Lamont, J.D. Gage, and J.D. Humphery. 2000. "Seabed Photography, Environmental Assessment and Evidence for Deep-Water Trawling on the Continental Margin West of the Hebrides." *Hydrobiologia*. 441:173-183.

Stoker, M.S., K. Hitchen, and C.C. Graham. 1993. *The Geology of the Hebrides and West Shetland Shelves, and Adjacent Deep-Water Areas*. United Kingdom Offshore Regional Report. HMSO for the British Geological Survey, London.

Swift, R.J., G.D. Hastie, T.R. Barton, C.W. Clark, M.L. Tasker, and P.M. Thompson. 2002. *Studying the Distribution and Behaviour of Cetaceans in the Northeast Atlantic Using Passive Acoustic Techniques*. A Report to the Atlantic Frontier Environmental Network, 87pp.

Terlizzi, A., S. Fraschetti, P. Guidetti, and F. Boero. 2002. "The Effects of Sewage Discharge on Shallow Hard Substrata Sessile Assemblages." *Marine Pollution Bulletin*. 44:544-550.

Tizard, R.N. and J. Murray. 1882. "Exploration of the Faroe Channel during the Summer of 1880, in H.M.'s Hired Ship "*Knight Errant*"." *Proceedings of the Royal Society of Edinburgh*. 11:638-719.

Turrell, W.R., G. Slesser, R.D. Adams, R. Payne, and P.A. Gillibrand. 1999. "Decadal Variability in the Composition of Faroe Shetland Channel Bottom Water." *Deep-Sea Research 1*. 46:1-25.

UKOOA. 2000. UKbenthos. *Database of Offshore Environmental Benthic Surveys in the UK Sector of the North Sea*. Prepared by Heriot Watt University for UK Offshore Operators Association. CD

Walsham P., L. Webster, R.J. Fryer, A.D. McIntosh, and D.C. Moore. 2002. *The Effects of Oil Exploration and Production in The Fladen Ground: Composition and Concentration of Hydrocarbons in Sediment Samples Collected in 2000 and their Comparison with Sediment Samples Collected in 1989*. FRS Report No 14/02 Fisheries Research Services, Aberdeen, 17pp. plus figures and tables.

Wilson, J.B. 1979. "The First Recorded Specimens of the Deep-Water Coral *Lophelia pertusa* (Linnaeus 1758) from British Waters." *Bulletin of the British Museum (Natural History), Zoology Series*. 36:209-215.

Wynn, R.B., D.G. Masson, and B.J. Bett. 2002. "Hydrodynamic Significance of Variable Ripple Morphology across Deep-Water Barchan Dunes in the Faroe-Shetland Channel." *Marine Geology*. 192:309-319.

Wyville Thomson, C. 1874. *The Depths of the Sea*. London: Macmillan & Co., 527pp.

Terra Nova Environmental Effects Monitoring Program: From Environmental Impact Statement Onward

DeBlois, E.M. (edeblois@jacqueswhitford.com), C. Leeder, K.C. Penney (Jacques Whitford Environment Ltd., St. John's, Newfoundland, Canada), Mary Murdoch (Jacques Whitford Environment Ltd., Fredericton, New Brunswick, Canada), M.D. Paine (Paine, Ledge and Associates, North Vancouver, British Columbia, Canada), F. Power, and U.P. Williams (Petro-Canada East Coast Operations, St. John's, Newfoundland, Canada)

ABSTRACT: In 1996, Petro-Canada submitted an Environmental Impact Statement (EIS) for the Terra Nova oil field development outlining anticipated project effects. As a condition of project approval, an Environmental Effects Monitoring (EEM) program was designed with input from stakeholders and national and international specialists. The primary purpose of the EEM program was to test effects predictions made in the EIS. As such, the program was established to detect change in the quality of the receiving environment through examination of sediment quality, water quality, and ultimately biological resources. Sediment toxicity, hydrocarbon content, metals concentration, and benthic infaunal community structure were assessed as part of the program. Hydrocarbon, metal, and chlorophyll concentrations were measured on water samples. Metals and hydrocarbon body burden, taste and morphometrics, and life history characteristics were assessed for Icelandic scallop and American plaice. Mixed-function oxygenase (MFO) induction assessment, haematology, and histopathology were performed on American plaice tissue samples. The strength of the Terra Nova EEM program results from these many supporting program elements. A weight of evidence approach was used to assess project effects to the receiving environment. Since its inception in 2000, the program has proven effective at detecting very small inter-annual changes in the marine environment unrelated to project activity and therefore provides a powerful tool for detecting and identifying project effects outlined in the EIS.

INTRODUCTION

Project Area

The Terra Nova oil field is situated on the Grand Banks of Newfoundland (Canada), approximately 350 km east-southeast of St. John's (Fig. 1). Petro-Canada acts as operator of the development on behalf of the owners (Petro-Canada, Mobil Oil Canada Properties, Husky Energy Ltd., Norsk Hydro Canada Oil and Gas Inc., Murphy Oil Company Ltd., Mosbacher Operating Limited, and Chevron Canada Resources).

The oil field is being developed using a floating production, storage and offloading (FPSO) facility and a semi-submersible drilling rig (Fig. 2). A total of 24 wells are being drilled through 7 subsea templates, located in 5 drill centres (Fig. 3). Trenched and bermed flowlines connected to flexible risers link the subsea installations to the FPSO. First oil was achieved on January 20, 2002.

EEM Background

EEM programs are follow-up surveys that assess predictions on biological effects made as part of EIS and determine the zone of influence of project contaminants. If effects exceed predications made in the EIS, then additional mitigation measures are considered. The Terra Nova EIS ranked potential effects as negligible, minor, moderate, and major. Using this scale, minor effects on benthos were predicted within 1 to 10 km from the Terra Nova development as a result of development drilling, but minor to major effects were expected a few

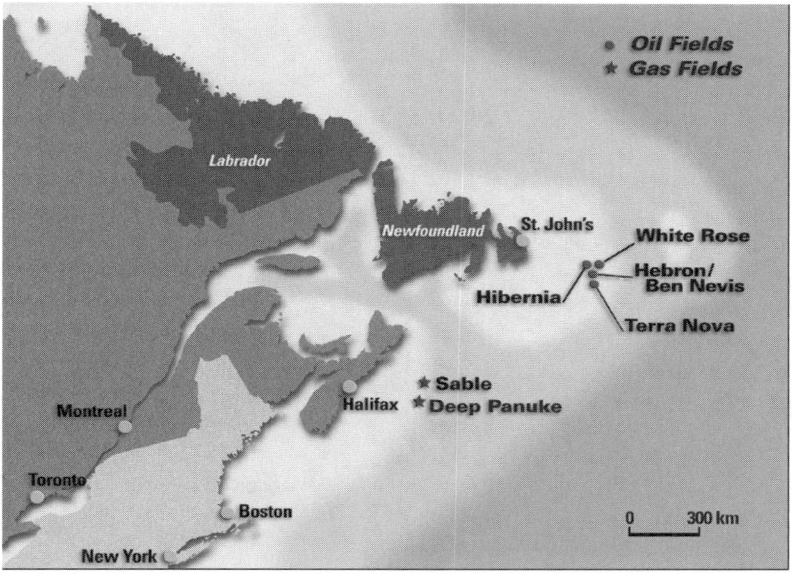

FIGURE 1. Terra Nova and other oil field locations on the Grand Banks.

FIGURE 2. Floating production, storage, and offloading (FPSO) facility and drilling rig.

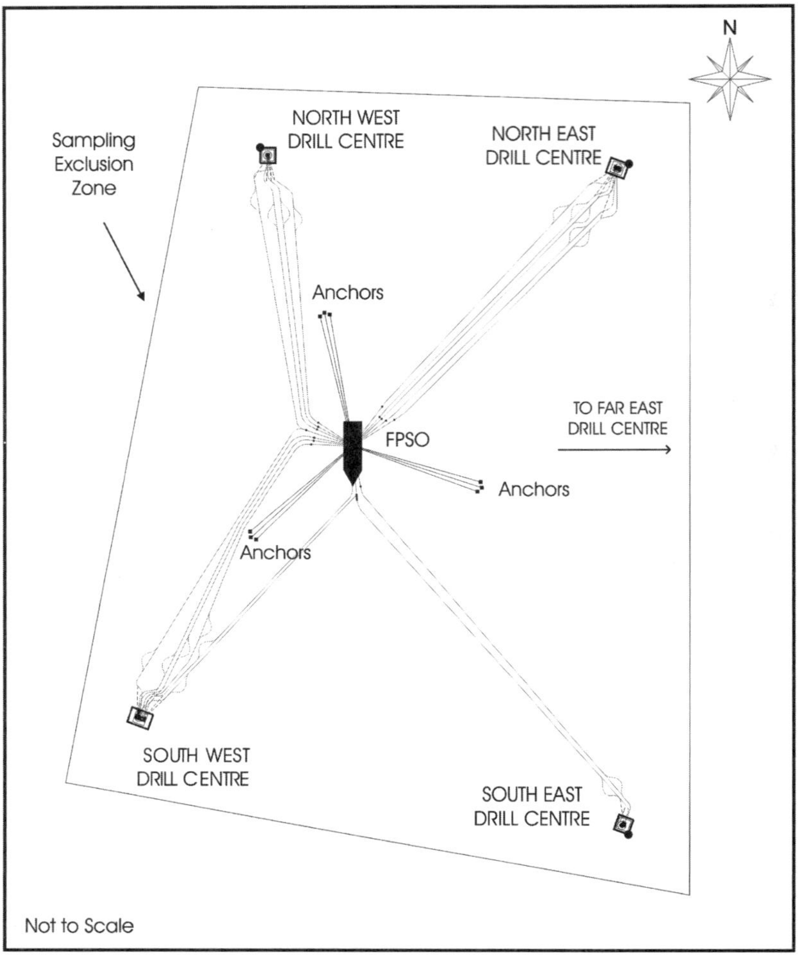

FIGURE 3. Drill centres at Terra Nova where development activity occurred (1997–2001). As of August 2001, construction had not occurred at the Far East drill centre, located approximately 5 km east of the sampling exclusion zone.

hundred metres from drill centres. From dispersion model results, the zone of influence for drill mud contaminants was expected to extend to 5 km from source (Seaconsult, 1998). Main drill mud contaminants at Terra Nova include hydrocarbons in the C_{10}–C_{21} range from the synthetic-based mud, PureDrill IA-35, and barium from both synthetic and water-based drilling muds (PureDrill IA-35 is a synthetic isoalkane (saturated parrafin) fluid manufactured from a petroleum feed stock which consists of linear and branched chain aliphatic hydrocarbons from C_{10}–C_{21} and like most synthetic-based fluids it is free of PAHs). Negligible effects were anticipated on fish health and fish taint; minor effects on water quality were expected within a few hundred meters of drill centres. Finally, minor effects on both water quality and plankton were predicted as a result of produced water discharge. However, those effects were not anticipated at Terra Nova until release of produced water at the site.

Petro-Canada has carried out an EEM program at Terra Nova in 2000 and 2001. Baseline data were collected prior to development in 1997. The following paper describes the Terra Nova EEM program and provides a general overview of results from 1998 to 2001.

MATERIALS AND METHODS

Data Collection and Laboratory Analysis

EEM sediment samples were collected at 49 stations located along transect lines centred on the FPSO (Fig. 4). The location of sediment stations during baseline was similar (33 common stations) to EEM station locations. However, construction activity in 2000 and 2001 precluded sampling in the immediate vicinity of drill centres and required relocation of stations to an area outside the core of construction activities. Sediment samples were processed for hydrocarbon concentrations, metal concentrations (a suite of 24 metals), particle size, sulphur, sulphide and ammonia. Toxicity tests that characterise whether sediments are toxic to bacteria (*Vibrio Fischeri*) and a marine amphipod (*Rhepozynius abronius*) were performed. In addition, benthic invertebrate infaunal species were identified and enumerated.

Water samples and conductivity, temperature, depth (CTD) data, including fluorometric readings, were collected at 16 stations in a study area located within approximately 5 km of the FPSO, and at eight stations located in two reference areas approximately 20 km to the southeast and southwest of Terra

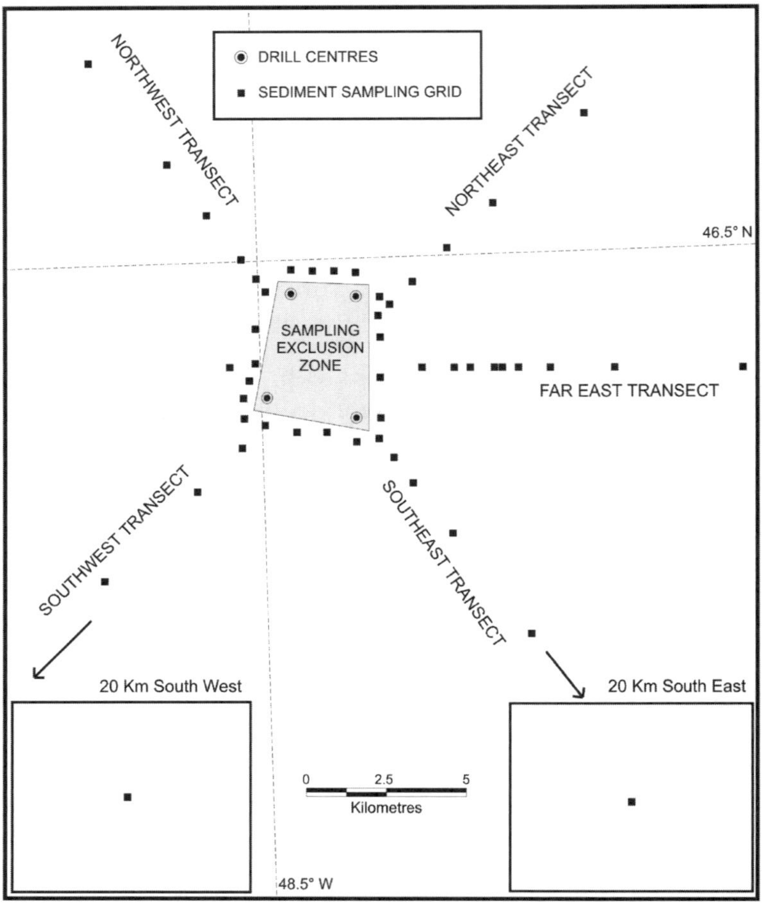

FIGURE 4. Sediment sampling stations.

Nova (Fig. 5). Water samples were analysed for hydrocarbons and metals (a suite of 11 metals), total suspended solids (TSS), and phytoplankton pigment concentration.

Samples of a common flatfish species (American plaice; *Hyppoglossoides platessoides*) and a commercial bivalve species (Iceland scallop; *Chlamys islandica*) were collected in the Study Area and in the Southeast Reference Area (Fig. 5). Both these species were identified as important commercial fisheries species during stakeholder consultation (Petro-Canada, 1996).

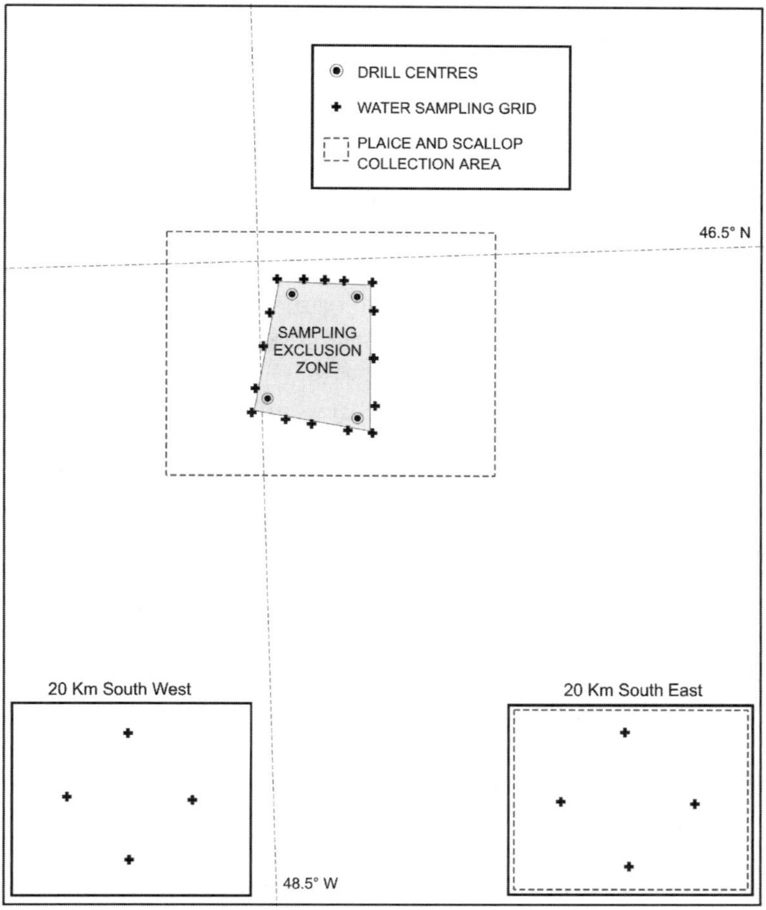

FIGURE 5. Water, plaice, and scallop collection areas.

American plaice and Iceland scallop size, shape, fecundity and maturity status (morphometric and life history characteristics) were measured. Tissue samples (liver and fillet for American plaice, and viscera and muscle for scallop) were analysed for hydrocarbon and metal (25 metals) body burden. Fillet and muscle samples were tested for taint using both triangle and hedonic scaling taste tests. Mixed-function oxygenase (MFO) induction was assessed in liver samples for American plaice. Tissue histopathology on both liver and gill, and haematological analyses were also performed on American plaice.

Data Analysis

The sediment portion of the Terra Nova EEM program followed a gradient design (see Figure 4 for spatial lay-out of stations). The aim of such a design is to assess attenuation or accentuation of a response variable (e.g., hydrocarbons, benthic community structure) with distance from source. In the presence of project induced contamination, the expectation is that contaminant concentration, for instance, will decrease as one moves away from the development. For Terra Nova, regression analysis of response variables on distance from the nearest drill centre was used to assess the presence of gradients within years. Repeated-measures analysis of covariance comparing regressions of response variables on distance was used to assess change in gradients across years. These cross-year comparisons to baseline data are particularly important when gradients occur naturally, as is the case for Terra Nova.

Both the water quality and commercial fish components of the program followed a Control-Impact (CI) design, where response variables in the vicinity of the development are compared to response variables at a more distant site (see Fig. 5 for spatial lay-out). Statistical tests for these designs included t-test (within years) and analysis of variance (across years) comparing conditions at the Control and Impact sites, and in the case of multi-year comparisons, Before and After (BA) development. These types of designs are commonly referred to as BACI designs (Green, 1979).

Principal component analysis was used to summarise metals concentrations ($\log_{10}$ transformed) and simplify analyses for most metals. Those metals judged to be of particular relevance (e.g. Barium) were examined separately. Non-metric multidimensional scaling (NMDS) was used to summarise multivariate benthic community data (Clarke, 1993).

RESULTS AND DISCUSSION

To date, few project-related effects have been noted at Terra Nova. Results for the 2000 and 2001 EEM programs were similar and comparable to results obtained during baseline data collection. No project related effects were identified for sediment particle size, barium concentration and concentrations of other metals. A gradient in particle size and metals concentration, with finer sediment, higher organic carbon and metals concentrations near the development was observed in all sampling years, including baseline. Natural, near-surface fines and organic carbon content and metal concentrations could be higher in areas where sub-surface oil occurs. There is documented evidence of natural organic carbon enrichment in these areas (Spies and DesMarais, 1983; Braddock et al., 1995).

Changes in sediment hydrocarbon concentration relative to baseline and accentuation of this effect from 2000 to 2001 were noted (see Fig. 6 for concentrations of C_{10}–C_{21} hydrocarbons). Contamination dropped to low levels (1 mg/kg) at roughly 3 km from drill centres. All chromatograms within the contaminated area revealed peaks characteristic of PureDrill IA-35 and in all these cases, hydrocarbons in the PureDrill IA-35 range (C_{10}–C_{21}) accounted for the bulk of hydrocarbons measured. Maximum C_{10}–C_{21} concentration was 28.6 mg/kg (97% of total petroleum hydrocarbons measured at this station). This concentration of PureDrill IA-35 is orders of magnitude lower than levels expected to have biological effects. Amphipod toxicity tests on sediment samples contaminated with PureDrill IA-35-on-cuttings indicated that toxicity (reduced survival to below 80%) occurred only at hydrocarbon concentrations greater than 1900 mg/kg (Payne et al. 2001). (Note that Figure 6 does not indicate C_{10}–C_{21} hydrocarbon levels for 1997, when the detection limit for hydrocarbons was higher (30 mg/kg) than any concentrations measured in 2000 and 2001. The working assumption for 1997 data has been that levels would compare to levels noted at more distant stations in 2000 and 2001 (i.e. approximately 1 mg/kg).)

Amphipod survival in toxicity tests conducted on Terra Nova sediments was high in all three sampling years and there has been no evidence of toxicity. This would be expected from Payne et al. (2001). Toxicity of sediments to bacteria has increased since 1997, but increases were more closely related to strontium and total inorganic carbon (TIC) concentrations

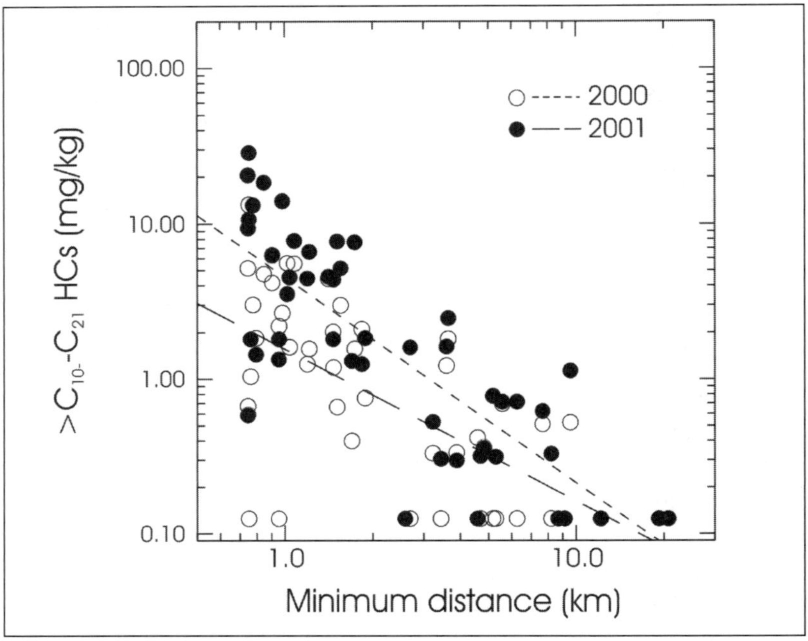

FIGURE 6. Relationship between sediment $>C_{10}$-C^{21} concentrations and distance to the nearest of the four drill centres (2000, 2001).

than to concentrations of barium and hydrocarbon, which are the major contaminants in drilling muds. Since strontium and TIC are high where shell fragments are abundant, the noted toxicity response could result from decomposition of organic material (for instance dead bivalves) in these sediments. These relationships are examined in more detail in Paine et al. (2003).

Evidence of effects on benthic infauna was minimal. There were no noted differences in distance gradients among years for total abundance, biomass, richness and diversity. Distance gradients for abundances of Amphipoda and Echinodermata, taxa known to be sensitive to contaminants in drill muds, also remained unchanged since 1997. However, the relative abundance of Spionid polychaetes decreased with increasing distance from the drill centres in 2000 and 2001, and this gradient was weak or absent in 1997. Abundances of bivalves were lower near drill centres in 1997, but this spatial pattern was reversed by 2001. The relative abundances of Syllid and Cirratulid polychaetes increased with increasing distance from the drill centres in 2000 and 2001 (reliable information on these later two taxa is lacking for 1997). Collectively, these results may be evidence of positive effects (enrichment) of project activities on Spionids and Bivalves, and negative effects on Syllids and Cirratulids. However, only Spionid and Syllid abundances were correlated with hydrocarbon concentrations, and this, potentially through a common relationship with total organic carbon (TOC). Spionids dominated sediments with higher TOC content, where hydrocarbon concentrations were usually higher. Syllids dominated in sandier sediments with lower TOC and hydrocarbon levels.

In general, the benthic infauna response noted at Terra Nova was ambiguous when compared to responses noted at other developments. Peterson et al. (1996) reviewed more than 50 studies assessing benthic infauna abundance and community changes in the vicinity of contaminant sources including oil spills, natural oil seeps, sewage discharges, industrial discharges, as well as within some experimental and mesocosm studies. These studies consistently showed increases in polychaetes in response to enrichment and decreases in echinoderms and amphipods in response to toxicants. One study (Steichen et al., 1996) showed decreased bivalve abundance near natural oil seeps. Conversely, the Petro-Canada EEM program showed increases in bivalves near the drilling sites, no effect on amphipods, either in situ or in toxicity tests, and increases as well as decreases in the relative abundance of certain polychaete taxa. Intriguingly, positive effects (enrichment) on benthic infauna at Terra Nova were related to increased toxicity of sediments to laboratory bacteria, supporting the decomposition hypothesis noted above (see Paine et al., 2003, for details).

Water quality variables did not change markedly since 1997. However, surface or mid-depth effects from produced water discharge may be important in the future (Holdway, 2002). Barium contamination of scallop viscera tissue was noted in 2000. Median viscera concentration in the Study Area in 2000 was 11 mg/kg. Barium occurs naturally in marine sediments (baseline levels were in the range of 4 to 5 mg/kg), but could also originate from water-based drilling muds high in barium sulphate (barite). Barite has a low bioaccumulation factor and water-based drilling muds show little or no chemical toxicity to marine organisms (Neff, 1989a, b). However, at concentrations well above background, barite can have physical or irritation effects on tissues and membranes (Barlow and Kingston, 2001). Barium levels in scallop viscera in the Study Area were not elevated in 2001 and barium has generally not been detected in scallop muscle tissue (except for one in the Reference Area sample in 2000).

Increases in hydrocarbon concentrations in scallop tissue in the Study Area were also noted in both years. This was expected, since hydrocarbon contamination was noted in sediments. Median levels in 2000 were 21.5 mg/kg and 110 mg/kg in muscle and viscera samples, respectively. Levels decreased in 2001 to 18 mg/kg and 54 mg/kg in muscles and viscera samples. The biological effects of tissue contamination of drilling mud hydrocarbons have not been investigated in detail in either laboratory or field studies. In laboratory experiments, chronic exposure to whole synthetic-based drilling muds (SBMs), including PureDrill IA-35, reduced scallop growth, reproductive development, and nutrient storage capacity (Cranford et al., 2000). However, effects of treated drill cutting (less than 15% SBMs-on-cutting at Terra Nova) were not explored and no relationship between hydrocarbon body burdens and effects was provided. Since hydrocarbon contamination in scallop tissues at Terra Nova did not result in scallop taint, nor did it result in changes in scallop morphometric and life history characteristics, contamination may have been too mild to result in biological effects.

Hydrocarbons were not detected in American plaice tissue samples in 2001. In 2000, low levels of hydrocarbons were detected in only 1 of 5 Study Area liver samples and 1 of 21 Study Area fillet samples. The mobility of American plaice will continue to limit the identification of project effects relative to effects from other sources. Taste tests in both years showed that fillet samples were not tainted. Fish health assessment, including MFO induction, haematology, and histopathology, showed that plaice caught in the vicinity of the development have continued to be similar in health to those caught in the Reference Area. More details on American plaice health assessment are provided in Mathieu et al. (2004, this publication).

Summary of Project Related Effects

Changes in sediment hydrocarbon concentration relative to baseline, and accentuation of this response from 2000 to 2001 were the most obvious project effects at Terra Nova. Sediment contamination extended to approximately 3 km from the development. Evidence of effects on invertebrate communities was minimal. The project may be having minor positive effects (enrichment) on Spionid polychaetes and bivalves and minor negative effects on Syllid and Cirratulid polychaetes. Hydrocarbon contamination was noted in scallop tissue in both years; barium contamination was noted in viscera in 2000 only. Tissue contamination in scallop did not translate into biological effects, as measured through taint tests and comparison of morphometric and life history characteristics.

Summary of Other Relevant Findings

The EEM also revealed a number of other findings that could not reasonably be linked to project activity. The more interesting of these findings were:

- Positive correlation between Microtox toxicity, strontium, and TIC concentrations (potential indicators of sediment shell content and organic decomposition).
- Positive correlations between Microtox toxicity and measures of benthic community composition and richness. (see Paine et al., 2003, for details on this and the previous item).
- Strong correlations between sediment gravel content and most benthic invertebrate community variables (perhaps a habitat heterogeneity effect).
- Higher concentrations of most metals in scallop viscera in the Study Area than in the Reference Area since baseline, but higher concentrations in adductor muscle in the Reference Area (potentially indicative of differences in the bioavailability of metals in these two areas).

CONCLUSIONS

Results to date are consistent with drill cuttings dispersion modelling results (Seaconsult 1998) and within predictions made in the Terra Nova EIS (Petro-Canada 1996):

- The zone of influence for drill cuttings was expected to extend to approximately 5 km from drill centres. Hydrocarbon concentrations at Terra Nova fell to low levels (1 mg/kg) at 3 km from drill centres.
- Minor to major effects were expected on benthic infauna. To date, however, the EEM program indicates a mix of mild inhibitory or stimulatory responses for a restricted number of taxa. Also, the evidence linking these responses to the project is weak.
- Discharges of drilling muds and cuttings were expected to have minor effects on water quality. No effects were noted on water quality.
- Negligible effects were expected on fish taint and fish health indices (MFO, haematology, histopathology). Tissue contamination was noted in scallop. However, there was no evidence to suggest that this contamination was having a biological effect. Tissue contamination in one sample of American plaice tissue in 2000 could not reasonably be attributed to the project. No taint was found in American plaice and plaice from the Study and Reference Area were in similar health.

The absence of marked project effects should not be attributed to any limitations of the Terra Nova EEM program. The program had sufficient power and sensitivity to detect a fines and strontium/TIC effect on Microtox, a strong particle size effect on benthic community structure, extremely mild benthic invertebrate responses, and interesting differences in scallop tissue metal concentrations (consistent since baseline data collection). Payne (2003) states that EEM programs being carried out in the Newfoundland region with respect to oil and gas activities equal or surpass those being carried out in the North Sea, the Gulf of Mexico, and Australia. The Terra Nova EEM should therefore be able to serve as a good model for effects monitoring programs in other offshore areas.

ACKNOWLEDGEMENTS

We would like to acknowledge the Canada-Newfoundland Offshore Petroleum Board, the Department of Fisheries and Oceans, Environment Canada, those interested parties who participated in stakeholder consultation, and all other individuals and organisations that have provided feedback during the design phase of this EEM and/or continue to provide feedback on results and interpretations for this program. We would also like to thank Beverley Best, Paula Dalton, and Tony Clemens for assistance in the production of this document.

REFERENCES

Barlow, M.J. and P.F. Kingston. 2001. "Observations on the Effects of Barite on the Gill Tissues of the Suspension Feeder *Cerastoderma edule* (Linne) and the Deposit Feeder *Macoma balthica* (Linne)." *Marine Pollution Bulletin*. 42(1):71-76.

Braddock, J.F., J.E. Lindstrom, and E.J. Brown. 1995. "Distribution of Hydrocarbon Degrading Microorganisms in Sediments from Prince William Sound, Alaska Following the Exxon Valdez Oil Spill." *Marine Pollution Bulletin*. 30:125-132.

Clarke, K.R. 1993. "Nonparametric Multivariate Analyses of Changes in Community Structure." *Austrailian Journal of Ecology*. 18:117-143.

Cranford, P.J., S. Armsworthy, and K. Querbach. 2000. *Chronic Toxicity of Synthetic Oil-Based Drilling Mud to Sea Scallops*. Department of Fisheries and Oceans, Bedford Institute of Oceanography, Dartmouth, NS, 7 pp.

Green, R.H. 1979. *Sampling Design and Statistical Methods for Environmental Biologists*. John Wiley and Sons, Toronto, ON.

Holdway, D.A. 2002. "The Acute and Chronic Effects of Wastes Associated with Offshore Oil and Gas Production on Temperate and Tropical Marine Ecological Processes." *Marine Pollution Bulletin*. 44(2002):185-203.

Mathieu, A., B. French, M. Dawe, M. Myers, E.M. DeBlois, F. Power, and U. Williams. 2005. "Bioindicator Studies on Fish Health around the Terra Nova Development Site on the Grand Banks." In S.L. Armsworthy, P.J. Cranford, and K. Lee (Eds.), *Proceedings of the Offshore Oil and Gas Environmental Effects Monitoring Workshop: Approaches and Technologies, Bedford Institute of Oceanography, Dartmouth, Nova Scotia, May 26–30, 2003*, Battelle Press. This volume.

Neff, J.M., M.H. Bothner, N.J. Maciolek and J.F. Grassle. 1989a. "Impacts of Exploratory Drilling for Oil and Gas on the Benthic Environment of Georges Bank." *Marine Environmental Research.* 27:77-114.

Neff, J.M., R.J. Breteler and R.S. Carr. 1989b. "Bioaccumulation, Food Chain Transfer and Biological Effects of Barium and Chromium from Drilling Muds by Flounder *Pseudopleuronectes americanus* and Lobster *Homarus americanus." Proceedings of the 1988 International Conference on Drilling Wastes.* pp. 439-459. Elsevier, London, UK and New York, NY.

Paine, M.D., E.M. DeBlois, S. Whiteway, F. Power, and U. Williams. 2003. "Can *in situ* invertebrate communities directly or indirectly affect toxicity test results?" (Abstract only). *Workshop on the Offshore Oil and Gas Environmental Effects Monitoring, Bedford Institute of Oceanography, Dartmouth, Nova Scotia, May 26–30, 2003*, Environmental Science Research Fund Report. In press.

Payne, J.F., L.L. Fancy, C.D. Andrews, J.D. Meade, F.M. Power, K. Lee, G. Veinott and A. Cook. 2001. "Laboratory Exposures of Invertebrate and Vertebrate Species to Concentrations of IA-35 (Petro-Canada) Drill Mud Fluid, Production Water and Hibernia Drill Mud Cuttings." *Canadian Manuscript Report of Fisheries and Aquatic Sciences.* No. 2560. April 2001. iv + 27 pp.

Payne, J.F., 2003. "Adequacy of Monitoring Programs for Production and Exploratory Wells on the Grand Banks of Newfoundland" (Abstract only). *Workshop on the Offshore Oil and Gas Environmental Effects Monitoring, Bedford Institute of Oceanography, Dartmouth, Nova Scotia, May 26–30, 2003*, Environmental Science Research Fund Report. In press.

Peterson, C.H., M.C. Kennicutt, R.H. Green, P. Montagna, D.E. Harper Jr., E.N. Powell, and P.F. Roscigno. 1996. "Ecological Consequences of Environmental Perturbations Associated with Offshore Hydrocarbon Production: A Perspective on Long-Term Exposures in the Gulf of Mexico." *Canadian Journal of Fisheries and Aquatic Sciences.* 53(11):2637-2654.

Petro-Canada. 1996. *Development Application: Terra Nova Development. Environmental Impact Statement.* Prepared by LGL Limited for Petro-Canada. St. John's, NL. 624 pp. + Appendices.

Petro-Canada. 1997. *Development Application: Terra Nova Development. Environmental Impact Statement—Addendum.* Prepared by LGL Limited for Petro-Canada. St. John's, NL. 111 pp. + Appendices.

Petro-Canada. 1998a. *Terra Nova Baseline Characterization Data Report.* Prepared by Jacques Whitford Environment Limited for Petro-Canada, St. John's, NL. 98 pp. + Appendices.

Petro-Canada. 1998b. *Environmental Effects Monitoring Program.* Document No. TM-IM-EV02-X00-001). Prepared by Jacques Whitford Environment Limited for Petro-Canada, St. John's, NL. 99 pp. + Appendices.

Seaconsult. 1998. *Distribution of Well Cuttings and Produced Water for the Terra Nova Development.* Prepared for Petro-Canada, St. John's, NL by Seaconsult Marine Research Ltd., Vancouver, BC. 40 pp. + Appendices.

Spies, R.B. and D.J. DesMarais. 1983. "Natural Isotope Study of Trophic Enrichment of Marine Benthic Communities by Petroleum Seepage." *Marine Biology.* 73:67-71.

Steichen, D.J., S.J. Holbrook, and C.W. Osenberg. 1996. "Distribution and Abundance of Benthic and Demersal Macrofauna within a Natural Hydrocarbon Seep." *Marine Ecology Progress Series.* 138:71-82.

Monitoring Explosive-Based Winter Seismic Exploration in Waterbodies, NWT 2000–2002

Cott, P. (cottp@dfo-mpo.gc.ca) and B. Hanna (Department of Fisheries and Oceans, Yellowknife, NWT, Canada)

ABSTRACT: During the winter of 2000/2001, there was a resurgence of oil and gas exploration in the Mackenzie Delta area of the Northwest Territories. Explosives were the primary energy source for seismic exploration crossing waterbodies not frozen to the bottom. Industry initially followed burial depths outlined in the Department of Fisheries and Oceans, *Guidelines for the Use of Explosives in or Near Canadian Fisheries Waters* (the Guidelines), when setting charges under waterbodies. The instantaneous pressure change (IPC) in the water column resulting from charge detonation was monitored and revealed that many charges produced IPC higher than DFO's 100 kPa threshold guideline for the protection of fish. DFO required additional monitoring and subsequently determined that 50% of the remaining charges monitored and recorded (n = 429) exceeded 100 kPa. As a result, DFO outlined a series of requirements for 2001/2002 winter seismic exploration programs. These included program-specific testing of charge size/burial depth combinations prior to initiating production seismic programs over waterbodies. The test results indicated that 10 of 11 explosive-based seismic programs required increased burial depths for production shot holes; up to five-times the distance outlined in the Guidelines in order to not exceed the 100 kPa threshold. Despite using greatly increased burial depths for a given charge size in most cases, approximately 1 of 12 charges monitored and recorded (n = 507) still exceeded 100 kPa. Although the reasons for high IPC are not fully understood, it is clear that the IPC from the use of explosives in waters potentially bearing fish is not entirely predictable. Due to the potential negative impacts on fish, DFO no longer authorizes explosive-based seismic exploration programs under waterbodies not frozen to the bottom in the NWT.

INTRODUCTION

It has been well documented that the use of explosives in aquatic environments has the potential to negatively impact fish by rupturing the swim bladder and/or damaging other internal organs (Falk and Lawrence, 1973; Teleki and Chamberlain, 1978; Wright, 1982; Alaska Department of Fish and Game, 1991; CAPP, unpublished); hence, the Department of Fisheries and Oceans developed the *Guidelines for the Use of Explosives in or Near Canadian Fisheries Waters* (hereafter the Guidelines). The Guidelines recommend burial depths for given charge sizes for certain generalized substrate types. These charge size/burial depth combinations were intended to keep peak pressures of the Instantaneous Pressure Change (IPC) in the water column below a critical threshold of 100 kPa, a level believed to cause fish injury or death. The killing of fish by any means other than fishing is prohibited under Section 32 of the *Fisheries Act* unless authorized by the Department of Fisheries and Oceans (Government of Canada, 1985).

Alaska has many similarities to the Northwest Territories (NWT) in both environment and types of hydrocarbon exploration and development. Current mitigation for seismic exploration in Alaska prohibits the use of explosives in fish-bearing waters, and charges cannot be set under or near fish-bearing waters if the explosive produces IPC values greater than 2.5 pounds per square inch (17.2 kPa) unless the waterbody and its substrates are solidly frozen (Alaska Department of Natural Resources, 2003). The last explosive-based seismic program near waterbodies approved in Alaska was in the late 1980s, however it was never conducted as the company withdrew the application (A. Ott, Alaska Department of Fish and Game, pers. comm.).

In 2000, seismic exploration and exploratory drilling programs re-commenced in the NWT at a scale not seen since the 1970s and early 1980s. Seismic exploration projects were conducted throughout the Mackenzie Delta and surrounding area within the Inuvialuit Settlement Region (ISR), using both dynamite and vibrator energy sources. Over 11,000 km of 2-D and 3-D seismic lines were surveyed (Cott et al., 2003; Fig. 1).

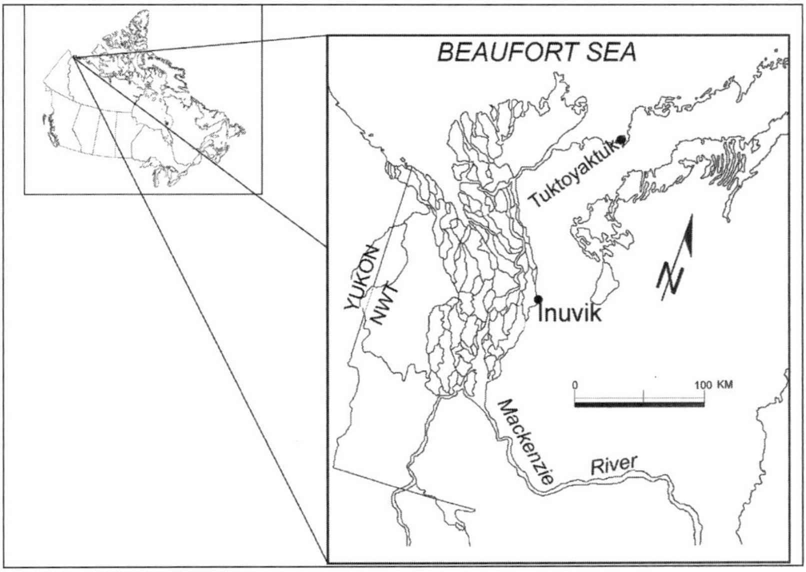

FIGURE 1. The Mackenzie Delta and surrounding area within the Inuvialuit Settlement Region, NWT where Instantaneous Pressure Change monitoring of explosive-based winter seismic programs 2000/2001 and 2001/2002 was conducted.

Most of the seismic exploration conducted under waterbodies not frozen to the bottom (including lakes, delta channels, and the near-shore Beaufort Sea) used dynamite as the energy source. Burial depths outlined within the Guidelines were followed when shooting under waterbodies not frozen to the bottom. In the winter of 2000/2001, industry experienced problems maintaining the desired depth of buried charges, particularly when using a Vibra-ram® (seismic drilling equipment using a vibrating rod designed to deploy explosive charges in soft substrates). Concerned with exceeding the instantaneous pressure change (IPC) threshold, industry voluntarily implemented an IPC monitoring program and notified DFO that in many cases, IPC was exceeding 100 kPa. This experience prompted DFO to increase requirements, such as extensive IPC monitoring, for seismic programs proposed for the following winter season.

MATERIALS AND METHODS

Monitoring of IPC was conducted in 2000/2001 and 2001/2002 during winter seismic exploration in the NWT. In both years, seismographs equipped with hydrophones were used to measure IPC. The most frequently used seismographs were the BlastMate™ and MiniMate™ series units manufactured by Instantel®. These units are portable and capable of a sampling rate greater than 16,000 s^{-1} (up to 65,536 s^{-1} on some models; Instantel, 2001). The sampling rate for units was set to the highest possible frequency in order to ensure a reading as close to the true peak pressure as possible (AquaScience, 2001a; R. Cyr, Explotech, pers. comm.). The waveform frequency of dynamite is very high, with peak pressure occurring within 1/1000th of a second of detonation. The hydrophones used in the monitoring program are capable of recording IPC in the water-column up to 346 kPa (AquaScience, 2001a).

Monitoring for IPC was conducted using three hydrophones placed under the ice in the water column; each suspended approximately 15 cm above the bottom. Hydrophones set during the 2000/2001 monitoring were placed above the shot hole (0 m), 5 m, and 15 m laterally away from the shot hole, measuring from the initial auger hole on the ice surface where the charge was loaded (AquaScience, 2001b). The reason for the placement of three hydrophones at different distances from the charge was to determine the rate of energy attenuation and potential area of exposure of IPC to fish.

For monitoring in the winter of 2001/2002, a minimum sampling frequency of 16,000 s^{-1} per hydrophone was specified. In addition, there was a change made in the hydrophone placement requirements. The three hydrophones were now placed directly over the shot hole (burial depth plus 0 m), burial depth plus 5 m, and burial depth plus 10 m. Trigonometry was used to calculate the appropriate lateral distance on the ice surface. This method enabled standardized hydrophone placement laterally from the shot hole regardless of the burial depth achieved (Fig. 2).

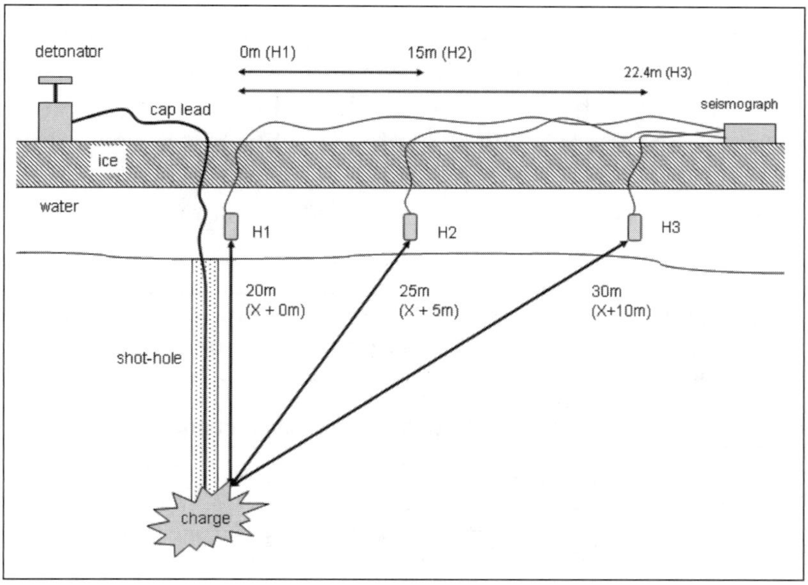

FIGURE 2. Hydrophone placement for instantaneous pressure change (IPC) monitoring using a 20 m burial depth example. Hydrophone 1 (HP1) is placed directly above the shot-hole to monitor burial depth of 20 m, HP2 is placed 15 m laterally from shot-hole to monitor burial depth + 5 m and HP3 is placed at 22.4 m laterally from shot-hole to monitor burial depth + 10 m.

It should be noted that pre- and post drilling, and pre- and post detonation water quality monitoring for turbidity, suspended solids and dissolved oxygen, and long-term water quality monitoring of waterbodies exposed to seismic activity was also conducted (Cott et al., 2003).

RESULTS AND DISCUSSION

Since it was not feasible to leave potentially shallow buried charges un-detonated due to safety concerns, DFO issued authorizations under section 32 of the *Fisheries Act* to allow for the completion of seismic programs experiencing high IPC values in the winter of 2000/2001. The authorizations stipulated monitoring 20% of the un-detonated charges remaining under waterbodies. Determination of charge burial depth was based on the amount of cap lead on the ice surface (i.e. burial depth of charge = total cap lead used – (ice depth + water depth + remaining cap lead on the ice surface)).

Of the 436 charges monitored and recorded for IPC in 2000/2001, 50.0% exceeded 100 kPa and 74.6% exceeded 50 kPa directly above the shot-hole (Table 1; Fig. 3). After several unsuccessful attempts were made to rectify the shortcomings of the Vibra-ram®, DFO recommended that it no longer be used when conducting seismic activities under waterbodies

TABLE 1. Summary of instantaneous pressure change data (IPC) data from monitoring explosive-based seismic exploration, by hydrophone (H) in the Mackenzie Delta area, NWT, 2000–2002. *Reasons for "no record" include an iced hydrophone, IPC value below the trigger setting on the monitoring equipment (< 5 kPa), or an undetonated charge.

IPC (kPa)	H-1 recorded and monitored % (n)	H-1 monitored % (n)	H-2 monitored % (n)	H-3 monitored % (n)
Winter 2000/ 2001	(n = 429)		(n total = 436)	
> 100	50.0 (214)	49.1 (214)	25.2 (110)	13.1 (57)
100 - 50	24.7 (106)	24.3 (106)	10.1 (44)	2.3 (10)
< 50	25.4 (109)	25.0 (109)	36.0 (157)	33.5 (146)
*no record	N/A	1.6 (7)	28.7 (125)	51.1 (223)
Winter 2001/2002	(n = 507)		(n total = 626)	
> 100	8.7 (44)	7.0 (44)	2.6 (16)	1.1 (7)
100 - 50	7.1 (36)	5.8 (36)	3.3 (21)	1.6 (10)
< 50	84.2 (427)	68.2 (427)	70.3 (440)	68.7 (430)
* no record	N/A	19.0 (119)	23.8 (149)	28.6 (179)

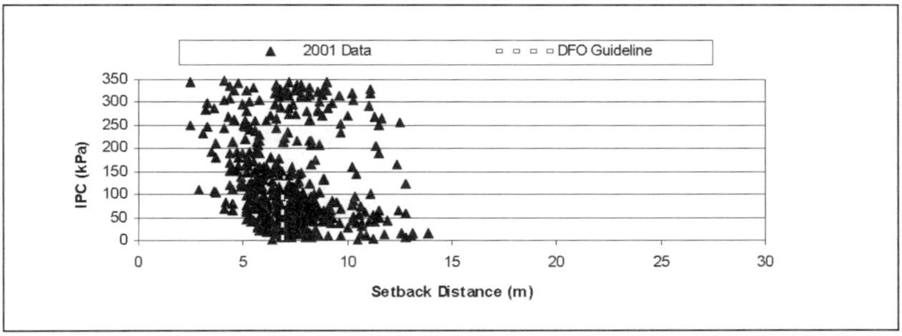

FIGURE 3. Instantaneous pressure change (IPC) values (kPa) versus burial depth (m) for 2.0 kg charges recorded directly above the shot-hole with hydrophone 1 for waterbodies not frozen to the bottom in the Mackenzie Delta area, NWT (January–April 2001).

not frozen to the bottom in the NWT. However, a great deal of variation in IPC values between program areas suggested that the reasons for excessive pressures were not limited to equipment problems or inadequate burial depths (Cott et al., 2003).

It should also be noted that a summer, open-water seismic test program using explosives on Parson's Lake, NWT, resulted in IPCs over 100 kPa and confirmed fish kills (Golder Associates Limited, 2000). This included charges buried to depths recommended in the Guidelines, indicating that the problem with high IPC values is not limited to winter conditions.

As a result of the previous year's experience, DFO outlined new requirements for conducting and monitoring seismic programs in the NWT. These requirements included the use of drilling technology proven to be effective in northern environments under winter conditions, proven mitigation to minimize the exposure of fish to critical IPC thresholds, contingency plans in place for unexpected events such as shallow buried charges, an assessment of alternative seismic exploration methods, IPC monitoring, long-term aquatic monitoring, and baseline studies on certain waterbodies likely to be encountered in the program areas (Cott et al., 2003). Also, it appeared that the Guidelines were being used too prescriptively, and that adjustments were required to account for site-specific conditions.

By the fall of 2001, industry had not presented the requested contingency plans and proven mitigation methods to DFO for review. As a result of these deficiencies, DFO issued a follow-up letter stipulating that charges should not exceed 2 kg in size, and that a conservative charge size/burial depth combination of 2 kg/15 m was being implemented in order to avoid exceeding the 100 kPa threshold. This charge size/setback combination is approximately double what is outlined in the Guidelines when using the rock substrate value (Wright and Hopky, 1998). As rock offers the most conservative value in the Guidelines, it was thought that that this precautionary approach would assist in providing an additional margin of safety to avoid or reduce impacts to fish.

Some proponents felt the 15 m minimum burial depth was too conservative, and that acceptable IPC values could be achieved with shallower depths. As such, a request was made to conduct field tests to demonstrate that shallower burial depths could achieve low IPC. DFO agreed to allow testing, and under *Fisheries Act* authorizations outlined IPC monitoring requirements for both tests and subsequent production seismic operations. The *DFO Western Arctic Area 2001/2002 Protocol for Monitoring the Use of Dynamite in Waterbodies not Frozen to the Bottom* was developed to assist proponents in developing monitoring programs, and to ensure consistency between seismic projects. The protocol was distributed to industry in February 2002. To take into account likely variations in subsurface geology, program-specific tests were used to determine appropriate combinations for individual program areas. Charge size/burial depth combinations for full production seismic exploration would be determined, pending the outcome of each program's test. The charge size/burial depth combinations using the rock substrate values outlined within the Guidelines would act as the minimum to begin testing.

Upon conclusion of the test programs, the burial depth for production seismic exploration on 10 out of the 11 programs had to be increased beyond double the Guideline levels, in some cases up to five times deeper (e.g. 24 m burial depth for 2 kg charge) in order to maintain IPC at levels that were consistently below 100 kPa. Also, through monitoring of production seismic operations, it was determined that the burial depths had to be increased for some programs as they encroached on different waterbodies in their program areas. During one program, the IPC on several charges surpassed 100 kPa in one lake after achieving safe IPC over much of their program area. In this case, the program was suspended by DFO until additional testing was conducted and a revised and acceptable charge size/burial depth combination was determined.

Despite the burial depths equal to, or in most cases, well beyond those outlined in the Guidelines, 8.7% of detonations monitored and recorded still produced IPCs in excess of 100 kPa, and 15.8% exceeded 50 kPa (n = 507) (Table 1; Fig. 4). The majority of charges used in production seismic were 2 kg, although other charge sizes were used experimentally and in limited production shooting.

Data derived from seismograph units that measured at frequencies of 16,000 s^{-1} and 65,000 s^{-1} were analysed together. This was acceptable as measurements taken at a frequency of 16,000 s^{-1} may be an under-representation of peak pressure. A higher sampling rate can more accurately delineate the peak pressure (R. Cyr, Explotech, pers. comm.; AquaScience, 2001a). Studies have indicated that seismograph units sampling at the

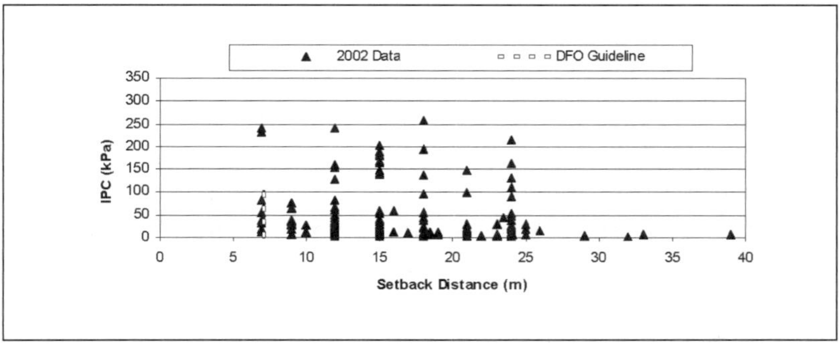

FIGURE 4. Instantaneous pressure change (IPC) values (kPa) versus burial depth (m) for 2.0 kg charges recorded directly above the shot-hole with hydrophone 1 for waterbodies not frozen to the bottom in the Mackenzie Delta area, NWT (January–April 2002).

higher frequencies record IPC values up to 25% higher than those sampling at lower frequencies, e.g. 65,000 s^{-1} versus 16,000 s^{-1} (Instantel, 2003; Explotech, unpublished data; T. Linton, Texas A and M University, unpublished data). If this is indicative of the differences between the two sampling rates it implies that the data sampled at 16,000 s^{-1} may be a gross underestimation of IPC peak pressure.

IPC was highly variable from waterbody to waterbody and within waterbodies even when the substrate appeared similar (similarity based on drillers' logs). This variability has been noted in other studies (T. Linton, Texas A and M University, unpublished data), and is possibly due to variations of the substrata, which further complicates the process of making accurate IPC predictions when applying the general substrate category values listed in the Guidelines (Cott et al., 2003).

No dead fish were observed in lakes by government regulators or industry during post-program site inspections. Due to the time between the seismic activity and ice melt (> 2 months), observed fish mortality was not anticipated as scavengers would have consumed any dead fish. Also, small-bodied fish are most susceptible to impacts from high IPC and would be very difficult to observe during aerial reconnaissance (Alaska Department of Fish and Game, 1991). Even during open water conditions, less than 50% of pressure-killed fish may actually be observed, due to injured fish swimming out of the immediate area prior to expiring or initially sinking to the bottom and surfacing at a later date (Teleki and Chamberlain, 1978).

Water quality monitoring from pre- and post drilling and pre- and post blasting showed that dissolved oxygen and turbidity exhibited a short-lived temporal and spatial increase (< 3 meters, < 10 minutes), or no change at all. The rapid settling of suspended sediments was observed with spot surveys using underwater video. Preliminary results from long-term monitoring suggest that seismic operations using explosive charges set several meters below the substrate of a waterbody do not adversely alter selected water quality parameters such as dissolved oxygen and turbidity. Similar findings were reported from water quality monitoring of the seismic operations during open-water conditions at Parsons Lake (Golder Associates Limited, 2000) and during underwater construction blasting in Lake Erie (Teleki and Chamberlain, 1978).

As a result of the data collected through the monitoring programs in 2000 to 2002, DFO released a position statement on February 14, 2003 that explosive-based seismic exploration in areas of waterbodies not frozen to the bottom in the NWT will not be authorized by DFO (Cott et al., 2003).

Lessons Learned

It became obvious that there were many potential reasons for the unpredictability of IPC values when conducting exploration under areas of waterbodies not frozen to the bottom. Differences were noted between program areas, within program areas, and between sequential shot holes under individual waterbodies. Possible reasons for high and/or unpredictable IPCs may include differences in substrate type, substrata, ice thickness, water depth, as well as equipment failure, human error, and subjectivity regarding substrate type on the part of drilling operators when working with different types of equipment. Most of these variables are difficult to predict or control in the field, and many are also a challenge to measure.

Conducting tests to determine acceptable charge size/burial depth combinations prior to initiating production seismic programs proved to be a prudent decision. The mean IPC level for production seismic exploration was approximately 43% lower than what was recorded during the testing phase as a result of using site-specific burial depth/charge size combinations determined for each production area. This was a successful example of adaptive management. If explosive-based seismic exploration programs under waterbodies are proposed elsewhere in Canada, it is imperative that project-specific testing to determine safe charge size/burial depth combinations be conducted to ensure that the potential for negative impacts to fish and fish habitat is minimized as much as possible.

A more conservative threshold of < 50 kPa should be utilized, rather than the 100 kPa in the Guidelines, which is essentially an LD_{50} where mortality of 50% of fish species exposed to that IPC would be expected (Wright and Hopky, 1998; D. Wright, DFO, pers. comm.). The < 50 kPa threshold should be used until such time that further research determines the correct threshold necessary to protect all life stages of fish. A lower threshold is also supported based on the results of IPC monitoring required by DFO biologists in the Pacific Region on a pile-driving operation after fish kills were observed. The IPC responsible for the fish kills were < 65 kPa (C. Salomi, DFO, pers. comm.; Cott et al., 2003). This would be closer to the guidelines used in Alaska; an area that has had much more experience with seismic exploration than the NWT.

Based on the results of short term and long term monitoring for dissolved oxygen (DO) and turbidity, the disturbance associated with deploying and detonating charges is thought to be negligible, and is not anticipated to have any adverse impact on fish and fish habitat in the Mackenzie Delta area.

Methods that were used to confirm burial depth of charges included keeping track of the number of drill stem sections and the total amount of cap-lead used. These methods are not precise and are subject to error such as slack cap-lead frozen under the ice giving an overestimation of burial depth. It is essential that a method to accurately measure and confirm burial depth be developed.

It is likely beneficial to allow time to elapse between charge burial and detonation. This is commonly referred to as "sleep time". If the detonation of a charge is delayed, the shot-hole has the opportunity to collapse on itself and re-consolidate; depending on substrate, the amount of energy escaping up the shot-hole may be reduced (R. Prudholm, Conoco, pers. comm.). The length of sleep time would be dependent on the type of substrate and its consolidation properties. The collapse of a shot-hole immediately upon drill stem removal was observed by using underwater video. Reducing the amount of energy transmitted up the shot-hole into the water column has the potential to lower the IPC value, likely minimizing impacts to fish. However, prolonged sleep time does not negate the need for adequate charge burial depth. Some extremely high IPCs (> 200 kPa) resulted from charges that "slept" for over a month.

Some of the most difficult variables to control are human error and subjectivity. In some cases, this relates directly to the equipment type being used by the individual operator. For instance, the operator's "feel" for what substrate is being encountered can be influenced by how powerful or heavy the particular drill is, and how the equipment is handled by different individuals. In one instance, an operator determined that consolidated material had been reached after going through 3 m of water. When this was checked by the person monitoring IPC, the water depth was actually 13 m. Mistakes happen. This is why it is imperative to develop and have on site mitigation that can minimize the effects of high IPC. Also, it is important that monitoring results are clear regarding "no-record" events. To accurately interpret data, the reason for "no-record" events must be established, distinguishing those that occur because the IPC is below recordable levels from those that are due to equipment or charge failure.

With all other variables being equal, deeper burial depths are an effective mitigation measure against high IPC. Unfortunately, all of the other variables are never equal, making it impossible to predict IPC strictly from a charge size/burial depth combination. The only completely effective mitigation measure currently available is to avoid conducting explosive-based seismic operations under fish bearing waters.

CONCLUSIONS

Monitoring of seismic programs in the ISR was conducted in a stepwise, exploratory fashion and yielded a significant amount of new information for management purposes. We now recognize that explosive-induced IPCs are very difficult to predict with any accuracy in waterbodies, and that the charge size/burial depth combinations suggested in the DFO *Guidelines for the Use of Explosives in or Near Canadian Fisheries Waters* may not provide the required level of protection to fish.

The following recommendations are based on lessons learned during seismic exploration monitoring in the winter seasons of 2000/2001 and 2001/2002.

General Recommendations and Areas for Further Study

1. Seismic exploration should not be conducted under waterbodies not frozen to the bottom in the NWT due to the unpredictability of IPC and absence of proven mitigation to suppress the negative effects of a detonated charge.
2. The existing Guidelines should be re-investigated using data obtained from modern monitoring equipment that is the current industry standard.
3. Guidelines should be used as intended, as "guidelines", and be adjusted to site-specific conditions accordingly, not applied as mitigation.
4. Additional research should be conducted to determine what the exact IPC threshold should be to offer protection for all affected life stages of fish species likely to be impacted.
5. Ice profiling on waterbodies should be used as a tool to determine the extent of bottom-fast ice. If the bottom-fast ice zones are determined, it may be possible to infill/undershoot the non-bottom fast portions without large gaps being left in the seismic data, while providing the necessary protection for over wintering fish and incubating eggs. Also, when the bottom fast ice zone of a lake is accurately delineated, it allows the use of vibrator-based energy sources.
6. The results of a recent fish deterrent study in the Mackenzie Delta area of the NWT demonstrated that the deterrents tested in winter conditions did not successfully move fish out of a designated area, and therefore should not be used as mitigation (Racca et al., 2004). Further investigation with other fish deterrent options is needed.

Recommendations for Project Requirements and Monitoring of Explosive-based Seismic Exploration Programs in Fish-bearing Waters

Steps need to be taken to minimize human error as much as possible.

1. Equipment must be proven to be appropriate for the operating environment. The use of new technology, as it becomes available, is encouraged, but it must be tested to ensure it can operate as intended in the environment selected (e.g. Canada's Western Arctic in winter conditions) before it is used for the full seismic program. This applies to all equipment from drill rigs to dissolved oxygen probes.
2. Proven mitigation to minimize the impact on fish from the effects of high IPC should be available on site in the event that an unforeseen event occurs, such as a shallow buried charge. Currently no such mitigation has been proven to be effective in the NWT.
3. For any explosive-based seismic program, a protocol must be developed that clearly indicates what is expected, how monitoring is to be conducted, what and how information is to be recorded, and when the results are to be submitted. The protocol should be designed well in advance of the proposed seismic exploration program, and be a joint effort between industry and regulators. Maintaining effective communication with the people conducting the field monitoring is essential.
4. Initial testing should be conducted to determine site-specific charge size/burial depth combinations.
5. Charge burial depth must be accurately measured and confirmed.
6. Substrate type must be carefully recorded. Drill logs completed at the end of the day do not provide an accurate assessment.

7. A maximum threshold of < 50kPa should be set for testing and production seismic operations.

8. Two hydrophones should be utilized for monitoring each shot hole as a contingency to minimize "no-record" events. However, if "no-record" events do occur, the reasons should be well documented (e.g. charge not detonating). See point 4 above. Additional hydrophones radiating from the shot hole could be used if monitoring IPC attenuation is desired.

9. Monitoring equipment should be capable of monitoring at the highest frequency available; currently 65,000 s^{-1} is standard. Also, the hydrophones must be compatible with the monitoring equipment being used.

10. A pre-determined number of production holes should be monitored to confirm the adequacy of site-specific charge size/burial depth combinations for the entire project area. Shot holes should be monitored sequentially rather than monitoring a certain number of charges after everything else has been detonated. They should also be distributed in a manner that ensures the shot holes monitored sufficiently represent the variation found within the program area.

11. When designing a program to monitor activities of industry, it is important that the requirements be practical and considers the technical and environmental conditions in which industry is bound to operate. Involving industry in the development of such monitoring programs is a good way to ensure that the requirements set forth are actually feasible. It is important to have all monitoring requirements clearly defined to ensure consistency and compliance, and to avoid the intent of the requirement from being misinterpreted. From the authors' experiences, fostering good and cooperative working relationships with industry, other government agencies, and co-management partners are essential ingredients in having successful monitoring programs that assist in the effective management and protection of NWT's fisheries resources.

ACKNOWLEDGEMENTS

The authors would like to thank the following people and companies for their assistance and cooperation in monitoring seismic activities in the NWT and helping to conserve fisheries resources: Ron Allen, Julie Dahl, Jeff Stein, Ray Pierce, Gerry Leering, Kathleen Simms, Kelly Cott, Dave Tyson, Corino Salomi, Dennis Wright, and Sam Stephenson from DFO; John Korec and Laura Van Ham from the National Energy Board; Conrad Baetz and Rob Walker from Indian and Northern Affairs Canada; Ed McLean and Bob Bell and the Fisheries Joint Management Committee; all of our Alaskan colleagues and participants of the 2002 and 2003 Oil and Gas Regulators meetings; Anadarko Canada Corporation; Darwin Monita and Aquatics; Chevron Canada; Conoco Canada (formerly Gulf Canada, presently ConocoPhilips); Delta Trace Ltd.; Devon Canada Corporation (formerly Anderson Exploration); Encana Corporation (formerly AEC West Ltd.); Golder Associates; IMG—Golder Corporation; Inuvialuit Environmental and Geotechnical Inc.; Petro-Canada; Salmo Consulting Inc.; Shell Canada; Veri-Illuq Geophysical Ltd.; and WesternGeco. Thanks also goes to Colin Lake of Kestrel Biological Consulting for his help with the figures, Dale McGowan for map preparation, Rob Cyr of Explotech for providing useful information, Andy Majewski, Dennis Wright and two anonymous reviewers for manuscript review and providing excellent comments, and Shelley Armsworthy (editor) for all of her assistance. We extend our apologies to anyone who was missed above.

REFERENCES

Alaska Department of Fish and Game. 1991. *Blasting Standards for the Protection of Fish. Rational for Blasting Standards (11 AAC 95) Developed to Prevent Explosive Injury to Fish.* 35 p.

Alaska Department of Natural Resources. 2003. "Mitigation Measures, North Slope Foothills Areawide Oil and Gas Lease Sale." http://www.dog.dnr.state.ak.us/oil/ (accessed May 1, 2003)

AquaScience. 2001a. *TUK 3D Aquatic Monitoring and Mitigation Program.* Prepared for Anderson Resrources Ltd. Calgary, Alberta. 34 pp.

AquaScience. 2001b. *Hanson Harbour Aquatic Monitoring Program.* Prepared for WesternGeco. Calgary, Alberta. 34 pp.

Canadian Association of Petroleum Producers (CAPP). 2003. *Background Paper on Explosive Use beneath Waterbodies.* Draft report.

Cott, P. A., B.W. Hanna, and J.A. Dahl. 2003. *Discussion on Seismic Exploration in the Northwest Territories 2000–2003.* Canadian Manuscript Report of Fisheries and Aquatic Science. 2648:vi+36 p.

Department of Fisheries and Oceans. 2002. *Canada's Oceans Strategy.* Department of Fisheries and Oceans, Oceans Directorate. Ottawa. vi + 30 p.

Falk M.R. and M.J. Lawrence. 1973. *Seismic Exploration: Its Nature and Effect on Fish.* Environment Canada, Fisheries and Marine Service. Technical Report No. CEN T-73-9:v+51 p.

Golder Associates Limited. 2002. *Parsons Lake Seismic Research Program—Environmental Impact Monitoring.* Calgary, Alberta. 53 pp. + appendices.

Government of Canada. 1985. *Fisheries Act, R.S.C.* c.f-14. Amended list April 23, 1993.

Instantel. 2001. *MiniMate PlusTM Operators Manual.* Kanata, Ontario. x + 74 p.

Instantel. 2003. "Sampling Rate and Underwater Overpressure. Just for the Record." Volume 3:Q4. http://www.instantel.com/Newsletters/news_current.htm (last accessed March 20, 2004)

Racca R.G., D.E. Hannay, R.B. Murray, W.B. Griffiths, and M. Muller. 2004. *Testing Fish Deterrents for Use Under-Ice in the Mackenzie Delta Area.* Environmental Studies Research Funds Secretariat. Calgary. 28p.

Teleki, G.C. and A.J. Chamberlain. 1978. "Acute effects of underwater construction blasting on fishes in Long Point Bay, Lake Erie." Journal of Fisheries Research Board of Canada. 35:1191-1198.

Traxler, S.L., B.R. Murphy, and T.L. Linton. 1992. "Subsediment Seismic Explosions Do Not Injure Caged Fishes in a Freshwater Reservoir." Journal of Freshwater Ecology. 8(1):73-75.

Wright, D.G. 1982. *A Discussion Paper on the Effects of Explosives on Fish and Marine Mammals in the Waters of the Northwest Territories.* Canadian Technical Report of Fisheries and Aquatic Sciences. 1052: v + 16 p.

Wright, D.G. and G.E Hopky. 1998. *Guidelines for the Use of Explosives in or Near Canadian Fisheries Waters.* Canadian Technical Report of Fisheries and Aquatic Sciences. 2107: iv + 34 p.

Monitoring Seismic Effects on Marine Mammals—Southeastern Beaufort Sea, 2001–2002

Miller, G.W. (gmiller@lgl.com), V.D. Moulton, R.A. Davis, M. Holst (LGL Limited, King City, ON, Canada), P. Millman (Devon Canada Corporation, Calgary, AB, Canada), A. MacGillivray, and D. Hannay (JASCO Research Ltd., Sidney, BC, Canada)

ABSTRACT: Devon Canada conducted seismic surveys offshore of the Mackenzie Delta in the Canadian Beaufort Sea during 2001–02. Concerns were expressed during community consultations that the seismic program might affect marine mammals and the subsistence hunt for beluga whales. A monitoring and mitigation program emphasising community involvement was designed to assess the effects of the seismic operations. Monitoring consisted of observations from the seismic ship, aerial surveys, and acoustic measurements. Mitigation included: (1) delay of operations until late July/early August to avoid peak period of beluga occurrence, (2) airgun shutdowns if whales were detected ≤1000 m from the seismic vessel, (3) ramp up of the airgun arrays, and (4) avoidance of beluga harvesting areas. During the two-year program, there were nine airgun shutdowns for whales (all bowheads) within the 1000 m "safety radius". The combined vessel-based and aerial monitoring documented very limited avoidance by some ringed seals, and a relatively small (minimum 600 m) zone of avoidance by summering bowhead whales. There was a considerably larger zone of avoidance for beluga whales, perhaps as large as 10–20 km. The localised avoidance of the seismic vessel by some marine mammals did not affect the subsistence hunt, as successful beluga hunts occurred in 2001–02.

INTRODUCTION

Devon Canada Corporation (Devon) conducted a two-year marine seismic program offshore of the Mackenzie Delta in the southeastern Beaufort Sea during 2001 and 2002 (Fig. 1). Marine seismic projects emit strong pulsed sounds into the water. Given the known auditory and behavioural sensitivity of many marine mammals to underwater sounds (Richardson et al., 1995), marine seismic projects have the potential to affect them. The waters offshore of the Mackenzie Delta in the Beaufort Sea are occupied by ringed (*Phoca hispida*) and bearded (*Erignathus barbatus*) seals, and beluga (*Delphinapterus leucas*) and bowhead (*Balaena mysticetus*) whales. Subsistence hunters hunt seals and whales in the area, so there is also the potential for acoustic disturbance of marine mammals to have effects on the accessibility of marine mammals to hunters. In order to address concerns about the potential effects of seismic exploration on marine mammals, LGL Limited (LGL) conducted a monitoring and mitigation program on behalf of Devon. The 2001 monitoring program (late July–early October) consisted of three primary components: acoustic measurements, vessel-based observations, and aerial surveys. Monitoring of sound levels was important because strong sounds might cause physical damage to a marine mammal's hearing or other organs, and can have a negative effect on its behaviour. The two main physical effects that a strong sound might have on marine mammal hearing are temporary and permanent auditory impairment. Strong sounds can cause temporary hearing impairment (temporary threshold shift, TTS). TTS is a mild form of hearing impairment that only lasts for minutes, hours or (at most) days (Thomson and Davis, 2001). Exposure to very strong sounds can cause permanent hearing loss (permanent threshold shift, PTS), at least in terrestrial mammals. This type

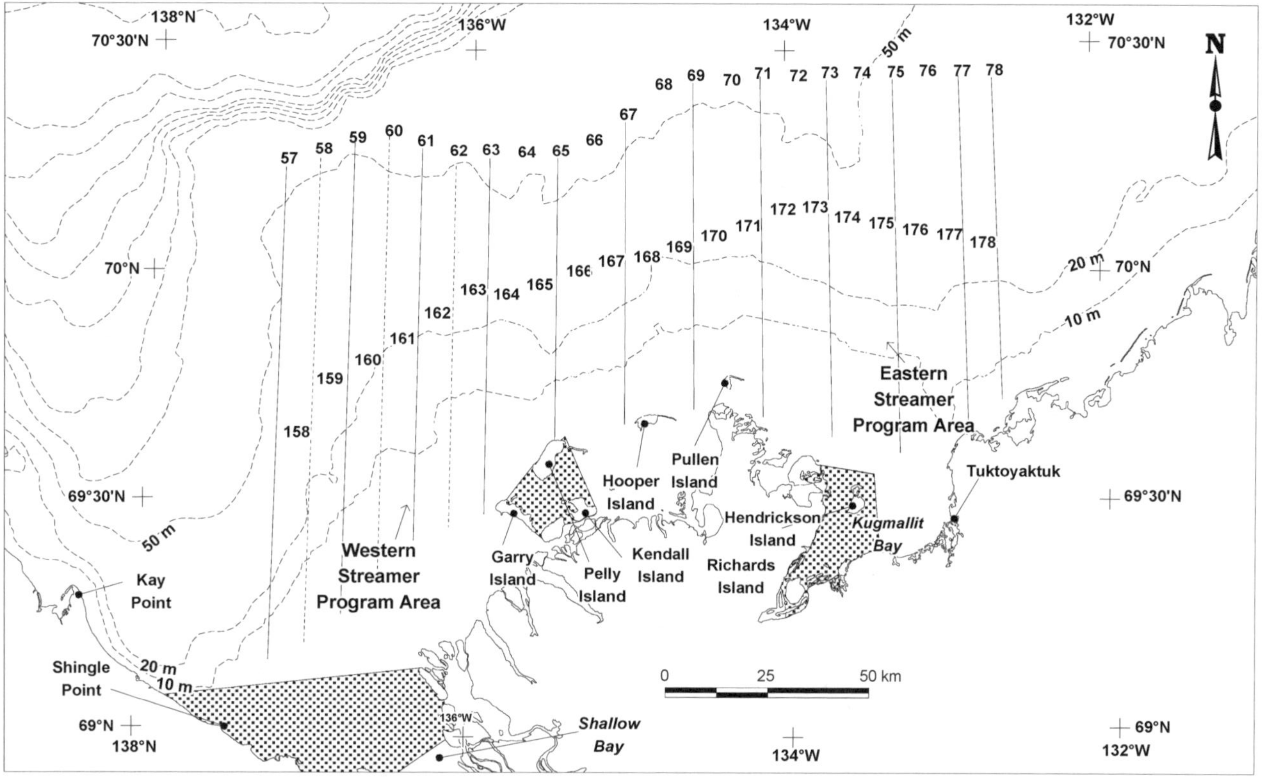

FIGURE 1. Study area showing program areas and aerial survey transects flown during 2001. Solid lines indicate standard "pre-seismic" aerial survey lines flown by LGL during 28 July–8 August 2001, prior to the start of full-scale seismic operations.

of damage results in a permanent decrease in the sensitivity of the hearing system at some or all frequencies (Thomson and Davis, 2001). There is no evidence that airgun pulses can cause PTS in marine mammals, and if they do, it would be limited animals exposed at very close distances.

Minimum levels of airgun sounds that cause TTS in whales have not been determined. However, in a beluga exposed to a single watergun pulse (similar to an airgun pulse), mild and brief TTS became evident when the received level was 226 dB re 1 μPa peak-to-peak or 186 dB re 1 $\mu Pa^2 \cdot s$ in energy terms (Finneran et al., 2002). That was probably equivalent to *c.* 205–210 dB re 1 μPa in the "rms over duration of pulse" units used elsewhere in this paper. TTS would be expected to occur at a somewhat lower received level in a beluga exposed to a series of pulses, but the levels of multiple pulses necessary to elicit TTS have not been measured.

The US National Marine Fisheries Service (NMFS) has concluded, provisionally, that whales should not be exposed to levels of impulse sound exceeding 180 dB re 1 μPa (rms) (NMFS, 2000). Limiting the exposure of whales to receive sounds exceeding that level is intended to protect them from hearing loss and other forms of physical injury. The NMFS 180 dB criterion was adopted as a mitigation standard for the 2001–2002 Monitoring and Mitigation Program, and was the basis for the establishment of the 1000 m "safety radius" around airgun operations and the airgun shutdown procedures for whales. Given the TTS information for belugas, it is unlikely that TTS, let alone PTS, would occur in belugas unless they were exposed to airgun sounds with received levels well above 180 dB re 1 μPa (rms). However, changes in marine mammal behaviour would be expected upon exposure to considerably lower received sound levels (Richardson et al., 1995).

The relatively brief (late July–late August) 2002 monitoring program consisted of vessel-based observations. The objectives of the monitoring program were to:

- address existing data gaps;
- mitigate the effects of the seismic program on marine mammals and subsistence hunting;

- answer questions about potential seismic effects on marine mammals in the study area; and
- address the concerns of communities and regulators.

The approach used in the monitoring program took into account the experience of studies conducted recently (1996–2001) during late summer and autumn in the Alaskan Beaufort Sea (e.g., Richardson [Ed.], 1999), as well as local concerns and sensitivity to the unique characteristics of the Canadian Beaufort Sea study area. Community consultation and peer review were key elements of the approach. A series of meetings with Community Corporations, Hunter's and Trapper's Committees (HTCs), and the public was held in Aklavik, Paulatuk, Holman, Sachs Harbour, Inuvik and Tuktoyaktuk in February 2001. After those meetings were held, there were additional meetings and consultations with the Inuvik HTC and the Fisheries Joint Management Committee (FJMC), as well as with representatives of Community Corporations and HTCs from the above-mentioned communities. The Monitoring and Mitigation Plan (LGL and JASCO, 2001) also received further review from the Environmental Impact Screening Committee, FJMC, Department of Fisheries and Oceans (DFO), and other key stakeholders. The Project Description and accompanying Monitoring and Mitigation Plan were approved by the Environmental Impact Screening Committee on 27 June 2001. A Geophysical Operations Authorization was granted by the National Energy Board (NEB) on 19 July 2001.

A number of mitigation measures were adopted to minimise the impact of the seismic exploration program on marine mammals and subsistence hunting. Key among these was the delay of the start of seismic operations until late July/early August. By then the beluga hunt had been completed and the period of peak occupation of the Mackenzie River estuary by belugas was over. Other mitigation measures included:

- airgun array shutdowns when whales were sighted within or closely approaching a 1000 m "safety radius" around the seismic vessel;

- gradual ramp up ("soft start") of the airgun arrays when resuming airgun operations after they had been shut down for a period of time; and
- avoidance of the Beluga 1A Management Zones (beluga harvesting areas) during vessel and aircraft operations.

More detailed descriptions of the overall monitoring program and its results can be found in Miller and Davis (Eds., 2002) and Miller and Moulton (2003). Details concerning the individual acoustic, vessel-based, and aerial monitoring programs conducted in 2001 are described, respectively, in MacGillivray et al. (2002), Moulton et al. (2002), and Holst et al. (2002).

MATERIALS AND METHODS

Seismic Operations

The seismic program consisted of streamer seismic surveys and associated bathymetric surveys conducted offshore of the Mackenzie Delta in the southeastern Beaufort Sea in two regions known as the Western Streamer and the Eastern Streamer program areas (Fig. 1). The offshore bathymetric surveys (2001 only) were conducted by Thales GeoSolutions (Canada) Ltd. from 15 July to 5 October. Those surveys employed two medium-sized (47 and 25 m) vessels equipped with side-scan sonars and single beam echosounders. The marine seismic operations were conducted by WesternGeco, from the 67 m Ice-Class 1, *Geco Snapper*. Seismic operations in 2001 were conducted on the Western Streamer program area from 6 August to 18 September, and on the Eastern Streamer program area from 19 September to 6 October. The 2002 seismic operations were conducted in the Eastern Streamer program area from 28 July to 21 August. In 2001, water depths within the areas where seismic operations occurred ranged from 6 to 31 m and averaged about 13 m. Seismic operations in 2002 were conducted in shallower waters averaging about 10 m (range 8–13 m).

The 3-D seismic program employed two identical 2250 in^3 sleevegun arrays. Each airgun array consisted of 24 airguns with volumes ranging from 40 to 150 in^3 that operated at an air pressure of 2000 psi. The two 2250 in^3 airgun arrays fired alternately at shot points spaced every 18.7 m (or ~8 seconds) along the seismic lines. The airgun arrays were operated at a depth of 5 m below the water surface. Three 4-km long streamers with hydrophones were towed ~150 m behind the vessel to record the seismic data. The streamers were separated laterally by a distance of 160 m, with a total resultant distance of 320 m between the two outside streamers.

Acoustic Monitoring

JASCO Research Ltd. (JASCO) prepared a pre-season report that outlined estimates of the anticipated sound levels at several ranges from the airgun array planned for use in the 2001 seismic program (Hannay and Austin, 2001). These estimates were used by LGL to derive the 1000 m "safety radius" that was used throughout the monitoring program. This radius corresponds to the approximate distance at which the received levels of airgun sounds are not expected to exceed 180 dB re 1 μPa. During the 2001 summer seismic program, JASCO carried out field studies to obtain calibrated recordings of the underwater sounds generated by the acoustic sources used by the seismic and bathymetric operations. JASCO prepared a report on the results of the acoustic measurements of the bathymetric sources (Hannay and MacGillivray, 2001a), and the preliminary results of the measurements of the airgun arrays were compared to the pre-season estimates of the airgun array sound levels (Hannay and MacGillivray, 2001b). Based on the latter report, LGL confirmed the suitability of the 1000 m safety radius, and that radius remained operative throughout the 2001–02 field seasons.

Acoustic monitoring was conducted from 19 August to 9 September 2001 while seismic operations were being conducted on the Western Streamer program area. Acoustic recordings were made from the 12-m vessel, *Suvukti*, using calibrated hydrophones to measure the characteristics and levels of the waterborne sound pulses generated by the seismic vessel's airgun arrays and by the sonars used by the bathymetric vessels. The hydrophones and recording systems were suitable for quantitative measurements of sounds at frequencies ranging from 10 Hz to 125 kHz. For the measurements of airgun sounds, the bandwidth was 10 Hz to 24 kHz.

Vessel-based Monitoring

Three marine mammal observers (LGL biologists and Inuvialuit monitors) were stationed on the seismic vessel *Geco Snapper* from 31 July–7 October 2001 and 26 July–24 August 2002. The vessel-based observers (and aerial observers; see below) participated in a one-day training session in Tuktoyaktuk to become familiar with the operational procedures and conditions, and reporting protocols. The bridge of the *Geco Snapper* served as the platform from which the vessel-based monitors watched for marine mammals. These observers monitored the occurrence and behaviour of marine mammals near the seismic vessel during all daytime and nearly all night-time periods when the airgun array was operating and for many periods when it was not operating. There was 24 h of daylight in late July and early August, but increasing amounts of darkness starting soon thereafter. By the end of the 2001 study period (early October), there were about 11.5 h of darkness per day. The observers' duties included watching for and identifying marine mammals; recording their numbers, distances, and reactions to the seismic operations; initiating mitigation measures when appropriate; and reporting the results. There was no requirement for suspension of seismic operations during times with poor visibility or at night, and seismic work continued during such periods.

Whenever airgun operations were underway, or when they were expected to start within 30 min, an observer watched continuously for marine mammals. To compare marine mammal sighting rates during periods with and without airgun operations, observers were normally on duty at times when the seismic vessel was underway but the airgun arrays were not operating. The observer scanned around the vessel using unaided vision and 7 × 50 Fujinon FMTRC-SX binoculars that were equipped with reticles to measure depression angle relative to the horizon. These reticle markings, together with the known height of the observers above sea level (eye level was about 10.28 m above sea level), enabled the observers to determine distances to the marine mammals that were sighted. While on watch, the observers kept systematic written records of the seismic activity and environmental conditions. Additional data were recorded when marine mammals were observed. For all records, the date, time, and observer on duty were recorded. Latitude, longitude, and information about seismic activity were available from the computer monitors located on the bridge. Operational activities that were recorded included the survey line number being shot and the type of seismic activity—ramping up, line shooting, seismic testing, shutdowns, and other. Environmental conditions that were recorded included ice cover within ~1 km of the seismic vessel (percent cover, primary type), wind force (Beaufort Scale), visibility (km), obstructions to visibility (e.g. snow, fog, darkness), and glare (severity, location relative to a clock face where 12 o'clock represented the heading of the vessel). Standardised codes were used for most of these records, but written descriptive comments were usually added as well.

Aerial Monitoring

In 2001, a chartered deHavilland Twin Otter was used to conduct aerial monitoring of a much larger region surrounding the area of seismic exploration. Aerial surveys were conducted to monitor the distribution, movements, and general activities of belugas, bowheads, and other marine mammals in and adjacent to the seismic exploration area. These surveys were conducted to provide the data needed to document potential displacement of marine mammals during seismic operations, and to evaluate any potential effects on subsistence hunting. The surveys were designed to avoid over-flying the Beluga 1a Management Zones that are known to be important to belugas and subsistence hunters. Aerial surveys began on 28 July, a few days before the *Geco Snapper* arrived in the area and about two weeks before the start of seismic production. The aerial surveys continued until the end of September.

The study area for the aerial surveys extended from about 20 km west of the western edge of the Western Streamer program area (137°12′W) to 20 km east of the eastern edge of the

Eastern Streamer program area (132°39′W), an east-west extent of ~180 km (Fig. 1). The transects that were flown extended from shallow nearshore waters to ~50 km north of the seismic exploration area. Weather permitting, aerial surveys were flown two or three times a week. Aerial surveys flown before 8 August were classified as "pre-seismic" control surveys. A complete or partial survey of the planned transects was conducted on 22 days from 28 July to 30 September. Surveys were attempted on an additional 16 dates but poor weather conditions prevented the completion of any transects. On some days, surveys were not attempted because poor weather conditions were reported offshore.

The pre-seismic control surveys (total of 2635 km of transects flown during good sighting conditions) covered alternate transects across the entire study area (Fig. 1). The surveys conducted after seismic operations began (total of 9831 km of transects flown during good sighting conditions) included coverage of consecutive transects centred on the then-current seismic operations area. Survey coverage from 8 August–17 September included transects centred on the Western Streamer program area. Survey coverage after 18 September was centred on the Eastern Streamer program area.

At any given time, aerial surveys were conducted by two LGL biologists plus an Inuvialuit observer. Two of the observers surveyed full-time for marine mammals. A third observer also functioned as a data recorder, entering observations of environmental conditions and marine mammals into the on-board computer.

The aerial surveys used line-transect methods and were flown in a Twin Otter (a twin-engine high-winged aircraft) specially modified for survey work. The special features included an external GPS antenna, bubble windows at the rear observer positions, and an AC power supply (an inverter provided 110 V 60-cycle power). Two pilots were present for takeoffs, landings, and ferry flights. Before arriving at the start of the flight's first transect, the co-pilot moved to the rear of the aircraft to allow a primary observer to use the co-pilot's seat. Surveys were conducted at an altitude of 305 m (1000 ft) above sea level and a nominal groundspeed of 222 km/h (120 knots).

Community Liaison and Reporting

Inuvialuit participation in the monitoring program provided the community with a local link to the program. Devon also kept local hunters updated by broadcasting operational information over the Trapper Radio. Twice-daily broadcasts were made by an Inuvialuit who outlined Devon's planned marine and aerial activities.

Bi-weekly reporting of the monitoring program was made to the FJMC, DFO, and NEB throughout the seismic program. The final report of the monitoring program was shared with the regulators along with the Inuvialuit Game Council and each community HTC.

RESULTS AND DISCUSSION

Sound Levels Received by Marine Mammals

The physical characteristics of the local environment had a strong effect on sound transmission. The study area was located near the Mackenzie River estuary where cold saline water mixes with the warmer fresh water from the river. As a result, the physical parameters that affect underwater sound transmission such as salinity, water temperature, depth, and bottom type were highly variable, and sound transmission varied greatly across the area.

The maximum ranges from the acoustic sources (side-scan sonars, echosounders and airgun arrays) at which received levels would be likely to diminish to 180 dB re 1 μPa (rms) were estimated based on the acoustic data obtained during the 2001 field work. The measurements showed that the effective source level of the 24 kHz echosounder was less than 160 dB re 1 μPa (rms) in the horizontal direction. Thus, the received levels from this acoustic source were less than 180 dB re 1 μPa (rms) at all horizontal distances from the source (Fig. 2). Stronger sounds would occur in the echosounder beam below the vessel.

The 114 kHz side-scan sonar transducers were also directional with an approximate beamwidth of 2°, in this case directed to the side. Only pulses received directly broadside of the instrument were used to estimate the 180 dB radius. For the side-scan sonar the source level was estimated to be 226 dB re 1 μPa (rms). The maximum range from the side-scan sonar to rms level 180 dB re 1 μPa was estimated to be 99 m (Fig. 2).

The *Geco Snapper*'s airgun arrays were observed to be directional, with the strongest horizontal signals directed immediately broadside to the arrays (Fig. 2). On four different dates, in water depths ranging from 8.2 to 16.5 m, the estimated maximum range from the airgun array to the 180 dB re 1 μPa (rms) level (as measured at bow and stern aspects) ranged from 372 to 587 m. For measurements broadside to the array, the estimated maximum distances ranged from 665 to 1370 m. The strongest received levels occurred within about a 15° broadside zone on either side of the airgun array, and were there-

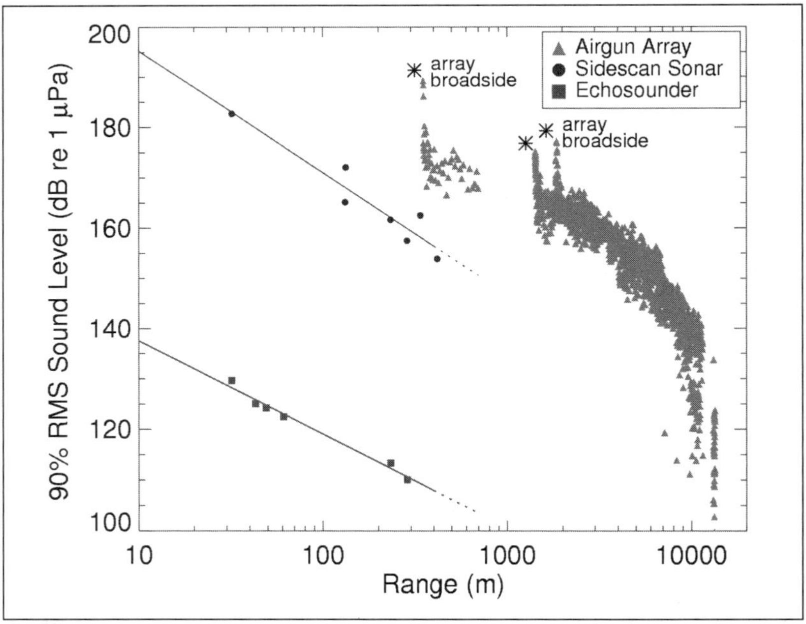

FIGURE 2. Received levels vs. range for the acoustic sources measured during 19 August to 9 September 2001.

fore typical of only about 8% (30° of 360°) of the possible range of directions about the airgun array.

Beluga Whales

The results from the 2001 aerial monitoring indicated that belugas were most abundant in the study area prior to the start of the seismic operations. Sightings of belugas during the pre-seismic period (first two weeks of the study) were widely distributed across the study area (Fig. 3A). In accord with the usual seasonal pattern (DFO 2000; Richard et al. 2001), beluga numbers declined sharply during the first few weeks of the study as belugas moved offshore, abandoning their summer concentration areas in the Mackenzie River estuary. During the seismic operations period when the aerial surveys were conducted over the Western Streamer program area, belugas were well distributed across the western portion of the study area (Fig. 3B). Sighting rates declined to very low levels by mid-September. By the time the seismic operations moved to the Eastern Streamer program area in late September, there were few belugas left in that portion of the study area.

Very few belugas were sighted during the 1561 h of daylight observations conducted from the *Geco Snapper*. In 2001 (1053 h of daylight observations) there were only seven vessel-based sightings of belugas, and in 2002 (508 h of daylight observations) there was only one vessel-based sighting. Five of these sightings were during "no gun" periods and the remaining three sightings were during line seismic (two sightings) or testing periods (one sighting). During line seismic operations, both sightings of belugas were at a distance of 1540 m. The single sighting during testing was at a distance of 2508 m.

In contrast to the vessel-based results, there were numerous sightings of belugas during the 2001 aerial surveys (Fig. 3). During the period when observers were present on the *Geco Snapper* (2 August–6 October), there were 123 aerial sightings of belugas. The closest aerial "transect" beluga sighting during seismic periods was at a distance of 7.8 km from the seismic vessel. Within 20 km of the airgun operations, there were

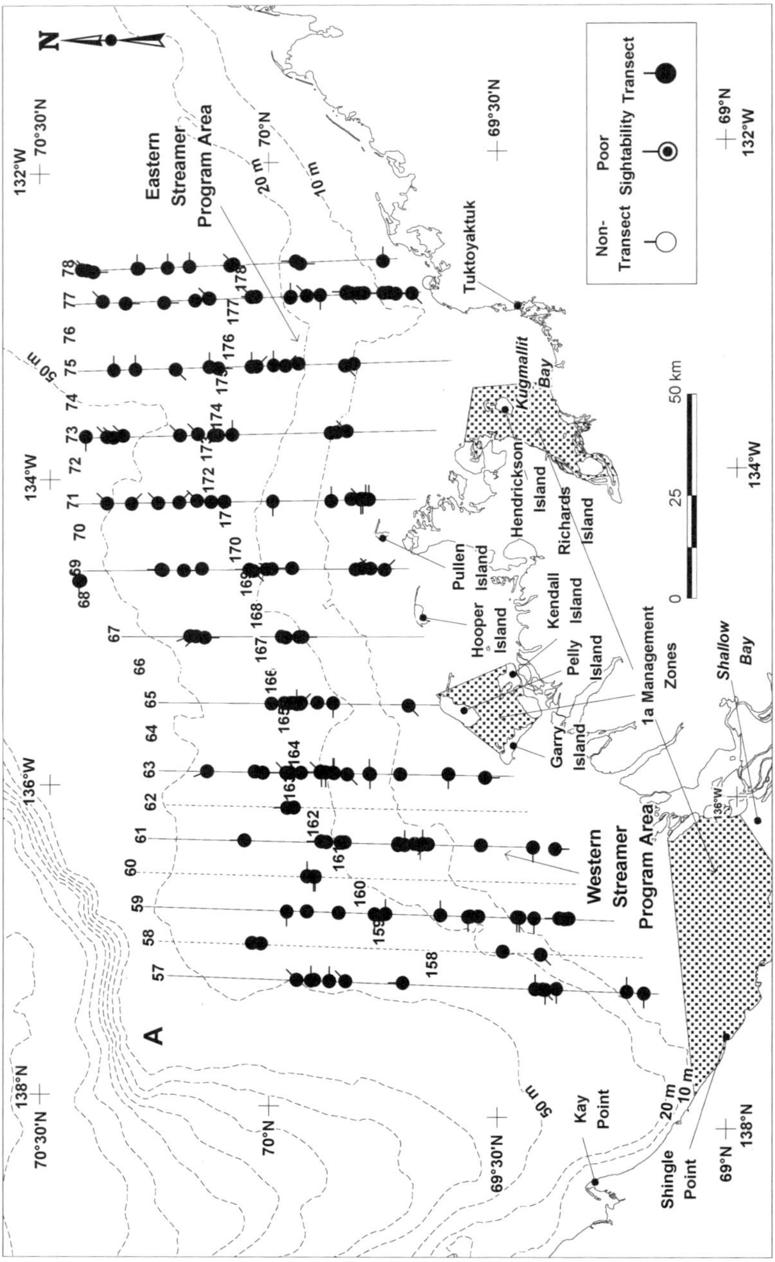
N
Eastern Streamer Program Area
Western Streamer Program Area
Tuktoyaktuk
Kugmallit Bay
Hendrickson Island
Richards Island
Pullen Island
Hooper Island
Kendall Island
Pelly Island
Garry Island
1a Management Zones
Shallow Bay
Kay Point
Shingle Point
Non-Transect
Poor Sightability
Transect
0
25
50 km
132°W 70°30'N
134°W
136°W
138°W 70°30'N
70°N
69°30'N
69°N 132°W
69°N 138°N
20 m
10 m
50 m
A

A

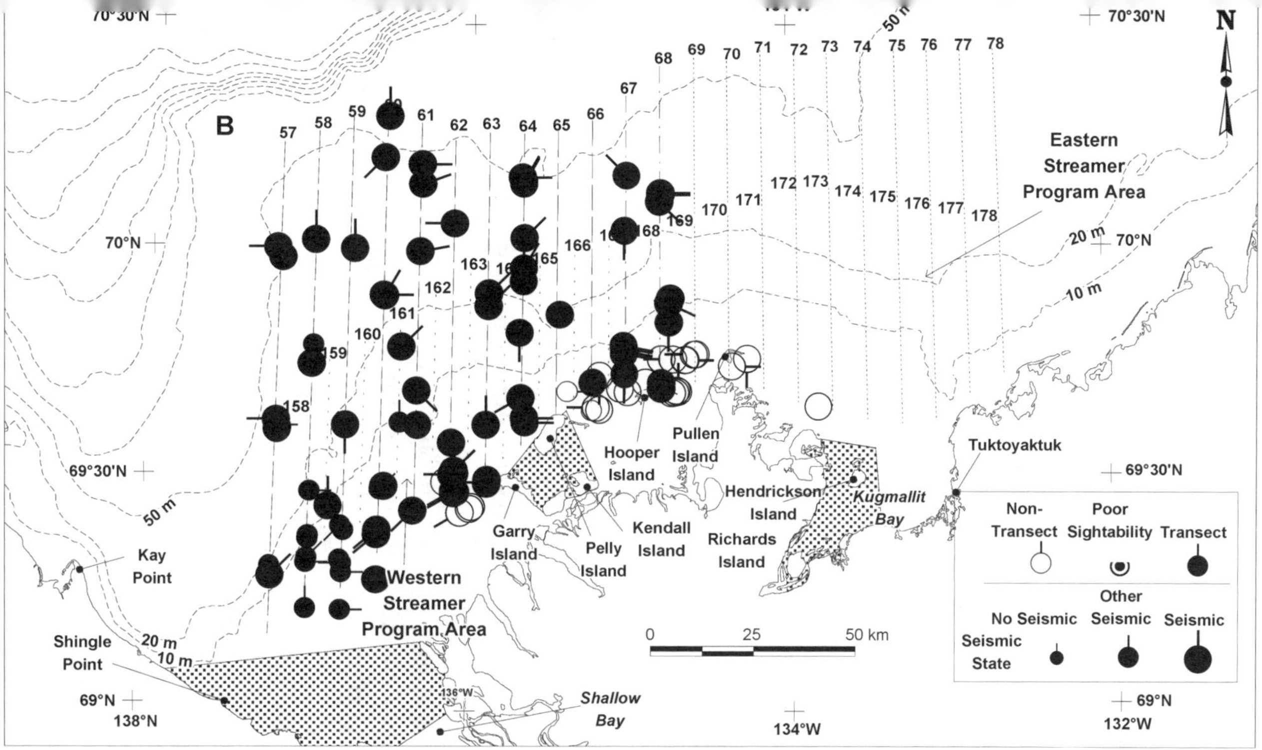

FIGURE 3. Beluga sightings during aerial surveys in the southeastern Canadian Beaufort Sea study area during **(A)** "pre-seismic" (28 Jul.–8 Aug.), and **(B)** "seismic operations" (9 Aug.–29 Sept.) periods, based on 2001 LGL/Devon aerial surveys. Large symbols denote seismic sightings, small symbols denote "no-seismic" sightings, and medium symbols denote sightings excluded from seismic and no-seismic categories (e.g. seismic testing, ramp up). Sightings during transect surveys are denoted by solid circles, sightings during non-transect surveys are shown by open circles, and sightings during poor sighting conditions are shown as "Poor Sightabil-

seven such aerial sightings at distances of 7.8 to 18.3 km (mean distance 12.8 km).

We examined the expected vs. observed numbers of beluga sightings in eight concentric 10-km bands (from 0–10 to 70–80 km) around the most recent shotpoint during seismic periods. Sighting rates were significantly different among the eight distance bands considered ($c^2 = 25.12$, $df = 7$, $P = 0.001$). Sighting rates at distances of 10–20 km from the airgun array were significantly lower than those in areas 20–30 km from the airgun array, where sighting rates were unexpectedly high. These findings suggest that some belugas may have been avoiding the area of seismic operations at distances up to 20 km. The low number of beluga sightings recorded by the vessel-based observers is consistent with the apparent avoidance of the seismic operations area indicated by the aerial data. The combined results of the vessel and aerial monitoring suggest that belugas may have avoided the area of seismic operations by 10–20 km. These long distance effects can only be determined through techniques such as aerial surveys or satellite tagging.

The belugas sighted closest to the vessel (1540–2500 m) would have been exposed to airgun sound levels of about 165 dB re 1 μPa (rms) (Fig. 2). The few aerial sightings during seismic periods at 7.8–18.3 km would have been exposed to underwater sound levels ranging from ~150 to 130 dB re 1 μPa (rms), respectively, based on measurements made during the acoustic program (Fig. 2).

The sound signature of the *Geco Snapper* underway at 4–5 knots (7.4–9.3 km/h), exclusive of its airgun sounds, was determined. The broadband source level (~180 dB re 1 μPa · m rms) was much lower than the source level of the airgun array. The observed avoidance of seismic operations by belugas was much greater than would be expected if the whales were only reacting to vessel noise, based on other observations in the Canadian Beaufort Sea (Richardson et al., 1985; 1995).

Bowhead Whales

During 2001, bowheads occurred in the offshore Mackenzie Delta study area in substantial numbers from late August until mid September (Fig. 4). This pattern was evident from both the vessel-based and aerial sightings. The four-week occupation of the study area by bowheads is typical of patterns observed in other years (Richardson et al., 1987; Moore and Reeves, 1993), and suggests that there were good feeding opportunities for bowheads in these waters during that period. During the seismic operations period when the aerial surveys were flown over the Western Streamer program area, bowheads were well-distributed across the western portion of the study area, but sighting rates declined to low levels by mid September. By the time the seismic operations moved to the Eastern Streamer program area in late September, there were few bowheads in that portion of the study area, presumably because they had already begun their westward migration into Alaskan waters (Fig. 4). The pattern of 2001 sightings from the *Geco Snapper* was similar (Fig. 5). Bowheads were sighted throughout the deeper (> 10 m) portions of the Western Streamer program area, but there was only one sighting during operations in the Eastern Streamer program area.

In 2001, a total of 259 bowheads in 166 groups were observed from the *Geco Snapper* during 1053 h of daylight observations. Sighting rates during daylight hours were higher when no guns were operating than during periods with airgun operations. Considering the 4-week period (23 August–19 September) when bowheads were most abundant in the study area, the bowhead sighting rate during "no guns" periods (0.85 bowheads/h) was about twice as high as that recorded during line seismic periods (0.40 bowheads/h). Average sighting distances from the vessel were significantly lower during no guns vs. line seismic periods (mean radial distances 1368 vs. 1957 m; Mann-Whitney $U = 1584.0$, $n = 86, 63$, $P < 0.001$). The locations of bowhead sightings relative to the seismic vessel (considering relative bearing and distance) were significantly different during no guns vs. line seismic periods, based on Hotelling's two-sample test (Batschelet, 1981; $T^2 = 26.63$, $n =$

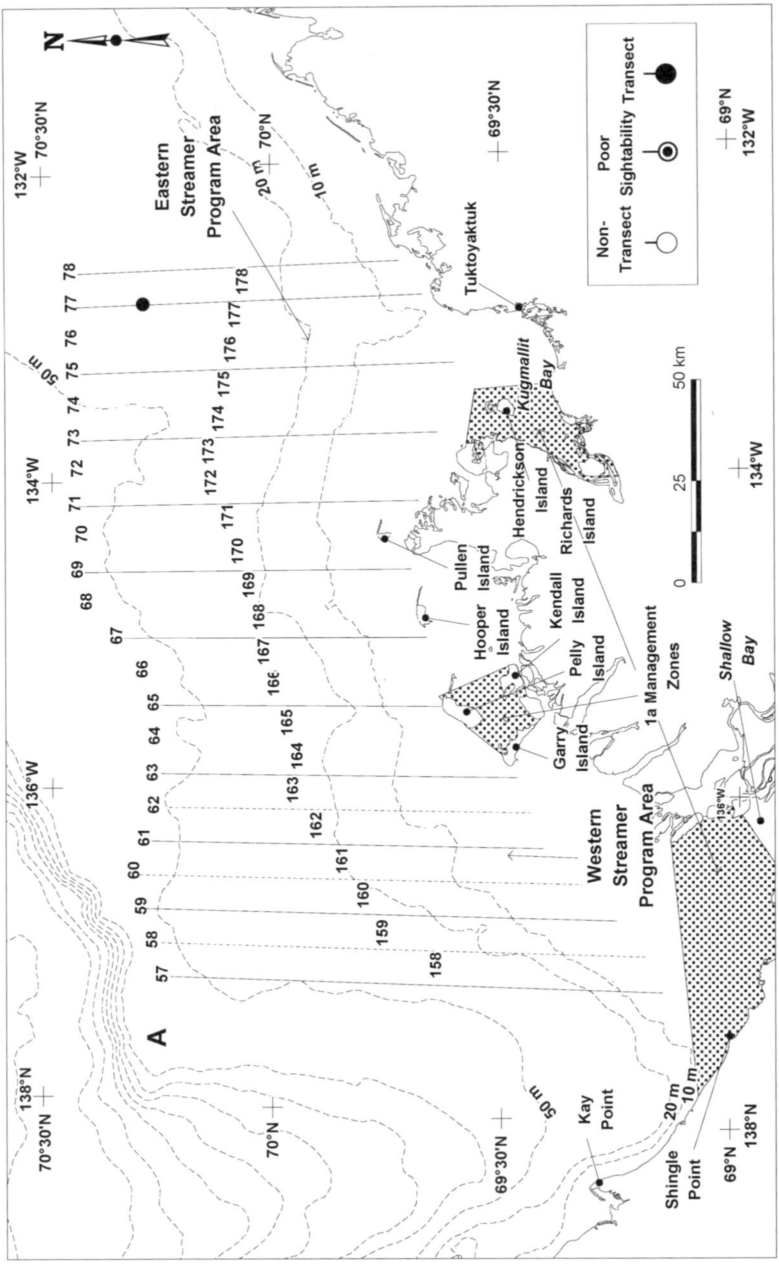
N
132°W
70°30'N
Eastern
Streamer
Program Area
20 m
70°N
10 m
69°30'N
50 m
Tuktoyaktuk
Kugmallit
Bay
Hendrickson
Island
Richards
Island
Pullen
Island
Hooper
Island
Kendall
Island
Pelly
Island
Garry
Island
1a Management
Zones
Shallow
Bay
Western
Streamer
Program Area
134°W
136°W
138°W
69°N
Kay
Point
Shingle
Point
A
0
25
50 km
Non-
Transect
Poor
Sightability
Transect
57
58
59
60
61
62
63
64
65
66
67
68
69
70
71
72
73
74
75
76
77
78
158
159
160
161
162
163
164
165
166
167
168
169
170
171
172
173
174
175
176
177
178

A

B

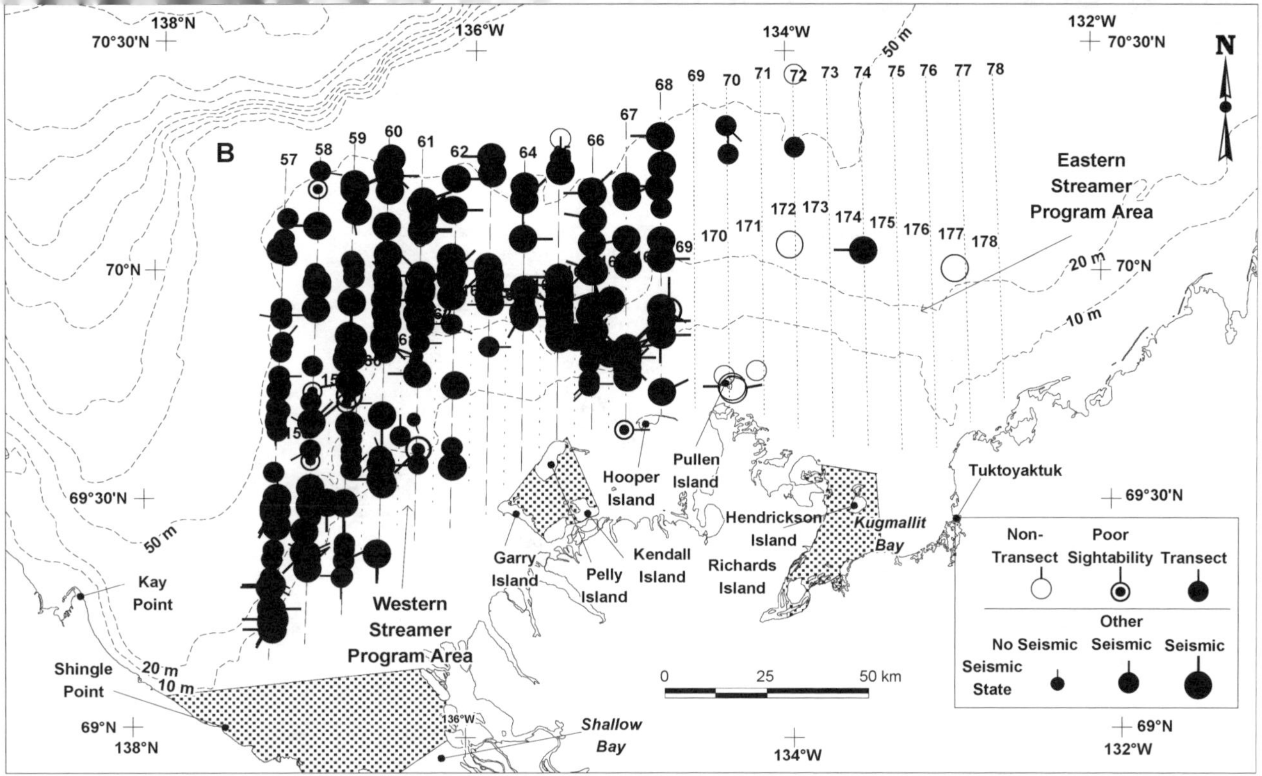

FIGURE 4. Bowhead whale sightings during aerial surveys in the southeastern Canadian Beaufort Sea study area during **(A)** "pre-seismic" (28 Jul.–8 Aug.), and **(B)** "seismic operations" (9 Aug.–29 Sept.) periods, based on 2001 LGL/Devon aerial surveys.

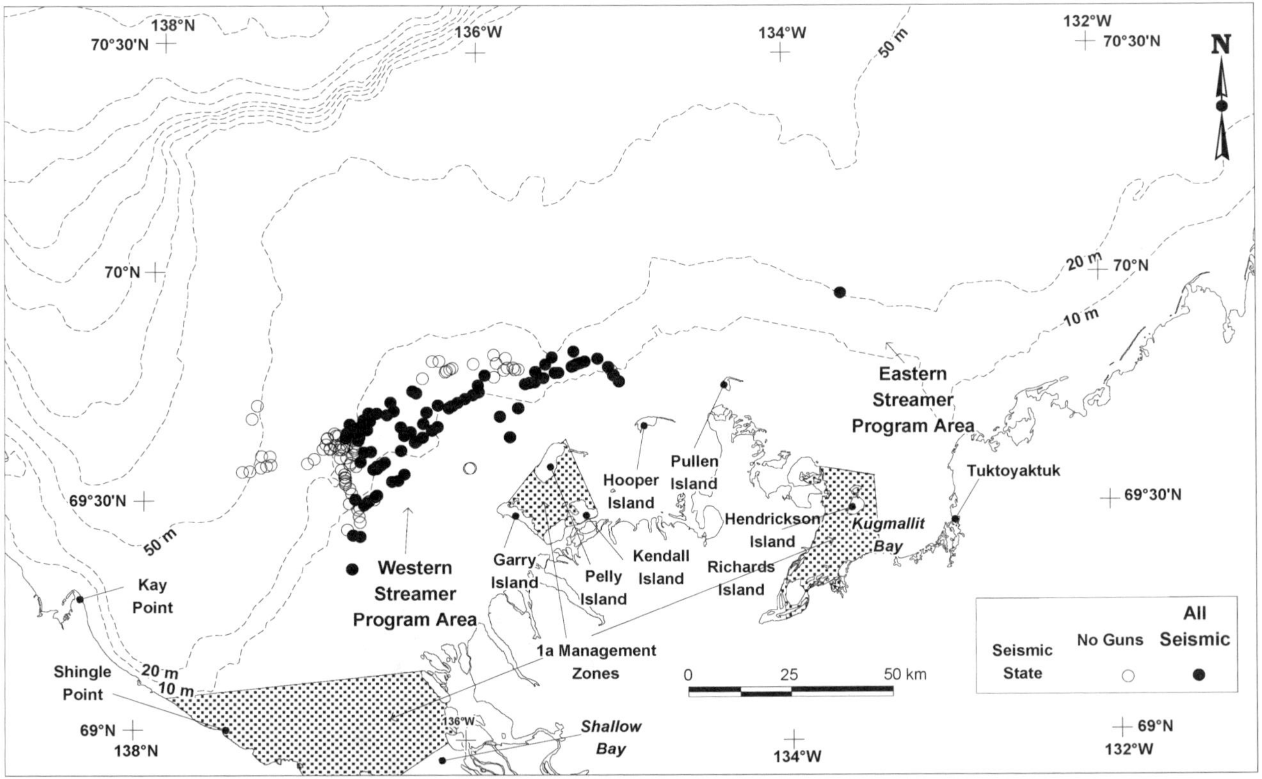

FIGURE 5. Bowhead whale sightings during LGL/Devon vessel-based monitoring in the southeastern Canadian Beaufort Sea study area, 2 Aug.–6 Oct., 2001. Solid symbols denote sightings during airgun operations, and open symbols denote sightings during periods without airgun operations.

86, 63, $P < 0.005$), indicating that sightings tended to be clustered closer to the vessel at times without airgun operations. The observed difference in sighting rates and the significant differences in sighting distances and location relative to the vessel suggest that bowheads did tend to avoid close approach to the area of seismic operations. However, the still substantial number of sightings by vessel-based observers during seismic periods and the relatively small (600 m) but significant difference in sighting distances with vs. without airgun operations, suggests that the avoidance was localised. At a minimum, the distance by which bowheads avoided seismic operations was about 600 m greater than the average distance by which they avoided general vessel operations. However, the lower sighting rates recorded during seismic operations suggest that some bowheads avoided the seismic operations by larger distances and thereby stayed out of visual range of the marine mammal observers on the *Geco Snapper*. Thus, the 600 m figure is no doubt an underestimate.

The 2001 aerial surveys did not document bowhead avoidance of the seismic operations area. The area of avoidance around the seismic operations area was apparently too small to be evident from the broad-scale aerial surveys that were flown, especially considering the small amount of surveying done when no seismic surveying was underway. During airgun operations, the closest bowhead sighting during systematic aerial surveys was at a distance of 5.3 km. Within 20 km of the airgun operations there were 30 such aerial sightings at distances of 5.3 to 19.9 km.

The bowheads that surfaced closest to the vessel (323–614 m) would, when underwater, have been exposed to sound levels of about 180 dB re 1 μPa (rms) before the airguns were shutdown (Fig. 2). In 2001, there were seven shutdowns of the airgun array in response to sightings of bowheads within 1 km of the seismic vessel. The average distance to bowheads seen from the vessel during line seismic operations was 1957 m. The underwater sound level at that distance was about 170 dB re 1 μPa (rms) (Fig. 2). The many bowheads seen (from the aircraft) at distances ranging from 5.3 to 19.9 km would, when underwater, have been exposed to sound levels ranging from ~150 to 130 dB re 1 μPa (rms), respectively (Fig. 2).

During 2002, marine mammal observers aboard the *Geco Snapper* conducted 508 h of daylight observations. There were 76 bowheads seen from the *Geco Snapper* during those observations (Fig. 6). Bowhead sighting rates during 2002 were similar during no guns vs. line seismic periods, in contrast with 2001. The distances at which bowheads were initially sighted were, on average, significantly smaller during non-seismic than line seismic periods (mean distances 1215 vs. 1931 m; Mann-Whitney $U = 81.5$, $n = 51, 6$, $P = 0.030$). The locations of bowhead sightings relative to the seismic vessel were not significantly different during non-seismic vs. line seismic periods (Hotelling's $T^2 = 0.54$, $n = 50, 6$, $P > 0.25$). The 2002 data indicate that there was some displacement of bowheads from the immediate area where seismic operations occurred. During the 2002 seismic program, the airguns were shut down twice because of bowhead whales within or about to enter the 1000 m safety radius. The first shutdown was during seismic testing after the completion of a production line. The second shutdown occurred along a production line.

The combined 2001–2002 vessel-based data (335 bowheads in 232 groups seen during 1561 h of daylight observations) also indicate that there was displacement of bowheads from the immediate seismic operations area. This displacement was evident in higher sighting rates and significantly shorter sighting distances (mean distances 1311 vs. 1954 m; Mann-Whitney $U = 2609$, $n = 137, 69$, $P < 0.001$), and significant differences in bowhead location about the vessel (Hotelling's $T^2 = 35.1$, $n = 136, 69$, $P < 0.001$), during non-seismic vs. line seismic periods. However, apart from displacement, the behaviour of whales observed during 2001-2002 was similar during non-seismic vs. seismic periods.

Seals

Seals were frequently observed during both aerial and vessel-based monitoring during 2001. There were 153 seals recorded during the aerial surveys (Fig. 7) and 196 seals seen from the

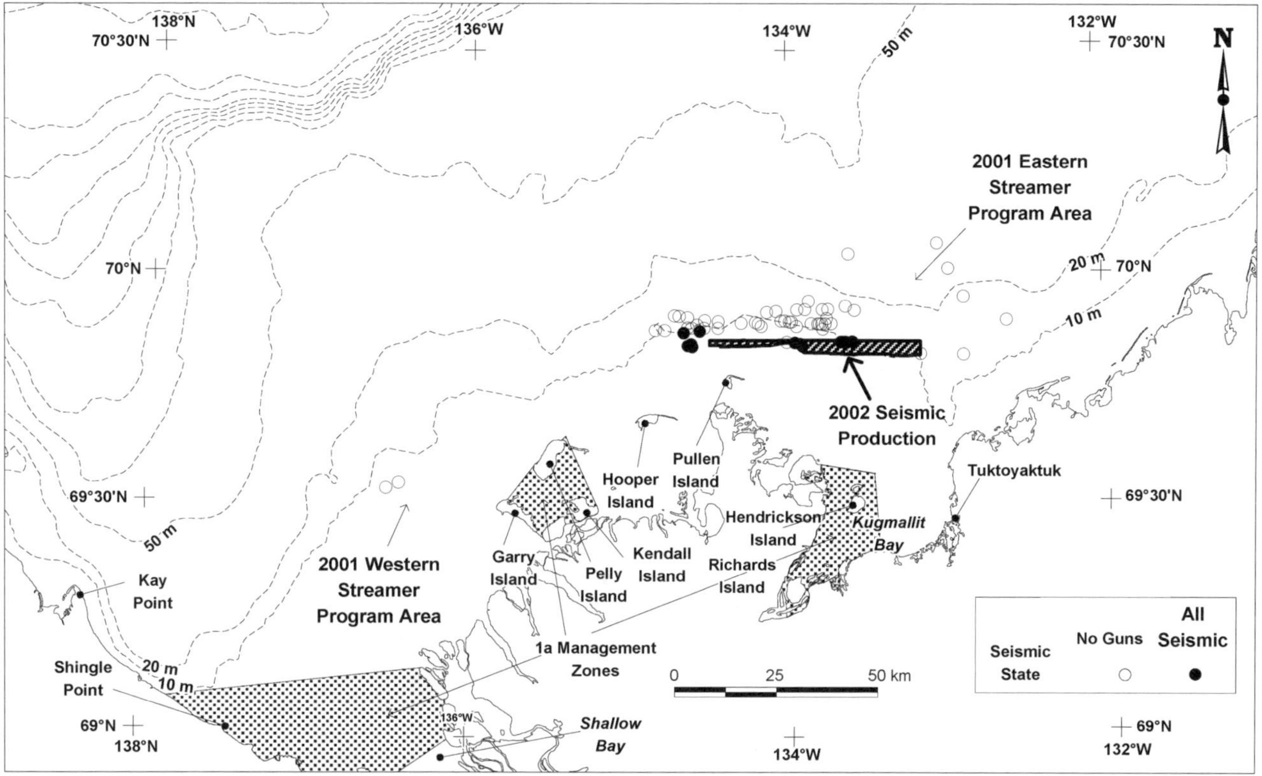

FIGURE 6. Bowhead whale sightings during LGL/Devon vessel-based monitoring in the south-astern Canadian Beaufort Sea study area, 27 Jul.–23 Aug. 2002.

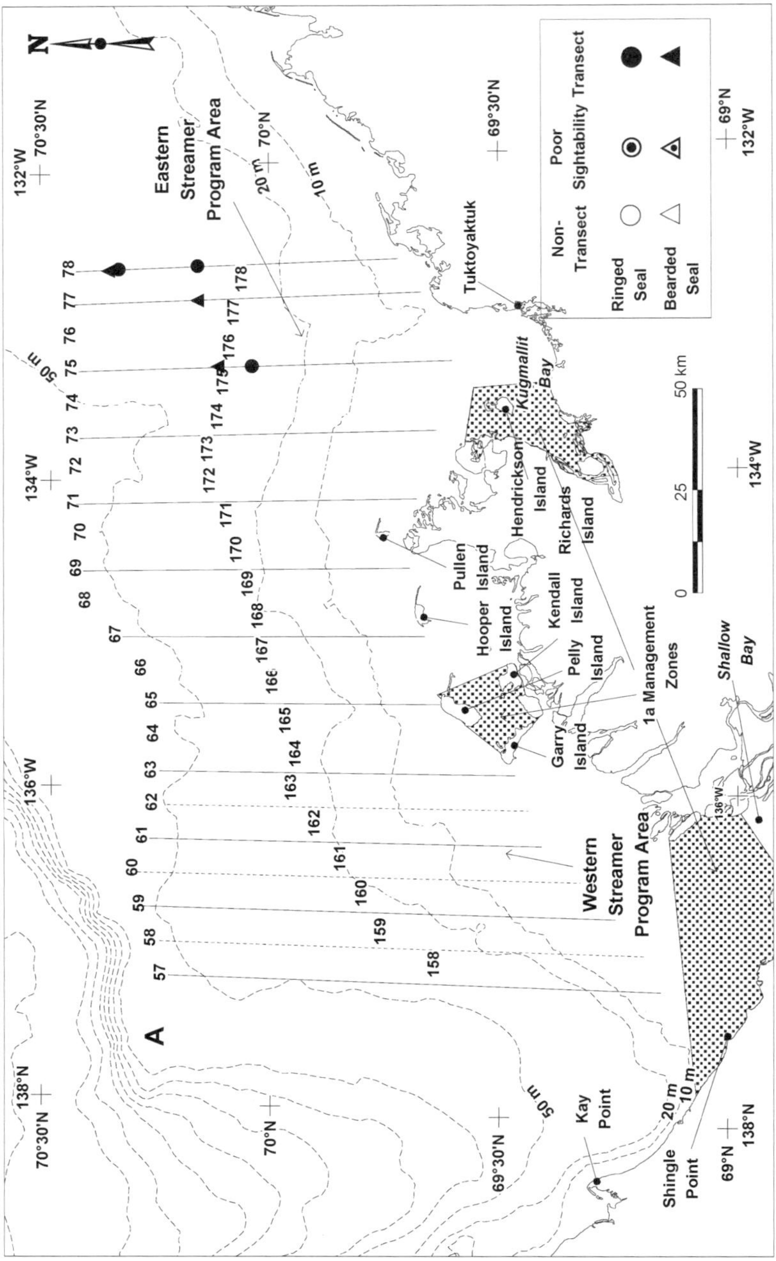
N
A
132°W
70°30'N
134°W
136°W
138°N
70°30'N
70°N
69°30'N
69°N
138°N
134°W
69°N
132°W
69°30'N
70°N
Eastern
Streamer
Program Area
Western
Streamer
Program Area
20 m
10 m
50 m
Tuktoyaktuk
Kugmallit
Bay
Hendrickson
Island
Richards
Island
Pullen
Island
Hooper
Island
Kendall
Island
Pelly
Island
Garry
Island
1a Management
Zones
Shallow
Bay
Kay
Point
Shingle
Point
0
25
50 km
Non-
Transect
Poor
Sightability
Transect
Ringed
Seal
Bearded
Seal
57
58
59
60
61
62
63
64
65
66
67
68
69
70
71
72
73
74
75
76
77
78
158
159
160
161
162
163
164
165
166
167
168
169
170
171
172
173
174
175
176
177
178

A

B

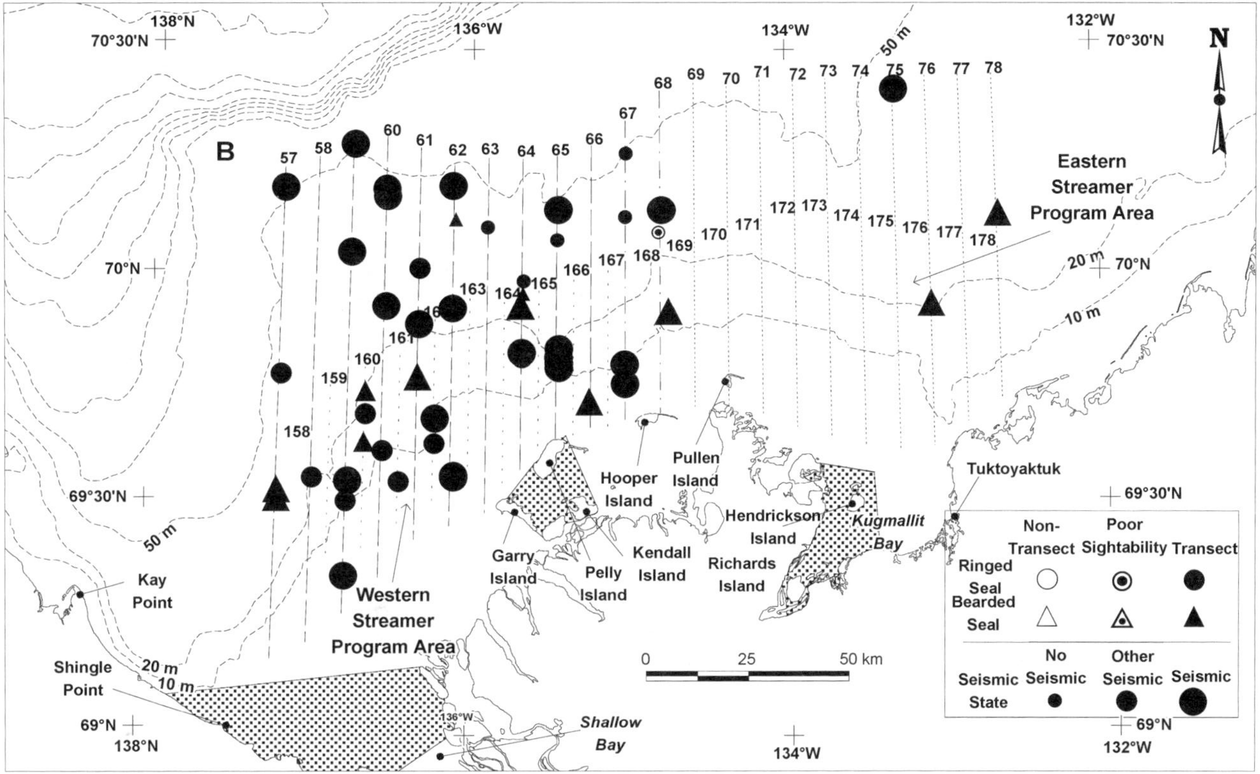

FIGURE 7. Ringed and bearded seal sightings during aerial surveys in the southeastern Canadian Beaufort Sea study area during **(A)** “pre-seismic” (28 Jul.–8 Aug.), and **(B)** “seismic operations” (9 Aug.–29 Sept.) periods, based on 2001 LGL/Devon aerial surveys. Circles denote ringed seal sightings; triangles denote bearded seal sightings.

Geco Snapper. Most of the sightings were of ringed seals. Of the seals identified during aerial surveys, 134 (88%) were ringed seals and 19 (12%) were bearded seals. Of 179 seals that were identified from the *Geco Snapper*, 176 (98%) were ringed seals and the remaining 3 (2%) were bearded seals.

During the period when the aerial surveys were flown over the Western Streamer program area, seals were well distributed across the western portion of the study area (Fig. 7). By the time the seismic operations moved to the Eastern Streamer program area in late-September, there were few seals left in that portion of the study area. These lower numbers may reflect a westward migration out of the area, as indicated by recent satellite telemetry data (L. Harwood, DFO, pers.comm.; www.permafrost.com/seals). Within the Western Streamer program area, ringed seal sightings were particularly abundant along a band that roughly approximated the 10 m depth contour.

Seals were difficult to see from the aircraft at the altitude at which the surveys were flown and during the moderate to high sea state conditions that were often encountered. The best information concerning seal reactions to the seismic program was available from the vessel-based monitoring. During 2001, daylight-sighting rates from the *Geco Snapper* were similar during all seismic states, including periods with no airgun operations. The mean distances at which seals were initially sighted were significantly closer during non-seismic than line seismic periods (mean radial distances 150 vs. 210 m; Mann-Whitney $U = 2186.5$, $n = 72, 97$, $P < 0.001$). The operation of the airgun array had minor and variable effects on the behaviour of some seals within a few hundred metres of the seismic vessel and possibly (to a limited extent) the distribution of seals around the seismic vessel. Nonetheless, seals were observed throughout the season in the general area where seismic operations were occurring. These results indicate that avoidance of seismic operations by seals was very localised. Seals sighted during line seismic operations at the average distance of 210 m from the *Geco Snapper* would have been exposed to sound levels of about 190 dB re 1 μPa (rms) (Fig. 2).

A total of 189 seals (99% ringed seals) were seen from the *Geco Snapper* during 2002. Seal sighting rates were higher during non-seismic periods (0.40 seals/h) than line seismic (0.21 seals/h) operations. Seals tended to be observed at smaller distances from the seismic vessel during line seismic vs. during non-seismic periods (mean distances 153 vs. 236 m; Mann-Whitney $U = 860$, $n = 16, 142$, $P = 0.056$). This marginally significant result is contrary to the results from 2001, when sighting distances were significantly smaller during non-seismic (150 m) vs. line seismic (210 m) periods. The locations (based on relative bearing and distance) of seal sightings relative to the seismic vessel were not significantly different during periods of no guns vs. line seismic periods (Hotelling's $T^2 = 2.80$, $n = 142, 16$, $P > 0.25$). The 2002 data suggest that airgun array operations may have had minor and variable effects on the behaviour of some seals within a few hundred metres of the seismic vessel.

Considering combined 2001–2002 data, there were 385 seals seen during 1561 h of daylight observations over the two seasons. Overall, seal sighting rates were higher during non-seismic periods (0.30 seals/h) than during line seismic (0.17 seals/h) operations. The mean distances at which seals were initially sighted were similar during non-seismic vs. line seismic periods (mean distances 207 vs. 202 m; Mann-Whitney $U = 10982.5$, $n = 214, 113$, $P = 0.086$), as were the locations (bearing and distance) of seal sightings relative to the seismic vessel (Hotelling's $T^2 = 0.441$, $n = 214, 113$, $P > 0.25$). Thus, there was no evidence that seals were displaced from the vessel by seismic operations, based on the 2001–2002 combined data. The behaviour of seals observed during 2001–2002 was similar during non-seismic and seismic periods, insofar as could be determined from the vessel-based observations. These results are generally similar (with minor variations) to results from observations of ringed seals near seismic vessels in the Alaskan Beaufort Sea (Harris et al., 2001; LGL Ltd., unpubl. data).

CONCLUSIONS

The 2001–02 marine mammal monitoring and mitigation program conducted off the Mackenzie Delta in the Canadian Beaufort Sea successfully determined the reactions of marine mammals to the marine seismic operations. The combined vessel-based and aerial monitoring documented very limited avoidance by some ringed seals, a relatively small zone of avoidance by summering bowhead whales, and a considerably larger zone of avoidance for beluga whales.

The mitigation measures undertaken during the monitoring program proved to be effective. Delay of the start of seismic operations until late July/early August prevented potential impacts on the beluga hunt, which was completed by the time seismic operations began. The delayed start also mitigated against disturbance of the portion of the beluga population that concentrates in the Mackenzie Estuary during late June and July. The period of peak occupation of the Mackenzie Estuary by belugas was over by the time the seismic surveys began.

During combined 2001–2002 operations (777.6 h of daylight observations with airgun operations), nine airgun array shutdowns were initiated by the vessel-based observers in response to bowhead whales appearing within the 1000 m "safety radius". The size of the safety radius and the shutdown procedures were designed to protect the hearing of whales that closely approached the airgun operations. Gradual ramp up of the airguns when resuming airgun operations allowed marine mammals disturbed by seismic pulses to move away from the source vessel before the pulses reached their maximum strength. These mitigation measures all served to minimise the potential disturbance of marine mammals. The aerial surveys provided evidence of reactions of belugas to seismic operations at distances beyond the effective visual range of the monitors on the seismic vessel. The reduced rate at which vessel-based observers detected bowheads during times with vs. without airgun operations (at least in 2001) suggested that some bowheads were also reacting beyond visual range from the ship. However, the results for bowheads were consistent with data from the early 1980s (Richardson et al., 1986) in indicating that reaction distances of summering bowheads exposed to a seismic program were generally on the order of a few kilometres. The acoustic monitoring provided important information on the propagation of survey sounds through the study area.

ACKNOWLEDGMENTS

The monitoring program was funded by Devon Canada Corp. via a contract to LGL and JASCO. We thank Doug Schmidt, of Devon Canada Corp., for his support and assistance during the planning and conduct of the project. Dave Valescu, Jerry Milot, and Raymond Young (Devon's Client Representatives), and Steven Bruce and Reginald Morais (of WesternGeco) were very helpful to and supportive of the marine mammal monitors on the *Geco Snapper*. Stuart Porteous, Peter (Speedy) Dawson, Steve Gowan and Kevin McGill (all of WesternGeco) graciously assisted the marine mammal monitors by providing navigation data. We also thank Joanne Nieuwenhuis of Devon Canada Corp. and Hal Burton of Accu-Audit Ltd. for providing navigation data after the end of the field season.

Kevin Allen, Lucky Pokiak, and Roy Smith were the Inuvialuit observers on the vessel. Arturo Serrano (of LGL) also served on the vessel-based marine mammal monitoring crew. We thank the many people on the crew of the *Geco Snapper,* including the captain (Craig Feeney) and officers (Sean Sheppard and Shannon Aylward), who extended their hospitality and friendship during the field season.

We thank Dave Bowie, Erin McEvoy, Matt Gacek, Steve King, Sandi Loutitt, Dave Sauder, and Greg Smith of Kenn Borek Air for flying the survey aircraft during July–September, 2001. Steve Penikett of Kenn Borek Air provided logistical support. Kevin Allen, Bertram Pokiak, Lucky Pokiak, and Roy Smith were the Inuvialuit observers on the aircraft. Jon Gorniak and Karen Truman (of LGL) were also key personnel on the aerial survey crew.

We experienced logistical support from many people at the Tuk Base Camp, including Randi Cofield (Devon Canada Corp.), and Keith Rosindell, Mike Dizep, Anita Canuel, Charity Ward, and Mark Potter (all of WesternGeco).

We also thank Art Adey of EGT Ltd., and Inuvialuit guides and observers Boogie and Lucky Pokiak for their assistance with the acoustic monitoring fieldwork. We are grateful to several people at LGL, including Ted Elliott, (processing and analysis of data), Dr. W. John Richardson (review of the manuscript), and Anne Wright (word processing).

REFERENCES

Batschelet, E. 1981. *Circular Statistics in Biology.* Academic Press, London, UK.

DFO (Department of Fisheries and Oceans Canada). 2000. *Eastern Beaufort Sea Beluga.* DFO Science Stock Status Report E5-38 (2000). Available at www.dfo-mpo.gc.ca/csas.

Finneran, J.J., C.E. Schlundt, R. Dear, D.A. Carder, and S.H. Ridgway. 2002. "Temporary Shift in Masked Hearing Thresholds in Odontocetes after Exposure to Single Underwater Impulses from a Seismic Watergun." *Journal of Acoustic Society of America.* 111(6):2929-2940.

Hannay, D. and M. Austin. 2001. *Pre-Season Acoustic Modeling for Prediction of Expected Underwater Noise Levels Generated by Anderson Resources Ltd, Summer-Autumn 2001/2002 3D Marine Seismic Program.* Report from JASCO Research Ltd, Victoria BC for Anderson Resources Ltd., Calgary, AB.

Hannay, D. and A. MacGillivray. 2001a. *Preliminary Acoustic Level Measurements from Echosounder and Side Scan Sonar for Bathymetry Survey Component, Anderson 2001/2002 Seismic Project, Tuktoyaktuk, NWT.* Rep. from JASCO Research Ltd, Victoria BC for Anderson Resources Ltd., Calgary, AB.

Hannay, D. and A. MacGillivray. 2001b. *Preliminary Acoustic Level Measurements of Airgun Sources from Anderson 2001/2002 Seismic Project, Tuktoyaktuk, NWT* Rep. from JASCO Research Ltd, Victoria BC for Anderson Resources Ltd., Calgary, AB.

Harris, R.E., G.W. Miller, and W.J. Richardson. 2001. "Seal Responses to Airgun Sounds during Summer Seismic Surveys in the Alaskan Beaufort Sea." *Marine Mammal Science* 17(4):795-812.

Holst, M., G.W. Miller, V.D. Moulton, and R.E. Elliott. 2002. "Aerial Monitoring, 2001." In G.W. Miller and R.A. Davis (Eds.), *Marine Mammal and Acoustical Monitoring of Anderson Exploration Ltd.'s Open-Water Seismic Program in the Southeastern Beaufort Sea*, pp. 5-1 to 5-42. Final report by LGL Ltd., King City, ON, and JASCO Research Ltd., Victoria, BC for Devon Canada Corporation, Calgary, AB.

LGL and JASCO. 2001. *Marine Mammal and Acoustic Monitoring Plan for the Proposed Anderson Resources Ltd. Southern Beaufort Sea Marine Seismic Program, Summer 2001.* Rep. from LGL Ltd., King City, ON, and JASCO Research Ltd., Victoria, BC for Anderson Resources Ltd., Calgary, AB.

MacGillivray, A., D. Hannay, and S. Carr. 2002. "Acoustic Monitoring, 2001." In G.W. Miller and R.A. Davis (Eds.), *Marine Mammal and Acoustical Monitoring of Anderson Exploration Ltd.'s Open-Water Seismic Program in the Southeastern Beaufort Sea*, pp. 3-1 to 3-51. Final report by LGL Ltd., King City, ON, and JASCO Research Ltd., Victoria, BC for Devon Canada Corporation, Calgary, AB.

Miller, G.W. and R.A. Davis (Eds.). 2002. *Marine Mammal and Acoustical Monitoring of Anderson Exploration Ltd.'s Open-Water Seismic Program in the Southeastern Beaufort Sea*. Final report by LGL Ltd., King City, ON, and JASCO Research Ltd., Victoria, BC for Devon Canada Corporation, Calgary, AB.

Miller, G.W. and V.D. Moulton. 2003. *Vessel-Based Marine Mammal Monitoring of Devon Canada Corporation's 2002 Open-Water Seismic Program in the Southeastern Beaufort Sea*. Final report by LGL Ltd., King City, ON, for Devon Canada Corporation, Calgary, AB.

Moore, S.E. and R.R. Reeves. 1993. "Distribution and Movement." In J.J. Burns, J.J. Montague and C.J. Cowles (Eds.), *The Bowhead Whale*. pp. 313-386. Spec. Publ. 2, Soc. Mar. Mamm., Lawrence, KS.

Moulton, V.D., G.W. Miller, and A. Serrano. 2002. "Vessel-Based Monitoring, 2001." In G.W. Miller and R.A. Davis (Eds.), *Marine Mammal and Acoustical Monitoring of Anderson Exploration Ltd.'s Open-Water Seismic Program in the Southeastern Beaufort Sea*, pp. 4-1 to 4-56. Final report by LGL Ltd., King City, ON, and JASCO Research Ltd., Victoria, BC for Devon Canada Corporation, Calgary, AB.

NMFS. 2000. Small Takes of Marine Mammals Incidental to Specified Activities; Marine Seismic-Reflection Data Collection in Southern California/Notice of Receipt of Application. Fed. Regist. 65(60, 28 Mar.):16374-16379.

Richard, P.R., A.R. Martin, and J.R. Orr. 2001. "Summer and Autumn Movements of Belugas of the Eastern Beaufort Sea Stock." *Arctic* 54(3):223-236.

Richardson, W.J. (Ed.). 1999. *Marine Mammal and Acoustical Monitoring of Western Geophysical's Open-Water Seismic Program in the Alaskan Beaufort Sea*, 1998. LGL Rep. TA2230-3. Rep. from LGL Ltd., King City, ON, and Greeneridge Sciences Inc., Santa Barbara, CA, for Western Geophysical, Houston, TX, and US Nat. Mar. Fish. Serv., Anchorage, AK, and Silver Spring, MD.

Richardson, W.J., M.A. Fraker, B. Würsig, and R.S. Wells. 1985. "Behaviour of Bowhead Whales *Balaena mysticetus* Summering in the Beaufort Sea: Reactions to Industrial Activities." *Biol. Conserv.* 32(3):195-230.

Richardson, W.J., B. Würsig, and C.R. Greene Jr. 1986. "Reactions of Bowhead Whales, *Balaena mysticetus,* to Seismic Exploration in the Canadian Beaufort Sea." *Journal of Acoustic Society of America* 79(4):1117-1128.

Richardson, W.J., R.A. Davis, C.R. Evans, D.K. Ljungblad, and P. Norton. 1987. "Summer Distribution of Bowhead Whales, *Balaena mysticetus,* Relative to Oil Industry Activities in the Canadian Beaufort Sea, 1980–84." *Arctic* 40(2):93-104.

Richardson, W.J., C.R. Greene Jr., C.I. Malme, and D.H. Thomson. 1995. *Marine Mammals and Noise.* Academic Press, San Diego, CA.

Thomson, D.H. and R.A. Davis. 2001. *Review of the Potential Effects of Seismic Exploration on Marine Animals in the Beaufort Sea.* Rep. from LGL Ltd., King City, ON, for ISR Oceans Program Co-ordinator, Fisheries and Oceans Canada, Yellowknife. NT.

Live Bottom Impact Reduction, Mitigation, and Monitoring for the Construction of a Subsea Natural Gas Pipeline in the Gulf of Mexico

Ellsworth, S. (sellsworth@ensr.com), J. Schmidt, D. Martin
(ENSR International, Anchorage, AK USA, Tallahassee, FL USA, and Ft. Lauderdale, FL USA, respectively), and R. Shaul
(Sea Byte, Inc., St. Petersburg, FL USA)

ABSTRACT: This paper describes successful methods undertaken to minimise impacts to sensitive benthic marine habitats during the development and construction of a submarine natural gas pipeline. The construction of a 91-cm diameter pipeline known as the Gulfstream Natural Gas System (Gulfstream) was proposed to transport natural gas from onshore sources in Mississippi and Alabama to peninsular Florida traversing approximately 676 km of seafloor in the eastern Gulf of Mexico and Tampa Bay. Seagrasses and live bottom (low-relief hard substrate dominated by sponges and soft corals) are prevalent on the West Florida Shelf and were determined to be the most ecologically important and sensitive marine habitats that could be impacted by the project. Project proponents were required to map the resources, predict the extent/severity of impacts during permitting, conduct post-construction impact assessments, mitigate for unavoidable impacts, and monitor recovery of impact areas and colonization of mitigation areas.

Impacts to seafloor resources could occur through barge anchoring, excavation where the pipeline must be lowered, and elsewhere by placement of the heavy pipe on the seafloor. Impacts to seagrasses were avoided completely, and potential live bottom impacts were reduced by over 93% through the use of horizontal direction drills (HDDs), rerouting, submarine plows for lowering the pipeline, dynamically-positioned burybarges to pull the plow, and buoys on laybarge anchor cables. Mitigation consisted of placing limestone boulders and limestone/concrete modules over 23.2 ha of seafloor.

Mitigation and impact monitoring included the establishment of semi-permanently marked transects in control and impact areas, and semi-permanent plots in mitigation and control areas. Monitoring is to take place over five successive years following construction, and consists of using still and video photography and diver observations. Performance criteria include percent biotic cover, richness, and diversity of the macrobenthic communities. The project was constructed in 2001–2002. Mitigation was installed and impact assessment surveys commenced in 2002 and completed in 2003.

INTRODUCTION

The Gulfstream Natural Gas System (Gulfstream) was designed and permitted from 1999 to 2001, constructed in 2001 and 2002, and commissioned into service in June, 2002. As permitted, the system brings natural gas from existing natural gas treatment facilities and pipelines in coastal Alabama and Mississippi, across the eastern Gulf of Mexico to a landfall in Tampa Bay, and then on across much of peninsular Florida to provide fuel to various power plants (Fig. 1). The pipeline was designed to transport up to 1.1 billion cubic feet of natural gas per day (1.1 BCFD), and includes approximately 676 km of pipeline in the Gulf of Mexico and adjacent coastal waters.

Live bottom consists of low relief hard bottom (substrate) areas that are colonized by a community of benthic epi-fauna dominated by algae, sponges, and corals. Live bottom is sensitive to environmental perturbation and is considered ecologically important for the productivity of the marine

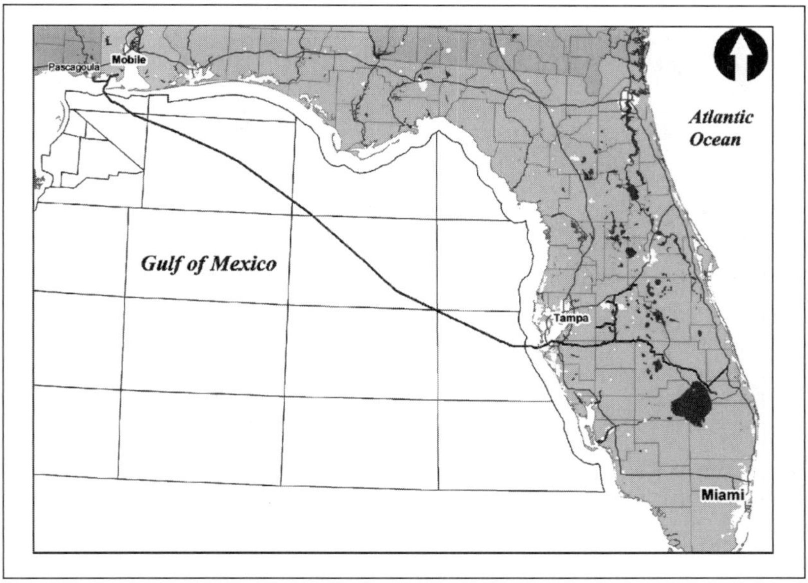

FIGURE 1. System map of the Guflstream Natural Gas System.

environment, particularly as habitat for finfish and shellfish. It is therefore protected by regulatory agencies with jurisdiction over pipeline construction. The Minerals Management Service issued a stipulation that applies to all leaseholders or potential leaseholders in that area, which states that video surveys are to be conducted for the presence of live bottom. The stipulation area is indicated as the hatched area in Figure 2.

This paper details the successful efforts of the project proponent in mapping live bottom and minimising potential impacts to live bottom from construction of the project. Live bottom mapping, and quantification and reduction of potential live bottom impacts, all had to be done during the permitting process, and therefore before actual pipeline construction. The live bottom impacts discussed in this paper are therefore considered to be potential impacts. Estimates of the potential impact were based on resource mapping, engineering design, pipeline construction experience, modelling, and the ecology of live bottom communities.

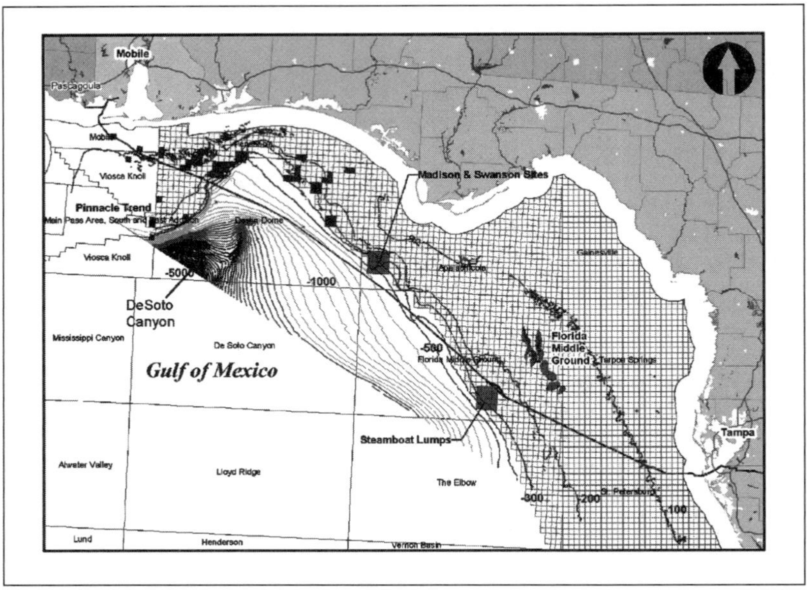

FIGURE 2. Bathymetry, the live bottom stipulation area (hatched), and known areas of hard bottom (darker blocks) in relation to Gulfstream Natural Gas System.

A mitigation and monitoring plan was also developed and approved by the agencies during the permitting phase. This plan mandates that the project proponent monitor the effects of pipeline construction, including measuring and mapping of the areal extent of live bottom impacts. The first year of monitoring is underway; however, the data has not yet been analysed sufficiently to comment on the accuracy of the impact estimates.

MATERIALS AND METHODS

Route Selection

The need for natural gas in peninsular Florida was apparent to several pipeline companies. Gulfstream's proponent selected an offshore route to transport natural gas from the major sources in Mississippi and Alabama to Florida as fewer sensitive areas and private properties would be traversed than with an onshore route. The end points of the pipeline route were then generally set by the location of the sources in Mississippi and Alabama and the markets in Florida. Between those locations the major, stationary, sensitive resources were identified as seagrasses and hard bottom (of which live bottom is one type). Known areas of these resources were then mapped. The proponent also wanted to stay in waters shallower than 244 m for engineering and cost reasons. A route was selected that was within the selected water depth and that avoided known and prominent hard bottom areas such as Florida Middle Ground and Pinnacle Trend (Fig. 2).

Seagrasses are very abundant in the shallow waters along most of the western coast of peninsular Florida, making landfall of a pipeline without significant disturbance of seagrass beds very difficult. Impacts to seagrasses were avoided in this project by coastal entry via Tampa Bay. There are no seagrasses in the deep waters at the mouth of the bay and a much narrower band of seagrasses is present along the bay shoreline than is the case for the peninsula shoreline. Pipeline landfall was made on the southern shoreline of the bay with the use of a horizontal directional drill (HDD) that drilled a section of the pipe under the seafloor, surfacing on an existing spoil island. Another HDD was used to install the pipeline under the seafloor (and the seagrasses) between the island and the mainland, avoiding all seagrass impacts. Live bottom could not be avoided, as it was widespread throughout the area indicated as live bottom stipulation area in Figure 2. However, three subsequent route modifications (re-routes) did reduce the length of route centerline within live bottom by 10.8 km.

Resource Mapping

Using their Integrated Video Mapping System, Sea Byte, Inc. mapped benthic habitats. A video camera was towed behind a vessel along pre-determined transects. In water depths less than 15.2 m, the camera was on the helmet of a diver who was towed behind the vessel. In deeper waters, the camera was mounted on a mechanical fish. In the area of high relief on the shelf edge west of DeSoto Canyon, the camera was mounted on a remotely operated vehicle. Digital video was wired from the camera to an on-board computer where it was integrated with the global positioning system data and mapping in the form of a Geographic Information System.

The level of effort in the video mapping increased throughout the permitting process. Widely spaced (152–762.0 m) transects were used early to define a route. When the route was set, transects with spacing as tight as 30.5 m were surveyed in live bottom areas where impact was expected to occur to more precisely define the limits of the live bottom and the impacts. Stations were also randomly established in mapped live bottom to take still photographs for quantification of biotic cover and composition. In the deeper waters (15–91 m) each station was 30.5 × 30.5 m. Nine stations were established along the route in water depths of 15–30 m, three stations were established in water depths of 30–61 m, and three stations were established in water depths of 61–91 m (one of which was in the pinnacle area, i.e. the shelf edge on the western side of the DeSoto Canyon, which has pinnacle-like spires that rise above the seafloor). Within each station, 100 still photographs were taken at random locations and analysed for percent biotic cover according to the method described by Bohnsack et al. (1979). The scale and dimensions of the photographs (0.56 m^2 of the seafloor) were maintained constant between photographs by holding the camera at the same height above the seafloor at each location (on a tetrapod in shallow water and on ROV in deeper water). A computer was used to generate 100 dots at random locations on each photograph (Fig. 3). Biotic cover of each taxon was calculated as the number of dots overlaying any portion of an organism of that taxon. The photographs were also examined to quantify taxonomic richness by identifying any other species or other taxon that might be within the photograph but not overlaid by a dot, and therefore not recorded. The sampling strategy differed in state waters (water depths of 0–15 m) where 23-m photographic transects were established; however, the photographs were taken and analysed in the same manner. Transects were established in each of the 11 mapped live bottom areas located

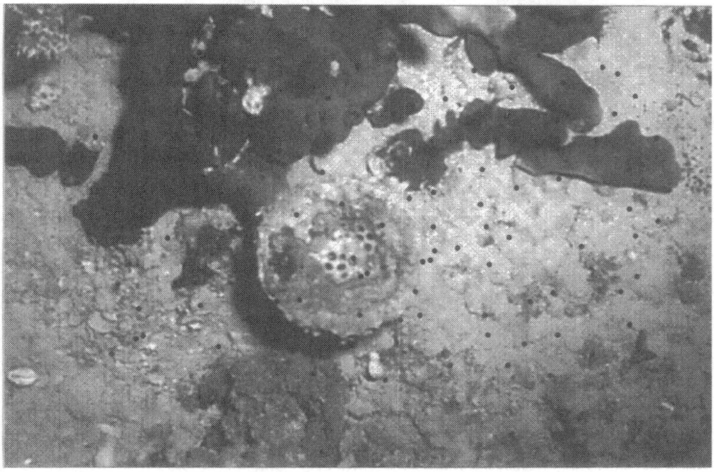

FIGURE 3. Digital photographs with random dots used to determine percent cover by taxon.

within 305 m of the proposed pipeline centerline (a total of 19 transects). Thirty photographs were taken along each transect.

Quantification of Impacts

Impacts to protected resources such as live bottom must be quantified during the permitting process for several reasons, the most important of which are: 1) resource protection agency must understand the potential impacts of the project before issuing permits and approvals, 2) alternative routes must be assessed to select the one that has the least associated environmental impact, 3) the potential effect of various impact reduction measures must be assessed in relation to efficacy and cost, and 4) the amount of mitigation that might be required to compensate for the temporary or permanent losses of live bottom due to the project must be determined. Impacts were quantified by estimating the amount and location of disturbances to the seafloor that could result under a given pipeline construction scenario, and then overlaying the locations of the disturbances over live bottom mapping produced for the project.

CHARACTERISTICS OF LIVE BOTTOM OBSERVED ALONG THE PIPELINE ROUTE

Live bottom was found to be widely distributed in the surveyed area, but was most abundant along the route on the West Florida Shelf outside Tampa Bay in water depths of 15.2 to 30.5 m. Surveys on the West Florida Shelf (Sea Byte, Inc., 2001a; 2001b) were conducted from the shoreline out to a depth of 100 m, therefore the presence or absence of live bottom in areas of greater water depth than that cannot be ascertained. However, no live bottom was observed along the pipeline centerline (or any survey transects) in water depths greater than 91.4 m, indicating that may be the limit of live bottom distribution in this area. The portion of the original pipeline route that was found to be within live bottom varied from 12 to 38% with water depth (Table 1). This frequency of occurrence agrees closely with reports found in the literature. Parker et al. (1983) reported that an estimated 38% of the seafloor between Key West and Pensacola, Florida, in water depths of less than 30.5 m, was live bottom. Live bottom was also observed along an approximately 0.8-km portion of the route, where it crosses the shelf break on the western side of DeSoto Canyon (Pinnacle area). Three significant re-routes reduced the total length of pipeline centerline within live bottom by 20% from 54.4 to 43.6 km (6.8, 3.5, and 0.6 km, for a total of 10.8 km).

Biotic cover observed in live bottom within the study area was found to vary with water depth (Table 2). The epi-faunal communities in the mapped live bottom were found to be

TABLE 1. Percent of the seafloor along the pipeline route that was found to be within mapped live bottom before and after re-routing.

Route	Parameter	Route Segment (Water Depth Range)				
		0 - 15 m	15 - 30 m	30 - 61 m	61 - 91 m	Pinnacle Area
Initial	% LB[1]	20 %	38 %	15 %	12 %	17 %
Initial	Length[2]	36.4 km	38.9 m	114.3 km	85.6 km	0.8 km
Final	% LB[1]	< 2 %	28 %	15%	12 %	17 %
Final	Length[2]	38.0 km	38.8 m	114.3 km	86.6 km	0.8 km

[1]% LB is percent of route segment centerline that is within live bottom.
[2]Length is total length of centerline in route segment.

TABLE 2. Percent biotic cover in live bottom found along the pipeline route.

Taxon	Percent and Relative Percent Biotic Cover by Route Segment /Water Depth Range									
	0 - 15 m		15 - 30 m		30 - 61 m		61 - 91 m		Pinnacle Area	
	% Cov[1]	Rel[2] Cov	% Cov	Rel Cov	% Cov	Rel Cov	% Cov	Rel Cov	% Cov	Rel Cov
Algae	2.6	20	5.0	41	7.3	28	2.7	19	1.5	19
Sponges	8.3	65	3.9	32	13.8	55	3.2	23	1.7	22
Hydrozoans	0.1	1	0.1	1	0.0	0	0.0	0	0.3	4
Soft Corals	1.4	11	1.4	12	1.1	5	4.6	34	4.0	50
Hard Corals	0.3	2	1.4	12	0.1	1	0.0	0	0.0	0
Bryozoans	0.0	0	0.0	1	0.6	3	1.4	10	0.0	0
Ascidians	0.1	1	0.1	1	1.6	5	1.1	8	0.0	0
Echinoderms	0.0	0	0.1	1	0.5	2	0.8	5	0.4	5
Total Cover	12.8	100	12.0	100	25.0	100	13.8	100	7.9	100

[1]% Cov is average percent biotic cover by taxonomic grouping determined from 0.707-m^2 photographic plots.
[2]Rel Cov is relative percent biotic cover by taxonomic grouping determined from 0.707-m^2 photographic plots.

dominated by algae, sponges, and soft corals (Table 2). Algae and sponges appeared to decrease in relative dominance, and soft corals appeared to increase in relative abundance, in depths greater than 61.0 m (Table 2).

Most of the live bottom observed on the West Florida Shelf (eastern portion of the pipeline) had 0.3–0.9-m relief. Approximately 88% of that live bottom was not on exposed hard substrate, but rather on hard substrate covered with a thin veneer of sand. Live bottom in this area appears to be ephemeral and subject to shifting sediments. This was evidenced by the fact that one live bottom area encompassing approximately 0.12 ha mapped in early surveys (June, 1999) were found to be covered by sand waves and shell hash (no evident live bottom) in subsequent (February, 2001) surveys. Live bottom organisms colonize any exposed hard substrate; however, the shifting nature of the seafloor sediments frequently covers existing live bottom resulting in its demise. Sediment movement may also result in the exposure of substrate for colonization.

The live bottom on the shelf edge on the western side of the DeSoto Canyon is different in that it has very substantial relief and is apparently very enduring. Live bottom in this area is found on pinnacle-like spires that rise above the seafloor anywhere from 3.0 to 18.3 m with slopes from 45% to nearly vertical. These formations and communities appear similar to that reported in the Pinnacle Trend Area (Fig. 2) of the Central Gulf (Continental Shelf Association, 1992).

QUANTIFYING AND REDUCING IMPACTS

Assessment of the pipeline route and pipeline installation methodologies indicated that live bottom could potentially be impacted by the following pipeline construction activities: 1) trenching of the pipeline where it is required (trenching impact), 2) laying the pipeline on the seafloor and leaving it exposed where that is permissible and proposed (pipelay impact), 3) deployment and retrieval of the anchors that moor the construction barges (anchor scar impact), and 4) the sweeping of anchor cables across the seafloor as the barge is winched ahead with the anchors in place (anchor cable sweep impact). The methods by which each of these impacts were quantified and then reduced are described below.

Trenching Impacts

By regulation, interstate natural gas pipelines in the Gulf of Mexico in water depths of 61.0 m or less, must be lowered into the seafloor so that the top of the installed pipe is 0.9 m or more below the original grade of the seafloor. In areas with water depths exceeding 61.0 m, the pipeline may be installed directly on the seafloor and left exposed on the seafloor with no trenching taking place. The intent of the regulatory requirement for lowering the pipeline into a trench in shallow waters is to protect the pipe from damage due to ship grounding and anchoring, and to protect mobile fishing gear from being damaged by the pipe.

Regulations do not mandate that submarine pipelines be covered with sediment. In fact, most subsea pipelines in US waters shallower than 61.0 m are left exposed within the trench. It is commonly held that these exposed pipelines do become buried within the trench due to the natural movement of sediments over the course of months or years; a process called self-burial. In the case of the Gulfstream Natural Gas System, the state of Florida was concerned with having pipe exposed even in a trench, within the busy harbour of Tampa Bay. The project proponent therefore agreed to backfill the installed, trenched pipeline within the limits of Tampa Bay proper (from the barrier islands landward).

Impacts associated with pipeline installation are generally considered to be greater and more deleterious where trenching takes place, than where the pipeline is to be installed and left exposed on the seafloor. This is because a much wider area of the seafloor is disturbed by the trench and by the placement of the trenched sediments on previously undisturbed seafloor (direct impacts). Significant re-deposition (indirect impacts) of sediments suspended in the water column during the trenching process can also occur.

Most available and applicable trenching techniques were investigated including plowing, jetting, and mechanical dredging. Potential direct live bottom impacts were quantified by using engineering methods and previous pipeline construction experience to predict the width of direct seafloor disturbance (width of trench and any spoil mounds) for each method/equipment type and overlaying that prediction as an impact corridor over the centerline and live bottom mapping. Potential indirect impacts due to trenching were estimated by modelling the resulting plume and depths of re-deposition with distance from the activity. Sediment suspension and re-deposition were modelled (ENSR, 2000) using the results of grain-size analyses of sediment samples collected along the route, engineering estimates of trench shape and volume, and several sediment transport models (e.g. DREDGE, STFATE, EFDC). Results of the University of South Florida's Tampa Bay transient three-dimensional circulation model were used to establish local flow patterns (ENSR, 2000). Plowing is a relatively new innovation that was developed in the North Sea. Most plows trench the pipeline after the pipe is installed on the seabed; i.e. they are post-lay plows. The plow is a large machine that sits over the pipeline with its weight supported by ski-like skids (Fig. 4). The plow is pulled along by the bury-barge at slow speeds (1–2 knots) and pushes the sediments out from under the pipe, creating a trench into which the pipe

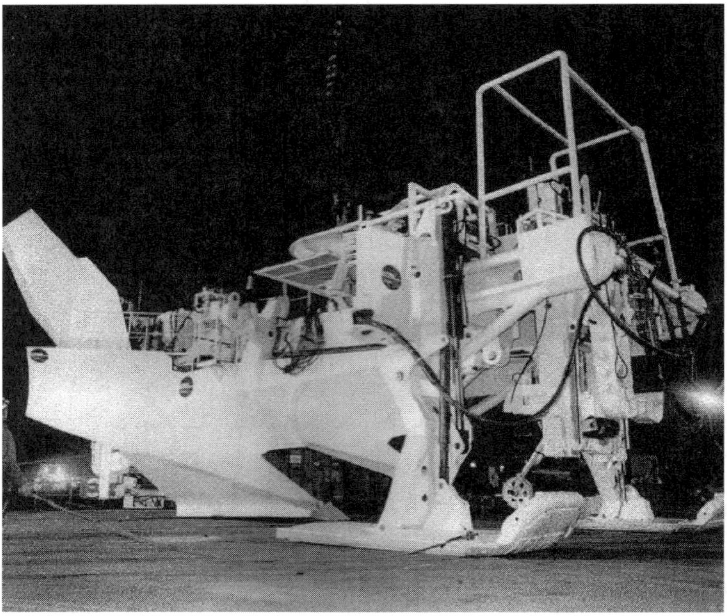

FIGURE 4. A post-lay submarine plow.

settles. Two or three passes are usually required to lower the pipe to the required depth. The plow is pulled so slowly that re-suspension of sediments is negligible in sandy substrates. The trench was expected to be 7.6-m wide with 7.6-m wide spoil mounds on either side.

Jetting is the most common method of installing subsea pipelines. The jet sled is pulled along the pipe in a similar fashion to the plow. However, the sediments below the pipe are fluidized and/or suspended by a series of pressurized water and air jets as opposed to being pushed to the side by the plow. This results in significant amounts of re-suspended sediments being deposited along the trench and for a distance perpendicular to the trench, depending on local currents and sediment grain size. Modelling conducted with input on local currents and sediments along the route indicated that re-suspended sediments would be deposited to depths that could be significantly detrimental to live bottom community out a distance of approximately 76.2 m from the centerline (indirect impact). The direct impact due to jetting was expected to be a trench 12.8 m wide.

Mechanical dredging involves the construction of a trench using a barge-mounted excavator and generally a clamshell bucket. The pipe is installed within the excavated trench. This is generally practicable only in relatively shallow areas. We estimated that a mechanically dredged trench would be 27.4 m wide with a 30.5-m spoil mound on one side. Mechanical dredging would also result in indirect live bottom impact due to the deposition of re-suspended sediments out approximately 15.2 m from the activity.

Mechanical dredging was deemed impractical due to water depths. The results of the comparison of the potential live bottom impact associated with jetting and plowing are presented in Table 4. The selection of a submarine plow as the equipment for trenching the Gulfstream Natural Gas System, as opposed to the standard method for pipeline trenching in the Gulf of Mexico (jetting), reduced the live bottom impact by an estimated 108.7 ha or 60%.

TABLE 4. Estimated potential live bottom impact associated with pipeline trenching for the Gulfstream Natural Gas System.

Trenching Method	Estimated Potential Live Bottom Impact		
	Direct Impact	Indirect Impact	Total Impact
Jetting	62.0 ha	121.3 ha	183.3 ha
Plowing	74.7 ha	0.0 ha	74.7 ha

Pipelay Impacts

Potential live bottom impacts associated with installation of the pipe on the seafloor in areas where it would not be trenched (in water depths > 61.0 m) were estimated by overlaying a corridor the approximate width of the concrete-coated pipe (122 cm) along the route and over the live bottom mapping. Pipelay impacts were reduced by re-routing, however, the achieved reduction was not large, as the pipeline is only exposed in the deeper waters where live bottom is less abundant. The most significant re-routes were instituted in shallow water where the pipeline was trenched. Re-routing in the area where the pipe was to be left exposed reduced the estimated potential trenching impact from 1.4 to 1.3 ha, a reduction of 3%.

Anchor Scar Impacts

Typical barges used to install the pipe on the seafloor (laybarges) and to lower or trench the pipe (burybarges) are moored in place with a set (array) of anchors. The anchors are attached to the barge with steel cable. A barge is moved along its course by winching in the bow anchors cables and letting cable out on the stern anchors. Large laybarges and bury barges used in deep water generally use an array of 12 anchors, while smaller barges used in shallower water use eight or ten anchors.

Almost all anchors used in this manner are embedment anchors. Embedment anchors are lowered to the seafloor from anchor-handling vessels; the anchor cable is then winched in, dragging the anchor along the seafloor until the anchor flukes dig into the seafloor sufficiently to hold the barge in place and to allow the barge to be moved by winching in the cable. We estimated the size of an anchor scar, i.e. the area of seafloor that would be impacted on average by deployment of a single anchor, to be 31.6 m^2, based upon the measurements of anchor scars reported in the literature (Centaur Association, 1984) and by engineering formulas (fluke length to drag length ratios). After discussions with marine contractors we assumed that: 1) twelve anchors would be used in the deeper (> 15.2 m) portions of the route and would be re-set at an average interval of 0.8 km, and 2) eight anchors would be used on the smaller barges in shallow waters, which would be re-set at an average interval of 146.3 m. In water depths greater than 61.0 m, a single pass with the laybarge to lay the pipe is all that is needed. Where trenching is required one pass is needed to lay the pipe and usually two or three more passes are needed to lower/bury the pipe. Potential seafloor and live bottom impacts were quantified by water depth as follows:

Seafloor impact = (scar area) × (number of anchors in a set) × (length of route within that water depth / anchor set interval) × (required number of barge passes)

Live bottom impact = (seafloor impact in depth interval) × (percent of seafloor in anchor corridor within that depth interval which was mapped as live bottom)

Potential live bottom impacts due to anchor scarring were reduced largely by the use of dynamically positioned barges (DP barges), and by the development and implementation of an anchor handling plan. DP barges use a series of thrusters on the bow and stern to hold the barge in place, and to move the barge along the route, so that anchors are not required. DP barges are a fairly recent innovation and are not yet widespread. Not all contractors have this technology. There are very few DP vessels in the world that can handle a pipeline of the diameter of this one. The selected pipeline installation contractor was able to employ a DP burybarge, but anchored laybarges had to be used to install the pipe. Use of DP burybarges precluded any anchor scar impacts from trenching and reduced potential live bottom impacts due to anchor scarring from 5.7 ha to 2.0 ha or approximately 65% (3.7-ha reduction).

Anchor scar impacts were reduced an additional 0.3 ha by development of an anchor-handling plan. The plan involved, among other things, ensuring anchor-handling techniques were implemented to minimise the size of anchor scars, and pre-installation identification of some locations where anchors could be placed without impacting live bottom. The latter was impossible in some areas, and impracticable in others, due to the widespread distribution and abundance of live bottom within the anchor corridor. However, it was accomplished in some small areas and resulted in a reduction of the estimated or potential live bottom impact. To the west of DeSoto Canyon, the presence of live bottom in the anchor corridor was restricted to a 0.8-km section of the route, and the live bottom was found in discrete pinnacle-like formations.

Pre-determined anchor locations were identified for this area that would allow pipelay to occur with no live bottom impacts.

Anchor Cable Sweep Impacts

Laybarges and bury barges that are moored and moved along the pipeline centerline with anchors can impact the seafloor and any objects on the seafloor such as live bottom, by the sweeping motion of the anchor cables. The barge is pulled ahead by winching in the bow anchor cables and letting out the stern anchor cables, with the anchors remaining in place (Fig. 5). As this is accomplished, the portion of the two-inch diameter steel cable that lays on the seafloor is swept in an arc across the seafloor and about the anchor (Fig. 5). Because cable sweep has been known to damage tropical coral reefs, we assumed that it could damage live bottom, particularly in those areas of live bottom with more relief. The area of seafloor that would be swept by anchor cables was estimated, by water depth interval (0–15.2 m, 15.2–30.5 m, 30.5–61.0 m, 61.0–91.4 m) (see example in Fig. 5), with engineering analyses, for each of the eight anchors used on barges in shallow (< 15.2 m), and for each of the 12 anchors used on barges in deeper waters. Potential seafloor and live bottom impacts were quantified by water depth as follows:

Seafloor impact = (sweep area of anchor array during one setting in that water depth range) × (length of route within that water depth / anchor set interval) × (required number of barge passes)

Live bottom impact = (seafloor impact in depth interval) × (percent of seafloor in anchor corridor within that depth interval which was mapped as live bottom)

Potential impacts to live bottom from anchor cable sweep were largely minimised by two methods. The use of DP barges minimised the number of required moorings and associated cable sweep, and the use of mid-line buoys on anchor cables reduced the length of cable that lies on the seafloor and thus the area of seafloor that would be swept by the remaining moorings. DP burybarges were available; a DP laybarge of

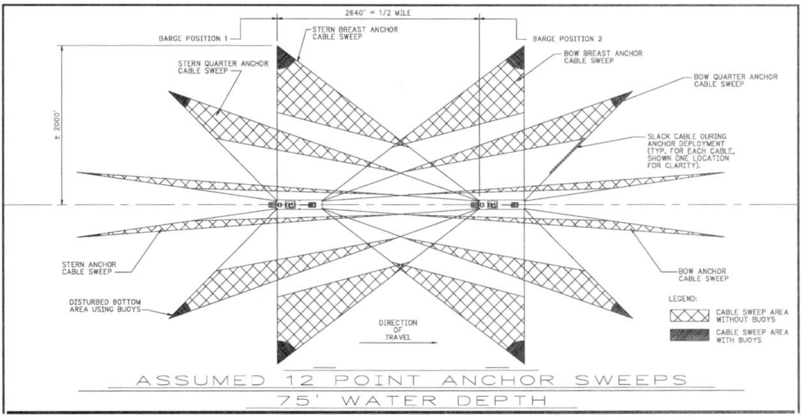

FIGURE 5. Schematic of seafloor area assumed to be impacted by anchor cable sweep with a 12-anchor array in water depths of 50–100 ft (15–30 m).

sufficient size to handle 91.4-cm pipe was not available. The use of DP burybarges reduced the number of barge passes with anchoring from three passes (one pass with laybarge and two passes with burybarge) to one pass, with a concomitant reduction of potential live bottom impact from 2,266.8 to 716.5 ha, a 68% reduction. The use of mid-line buoys was estimated to further reduce the potential live bottom impact to 108.7 ha, an additional 27% reduction (total reduction of 95%).

Total Impact Reduction

Reduction in potential live bottom impact achieved through the implementation of the measures described above, are summarised in Table 5. Total potential impacts were reduced by 93%, from 2,457.2 to 186.4 ha.

The greatest potential impact reductions on both a hectare basis and a percentage basis were achieved through the use of mid-line buoys, which significantly reduced the area of live bottom that would be swept by anchor cables. However, the impact associated with anchor cable sweep was assumed to be less severe and more temporary than impacts due to anchor scars or trenching. Cable sweep was expected to result mostly in impact to the benthic organisms, which could subsequently

TABLE 5. Total reduction in potential live bottom impact associated with construction of the Gulfstream Natural Gas System due to the implementation of mitigative measures by the project proponent.

Impact Type	Live Bottom Impact			
	Initial	Reduced	Difference	% Reduced
Trenching	183.3 ha	74.7 ha	108.7 ha	60 %
Pipelay	1.4 ha	1.3 ha	0.04 ha	3 %
Anchor Scar	5.7 ha	1.7 ha	4.0 ha	70 %
Cable Sweep	2,266.8 ha	108.7 ha	2,158.1 ha	95 %
Total Impact	2,457.2 ha	186.4 ha	2,270.8 ha	93 %

recover or re-colonize the area. Anchor scars and trenching would remove the hard substrate that forms the habitat. These substrate voids may be re-colonized on the exposed substrate, however moving sediments may well fill the void first, effectively preventing re-colonization.

Habitat Replacement Mitigation and Monitoring

Regulatory and resource management agencies required the project proponent to replace some of the live bottom habitat that would be impacted by the construction of the subsea pipeline, as mitigation. In state waters, habitat replacement was required for all live bottom impacts. In federal waters, habitat replacement was required only for the most productive and sensitive live bottom, which was considered to be live bottom on exposed rock.

The surface area of required replacement habitat was calculated with the National Oceanic and Atmospheric Administration (NOAA) Habitat Equivalency Model (NOAA, 1995), which was developed for seagrass mitigation, but has been used previously for several coral reef mitigation projects. Literature on the growth of live bottom organisms was reviewed and discussed with resource agency staff. It was concluded that impacts that largely affected the organisms and not the habitat (cable sweep impacts, sediment deposition), may have a duration of ten years, and that those impacts that damage the habitat (pipelay, anchor scars, and trenching) may have a duration as long as 100 years. Due to regulatory jurisdictions, replacement habitat requirements were calculated separately for state waters (3.2 ha) and federal waters (20.0 ha).

Habitat replacement in state waters (Gulfstream Natural Gas System, 2001) consisted of the deposition of limestone at six sites. Sites were selected that were: 1) located 305 to 610 m from the centerline so the mitigation was near the impact but not so near as to be impacted by construction, 2) located over hard substrate with 0.3 m or less sand veneer so that the replacement habitat would not self-bury in deep sediments, and 3) located in proximity to, but not on, existing live bottom communities that could act as a source for colonization by invertebrates. Each site was 73.5 × 73.5 m or 0.5 ha, and contained 16 or 17 limestone modules, each within 15.2 m of another module. Each module consisted of wire mesh 7.3 × 6.1 m or 44.6 m^2 covered with a single layer of limestone boulders with nominal diameters of 0.3–0.6 m. The limestone was mined from onshore Florida but was similar in origin, chemical composition, density, and porosity to the limestone found at existing West Florida live bottom, and therefore thought to be suitable substrate for colonization by live bottom epi-fauna. The limestone was placed on wire mesh designed to hold the boulders together until they were encrusted and cemented by epi-fauna.

Habitat replacement in federal waters consisted of 12 sites totalling 20.0 ha, with the same location parameters as listed above for the state waters sites (Continental Shelf Associates, 2001). Sufficient limestone boulders (approximately 111,600 boulders weighing 114,487 metric tonnes) with nominal diameters of 0.6–1.2 m were interspersed within that 20.0 ha to cover 9.0 ha of the seafloor. Additionally, 150 limestone/concrete modules, designed for improved fish habitat, were deployed at the mitigation sites (Fig. 6). Each module consisted of a 1.8 × 2.4 × 0.5 m concrete base in which 0.5–0.9 m diameter limestone boulders were embedded in a fashion that forms an inner cave with two exits.

Numerous monitoring efforts were also required as part of the approved mitigation and monitoring plan. Monitoring included: 1) measurement and quantification of actual impact

FIGURE 6. Limestone and concrete module used in habitat replacement in federal waters.

though a sampling program to verify impacts were within estimates provided by the proponent during permitting (size of anchor scars, cable sweeps, and trench width), 2) monitoring of the recovery of impacted areas, and 3) monitoring of the rate of colonization of replacement habitat by live bottom communities. Basic monitoring units consisted of: 10-m and 25-m transects, some of which were permanently marked and sampled repetitively and some that were randomly distributed during each sampling event, and permanent 0.707 m^2 plots. Monitoring is taking place over a five-year period. The transects are photographed with high-resolution video and sampled frame by frame. Still photography is used to sample the plots. Parameters measured include percent cover by taxon, density and cover of dominants, and species richness. Recovery and colonization of experimental transects and plots are compared by analysis of variance with reference transects and plots. The replacement habitat has been installed and the first annual sampling has taken place, however the data is not yet available.

DISCUSSION

Subsea pipelines cannot be installed without some impact to the seafloor. In some areas, all sensitive or highly productive (and protected) benthic habitats cannot be avoided. In the case of the Gulfstream Natural Gas System, all such habitats were avoided by prudent routing, except for live bottom, which is widely distributed on the West Florida Shelf. However, impacts can be greatly reduced. During the design and construction of the Gulfstream Natural Gas System, potential impacts to live bottom were reduced by approximately 93% by re-routing some sections of the pipeline centerline, and through selection of construction equipment and methodology. It behooves proponents of such large projects to reduce these impacts as much as they can for environmental reasons as well as cost. Although many of the measures used to reduce impacts come at considerable cost to the proponent, the costs are generally much less than the cost of the habitat replacement that is often required for the un-avoided impacts. This cost savings, as well as the ability to obtain the required permits (when all practicable avoidance efforts are undertaken) on schedule, makes avoidance a wise avenue to direct project dollars.

REFERENCES

Bohnsack, J.A. 1979. "Photographic Quantitative Sampling of Hard Bottom Communities." *Bulletin of Marine Science.* 29(2):242-252.

Centaur Associates, Inc. 1984. *Mitigation of seafloor conflicts between oil and gas pipelines and commercial trawl fisheries on California's outer continental shelf.* OCS Study / MMS 84-0058, USDI Minerals Management Service, Washington, D.C.

Continental Shelf Associates. 1992. Mississippi-Alabama shelf pinnacle trend habitat mapping study. OCS Study / MMS 92-0026. USDI Minerals Management Service, New Orleans. 3 vols.

Continental Shelf Associates. 2001. Revision 1 of the 17 May 2001 hard / live bottom mitigation plan – federal waters. Report prepared for Gulfstream Natural Gas System L.L.C. 19 pp.

ENSR. 2000. Evaluation of physical impacts associated with installation of the Gulfstream Natural Gas System in Florida waters. Report prepared for Gulfstream Natural Gas System by ENSR, Westford, MA.

Gulfstream Natural Gas System, L.L.C. 2001. Mitigation and monitoring plan for impacts to live bottom associated with the installation of the Gulfstream Natural Gas System in Florida state waters. Report prepared by ENSR International for Gulfstream Natural Gas System, L.L.C. 24 pp.

NOAA. 1995. Habitat equivalency analysis: an overview. Policy and Technical Paper Series No. 95-1. Damage and Assessment Program, National Oceanic and Atmospheric Administration.

Parker, R.O., D.R. Colby, and T.D. Willis. 1983. Estimated amount of reef habitat on a portion of the U.S. South Atlantic and Gulf of Mexico continental shelf. *Bulletin of Marine Science.* 33(4):935-940.

Sea Byte, Inc. 2001a. Gulfstream Natural Gas System benthic habitat survey: federal waters: April 2001. Report prepared by Sea Byte, Inc. and ENSR International for Gulfstream Natural Gas System.

Sea Byte, Inc. 2001b. Gulfstream Natural Gas System benthic habitat survey: Florida state waters: April 2001. Report prepared by Sea Byte, Inc. and ENSR International for Gulfstream Natural Gas System.

Aspects of Polycyclic Aromatic Hydrocarbons in Offshore Sediments in the Azeri Sector of the Caspian Sea

Boehm, P.D. (pboehm@exponent.com), J.S. Brown (Exponent Inc. Natick, MA, USA), C.L. Maxon, F.C. Newton, and Y. Galperin (Battelle Memorial Institute, Duxbury, MA, USA)

ABSTRACT: A detailed investigation of the composition and concentration of polycyclic aromatic hydrocarbons (PAH) (2 to 6 ringed parent and alkylated homologous series) and saturated hydrocarbons in sediments of Caspian Sea, 220 to 645 m depth in Azerbaijan waters, was conducted. Sediments in the study area consisted primarily of fine and very fine silts and average sediment organic carbon exceeded 3% of sediment dry weight. Significant levels of petroleum-related hydrocarbons, measured as 41 alkylated and parent polycyclic aromatic hydrocarbons (Σ PAH) were evident throughout the Absheron region. Σ PAH concentrations ranged from 830 to 3600 ng/g with a mean total concentration for all stations (n = 15) of 1700 ng/g. Most sediments showed a relatively uniform distribution of PAH, with variations in PAH levels being directly related to total organic carbon (TOC) and silt/clay content of the sediments. However, PAH inputs related to an underwater "mud volcano" in the eastern part of the study area were evidenced by significantly higher ΣPAH/TOC values and a higher proportion of low molecular weight PAH in these samples than in surrounding sediments. The background levels of PAHs in the Absheron region are attributed predominantly to petrogenic sources with minor contributions of pyrogenic (combustion related) PAH components. Assessment of potential impacts from future oil exploration and production activities need to take into account the relatively high pre-exploration, petrogenic PAH levels in the region's sediments.

INTRODUCTION

The oil and gas industry has undertaken extensive planning for the exploration and development of the offshore region of the Caspian Sea located in Azerbiajan. Though oil development in Azerbiajan dates back to the early part of the century, these new plans have targeted deepwater locations in the central Caspian area.

In response to requirements governing Environmental Standards and Practices of the Production Sharing Agreement (PSA) between Chevron Overseas Petroleum Azerbaijan Limited (COPAL) and the State Oil Company of the Azerbaijan Republic (SOCAR), COPAL conducted an environmental baseline survey (EBS) of the Absheron Prospect (Fig. 1). The completed EBS (Arthur D. Little, 1999) describes the chemical, physical, and biological properties of the Absheron Prospect sediments and the water column. Emphasis was placed on describing the sedimentary environment since sediments and their biota are relatively immobile and integrates the effects of depositional processes, including physical disturbances (Boesch and Rabalais, 1987). The benthic environment is a depositional sink for discharged drill materials and potential contaminants, and is recognised as an important marine habitat, which may be impacted by sedimentation of contaminated particles from petroleum exploration and development operations.

The general environmental features of the Caspian Sea and the Absheron Prospect area are described by Kosarev and Yablonska, 1994. The Absheron Prospect is located along the lower slope of the northern sector of the south Caspian Basin (Fig. 1). Prospect depths range from 220 to 645 m (Fig. 2),

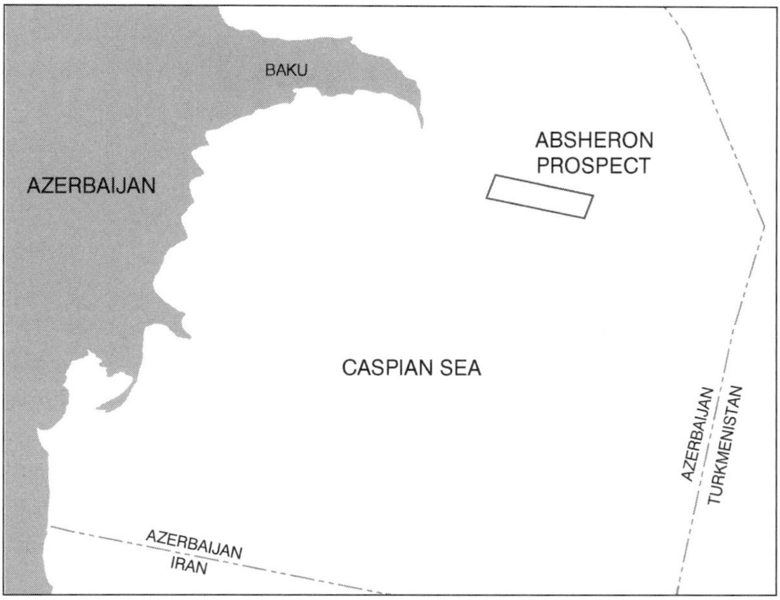

FIGURE 1. Location of Absheron Prospect area.

defining a topographically complex region of variable slope and configuration, including the presence of a mud volcano mound, evident in the eastern portion of the prospect. These mud volcanoes are common benthic features in the Caspian Sea and can be attributed to surface sediment mounds created by releases of hydrocarbon gases and other escaping subsurface formation fluids (Kosarev and Yablonska, 1994). A large sediment overburden typifies the entire south Caspian Basin, consisting of terrigenous minerals mixed with carbonates and fragmented remains of biogenic origin, primarily mollusc shells and silicious diatom skeletons. Regional sediments are typified by shelf-depth sands grading into silts and clays over offshore slopes and basins (Kosarev and Yoblanska, 1994).

Sediment chemical concentrations in the Absheron region are influenced by a history of onshore and offshore activities dating back over 100 years. Contaminants enter the Caspian Sea from a variety of onshore and offshore sources. Oil seeps from active mud volcanoes and through sea floor fissures elsewhere in the area, may provide additional sources of hydrocarbons to sur-

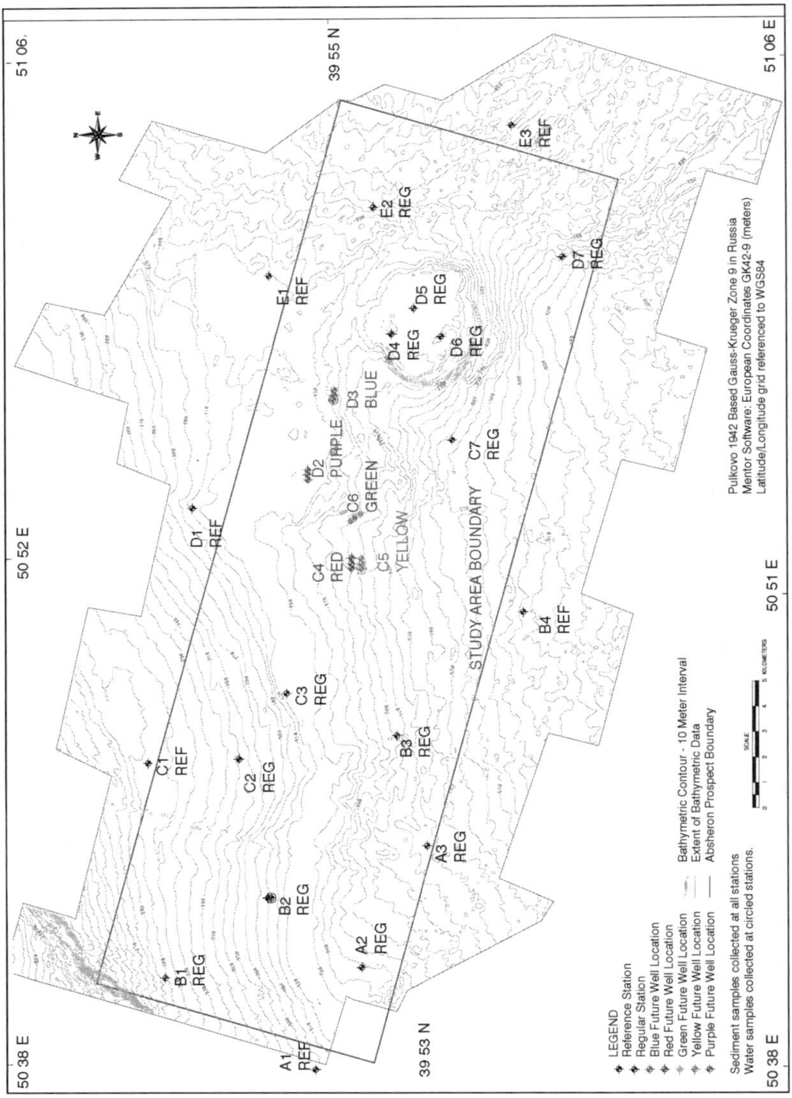

FIGURE 2. Absheron Prospect and reference area sampling stations.

rounding sediments. Although inputs from anthropogenic sources have been reduced in recent years, Baku Bay and surrounding coastal waters remain impacted by long-standing pollution sources (Kosarev and Yoblanska, 1994), and in turn, may influence contaminant concentrations in offshore sediments. Additionally, as in many coastal areas that are rich in fossil hydrocarbons, erosion of land based organic shales may contribute hydrocarbons to offshore sediments (Bence et al., 1996).

Among the important aspects of the EBS is the determination of the existing, pre-exploration levels of contaminants. In particular, because of their toxicological properties, specific attention has been paid to the concentrations and compositions of the petroleum-derived (i.e. petrogenic) polycyclic aromatic hydrocarbons (PAHs), as well as other sources of PAHs (Boehm, 1987; Neff et al, 1987; many others). PAHs are ubiquitous in the aquatic environment (e.g. Laflamme and Hites, 1978) and are derived from a variety of sources of both petrogenic and pyrogenic (i.e. combustion) sources (Blumer, 1976). Additionally, many areas that are targeted for oil exploration or transportation have records of natural oil seepages (Stuermer et al., 1982; National Research Council, 1985; Page et al., 1996). The most likely sources of existing PAH to Caspian Sea sediments are sewage and industrial effluents, atmospheric fallout of combustion particulates, ship ballast water, spilled petroleum, and natural seepage (Kosarev and Yoblanska, 1994).

The present study was an important component for determining the potential for environmental effects from PAH discharge and deposition from exploration and production activities. In addition, the measurements were aimed at the need to differentiate PAH from the pre-existing background or baseline, from those that might be added from oil development.

METHODS AND MATERIALS

Study Design

The EBS was designed to cover the complete range of depth, habitat, and longshore extent of the Absheron block (Figs. 1 and 2), with 24 sampling stations, including emphasis on the mud volcano habitat. Sampling stations, bathymetry, and special features of the prospect are shown in Figure 2. Survey sites were selected to encompass a representative range of depths, habitat types, special features, and topography, with inclusion of proposed well drilling sites. Stations were aligned roughly along five transects (Fig. 2, A through E), positioned along the longshore extent of the prospect and trending cross-slope to include representative depths and habitat types. Sampling stations were comprised of a total of 15 sites in the block (including five prospective well sites and five sites sampled in an earlier (1998) baseline survey), three sites in the vicinity of the mud volcano, and six reference sites (Fig. 2). Three replicate grab samples were collected at each of the five prospective well sites for the purpose of estimating small-scale variability in these potential impact areas. Single grabs were collected at the remaining 19 sites. Table 1 summarises the EBS sampling plan, including types and numbers of samples, sampling dates, depths, and geographic coordinates of all stations. The sampling program was designed to obtain benthic chemical and biological (see Arthur D. Little, 1999) data from a complete range of habitat types, encompassing variability related to depth, longshore position, topography, and proximity to well sites and the mud volcano. Refined descriptions of prospective well sites were included for the purpose of defining small-scale variability, and as a basis for future point source monitoring at these sites.

Sediments were analyzed for grain size characteristics, total organic carbon (TOC) content, petroleum hydrocarbons, including polycyclic aromatic hydrocarbons (PAH) and saturated aliphatic hydrocarbons (SHC), metals, and benthic biological communities. Only the sedimentological and PAH results are presented here.

Field Methods

Field operations were conducted from the vessel, *Svetlomor 2*, a 52-m oceanographic research vessel subcontracted by Azeri Fugro, Inc. A Starfix MN8 mobile Differential Global Positioning System (DGPS) using PCNav™ software was used to posi-

TABLE 1. Chevron Absheron baseline environmental survey sampling summary. Stations are shown in Figure 2.

Transect	Station	Sampling Date	Depth (m)	Geographic Coordinates Krassowsky 1940, Pulkovo 1942 Datum		No. of Sediment Samples
				Latitude (N)	Longitude (E)	
A	A1 (Ref)	20 Apr 99	398	39°55'24.3"	50°37'47.2"	1
	A2	21 Apr 99	470	39°54'19.5"	50°40'41.2"	1
	A3	21 Apr 99	510	39°52'55.1"	50°44'02.7"	1
B	B1	21 Apr 99	306	39°58'32.4"	50°40'22.9"	1
	B2	21 Apr 99	420	39°56'16.2"	50°42'35.7"	1
	B3	21 Apr 99	508	39°53'34.1"	50°47'06.0"	1
	B4 (Ref)	21 Apr 99	605	39°50'51.9"	50°50'32.1"	1
C	C1 (Ref)	21 Apr 99	300	39°59'17.8"	50°45'52.7"	1
	C2[1]	21 Apr 99	358	39°56'57.5"	50°46'27.0"	1
	C3[1]	21 Apr 99	430	39°55'56.1"	50°48'16.7"	1
	C4[2]	21 Apr 99	493 - 499	39°54'32.5"	50°51'52.1"	3
	C5[2]	21 Apr 99	505 - 507	39°54'19.8"	50°51'51.6"	3
	C6[2]	21 Apr 99	494 - 497	39°54'28.3"	50°53'08.6"	3
	C7[1]	24 Apr 99	570	39°52'22.5"	50°55'18.2"	1
D	D1 (Ref)	22 Apr 99	404	39°58'06.5"	50°53'15.1"	1
	D2[2]	24 Apr 99	440 - 442	39°55'28.4"	50°54'20.3"	3
	D3[2]	24 Apr 99	468 - 470	39°54'54.4"	50°56'35.3"	3
	D4[3]	24 Apr 99	460	39°53'40.5"	50°58'17.4"	1
	D5[3]	24 Apr 99	454	39°53'11.4"	50°59'00.5"	1
	D6[1,3]	24 Apr 99	484	39°52'36.3"	50°58'13.8"	1
	D7[1]	23 Apr 99	614	39°49'59.9"	50°00'25.1"	1
E	E1 (Ref)	23 Apr 99	450	39°56'16.2"	50°00'00.0"	1
	E2	23 Apr 99	500	39°54'03.2"	50°01'49.9"	1
	E3	23 Apr 99	550	39°51'43.8"	50°02'26.6"	1

Ref = Reference Station.
[1]Stations previously sampled in February, 1998.
[2]Proposed well location (location coordinates shown for middle replicate only).
[3]Samples collected in vicinity of mud volcano.

tion and navigate the vessel during all fieldwork. Accuracy of the navigation system was approximately ± 2 m. Positions of sampling stations were recorded in eastings and northings (Krassowsky 1940, Pulkovo 1942 datum) at the time of sample collection. Location data were converted to latitude/longitude coordinates in Krassowsky 1940, Pulkovo 1942 datum. Positions were printed out at the time of sample collection, recorded electronically, and stored on an onboard computer until subsequent transfer to the field database. Sediment samples were collected from the stern of the vessel using a stainless steel dual 0.2 m^2 vanVeen grab sampler. This sampler allowed collection of separate co-located 0.1 m^2 grab samples each for biological and chemical/physical testing without interference or cross contamination.

Chemistry and Grain Size

Prior to sediment subsampling, overlying water was decanted by gently tipping the grab and letting water escape through the 0.5 mm stainless steel mesh lids. The top 2 cm of sediment was removed and placed into pre-cleaned sample containers using a Kynar-coated scoop. The 2-cm deep scoop was used to gauge the depth of the collected sample. Subsamples for chemical analyses were collected from the center of the grab to prevent potential contamination from the grab sampler. The sample scoop was cleaned between each grab sample as previously described. Sediment aliquots were placed into 250-ml pre-cleaned I-Chem™ jars for hydrocarbon analyses, 25-ml paper containers for metals analysis, 500-ml plastic ziplock bags for total organic carbon (TOC), and 1 litre polyethylene jars for grain size analysis. Samples for hydrocarbons, metals, and TOC were stored frozen onboard ship and throughout subsequent transfer to analytical laboratories. Samples for grain size were stored at 4°C until analyzed. The processing of biological samples is presented elsewhere (Arthur D. Little, 1999).

Laboratory Methods

The polycyclic aromatic hydrocarbons (PAH) analyses were performed using gas chromatography with mass spectrometer in selected ion mode (GC/MS/SIM) (Boehm et al., 1998). A total of 41 parent (i.e. unsubstituted) and alkyl homologues were analyzed. These included alkylated homologues of the 2 to 6 ringed PAHs. The multiple isomers within each alkyl homologue grouping (e.g. C3-phenathrenes) were identified in the selected ion mode (SIM). The PAHs were quantified using relative response factors for the parent PAH in each series (e.g. phenanthrene for all of the alkylated phenanthrenes) according to the method published in the U.S. Federal Register and used in numerous studies (e.g. Bence et al., 1996; Boehm et al., 1997, 1998; Hyland et al., 1990; Steinhauer and Boehm, 1992). These analytical lists (Table 2) are consistent with those performed in sediment quality studies performed by the U.S. EPA

TABLE 2. Target polycyclic aromatic hydrocarbons (PAH) and saturated hydrocarbons.

PAH	Fig. 5 Ref #	Sediment DL* (ng/g)	PAH	Fig. 5 Ref #	Sediment DL* (ng/g)
Naphthalene (N)	1	0.5	C_1-Fluoranthene/Pyrene (FP1)	25	0.5
C_1-Naphthalenes (N1)	2	0.5	C_2-Fluoranthene/Pyrene (FP2)	26	0.5
C_2-Naphthalenes (N2)	3	0.5	C_3-Fluoranthene/Pyrene (FP3)	27	0.5
C_3-Naphthalenes (N3)	4	0.5	Benzo[a]anthracene (BAA)	28	0.5
C_4-Naphthalenes (N4)	5	0.5	Chrysene (C)	29	0.5
Acenaphthylene (AC)	6	0.37	C_1-Chrysenes (C1)	30	0.5
Acenaphthene (ACE)	7	0.39	C_2-Chrysenes (C2)	31	0.5
Biphenyl (B)	8	0.5	C_3-Chrysenes (C3)	32	0.36
Fluorene (F)	9	0.5	C_4-Chrysenes (C4)	33	0.36
C_1-Fluorenes (F1)	10	0.5	Benzo[b]fluoranthene (BBF)	34	0.5
C_2-Fluorenes (F2)	11	0.5	Benzo[k]fluoranthene (BKF)	35	0.5
C_3-Fluorenes (F3)	12	0.5	Benzo[e]pyrene (BEP)	36	0.5
Dibenzothiophene (D)	13	0.5	Benzo[a]pyrene (BAP)	37	0.5
C_1-Dibenzothiophenes (D1)	14	0.5	Perylene (PER)	38	0.5
C_2-Dibenzothiophenes (D2)	15	0.5	Indeno[1,2,3-c,d]pyrene (IP)	39	0.5
C_3-Dibenzothiophenes (D3)	16	0.5	Dibenzo[a,h]anthracene (DAA)	40	0.5
Phenanthrene (P)	17	0.5	Benzo[g,h,i]perylene (BGP)	41	0.5
Anthracene (A)	18	0.5			
C_1-Phenanthrenes/Anthracenes (P1)	19	0.5			
C_2-Phenanthrenes/Anthracenes (P2)	20	0.5			
C_3-Phenanthrenes/Anthracenes (P3)	21	0.5			
C_4-Phenanthrenes/Anthracenes (P4)	22	0.5			
Fluoranthene (FL)	23	0.5			
Pyrene (PY)	24	0.5			

Petrogenic PAH = N+N1+N2+N3+N4+AC+ACE+B+F+F1+F2+F3+D+D1+D2+D3+1/2(P+A) +P1+P2+P3+P4+C1+C2+C3+C4
Pyrogenic PAH = ½(P+A) +FL+PY+FP1+FP2+FP3+BAA+C+BBF+BKF+BEP+BAP+IP+DAA+BGP
*DL is laboratory detection limit. Fig. 5 Ref # is used as a key to identify chemical in Figure 5.

(Ozretich et al., 2000) and natural resource damage studies (Sauer and Boehm, 1991). Typical analytical precision (i.e. the same sample analyzed in replicate) is 10% to 50% for individual PAH analytes and 20% for ΣPAH.

Grain size was analyzed by a combination of wet and dry sieving of the larger fraction (4 Phi, i.e. ≥ 63 μm) and by electrical sensing particle zone counting for particles smaller than 4 Phi (i.e. < 63 μm). These two methods, requiring a separate subsample for each, are detailed by Buchanan (1984) and Little et al. (1984). Results were reported for 30 Phi sizes, as well as for four size categories: gravel, sand, silt, and clay. The sum of silt and clay (i.e. < 63 μm) was also calculated and is referred to as "fine sediments".

Total organic carbon (TOC) content was analyzed by combustion followed by infrared spectophotometry with a non-dispersive infrared detector. Samples were thoroughly homogenised with a stainless steel spatula and aliquoted into porcelain dishes where they were acidified and mixed to remove inorganic carbon. Samples were then dried for 12 hours at 70–75°C and ground into a homogeneous powder. Approximately 0.02 g of sample were placed into a platinum container and introduced into a high temperature total organic carbon analyzer that converts organic carbon into carbon dioxide through high temperature (> 500°C) combustion. The concentration of organic carbon (measured as carbon dioxide) was then determined by infrared spectrophotometry using a non-dispersive infrared detector.

RESULTS AND DISCUSSION

Physical Characteristics

Grain size and TOC results are summarised in Table 3. Absheron region contour plots for mean grain size and TOC distributions are shown in Figure 3.

Sediment Grain Size Characteristics

Sediments were characterised primarily by fine silts (17 stations) and very fine silts (5 stations). A single station in the mud volcano vicinity (Station D6) consisted of coarse silts. The average grain size from all stations was 0.0100 mm. The finest sediments were located near the prospective well sites, Stations C4 and C5. Sediment grain size was not significantly correlated with station depth.

Silt particles (0.004–0.063 mm) were the prevalent grain category, averaging 71% of sediment dry-weight, ranging from

TABLE 3. Sediment total organic carbon (TOC) and grain size summary.

	Average	Standard Deviation	Range	Highest Stations	Lowest Stations
Depth (m)	466.8	79.1	304 to 614	D7, B4 Ref	C1 Ref, B1
TOC	3.17	0.84	0.52 to 4.49	D4, D5	D6, B1
Mean Grain Size	0.01034	0.00693	0.00615 to 0.04124	D6, D5	C5 C4
Mean Phi	6.753	0.568	4.6 to 7.378	C5 C4	D6, D5
Sorting	1.554	0.506	1.012 to 3.375	A1, C3	D6, E3
Percent Sand	7.6	7.5	0 to 33.71	B1, D6	A1,B3
Percent Silt	70.93	9.53	45.25 to 91.74	B3, A1	D6,B4 Ref
Percent Clay	20.88	7.46	6.64 to 34.5	B4 Ref, E1 Ref	B3, D6
Skewness	-1.686	0.621	-2.597 to -0.302	B3, A1	B4, E1 Ref
Kurtosis	6.841	2.661	2.133 to 11.761	B4 Ref, E1 Ref	D6, A1

45% to 92%. Sorting indices below 1.5 represent various grades of well-sorted sediments. Fourteen of the 24 sites fell into this category, with a weak but significant tendency for better sorting at the shallower depths. Moderate sorting (Index = 1.5–2.0) characterised seven stations, while three stations (D5, D6, E3) had poorly sorted sediments (Index > 2.0). Those at the mud volcano site (D6) had an index of 3.4, classified as extremely poor sorting, indicating a high degree of sediment heterogeneity in this vicinity. Better sorting of sediments was evident in the western portion of the prospect where more even slope and bathymetry prevailed in contrast with the variable topography and slope associated with the mud volcano and areas to the east.

Organic Carbon

On average, TOC comprised 3.17% of sediment dry weight, but was highly variable in the study area, ranging from 0.52% to 4.49%. Seventeen of the stations had values exceeding 3%.

A general gradient of increasing TOC from west to east across the prospect area was apparent (Fig. 3). Lowest TOC concentrations were recorded at the shallow Station B1 and from Station D6, where sediments were coarsest. Highest values were recorded from the D-Transect at Stations D4 and D5, indicating a great deal of physical and chemical heterogeneity and

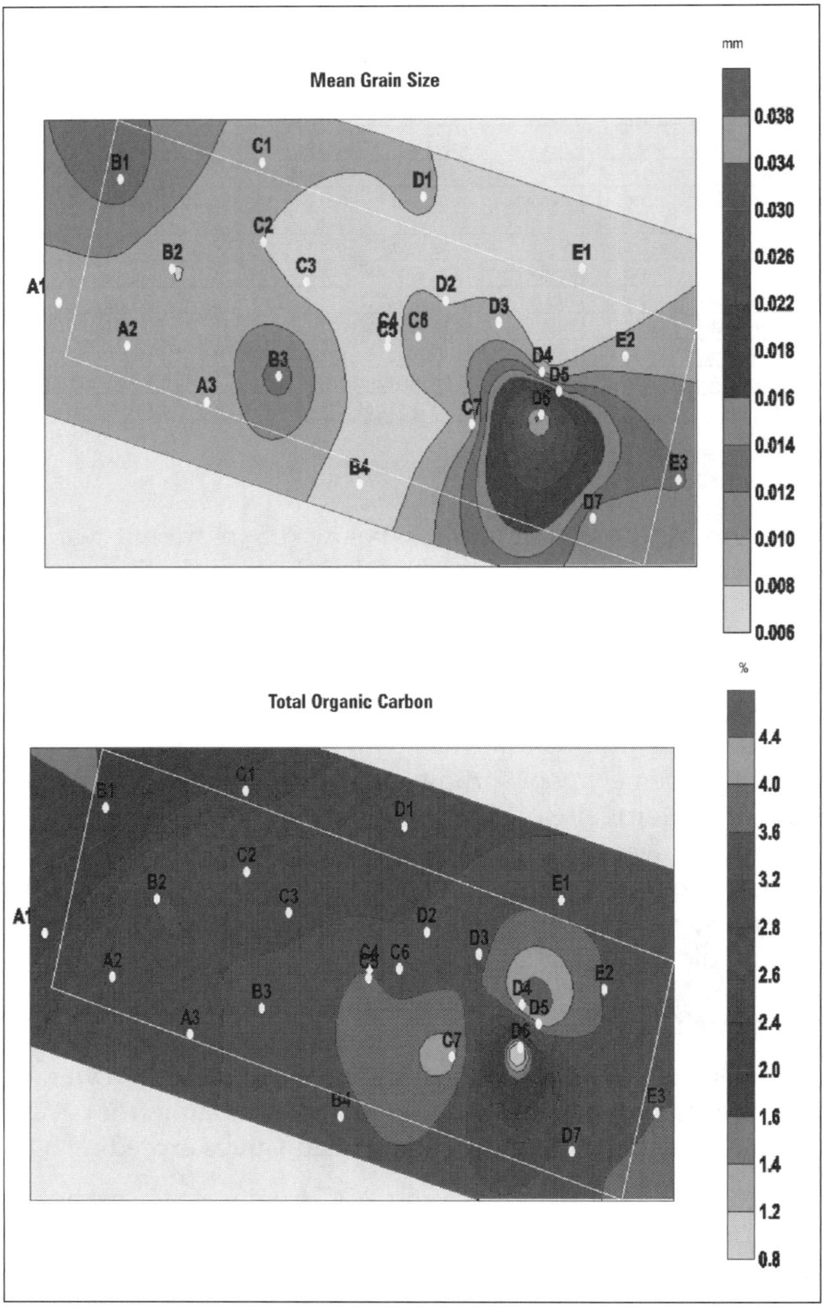

FIGURE 3. Distribution of sediment mean grain size and total organic carbon (TOC).

the possibility of carbon sources from seeps or gas charged sediments in this mud volcano area. TOC concentrations were positively correlated with depth and with finer sediments, as is typical for aquatic sediments. The distribution of TOC generally helps to define the sediment contaminant regime, as is seen below.

The TOC values in the mud volcano area varied considerably, with some of the highest TOC values observed just to the north of the stations and some of the lower values just to the south of the feature, in the region of coarser, more poorly sorted sediment (Fig. 3).

Chemical Characteristics: PAH Concentrations

Total target polycyclic aromatic hydrocarbons (ΣPAH) were measured at 15 stations. Spatial distributions of ΣPAH are shown in Figure 4. Comparisons of mean total PAH concentrations for the Absheron Prospect, mud volcano, and the reference stations (see Table 1) are presented in Table 4. Levels higher than 1 ppm (1000 ng/g) of PAH were evident throughout the study area's sediments. Total PAH concentrations ranged from 830 to 3580 ng/g (i.e. 0.83 to 3.58 ppm) with a mean total concentration of PAH for all stations of 1730 ng/g. The PAHs in the Absheron-Caspian Sea study area are contrasted with PAH (identical target analytes; i.e. 2–6 ringed parent and alkylated PAHs) assemblages from other offshore areas in Table 4. The PAH in the Caspian deepwater sediments are higher than many sediment levels, but are consistent with levels observed in areas of known natural fossil petrogenic inputs (e.g. Beaufort Sea, Alaska; see, Steinhauer and Boehm, 1992)).

Sediments located within the vicinity of the Absheron mud volcano (Stations D4, D5, D6) had higher ΣPAH concentrations compared to the remaining study area, with the exception of Station B3 sediments, which was highest in ΣPAH (3.6 ppm).

No significant differences in mean total PAH sediment concentrations were observed between Absheron Prospect and reference areas, when mud volcano stations (D4, D5, D6) were excluded ($P < 0.05$). The mud volcano area was generally

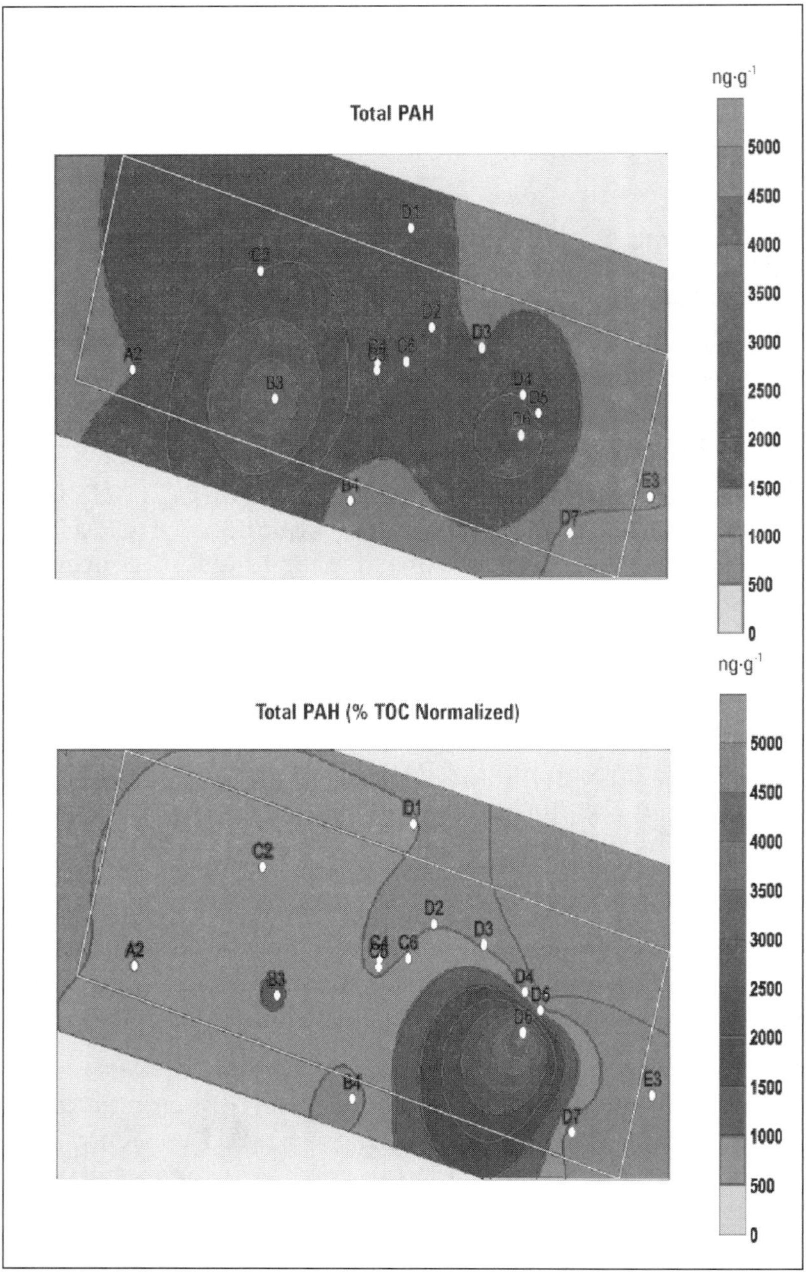

FIGURE 4. Spatial distribution of total PAH and TOC-normalised total PAH.

TABLE 4. Summary of PAH (ng/g) in sediments.

Parameter	No. of Stations	Minimum	Maximum	Mean	Standard Deviation
CASPIAN SEA STUDY AREA					
Reference Area					
Total PAH	3	1,040	1,590	1,260	290
Absheron Prospect					
Total PAH	9	834	3,560	1,760	747
Mud Volcano					
Total PAH	3	1,780	2,600	2,120	427
OTHER OFFSHORE REGIONS					
Beaufort Sea, AK (Boehm et al., 2002)					
Total PAH	48	16	10,400	951	1,585
Shelikof Strait, AK (Brown et al., 2003)					
Total PAH	94	120	1,080	459	151
Prince William Sound, AK (Page et al. 1995) (reference stations)					
Total PAH	26	--	--	517	560
Total PAH	15	--	--	435	243
Total PAH	12	--	--	619	315
Gulf of Alaska (Page et al. 1995) (reference stations)					
Total PAH	25	--	--	1481	1800
Gulf of Guinea, Nigeria (Brown and Boehm, unpublished)					
Total PAH	13	20	71	42	13
Santa Maria Basin, CA (Hyland et al., 1990)					
Total PAH	--	10	1,500	200	--

higher in ΣPAH, suggestive of a localised input of hydrocarbons on the sea floor in the vicinity of the mud volcano, both on an absolute basis and especially when expressed on an organic carbon basis (Fig. 4). This latter relationship strongly suggests a localised petrogenic input of PAHs (i.e. seepage) from the mud volcano itself, as an incremental input on top of an area-wide PAH input.

PAH concentrations, based on an equivalent PAH analyte list (i.e. ΣPAH; 2 to 6 ringed parent and alkylated compounds) for some other deep subtidal offshore areas that receive natural inputs of fossil hydrocarbons from offshore seeps and/or eroding onshore organic shale, are presented in Table 4. Some of these areas are subject to natural oil seepages and/or the input of fossil hydrocarbons from erosion of shales and coal beds. In general, the background PAH levels for the Caspian Sea sediments studied here are higher in overall PAH content. This finding appears to be rooted in three factors: the low energy-depositional nature of the study region as evidenced by the high silt clay content of the sediment; the proximity of the sampling stations to an apparent localised source of fossil PAHs (i.e. the "mud volcano" feature); the likely offshore transport of fossil PAHs, and related pollutant sources from nearshore/onshore activities.

PAH Composition and Source-Types

The basic compositional features of the PAH in the Absheron surface sediments are shown in Figure 5.

General sources of hydrocarbons are typically identified by evaluation of the ratios of selected constituents (i.e. diagnostic ratios) and parameters (e.g. Volkman et al., 1992; Douglas et al., 1992; Steinhauer and Boehm, 1992). Relatively high concentrations of low molecular weight PAH (2-, 3- and 4-ring PAH and their alkylated homologues) are typically associated with petrogenic sources. Increased concentrations of the parent high molecular weight PAHs in the absence of greater abundances of alkylated homologues (4-, 5-, and 6-ring PAH) generally suggest pyrogenic sources.

The PAH distributions (e.g. Fig. 5) exhibit the presence of a full suite of 2-, 3-, 4-, 5-, and 6-ring PAH compounds, and pronounced petrogenic distribution, with alkylated PAHs dominating the PAH assemblages. All sediment samples had PAH distributions with a higher content of petrogenic versus pyrogenic compounds, as evidenced in petrogenic PAH/pyrogenic PAH (Table 2) ratios of 4.1 to 8.9 with a grand mean value of 5.3. The observed values for this diagnostic ratio suggest that background sediments in the entire study area contain a high proportion of petrogenic PAH, originating from 1) natural seeps, 2) historic crude oil releases, and 3) terrestrial runoff of eroded petroleum sources rocks. Evidence for a localised input from the mud volcano has been presented above, and was based on the elevated PAH/TOC values.

The PAHs in the samples appear relatively unweathered with a full suite of the more weathering-susceptible naphthelenes. Similar results have been reported for surface sediment distribution from other offshore areas of high potential for

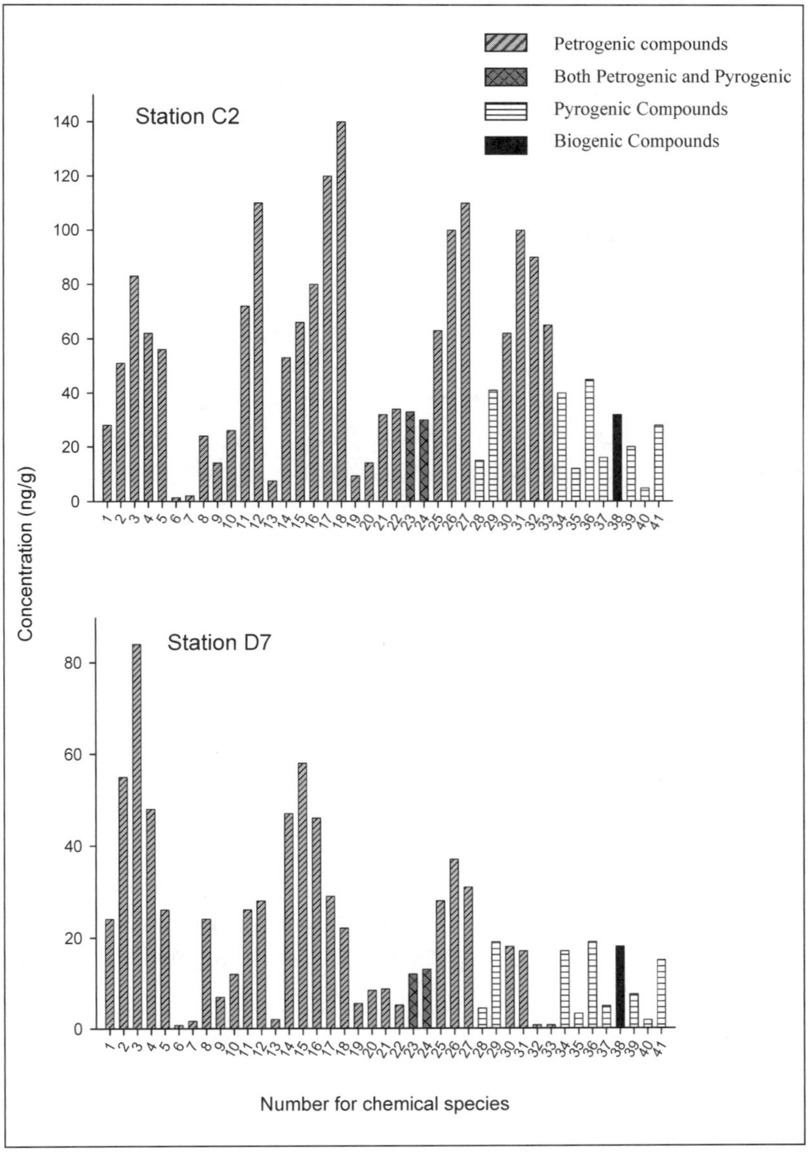

FIGURE 5. Distribution of PAHs for two representative sediment stations. Table 2 provides a key to the reference number for identifying chemical species.

recoverable petroleum resources and of high silt/clay (i.e. fine grained), high TOC content. All of these areas, including the Absheron study area of the Caspian Sea, exhibit a relatively "fresh" PAH assemblage indicative of either: 1) significant input of local oil seepages with preferential weathering of the saturated hydrocarbon fraction, and/or 2) inputs of eroding petroleum source rocks/shales from adjacent land areas (Boehm et al., 2001).

These findings indicate that subsequent monitoring of any chemical inputs and associated benthic effects from offshore exploration and oil and gas development in the offshore regions of the deep basins of this and presumably other areas of the Caspian Sea, need to take into account detailed PAH concentration distributions, localised sources from oil seepages, and the detailed compositional features of the PAH assemblages, to be able to distinguish the "environmental baseline" from potential future impacts.

ACKNOWLEDGEMENTS

This study was supported by Chevron Texaco and its subsidiary Chevron Azerbiajan Ltd.

REFERENCES

Little, Arthur D. 1999. *Absheron Prospect Environmental Baseline Study*—Final Report to Chevron Overseas Petroleum Azerbaijan Limited (COPAL), 300 pp.

Bence, A.E., K.A. Kvenvolden, and M.C. Kennicutt. 1996. "Organic Geochemistry of the Sediments of the Gulf of Alaska." *Organic Geochemistry.* 27:7-42.

Blumer, M. 1976. "Polycyclic Aromatic Hydrocarbons in Nature." *Scientific American.* 234:34-45.

Boehm, P.D. 1987. "Transport and Transformation Processes Regarding Hydrocarbon and Metal Pollutants in Offshore Sedimentary Environments." In D.F. Boesch and N.N. Rabalais (Eds.), Long *Term Environmental Effects of Offshore Oil and Gas Development*, pp. 233-286, Elsevier Applied Science, NY.

Boehm, P.D., L. LeBlanc, J. Trefry, P. Marajh-Whittemore, J. Brown, A. Schutzberg, A. Kick, and H. Costa. 1991. *Beaufort Sea Monitoring Program: Analysis of Trace Metals and Hydrocarbons from Outer Continental Shelf (OCS) Activities.* Report, Prepared for U.S. Department of the Interior, Minerals Management Service, Anchorage, Alaska.

Boehm, P.D., J.S. Brown, J. Trefry, R. Spies, S. Graham, C.A. Howell, S. Newton, J. Gold, and S. Metz. 1999. *Sediment Quality in Depositional Areas of Shelikof Strait and Outermost Cook Inlet, Draft Final Report.* Prepared by Arthur D. Little, Inc. for U.S. Department of the Interior, Minerals Management Service, Anchorage, Alaska. MMS Contract No. 1435-01-97-CT-30830.

Boehm, P.D., G.S. Douglas, W.A. Burns, P. J. Mankiewicz, D.S. Page, and A.E. Burns. 1997. "Application of Petroleum Hydrocarbon Chemical Fingerprinting and Allocation Techniques after the *Exxon Valdez* Oil Spill." *Marine Pollution Bulletin.* 34:599-613.

Boehm, P.D., D.S. Page, E.S. Gilfillan, A.E. Bence, W.A. Burns, and P.J. Mankiewicz. 1998. "Study of The Fates and Effects of The Exxon Valdez Oil Spill on Benthic Sediments from Two Bays in Prince William Sound, Alaska. 1. Study Design, Chemistry, and Source Fingerprinting." *Environmental Science and Technology.* 32:567-576.

Boehm, P.D., D.S. Page, W.A. Burns, A.E. Bence, P.J. Mankiewicz, and J.S. Brown. 2001. "Resolving the Origin of the Petrogenic Hydrocarbon Background in Prince William Sound, Alaska." *Environmental Science and Technology.* 35:471-479.

Boesch, D.F. and N.N. Rabalais (Eds.). 1987. *The Long-Term Effects of Offshore Oil and Gas Development: An Assessment and Research Strategy.* Report to the Interagency Committee on Ocean Pollution, Research, Development and Monitoring. Elsevier Applied Sciences Publishing, Barking, Essex, England. 708 pp.

Buchanan, J.B. 1984. "Sediment Analysis." In N.A. Holme and A.D. McIntyre (Eds.), *Methods for the Study of Marine Benthos,* pp. 41-65. IBP handbook No. 16. 2nd Edition, Oxford: Blackwell.

Douglas, G.S., K.J. McCarthy, D.T. Dahlen, J.A. Seavey, W.G. Steinhauer, R.C. Prince, and D.L. Elmendorf. 1992. "The use of Hydrocarbon Analyses for Environmental Assessment and Remediation." *Journal of Soil Contamination.* 1:1-197.

Hyland, J., D. Hardin, E. Crecelius, D. Drake, P. Montagna, and M. Steinhauer. 1990. "Monitoring Long-term Effects of Offshore Oil and Gas Development along the Southern California Outer Continental

Shelf and Slope: Background Environmental Conditions in Santa Maria Basin." *Oil and Chemical Pollution.* 6:195-240.

Kosarev, A.N. and E.A. Yablonska. 1994. *The Caspian Sea.* SPB Academic Publishing, The Hague, Netherlands. 259 pp.

LaFlamme, R.E. and R.A. Hites. 1978. "The Global Distribution of PAHs in Recent Sediments." *Geochimica et Cosmochimica Acta.* 42:289-304.

Little, D.I., M.F. Staggs, and S.S.C. Woodman. 1984. "Sample Pretreatment and Size Analysis of Poorly-Sorted Cohesive Sediments by Sieve Analysis and Electronic Particle Counter." In W.R. Parker and D.J.J. Kinsman (Eds.), *Transfer Process in Cohesive Sediment Systems,* pp. 47-74. Plenum Press, NY.

National Research Council. 1985. *Oil in the Sea.* National Academy Press. Washington, D.C.

Neff, J.N., N.N. Rabalais, and D.F. Boesch. 1987. "Offshore Oil and Gas Development Activities Potentially Causing Long-term Environmental Effects." In: D.F. Boesch and N.N. Rabalais (Eds.), *Long Term Environmental Effects of Offshore Oil and Gas Development*, pp. 149-173. Elsevier Applied Science, NY.

Ozretich, R.J., S.P. Ferraro, J.O. Lamberson, and F.A. Cole. 2000. "Test of ΣPolycyclic Aromatic Hydrocarbon Model at a Creosote-contaminated Site, Elliott Bay, Washington, USA." *Environmental Toxicology and Chemistry.* 19:2378-2389.

Page, D.S., P.D. Boehm, G.S. Douglas, A.E. Bence, W.A. Burns, and P.J. Mankiewicz. 1996. The Natural Petroleum Hydrocarbon Background in Subtidal Sediments of Prince William Sound, Alaska." *Environmental Toxicology and Chemistry.* 15:1266-1281.

Page, D.S., P.D. Boehm, G.S. Douglas, A.E. Bence, W.A. Burns, and P.J. Mankiewicz. 1995. "Identification of Hydrocarbon Sources in the Benthic Sediments of Prince William Sound and the Gulf of Alaska Following the Exxon Valdez Oil Spill." *In Exxon Valdez Oil Spill: Fate and Effects in Alaskan Waters*, pp. 41-83. Astm, Philadelphia, PA USA.

Sauer, T. and P. Boehm. 1991. "The Use of Defensible Analytical Chemical Measurements for Oil Spill Natural Resource Damage Assessment." In *Proceedings, Oil Spill Conference, American Petroleum Institute.*

Steinhauer, M.S. and P.D. Boehm. 1992. "The Composition and Distribution of Saturated and Aromatic Hydrocarbons in Nearshore Sediments, River Sediments, and Coastal Peat of the Alaskan Beaufort Sea:

Implications for Detecting Anthropogenic Inputs." *Marine Environmental Research.* 33:223-253.

Stuermer, D.H., R.B. Spies, P.H. Davis, D.J. Ng, C.J. Morris, and S. Neal. 1982. "The Hydrocarbon Chemistry of the Isla Vista Marine Seep Environment." *Marine Chemistry.* 11:413-426.

U.S. Federal Register. 40CFR Subchapter J, Part 300, Subpart L, Appendix C, par. 4.6.3 to 4.6.5 (40CFRPart 300, Subpart A).

Volkman, J.K, D.G. Holdsworth, G.P. Neill, and H.J. Bavor, Jr. 1992. "Identification of Natural, Anthropogenic, and Petroleum Hydrocarbons in Aquatic Sediments." *The Science of the Total Environment.* 112:203-219.

Seasonal and Spatial Trends of Marine Birds Along Support Vessel Transects and at Oil Platforms on the Grand Banks

Burke, C.M. (Memorial University of Newfoundland, NL, Canada), G.K. Davoren (University of Manitoba, MB, Canada), W.A. Montevecchi (mont@mun.ca) (Memorial University of Newfoundland, NL, Canada), and F.K. Wiese (University of Washington, Seattle, WA, USA)

ABSTRACT: Our primary goal is to ensure the well-being of seabirds and mammals on the Grand Banks through improved ecological understanding and environmental protection. Our major objective is to conduct year-round, scientifically robust surveys to document seabird and mammal diversity, distribution, and abundance in relation to hydrocarbon activity and oceanography. Surveys along a fixed support vessel route between St. John's, NL and offshore platforms are limited by narrow spatial coverage yet have wide temporal coverage and repeatability. Twenty-one surveys, including the first winter surveys to platforms, were conducted from 1999 through 2003. Oiled murres and dovekies, the species most susceptible to oil pollution, were recorded at the Hibernia platform during February and November when the risk of mortality due to thermal stress is greatest. Roosting gulls were recorded feeding at night on fish, apparently attracted by platform illumination. Marine mammals were observed at offshore platforms. Humpback and minke whales were the most common mammals and were seen within 5–10 meters of the Hibernia platform. The abundance of storm-petrels and shearwaters near Hibernia was highest during fall and spring migratory periods, when millions of these species move onto the shelf. These are critical periods for potential mortality from flaring, oiling, and collisions. It is environmentally and ethically essential to continue independent, scientifically rigorous surveys of seabirds in relation to ongoing hydrocarbon activities. Assessments could be greatly enhanced with independent, year-round scientifically valid observation procedures on platforms. Without such surveys, it is impossible to document seabird mortality associated with marine hydrocarbon activity in eastern Canada, and hence not possible to effectively mitigate negative environmental effects.

INTRODUCTION

Pelagic surveys of seabirds are recognised worldwide as a useful tool for examining the ecological effects of natural and anthropogenic influences on marine ecosystems (e.g. Veit et al., 1996; 1997; Veit and Montevecchi, 2003). However, the level of research directed at pelagic surveys in eastern Canadian waters is insufficient to provide the information necessary to detect changes in the pelagic diversity, distribution and abundance of marine birds and mammals.

The unique Low Arctic oceanographic regime of the Northwest Atlantic supports more than 40 million marine birds of regional and global significance including long-distance migrants from the High Arctic (dovekies, thick-billed murres; note the common and taxonomic names of all species mentioned in the text are given in Table 1) and from high latitudes in the South Atlantic (greater and sooty shearwaters) and millions of locally breeding seabirds (Leach's storm-petrels, common murres, gulls) (Montevecchi and Tuck, 1987; Montevecchi, 2000; Fig. 1). This oceanographic region also supports globally significant populations of marine mammals including humpback whales (Barlow and Clapham, 1997), fin whales (Hay, 1982), sperm whales (Braham, 1984), sei whales (Mitchell and Chapman, 1977), pilot whales (Nelson and Lien, 1996),

TABLE 1. Avian and mammalian species observed during offshore support vessel surveys along the survey route and at offshore platforms.

Seabird Species Common Name	Seabird Species Scientific Name	Transect	Platform
Northern fulmar	*Fulmarus glacialis*	yes	yes
Greater shearwater	*Puffinus gravis*	yes	yes
Sooty shearwater	*Puffinus griseus*	yes	no
Leach's storm-petrel	*Oceanodroma leucorhoa*	yes	yes
Northern gannet	*Morus bassanus*	yes	no
White-winged scoter	*Melanitta fusca*	no	yes
Merlin	*Falco columbarius*	yes	no
Red-necked phalarope	*Phalaropus lobatus*	no	no
Red phalarope	*Phalaropus fulicaria*	yes	no
Pomarine jaeger	*Sterocorarius pomarinus*	no	yes
Great skua	*Sterocorarius skua*	yes	no
Glaucous gull	*Larus hyperboreus*	yes	yes
Iceland gull	*Larus glaucoides*	yes	yes
Great black-backed gull	*Larus marinus*	yes	yes
Herring gull	*Larus argentatus*	yes	yes
Ring-billed gull	*Larus delawarensis*	yes	yes
Black-legged kittiwake	*Rissa tridactyla*	yes	yes
Common murre	*Uria aalge*	yes	yes
Common murre juvenile	*Uria aalge*	yes	no
Thick-billed murre	*Uria lomvia*	yes	yes
Dovekie	*Alle alle*	yes	yes
Black guillemot	*Cepphus grylle*	yes	yes
Atlantic Puffin	*Fratercula arctica*	yes	no
Fox Sparrow	*Passerella iliaca*	no	no

Mammal Species Common Name	Mammal Species Scientific Name	Transect	Platform
Humpback whale	*Megaptera novalanglia*	yes	yes
Minke whale	*Balaenoptera acutorostrat*	yes	yes
Harp seal	*Pagophilus groenlandicus*	yes	yes
Harbour seal	*Phoca vitulina*	yes	yes
Common dolphin	*Delphinus delphis*	yes	yes

FIGURE 1. Black-legged kittiwake flying in the Northwest Atlantic (photo: W.A. Montevecchi).

harbour porpoises (Gaskin, 1992), and harp, hooded, gray, and harbour seals (Bundy et al., 2000). Chronic oil pollution (Montevecchi and Tuck, 1987; Wiese and Ryan, 1999; 2003; Wiese, 2002) and increasing offshore hydrocarbon development (Wiese et al., 2001) in this region is encroaching on traditional seabird and mammal habitat. These events and their cumulative effects create disturbances not previously experienced by marine animals on the Grand Banks of eastern Canada.

Offshore platforms create artificial reefs that promote algal growth, attract crustaceans and fishes, and their avian and mammalian predators (Shinn, 1974; Duffy, 1975; Sonnier et al., 1976; Ortego, 1978; Weise et al., 2001). The vegetative growth and biologically attractive properties of these reefs are enhanced by direct discharges of human waste water directly below the platforms. The thermal and structural (barrier) properties of platforms are known to attract fishes, birds, and mammals (Klima and Wickham, 1971). The intense artificial lighting associated with platforms and flares also are highly attractive to seabirds, especially planktivorous and nocturnal species, such as Leach's storm-petrels (Wood, 1999). Once in platform vicinities, marine animals are exposed to mortality threats from spilled oil, incineration by flaring, and collision with the large illuminated structures. For these reasons, we investigated variability in the diversity, distributional, and abundance patterns of seabirds and marine mammals within and among seasons along a vessel-based, fixed transect route from St. John's to oil platforms on the Grand Banks from 1999 through 2003.

Seabirds are highly mobile, wide-ranging predators. Owing to the high energetic demands of endothermy in cold ocean regimes, avian and mammalian predators spend much of their time foraging. Therefore, the distributional patterns of marine birds are primarily influenced by the distributional patterns of their prey and by proximity to breeding sites (Davoren et al., 2002; 2003a) and generally reflect how individuals meet their energetic requirements for survival (Horne and Schneider, 1994). To meet these high energetic requirements, a high proportion of animals are found associated with a few large, dense prey aggregations (Heinemann et al., 1989; Hunt et al., 1991; Veit et al., 1993). Marine endotherms also tend to concentrate their foraging efforts in areas where prey aggregations are spatially and temporally persistent, thereby avoiding energetically costly search activities (Schneider et al., 1987; Cairns and Schneider, 1990; Hunt and Harrison, 1990; Coyle et al., 1992; Horne and Schneider, 1994, Decker and Hunt, 1996; Davoren et al., 2002; 2003b). Physical features of the pelagic environment, such as hydrographic regimes, combined with prey behaviour influence the distribution and density of prey (Mehlum et al., 1996; Ostrand et al., 1998) and the spatial and temporal persistence of prey patches (Cairns and Schneider, 1990, Hunt and Harrison, 1990; Coyle et al., 1992; Decker and Hunt, 1996; Irons, 1998). Oceanic structures, both natural (e.g. high bathymetric relief) and anthropogenic (e.g. offshore hydrocarbon platforms) concentrate prey in a persistent and predictable manner. These circumstances, combined with the energetic constraints of seabirds, suggest that these predators will concentrate at offshore platforms.

Energetic requirements and, thus, distributional patterns of marine birds vary among species due to dissimilar mobility constraints and flight efficiencies (Birt-Friesen et al., 1989). Auks are pursuit-diving birds and their wing design compromises aerial (high surface area) and underwater flight (low surface area; Livezey, 1988; Thompson et al., 1998) and results in high wing-loading (i.e. body mass to wing area ratio; Guillemette, 1994). Therefore, energy expenditure is elevated during flight compared to most other seabirds (Birt-Friesen et al., 1989). One strategy for auks to minimise energy expenditure is to reduce time spent flying. This results in more time spent sitting at sea, a behaviour that increases these birds' risk of mortality during winter when water temperatures are near or at freezing and birds are thermally stressed. Prolonged periods spent sitting on water also increases their chances of coming in contact with oil pollution making them more vulnerable than other avian species. In contrast, the wing design of surface-feeding birds, such as gulls, petrels, and shearwaters, optimises the efficiency of aerial flight and, thus, these birds have lower energetic requirements (metabolic rates) than pursuit-diving birds (Birt-Friesen et al., 1989; Croll and McLaren, 1993). These birds also tend to minimise thermal stress during the winter by roosting on land or other substrates at sea (e.g. offshore platforms). Therefore, not only are seabirds likely attracted to offshore platforms by elevated prey densities but they will also remain there due to flight constraints or accessibility to roosting sites. Overall, we predict that seabirds will aggregate at offshore hydrocarbon platforms on the Grand Banks throughout the year.

MATERIALS AND METHODS

Study Area

Research was carried out on the eastern Newfoundland Shelf, where globally and regionally significant seabird breeding sites occur (Montevecchi and Tuck, 1987). These include the world's largest colonies of Leach's storm-petrels (Sklepkovych and Montevecchi, 1989) and common murres (Nettleship and Evans, 1985), the only North American colonies of Manx shearwaters (Storey and Lien, 1985; Robertson, 2002) and common black-headed gulls (Montevecchi et al., 1987), three of North America's six northern gannet colonies including the world's southernmost colony at Cape St. Mary's (Montevecchi and Tuck, 1987; Nettleship and Chapdelaine, 1988), North America's largest colonies of Atlantic puffins (Rodway et al., 1996), the largest colonies of northern fulmars and Caspian terns in eastern Canada, and the world's southernmost breeding sites for thick-billed murres (Cairns et al., 1989; Lock et al., 1994). As well, tens of millions of birds that breed in the Arctic (dovekies, thick-billed murres) and at high latitudes in the South Atlantic (greater and sooty shearwaters), inhabit the region during their non-breeding seasons. Globally significant populations of marine mammals and fishes also inhabit the Grand Banks (Bundy et al., 2000).

Survey Design

The distribution and density of marine birds, as well as mammals, were quantified during monthly, fixed-line transect surveys from offshore support vessels traveling from St. John's to offshore oil exploration (Glomar Grand Banks, Bill Shoemaker) and production (Hibernia, Terra Nova) platforms on the Grand Banks (Fig. 2).

Surveys were carried out by dedicated, independent observers using scientifically standardised methods (method Ib, Tasker et al., 1984). While not free from biases (e.g. overestimation of densities of flying birds; Tasker et al., 1984), these methods have errors considered to be consistent among surveys. Observers underwent extensive training, which began by studying species identification with seabird and marine mammal guides, and then by accompanying experienced observers during at-sea surveys for up to two weeks prior to conducting surveys on their own. All marine birds and mammals were recorded by a single observer within a 300 m 90°-viewing arc from the bow to either the port or starboard side of the vessel. A single observer was determined to be sufficient due to low densities of seabirds encountered over the Newfoundland Shelf (Davoren et al., 2002). Species names, counts,

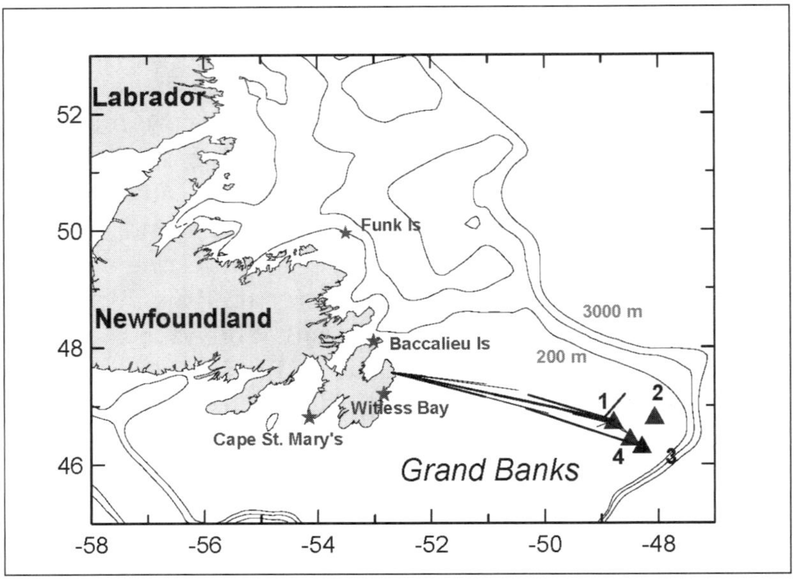

FIGURE 2. Monthly fixed transect surveys conducted from offshore support vessels to hydrocarbon platforms on the Grand Banks (total distance = 350 km). Seabird ecological reserves (*) and offshore platforms indicated: Hibernia (1), Bill Shoemaker drill rig (2), Terra Nova (3), Glomar Grand Banks drill rig (4).

and behavioural descriptions (on water, feeding, flying, and flight direction) were entered directly into a laptop computer. This laptop was interfaced with a hand-held Global Positioning System (Garmin III Plus) and counting software (D. Senciall, Birds and Beasty Counter, 1998, Fisheries and Oceans Canada, version 1.0) to append the GMT, latitude and longitude to each entry. Continuous observations were recorded while the vessel was moving between stations during daylight hours. Weather conditions (wind speed and direction, precipitation, sea state) and visibility were recorded at the onset of transects and every 30 min thereafter or whenever a change in ambient conditions impacted the detectability of seabirds and mammals. The 300 m survey width was maintained following Heinemann (1981) and counting was discontinued if visibility was < 300 m (e.g. during fog, high wind, and sea spray).

Stationary Vessels at Hydrocarbon Platforms

To detect fluctuations in the species composition and abundance within seabird aggregations near offshore platforms, systematic hourly scans were conducted throughout the day and night when vessels were stationed alongside platforms. Continuous 360° scans within a 500 m range of the support vessel were carried out for 15 min at 1 h intervals. Species, counts, and behavioural descriptions, as well as weather conditions, visibility, vessel distance from platform, and flare size were recorded on data sheets. Birds flying through the survey area were recorded separately to differentiate birds associated with offshore installations and those obviously flying rapidly through the study area. Protocols for systematic scans at offshore platforms were developed in 2001 subsequent to offshore vessel surveys in 1999 and 2000. Consequently, results from stationary scans at hydrocarbon platforms do not include those surveys conducted in 1999 and 2000. Data were collected at Hibernia during 11 out of 12 surveys and at Terra Nova during 4 surveys.

Physical Oceanographic Measurements

Sea surface temperature (SST) at 5 m depths were recorded every 5 min along the transect line using an Acoustic Current Doppler Profiler (ACDP) on support vessels. SST data were collected simultaneously with seabird observations whenever possible. When ACDP data were unavailable on survey vessels, SSTs were obtained from an alternate support vessel that had covered the transect within days of the survey. The data for the summer profile along the survey route was obtained from Fisheries and Oceans Canada (DFO) website (www.meds-sdmm.dfo-mpo.gc.ca/zmp/hydro_data-e.html). These data were related to seasonal species diversity, distribution, and abundance. These physical oceanographic variables are important to understanding the factors that influence distributions of seabirds at sea.

Mapping and Analysis

Seabirds and mammals were partitioned into six guilds. Birds were separated into: auks (common murre and thick-billed murres, dovekie, Atlantic puffin, black guillemot), gulls (herring, great black-backed, Iceland, and glaucous gulls), kittiwake (black-legged kittiwake; Fig. 1), fulmar (northern fulmar), storm-petrels (Leach's and Wilson's storm-petrels), and shearwaters (greater and sooty shearwaters). Marine mammal species were lumped into a single category, which included humpback and minke whales, harp and harbour seal and porpoise and dolphins.

The mean densities of birds on the water and flying and marine mammals observed along the survey route were calculated by averaging the number of individuals in each category per 5 km along the survey route using Matlab© routines provided by R. O'Driscoll (see O'Driscoll, 1998). Mean densities per 5 km were mapped using GIS software (Surfer 8.0, Golden Software) to examine distributional patterns relative to offshore platforms as well as shifting distributional patterns among seasons. SST data were also binned by 5 km means using Matlab© routines and were mapped. Raw count data along surveys and at offshore platforms were used to examine seasonal trends in species composition and abundance of birds and mammals. Species composition, mean abundance and distribution patterns were compared among years.

RESULTS

Seasonal Patterns

Striking seasonal fluxes were observed in seabird species diversity and abundance on the Grand Banks during surveys to offshore platforms. The numbers of seabirds recorded were highly variable between seasons, and mean numbers were highest during summer (Table 2). This pattern can be explained by dense summer concentrations of migrant shearwaters offshore, and breeding auks (common murres and Atlantic puffins) near coastal colonies in July and August. Sooty and greater shearwaters comprised nearly half (48%) of all birds recorded during summer surveys (Table 3; Fig. 3). These long-distance, southern hemisphere migrants were present from May through to October when warm SST ranged from 7 to 13.2°C. Leach's storm-petrels were also present from May to October and were most abundant offshore during fall.

TABLE 2. Seasonal mean ± standard deviation (sd) and range of seabird abundance per 5 km² blocks along transect route. All surveys pooled by season (1999–2003). N = total number of binned 5 km counts from all pooled surveys.

Summary Statistics	Winter (2 surveys)	Spring (4 surveys)	Summer (8 surveys)	Fall (7 surveys)
Mean ± sd	1.8 ± 3.5	1.83 ± 1.7	3.8 ± 22.7	1.7 ± 1.20
Range	(1 - 31)	(1 - 18)	(1 - 500)	(1 - 12.5)
N	85	232	569	409

TABLE 3. Mean ± standard deviation of seabirds (partitioned by guilds) and mammals per 5 km segments along the survey route during 12 surveys from October 2001 to June 2003.

Survey Dates	Auks	Fulmars	Gulls	Kittiwakes	Petrels	Shearwaters	Mammals
2001							
1 - 3 Oct	1.5 ± 0.8	1.1 ± 0.3	1.2 ± 0.3	3.2 ± 3.8	1.1 ± 0.3	1.0 ± 0.0	0.0
23 - 25 Oct	1.6 ± 0.7	1.3 ± 0.5	1.1 ± 0.2	4.0 ± 3.5	1.3 ± 0.6	1.8 ± 1.8	0.0
2002							
Feb	1.9 ± 2.2	1.1 ± 0.3	1.0 ± 0.0	1.1 ± 0.3	0.0	0.0	0.0
Apr	1.3 ± 0.5	1.6 ± 1.6	1.0 ± 0.0	1.4 ± 0.5	0.0	2.0 ± 0.0	1.3 ± 0.4
May	2.9 ± 4.5	1.1 ± 0.4	1.0 ± 0.0	1.2 ± 0.4	1.8 ± 2.2	0.0	1.2 ± 0.4
Sep	1.4 ± 1	1.0 ± 0.0	1.8 ± 1	1.0 ± 0.0	1.2 ± 0.3	1.3 ± 0.8	3.7 ± 2.2
Oct	1.8 ± 1.0	1.3 ± 0.6	2.2 ± 1.2	2.1 ± 1.6	1.1 ± 0.2	1.3 ± 0.6	1.3 ± 0.7
Nov	1.9 ± 1.0	2.0 ± 4.5	1.3 ± 0.7	1.6 ± 1.5	0.0	0.0	1.0 ± 0.0
2003							
Feb	2.4 ± 1.8	3 ± 7.5	1.0 ± 0.0	1.0 ± 0.0	0.0	0.0	1.0 ± 0.0
Apr	1.4 ± 0.7	2.4 ± 2.2	1.2 ± 0.4	1.9 ± 2.1	0.0	0.0	1.1 ± 0.2
May	1.8 ± 1.1	1.1 ± 0.4	1.2 ± 0.3	1.4 ± 0.5	1.1 ± 0.3	1.0 ± 0.0	1.0 ± 0.1
Jun	5.8 ± 9.2	1.1 ± 0.3	1.0 ± 0.1	1.3 ± 0.7	1.1 ± 0.2	1.3 ± 0.5	1.0 ± 0.0

During winter surveys when SSTs ranged between -1 to 0.5°C, auks (primarily dovekies and thick-billed murres) were the most abundant species.

Seasonal patterns in seabird species diversity and abundance were also obvious from observations at offshore platforms, and reflected trends along the survey route. Large aggregations of greater shearwaters were recorded at offshore platforms on the Grand Banks during late July/early August (Fig. 4). During October, black-legged kittiwakes formed large rafts in the water near the Hibernia platform (Table 4). Kittiwakes were also the most abundant seabirds along the

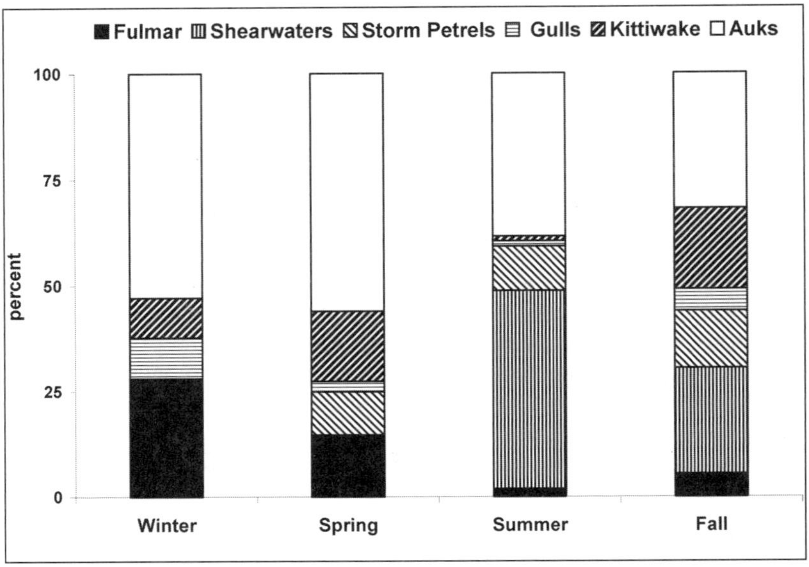

FIGURE 3. Seasonal diversity of seabird species grouped by guild during offshore support vessels surveys from 1999–2003.

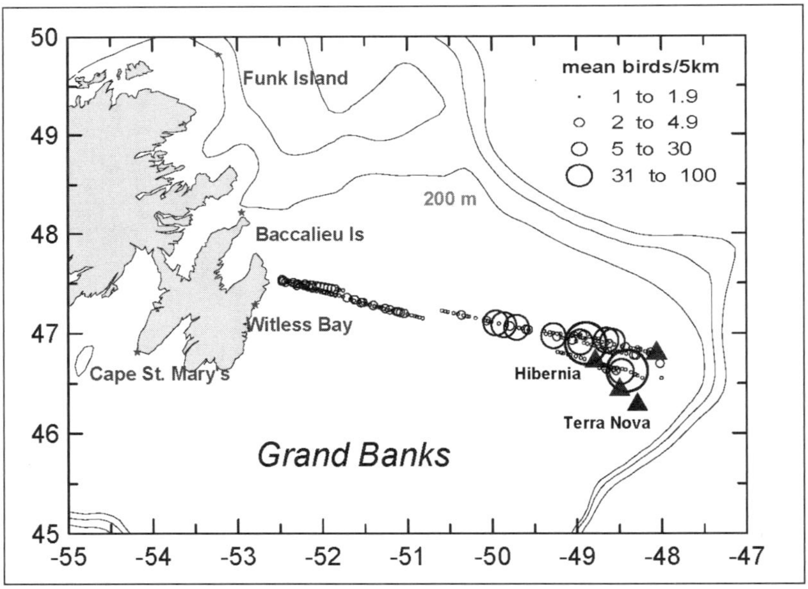

FIGURE 4. Distribution of all seabirds binned into 5 km segments from three surveys during July and August 2000.

TABLE 4. Mean ± standard deviation (range) of seabirds (partitioned by guilds) recorded during scans from support vessels positioned alongside offshore platforms. N = total number of scans completed during each survey.

Survey Dates	N	Fulmars	Shearwaters	Petrels	Gulls	Kittiwakes	Auks	Total Birds
2001								
1 - 3 Oct	7	0.0	0.0	2.1 ± 5.7 (0 - 15)	92.1 ± 84.8 (0 - 250)	2.9 ± 7.6 (0 - 20)	0.0	680
23 - 25 Oct	7	2.9 ± 5.9 (0 - 16)	0.0	0.0	31.7± 26.9 (0 - 61)	83.4 ± 148.5 (0 - 400)	1.1 ± 2.6 (0 - 7)	834
2002								
Feb	14	0.5 ± 1.3 (0-5)	0.0	0.0	0.7 ± 0.7 (0 - 2)	0.7 ± 1.3 (0 - 4)	4.1 ± 3.8 (0 - 12)	84
Apr	4	0.0	0.0	0.0	0.0	0.0	3 ± 1.4 (2-5)	12
May	10	0.0	0.0	0.0	0.5 ± 0.6 (0 - 5)	0.0	0.0	3
Sept	3	0.0	0.0	0.0	46.7 ± 5.8 (40 - 50)	0.0	0.0	140
Oct	18	0.5 ± 0.1 (0 - 3)	0.0	0.0	71.1 ± 67.1 (0 - 190)	4.4 ± 6.8 (0 - 20)	0.0	1369
Nov	8	0.9 ± 0.8 (0 - 2)	0.0	0.0	3.1 ± 6.1 (0 - 18)	40 ± 42 (0 - 115)	7.9 ± 6.9 (0 - 23)	415
2003								
Feb	11	0.5 ± 1 (0 - 3)	0.0	0.0	0.7 ± 1.3 (0 - 3)	0.8 ± 1.3 (0 - 4)	0.8 ± 0.1 (0 - 2)	32
Apr	9	2.2 ± 3.3 (0 - 15)	2.2 ± 3.3 (0 - 15)	0.0	0.6 ± 0.9 (0 - 2)	0.0	0.8 ± 0.1 (0 - 6)	40
May	7	0.0	0.0	0.0	0.3 ± 0.5 (0 - 1)	0.6 ± 1.5 (0 - 4)	1 ± 0.8 (0 - 2)	13
Jun	4	0.0	0.2 ± 0.5 (0 - 1)	0.0	0.6 ± 0.6 (0 - 1)	0.0	0.0	4

survey route during this period, with mean numbers ranging between 2.1 and 4 birds per 5 km segments. During September and October, large groups of great black-backed gulls roosting at the Hibernia platform during the day were observed foraging at night on fish attracted to the illuminated surface water under the structure. A total of 11 oiled birds including thick-billed and common murres, dovekies, great black-backed gulls, and kittiwakes were observed during 4 surveys over 41 scans at the Hibernia platform. Anecdotal information from support vessel personnel on several occasions provided further evidence of oiled birds near offshore platforms with increased incidents during winter (Fig. 5). A total of 18 marine mammals were observed over 8 days of observations in the immediate vicinity of Hibernia. Avian and mammalian occurrences at platforms were episodic in nature,

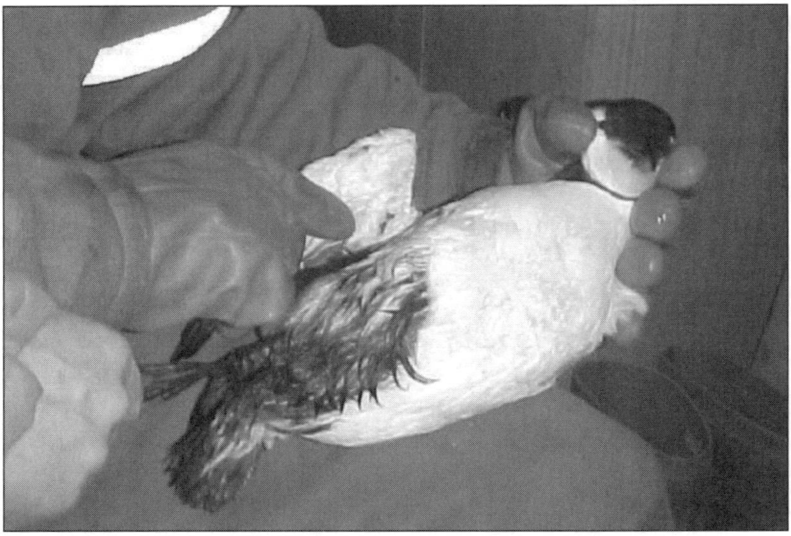

FIGURE 5. Oiled common murre retrieved by support vessel personnel near Hibernia during January 2003 (photo: crew of Maersk Nascopie).

and long-term dedicated observations will be required to adequately capture these events (Montevecchi et al., 1999).

Oceanographic Distributional Patterns

Seabirds were patchily distributed along the survey route, often forming considerable aggregations. Thick-billed and common murres were observed at densities that ranged from 1–50 birds per 5 km intervals (Fig. 6 f). These aggregations were closely associated with a branch of the Labrador Current. Densities of marine birds and mammals also were considerably higher in the vicinity of offshore platforms than along the transect routes leading to it (Figs. 6 a–f). This pattern of aggregation was previously documented by Wiese et al. (2001). The association of seabirds with offshore installations is readily observable and vessel and platform personnel have commented on this attraction. Seabird concentrations were more evident at Hibernia gravity-based platform than at the Terra Nova floating platform with an average of 71 birds per scan at Hibernia (3576 birds during 50 scans) and 6.8 birds at Terra

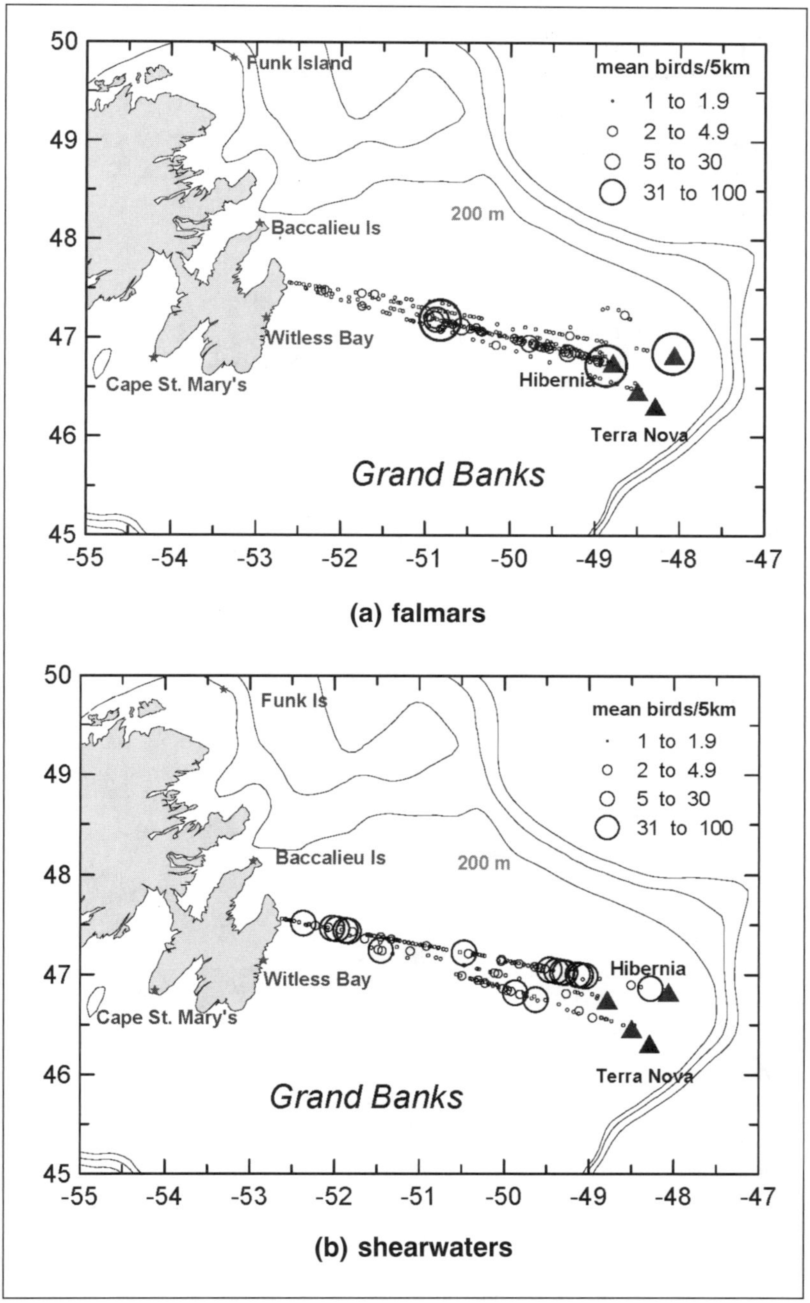

FIGURE 6. Distribution of seabirds: **(a)** fulmars, **(b)** shearwaters, **(c)** storm-petrels, **(d)** gulls, **(e)** kittiwakes, **(f)** auks binned into 5 km segments during surveys in 1999 and 2001–2003.

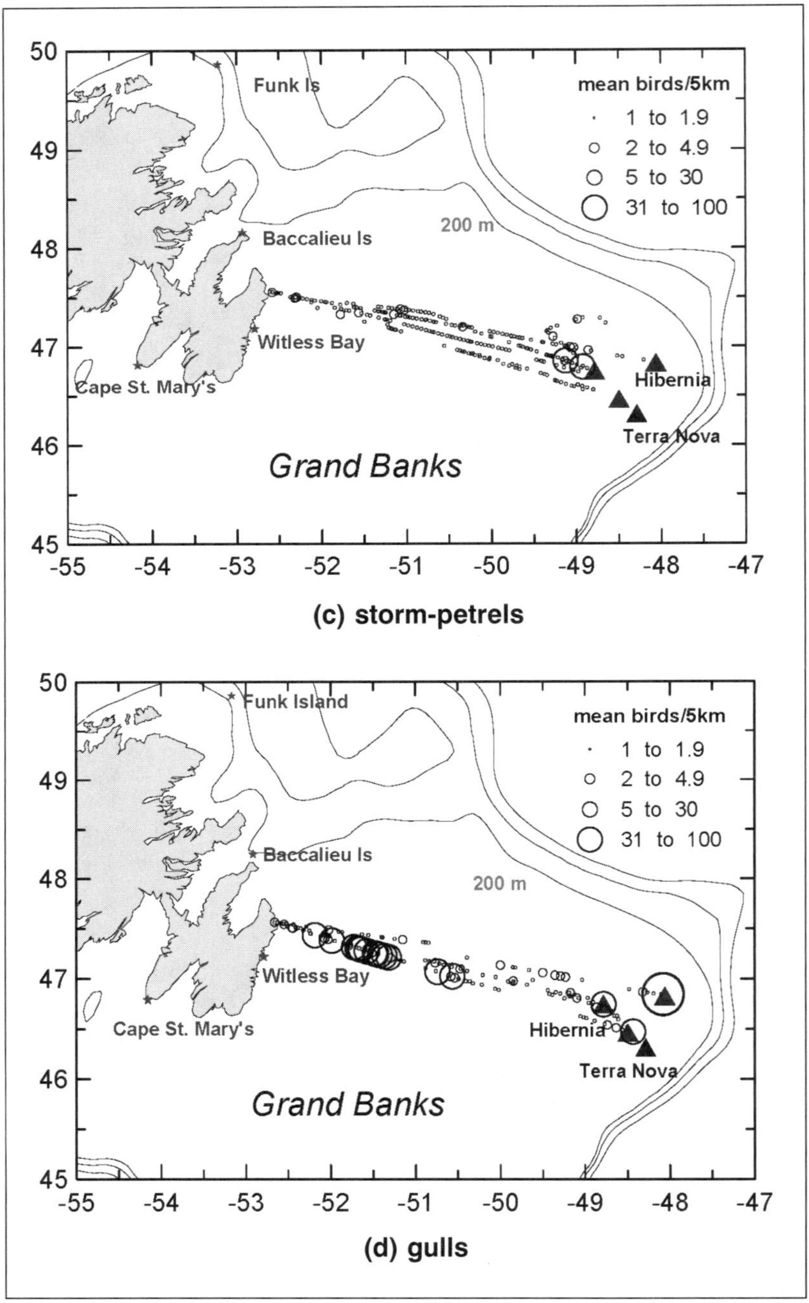

FIGURE 6. *(continued)*

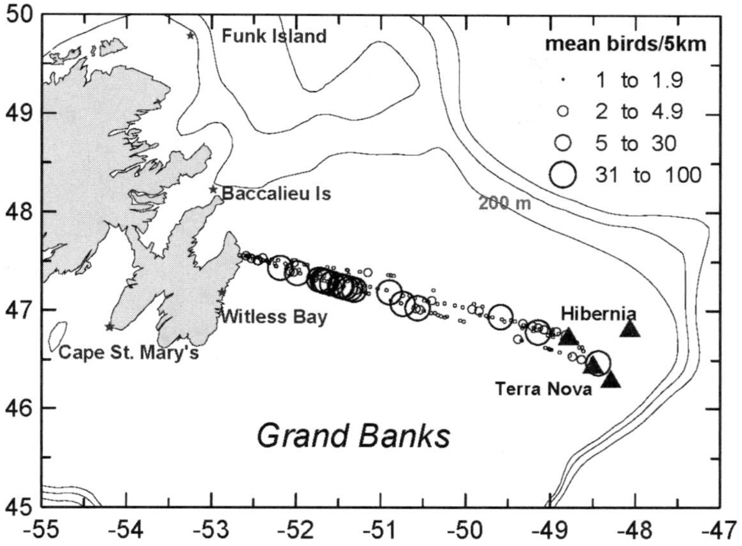

(e) kittiwakes

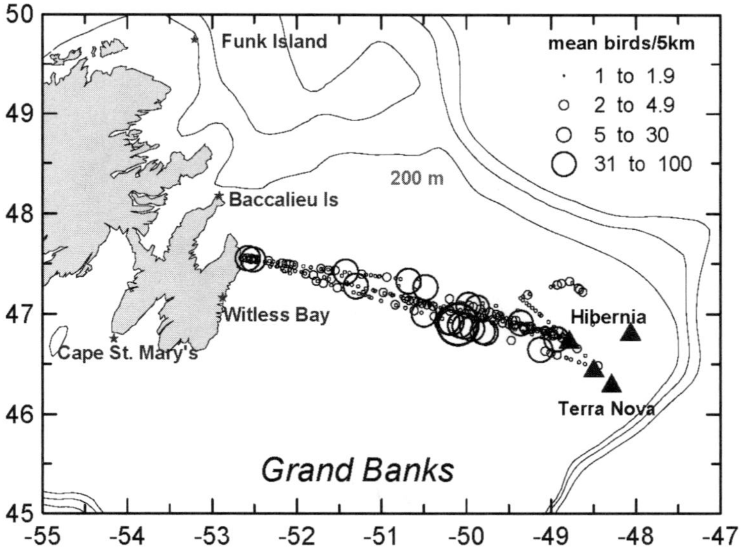

(f) auks

FIGURE 6. *(continued)*

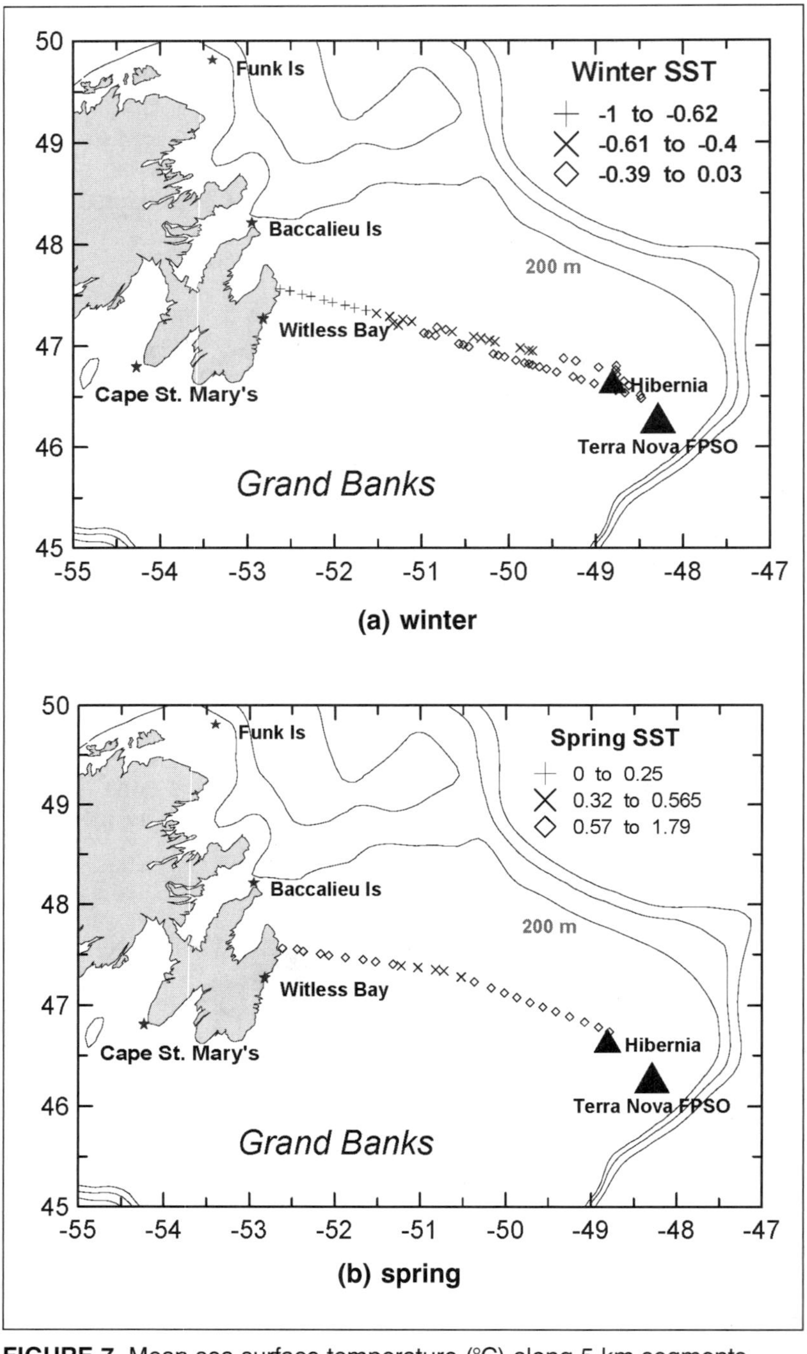

FIGURE 7. Mean sea surface temperature (°C) along 5 km segments during **(a)** winter, **(b)** spring, **(c)** summer (DFO data) and **(d)** fall.

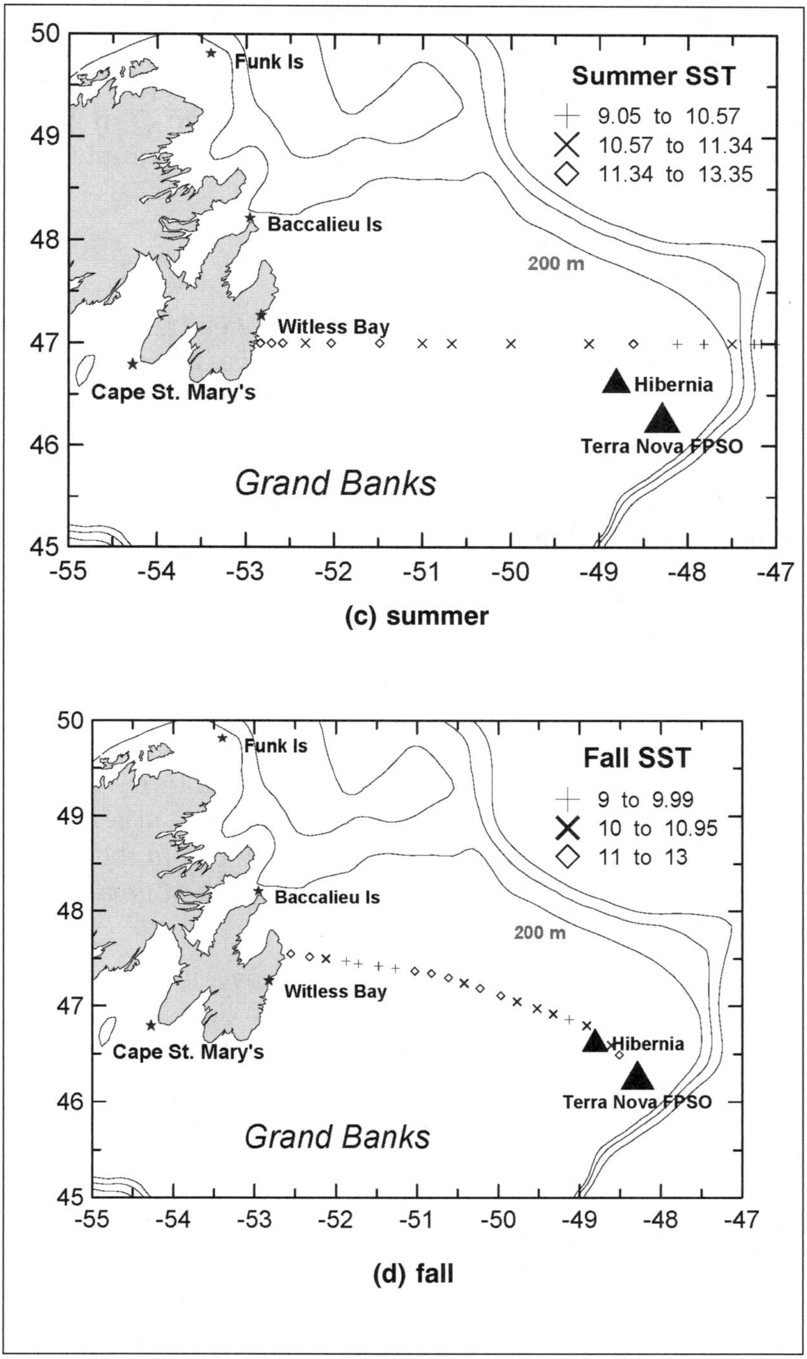

FIGURE 7. *(continued)*

Nova (89 birds during 13 scans). The physical features of the Hibernia platform (i.e. a fixed structure in place over years) present ideal conditions for artificial reef development, and the concrete icewall at the base of the platform (7 m above water) is an attractive roosting refuge for large numbers of gulls and kittiwakes.

DISCUSSION

Year-round surveys were carried out by well-trained, arm's length observers along offshore support vessel routes. Surveys, including the first winter observations along the support vessel route, and at the Hibernia platform, showed striking seasonal patterns in species occurrences and abundances. Thick-billed murres and dovekies, the species most vulnerable to oil pollution (Wiese and Ryan, 1999; 2003; Wiese, 2002), were most abundant during winter when thermal stresses, and hence risk of mortality associated with oil pollution are greatest. Oiled dovekies and thick-billed and common murres were documented at the Hibernia platform. Storm-petrels and shearwaters, species naturally attracted to light (e.g. Reed et al., 1985), were common during summer and fall. These periods represent vulnerable stages to mortality in flares in these species' annual cycles (Wood, 1999; Wiese et al., 2001). Fishes attracted to surface waters by lighting from the Hibernia platform were preyed on at night by flocks of great black-backed gulls that roosted at the platform.

It is no surprise that oiled murres and dovekies were documented at the Hibernia platform during our surveys, that large numbers of storm-petrels have been observed to die in the flare (Wood, 1999) or that they are easy prey for gulls when they fall in the water after impact with the platform. Marine mammals may also be influenced by hydrocarbon and produced waters in platform vicinities. Because there are no independent, dedicated observers on platforms (see Montevecchi et al., 1999; Wiese et al., 2001), it impossible to quantify mortality and cumulative effects on these species at offshore hydrocarbon platforms. Surveys at offshore platforms during fall when storm-petrels migrate offshore, are key times to quantify mortality rates associated with flaring. The timing of surveys is particularly crucial given the episodic nature of flaring mortalities.

Features of offshore platforms that attract marine animals can be modified and mitigated. For example, 1) waste water could be recycled rather than discharged directly below platforms, 2) lights on the platform could be reduced and shielded without causing safety concerns (Reed et al., 1985), 3) maintenance and refit (i.e. down time) of flare equipment could be carried out during periods of greatest risks to seabirds (e.g. September, October). Before and after comparisons could be made to attempt to assess influences of any manipulations that are carried out. There have been no known mitigation efforts for marine birds and mammals at the Hibernia platform in eastern Canada.

The year-round support vessel surveys conducted through Memorial University of Newfoundland were the only arm's length surveys for marine birds and mammals related to offshore hydrocarbon activities that have been carried out in eastern Canadian waters. These surveys have generated new and important information about the seasonal occurrences of marine birds and mammals along the support vessel route and in the vicinity of offshore hydrocarbon platforms on the Grand Banks. Funding for these surveys by the Environmental Studies Research Fund of the Energy Board of Canada was terminated in August 2003, with no anticipation of reinstatement. To not conduct independent scientific surveys for marine birds and mammals throughout and beyond the duration of hydrocarbon extraction on the Grand Banks is environmentally and ethically irresponsible.

Beyond the limited duration of these monthly support vessel surveys for seabirds and marine mammals, it has been impossible to station independent, dedicated, trained observers on offshore hydrocarbon platforms in eastern Canada. This environmental monitoring inadequacy is occurring despite a report prepared in 1999 for the Canadian Association of Petroleum Producers (CAPP) on monitoring marine birds and mammals in relation to offshore hydrocarbon activities that recommended a scientifically robust observation program on offshore platforms (Montevecchi et al., 1999). This report and recommendations were reviewed and supported by international scientific experts. Owing to the absence of dedicated, independent observers on platforms, the mortality of marine birds and disturbance of marine mammals that have occurred due to lighting, flaring, collisions and oiling at offshore hydrocarbon platforms in Canada's eastern ocean have not been documented.

CONCLUSIONS

Monthly transect surveys to offshore platforms were generating the only scientifically reliable indices of seabirds and mammals with respect to hydrocarbon developments in eastern Canada. The threat of seabird mortality due to flaring and collisions with offshore platforms is likely greatest during fall when tens of millions of small planktivorous and nocturnal seabirds (dovekies, Leach's storm-petrels) are moving in the Northwest Atlantic. The threat of oil pollution is highest during the winter when marine bird populations are dominated by mainly diving species (auks), water temperatures are lowest, and their populations expand into oil development and shipping areas (Lock et al., 1994). An estimated 300,000 seabirds die due to oiling off the coast of eastern Canada every year, which is equivalent to the mortality of the Exxon Valdez disaster on an annual basis (Wiese, 2002; Wiese and Robertson, 2003). Mortality associated with hydrocarbon developments cumulates with other sources of mortality from hunting and entrapment in fishing gear (Piatt et al 1984; Montevecchi and Tuck, 1987; Montevecchi, 2001; Wiese and Robertson, 2003).

Current levels of scientific effort directed at marine birds and mammals in relation to hydrocarbon developments in eastern Canadian waters are inadequate to document the dynamic distributional patterns of these animals and ongoing mortality. Yet even modest independent research efforts were curtailed due to lack of support from responsible government agencies. Whether or not these environmental and ethical challenges will be met depends on the commitments of individuals, both within and outside of government. The needs and circumstances are self-evident.

ACKNOWLEDGEMENTS

We are grateful to many people who have aided us with our research over the years. David Burley (CNOPB) helped develop the proposal for ESRF contract that supported the research surveys, and Kym Hopper-Smith (Energy Board of Canada) executed the contract logistics. Don Sutherland (Husky Oil) provided the initial funding of this research. Urban Williams (Petro-Canada) helped with the logistics for the support vessel survey to the Terra Nova floating platform. Special thanks to the crew members of the Maersk Nascopie, Maersk Norseman, and Maersk Bonavista for their support and enthusiasm. Dr. Richard O'Driscoll assisted with mapping analyses, and Tony Lock (CWS), Dave Taylor (Husky Oil) and Bruce Turner (CWS) provided advice on the research. Funding support for these surveys was provided by the Environmental Studies Research Fund—Energy Board of Canada, Husky Oil, the Natural Sciences and Engineering Research Council of Canada (NSERC) to W.A. Montevecchi and G.K. Davoren.

REFERENCES

Barlow, J. and P.J. Clapham. 1997. "A New Birth-Interval Approach to Estimating Demographic Parameters of Humpback Whales." *Ecology.* 78: 535-546.

Birt-Friesen, V.L., W.A. Montevecchi, D.K. Cairns, and S.A. Macko. 1989. "Activity Specific Metabolic Rates of Free-Living Northern Gannets and Other Seabirds." *Ecology.* 70: 357-67.

Braham, H.W. 1984. "The Status of Endangered Whales: An Overview." *Marine Fisheries Review* 46(4): 2-6.

Bundy, A., G. R. Lilly, and P. A. Shelton. 2000. (Eds.), *A Mass Balance Model of the Newfoundland-Labrador Shelf.* Canadian Technical Report of Fisheries and Aquatic Sciences. No. 2310.

Cairns, D.K. and D.C. Schneider. 1990. "Hot Spots in Cold Water: Feeding Habitat Selection by Thick-billed Murres." *Studies in Avian Biology.* 14:52-60.

Cairns, D.K., W.A. Montevecchi, and W. Threlfall. 1989. "Researcher's Guide to the Seabird Colonies of Newfoundland." *Memorial University of Newfoundland Occasional Papers in Biology.* Number 14.

Coyle, K.O., G.L. Hunt Jr., M.B. Decker, and T.J. Weingartner. 1992. "Murre Foraging, Epibenthic Sound Scattering and Tidal Advection Over a Shoal Near St. George's Island." *Marine Ecology Progress Series.* 83:1-14.

Croll, D.A. and E.B. McLaren. 1993. "Diving Metabolism and Thermoregulation in Common and Thick-billed Murres." *Journal of Comparative Physiology.* 163:160-166.

Davoren G.K., W.A. Montevecchi, and J.T. Anderson. 2002. "Scale Dependent Associations of Predators and Prey: Constraints Imposed by Flightlessness of Common Murres." *Marine Ecology Progress Series.* 245:259-272.

Davoren, G.K., W.A. Montevecchi, and J.T. Anderson. 2003a. "Distributional Patterns of a Marine and Its Prey: Habitat Selection Based on Prey and Conspecific Behavior." *Marine Ecology Progress Series.* 256: 229-242.

Davoren, G.K., W.A. Montevecchi, and J.T. Anderson. 2003b. "Search Strategies of a Pursuit-Diving Marine Bird and the Stability of Prey Patches." *Ecology*. 73:463-481.

Decker, M.B. and G.L. Hunt Jr. 1996. "Foraging by Murres (*Uria* spp.) at Tidal Fronts Surrounding the Pribilof Islands, Alaska, USA." *Marine Ecology Progress Series.* 139:1-10.

Duffy, M. 1975. "From Rigs to Reefs." *Louisiana Conservationist.* 27:18-21.

Gaskin, D.E. 1992. "Status of the Harbour Porpoise, *Phocoena phocoena*, in Canada." *Canadian Field-Nauralist.* 106:36-54.

Guillemette, M. 1994. "Digestive–rate Constraints in Wintering Common Eiders (*Somateria mollissima*): Implications for Flying Capabilities." *Auk.* 111:900-909.

Hay, K. 1982. "Aerial Line Transect Estimates of Abundance of Humpback, Fin, and Long-finned Pilot Whales in the Newfoundland-Labrador Area." *Report of the International Whalling Commission.* 32:475-486.

Horne, J.K, and D.C. Schneider. 1994. "Lack of Spatial Coherence of Predators and Prey: A Bioenergetic Explanation for Atlantic Cod

Feeding on Capelin." *Journal of Fish Biology.* 45(Supplement A):191-297.

Heinemann D., G.L. Hunt Jr., and I. Everson. 1989. "Relationship between the Distributions of Marine Avian Predators and their Prey, *Euphausia superba*, in Bransfield Strait and Southern Drake Passage, Antarctica." *Marine Ecology Progress Series.* 58:3-16.

Heinemann, D. 1981. "A Rangefinder for Pelagic Bird Censusing." *Journal of Wildlife Management.* 45:489-493.

Hunt, G.L. Jr. and N.M. Harrison. 1990. "Foraging Habitat and Prey Taken by Least Auklets at King Island, Alaska." *Marine Ecology Progress Series.* 65:141-150.

Hunt, G.L. Jr., J.F. Piatt, and K.E. Erikstad. 1991. "How do Foraging Seabirds Sample their Environment. *International Ornithological Congress.* 20:2272-2279.

Irons, D.B. 1998. "Foraging Area Fidelity of Individual Seabirds in Relation to Tidal Cycles and Flock Feeding." *Ecology.* 79:647-655.

Klima, E.F. and D.A. Wickham. 1971. "Attraction of Coastal Pelagic Fishes with Artificial Structures." *Transactions of the American Fisheries Society.* 1:86-99.

Livezey, B.C. 1988. "Morphometrics of Flightlessness in the Alcidae." *Auk.* 105:681-698.

Lock, A.R., R.G.B. Brown, and S.H. Gerriets. 1994. *Gazetteer of Marine Birds in Atlantic Canada.* Canadian Wildlife Service, Dartmouth, Nova Scotia.

Mehlum, F., G.L. Hunt Jr., Z Klusek, M.B. Decker, and N. Nordlund. 1996. "The Importance of Prey Aggregations to the Distribution of Brunnich's Guillemots in Storfjorden, Svalbard." *Polar Biology.* 16:537-547.

Mitchell, E.D. and D.G. Chapman. 1977. "Preliminary Assessment of Stocks of Northwest Atlantic Sei Whales (*Balaenoptera borealis*)." *Report of the International Whaling Commission.* 1:117-120.

Montevecchi, W.A. 2000. "Seabirds". In A Bundy, G. R. Lilly, and P. A. Shelton (Eds.), *A Mass Balance Model of the Newfoundland-Labrador Shelf*, Canadian Technical Report of Fisheries and Aquatic Sciences. No. 2310.

Montevecchi, W.A. 2001. "Interactions between Fisheries and Seabirds." In E. A. Schreiber and J. Burger (Eds.). *Biology of Marine Birds*, pp. 527-557. CRC Press, Boca Raton, FL.

Montevecchi, W.A. and L.M. Tuck. 1987. *Newfoundland Birds: Exploitation, Study, Conservation.* Nuttall Ornithological Club, Cambridge, MA.

Montevecchi, W.A., D.K. Cairns, A.E. Burger, R. D. Elliot, and J. Wells. 1987. "The Status of Common Black-headed Gulls, *Larus ridibundus* in Newfoundland and Labrador." *American Birds.* 41:197-203.

Montevecchi, W.A., A.W. Diamond, G.K. Davoren, F. Huettmann, F.K. Wiese, and J. Linke. 1999. *Seabird Attraction to Offshore Platforms and Seabird Monitoring from Offshore Support Vessels and Other Ships: Literature Review and Monitoring Designs.* Report for Canadian Association of Petroleum Producers (CAPP), St. John's, NL.

Nelson, D. and J. Lien. 1996. "The Status of the Long-finned Pilot Whale, *Globicephala melas*, in Canada." *Canadian Field-Naturalist.* 110:511-524.

Nettleship, D.N. and G. Chapdelaine. 1988. "Population Size and Status of Northern Gannets, *Sula bassanus*, in North America." *Journal of Field Ornithology*. 59:120-127.

Nettleship, D.N. and Evans, P.G.H. 1985. "Distribution and Status of the Atlantic Alcidae." In D. N. Nettleship and T.R. Birkhead (Eds.), *The Atlantic Alcidae*, pp. 53-154. Academic Press, Orlando.

Ortego, B. 1978. "Blue-footed Boobies at an Oil Production Platform." *Auk.* 95:762-762.

Ostrand, W.D., K.O Coyle, and G.S. Drew. 1998. "Selection of Foraging-Fish Schools by Murrelets and Tufted Puffins in Prince William Sound, Alaska." *Condor.* 100:286-297.

O'Driscoll, R.L. 1998. "Description of Spatial Pattern in Seabird Distributions along Line Transects Using Neighbour K Statistics." *Marine Ecology Progress Series.* 165:81-94.

Piatt, J.F., D.N. Nettleship, and W. Threlfall. 1984. "Net Mortality of Common Murres *Uria aalge* and Atlantic Puffins *Fratercula arctica* in Newfoundland 1951–1981." In D. N. Nettleship, G. A. Sanger, and P. F. Springer (Eds.), *Marine Birds: Their Feeding Ecology and Commercial Fishery Relationships*, pp. 196-206. Canadian Wildlife Service, Ottawa.

Reed, J.R., J.L. Sincock, and J.P. Hailman. 1985. "Light Attraction in Endangered Procellariiform Birds: Reduction by Shielding Upward Radiation." *Auk.* 102:377-382.

Robertson, G.J. 2002. "Current status of the Manx Shearwater (*Puffinus puffinus*) Colony on Middle Lawn Island, Newfoundland." *Northeastern Naturalist.* 9:317-324.

Rodway, M.S., H.M. Regehr, and J.W. Chardine. 1996. *Population Census of Breeding Atlantic Puffins at Great Island, Newfoundland in 1993–1994.* Canadian Wildlife Service Technical Report Series Number 263, Ottawa.

Schneider, D.C., G.L. Hunt, and K.D. Powers. 1987. "Energy Flux to Pelagic Birds: A Comparison of Bristol Bay (Bering Sea) and Georges Bank (Northwest Atlantic)." In J. P. Croxall (Ed.), *Seabirds: Feeding Ecology and Role in Marine Ecosystems*, pp. 259-277. Cambridge University Press, Cambridge, UK.

Shinn, E.A. 1974. "Oil Structures as Artificial Reefs." In *Proceedings of the International Conference of Artificial Reefs*, pp. 91-96. Houston, TX.

Sklepkovych, B.O. and W.A. Montevecchi. 1989. "The World's Largest Documented Colony of Leach's Storm-Petrels on Baccalieu Island, Newfoundland." *American Birds.* 43:36-42.

Sonnier, F., J Teerling, and J.D. Hoese. 1976. "Observations on the Offshore Reef and Platform Fish Fauna of Louisiana." *Copeia.* 1:105-111.

Storey, A.E. and O.J. Lien. 1985. "Establishment of a Breeding Colony of Manx Shearwaters in Newfoundland." *Auk.* 102:395-401.

Tasker, M.L., P. Hope-Jones, T. Dixon, and B.F. Blake. 1984. "Counting Seabirds at Sea From Ships: A Review of Methods Employed and a Suggestion for a Standardized Approach." *Auk.* 101:567-577.

Thompson, C.W., M.L. Wislon, E.F. Melvin, and D.J. Pierce. 1998. "An Unusual Sequence of Flight-Feather Molt in Common Murres and Its Evolutionary Implications." *Auk.* 115:653-669.

Veit, R.R., E.D. Silverman, and I. Everson. 1993. "Aggregation Patterns of Pelagic Predators and Their Principle Prey, Antarctic Krill, Near South Georgia." *Journal of Animal Ecology.* 62:551-564.

Veit, R.R. and W.A. Montevecchi. 2003. "Influences of Global Climate Change on Marine Birds." *Proceedings International Ornithological Congress (Beijing).* In press.

Veit, R.R., P. Pyle, and J.A. McGowan. 1996. "Ocean Warming and Long-Term Changes in Pelagic Bird Abundance within the California Current System." *Marine Ecology Progress Series.* 139:11-18.

Veit, R.R., J.A. McGowan, D.G. Ainley, T.R. Wahl, and P. Pyle. 1997. "Apex Marine Predator Declines 90% in Association with Changing Oceanic Climate." *Global Change Biology.* 3:23-28.

Wiese, F.K. 2002. *Seabirds and Atlantic Canada's Ship—Source Oil Pollution: Impacts, Trends, and Solutions.* World Wildlife Fund Canada, Ottawa.

Wiese, F.K. and G.J. Robertson. 2003. "Assessing Impacts of Chronic Oil Discharges at Sea on Seabirds: A General Oiled Seabird Mortality Model Applied to Eastern Canada." *Journal of Wildlife Management.* In press.

Wiese, F.K. and P.C. Ryan. 1999. "Trends of Chronic Oil Pollution in Southeast Newfoundland Assessed Through Beached-Bird Surveys, 1984–1997." *Bird Trends.* 7:36-40.

Wiese, F.K. and P.C. Ryan. 2003. "The Extent of Chronic Marine Oil Pollution in Southeastern Newfoundland Waters Assessed through Beached-Bird Surveys 1984–1999." *Marine Pollution Bulletin.* 46:1090-1101.

Weise, F.K., W.A. Montevecchi, G.K. Davoren, F. Heuttmann, A.W. Diamond, and J. Linke. 2001. "Seabirds at Risk from Offshore Platforms in the Northwest Atlantic." *Marine Pollution Bulletin.* 42:1285-1290.

Wood, D. 1999. "Hibernia." *Air Canada En Route.* 2:48-57.

APPENDIX
List of Papers Presented at the Workshop

The following titles of papers were presented either orally or as a poster presentation at the *Offshore Oil and Gas Environmental Effects Monitoring Workshop* held at the Bedford Institute of Oceanography in Dartmouth, Nova Scotia on May 26–30, 2003. Abstracts for these papers can be found in a separate report: Armsworthy, S.L., A. Muecke, and P.J. Cranford. 2004. *Workshop on the Offshore Oil and Gas Environmental Effects Monitoring, Bedford Institute of Oceanography, Dartmouth, Nova Scotia, May 26–30, 2003*, Environmental Science Research Fund Report (In press). Presentation titles may differ from paper titles in the book.

Oral Presentation Abstracts

Offshore Oil and Gas Environmental Effects Monitoring Investigations Conducted by the U.S. Minerals Management Service
Ahlfeld, T.E.—U.S. Department of the Interior/Minerals Management Service, Herndon, VA, USA

Overview of Environmental Risk Assessment and Monitoring: Past, Present, and Future
van Dalfsen, J.A., M.G.D. Smit, and C.C. Karman—TNO-Environment, Energy and Process Innovation, Department for Ecological Risk Studies, Netherlands

'Designing Out' EEM
Hurley, G.—EnCana Corporation, Halifax, NS, Canada

Evaluation of Offshore Drilling Cuttings Management Technologies using Multicriteria Decision-making
Worakanok, T.—Faculty of Engineering, Memorial University of Newfoundland, St. John's, NL, Canada

Adequacy of Monitoring Programs for Production and Exploratory Wells on the Grand Banks of Newfoundland
Payne, J.F.—Oceans and Environment Branch, Fisheries and Oceans Canada, St. John's, NL, Canada

A Concept for the Environmental Impact Factor for Drilling Discharges
Smit, M.G.D., R.G. Jak, and J.A. van Dalfsen—TNO-Environment, Energy and Process Innovation, Department for Ecological Risk Studies, Netherlands

The Efficacy of In-situ Remediation Techniques to Enhance Recovery of an Oil-contaminated Salt Marsh
Lee, K.—Centre for Offshore Oil and Gas Environmental Research, Bedford Institute of Oceanography Fisheries and Oceans Canada Dartmouth, NS, Canada

Prioritising Environmental Effects Monitoring (EEM) Programs: A Risk-based Strategy
Sadiq R.[1], B. Veitch[2], T. Husain[2], and N. Bose[2] — [1]Institute for Research in Construction, National Research Council, Ottawa, ON, Canada; [2]Faculty of Engineering and Applied Science, Memorial University of Newfoundland, St. John's, NL, Canada

A Community-based Approach to Environmental Effects Monitoring in the Beaufort Sea
Cott, K.J.[1], D.G. Cobb[2], and D.B. Chiperzak[3] — [1]Fisheries and Oceans Canada, Inuvik, NWT, Canada; [2]Fisheries and Oceans Canada, Winnipeg, MB, Canada; [3]Fisheries and Oceans Canada, Yellowknife, NWT, Canada

Development of Effect Monitoring, Risk Estimates, and Decision Making for the Deep-sea
Skadsheim, A., J.F. Børseth, A. Bjørnstad, E. Aas, and S. Sanni—RF-Akvamiljø, Randaberg, Norway

Environmental Effects Monitoring for Exploration Drilling of Single Wells
Buchanan, R.A.[1], N. Collins[2], A. Mathieu[3], and P. Stewart[4] — [1]LGL Limited, St. John's, NL, Canada; [2]CEF Consultants, Halifax, NS, Canada; [3]Oceans Ltd., St. John's, NL, Canada; [4]Envirosphere Consultants, Halifax, NS, Canada

Deciding When to Monitor Oil Spills: A Plan for Implementing Environmental Effects Monitoring During Production-related Oil Spills in Eastern Canada
Trudel, K.—S.L. Ross Environmental Research Ltd., Ottawa, ON, Canada

Sampling Design and Tools Used in SOEI's Offshore Environmental Effects Monitoring Program
Belford, S.L.[1], C. Ross[2], and S.M. Fudge[1] — [1]Jacques Whitford Environment Limited; [2]Sable Offshore Energy Inc. (ExxonMobil)

Sixteen Years of Harmonised Monitoring of Effects of Drilling Waste from Norwegian Offshore Installations. What Works and What Doesn't?
Bakke, T.[1] and I. Nilssen[2] — [1]Norwegian Institute for Water Research, Oslo, Norway; [2]Norwegian Pollution Control Authority, Oslo, Norway

Quantifying Post-exploration Drill Impacts in South Caspian Sea Sediments
Maxon, C.L.[1], R.D. Tait[2], and F.C. Newton III[1] — [1]Battelle Memorial Institute, Carlsbad, CA, USA; [2]ExxonMobil Production Company, Houston, TX, USA

Sediment Profile Imaging: Impact Assessment in the South Caspian Sea
Germano, J.D.[1], R.D. Tait[2], D.G. Browning[1], and F. Newton III[3] — [1]Germano & Associates, Inc., Bellevue, WA, USA; [2]ExxonMobil Production Company, Houston, TX, USA; [3]Battelle Memorial Institute, Carlsbad, CA, USA

Long-term Trends and Monitoring of the Macrobenthos at Sullom Voe Oil Terminal, Shetland, UK
Woodham, A.[1], C. Dalgleish[1], P.F. Kingston[2], and J.M. Mair[2] — [1]ERT Ltd., (Scotland), Heriot-Watt University, Edinburgh, UK; [2]Centre for Marine Biodiversity and Biotechnology, Heriot-Watt University, Edinburgh, UK.

Terra Nova Environmental Effects Monitoring Program: From EIS Onward
DeBlois, E.M.[1], C. Leeder[1], M. Murdoch[2], K.C. Penney[1], M.D. Paine[3], F. Power[4], and U.P. Williams[4] — [1]Jacques Whitford Environment Ltd., St. John's, NL, Canada; [2]Jacques Whitford Environment Ltd., Fredericton, NB, Canada; [3]Paine, Ledge and Associates (PLA), North Vancouver, BC, Canada; [4]Petro-Canada East Coast Operations, St. John's, NL, Canada

Environmental Effects of Mercury in Permitted Discharges from Offshore Platforms in the U.S. Gulf of Mexico
Neff, J.M.[1], J.P. Ray[2], J.P. Smith[3], and M.E. Parker[3] — [1]Battelle Memorial Institute, Duxbury, MA, USA; [2]Shell Global Solutions, Houston, TX, USA; [3]ExxonMobil, Houston, TX, USA

Assessment of Environmental Impacts from Drilling Muds and Cuttings Disposal, Offshore Brunei
Sayle, S.[1], M. Seymour[2], and E. Hickey[1] — [1]Adinin-Jacques Whitford; [2]Brunei Shell Petroleum

Lessons for Offshore Oil and Gas EEM from the Pulp and Paper and Metal Mining EEM Programs
Courtenay, S.C.[1], S.D. St-Jean[1], K.R. Munkittrick[2], K. Kim[3], and W.R. Parker[4] — [1]Gulf Fisheries Centre, Fisheries and Oceans Canada, Moncton, NB, Canada; [2]Department of Biology, University of New Brunswick, Saint John, NB, Canada; [3]Environment Canada, Dartmouth, NS, Canada; [4]Environment Canada, Fredericton, NB, Canada

Chronic Effects of Synthetic Drilling Muds on Sea Scallops (*Placopecten magellanicus*)
Armsworthy, S.L., P.J. Cranford, K. Lee, V. Burdett-Coutts, K. Querbach, and S. Magee — Fisheries and Oceans Canada, Dartmouth, NS, Canada

Distinguishing Between Artificial Reef Effects and Platform Effects on Benthos
Montagna, P.A.[1], S.C. Jarvis[1,3], and M.C. Kennicutt II[2] — [1]University of Texas Marine Science Institute, Port Aransas, TX, USA; [2]Geochemical and Environmental Research Group, Texas A&M University, College Station, TX, USA; [3]Emu Ltd, Marine Laboratory, Hayling Island, Hampshire, UK

Bioindicator Studies on Fish Health around the Terra Nova Development Site on the Grand Banks
Mathieu A.[1], B. French[1], M. Dawe[1], E. Deblois[2], F. Power[3], and U. Williams[3] — [1]Oceans Ltd., St. John's, NL, Canada; [2]Jacques Whitford Environment Ltd., St. John's, NL, Canada; [3]Petro-Canada, St. John's, NL, Canada

Gas Pipelines and the American Lobster, *Homarus americanus*
Cooper, R.A.[1], M. Clancy[2], and J.S. Cobb[3] — [1]University of Connecticut, Groton, CT, USA; [2] Boston University, Boston, MA, USA; [3] University of Rhode Island, Kingston, RI, USA

Long Term Lessons Learned from the *Exxon Valdez*: Acute Toxicity versus Chronic Embryotoxicity
Rice, S.D., J.W. Short, R. Heintz, and M.G. Carls —Auke Bay Laboratory, National Marine Fisheries Service, Juneau, AK, USA

Monitoring Oil Persistence on Beaches—
SCAT vs. Stratified Random Sampling Designs
Short J.W., M.R. Lindeberg, P.M. Harris, J.M. Maselko, J.J. Pella, and S.D. Rice —Auke Bay Laboratory, Alaska Fisheries Science Center, National Marine Fisheries Service, NOAA, Juneau, AK, USA

Monitoring Seismic Effects on Marine Mammals - Southeastern Beaufort Sea, 2001–2002
Miller, G.W.[1], P. Millman[2], R.A. Davis[1], V.D. Moulton[3], A. MacGillivray[4], D. Hannay[4], M. Holst[1], R.E. Elliott[1], A. Serrano[1], and S. Carr[4] — [1]LGL Limited, Environmental Research Associates, King City, ON, Canada; [2]Devon Canada Corporation, Calgary, AB, Canada; [3]LGL Limited, Environmental Research Associates, St. John's, NL, Canada; [4]JASCO Research Ltd., Victoria, BC, Canada

Effects of Seismic Energy on Snow Crab (*Chionoecetes opilio*)
Christian, J.R.[1], A. Mathieu[2], D. White[2], and R.A. Buchanan[1] — [1]LGL Limited, St. John's, NL, Canada; [2]Oceans Ltd., St. John's, NL, Canada

Sediment Quality in Depositional Areas of Shelik of Strait and Outermost Cook Inlet
Brown, J.S. and P.D. Boehm—Battelle, Waltham, MA, USA

Marine Coastal Monitoring Designs: Examples for Offshore Oil and Gas Environmental Effects Monitoring
Green, R.H.—University of Western Ontario, London, ON, Canada

Distribution of Suspended Particulate Drilling Wastes at the Hibernia Oilfield
Muschenheim, D.K., T.G. Milligan, A. Stewart, and B. Law— Bedford Institute of Oceanography, Fisheries and Oceans Canada, Dartmouth, NS, Canada

Quantifying Fine-grained Drill Waste in Scotian Shelf Suspended Sediments
Milligan, T.G., T. Tedford, D.K. Muschenheim, and C. Hannah—Bedford Institute of Oceanography, Fisheries and Oceans Canada, Dartmouth, NS, Canada

Scallops as Sentinel Organisms for Monitoring Environmental Effects of Offshore Oil and Gas Operations
Cranford, P.J.[1], S.L. Armsworthy[1], and K. Lee[2] —Bedford Institute of Oceanography, Fisheries and Oceans Canada, Dartmouth, NS, Canada: [1]Marine Environmental Sciences Division; [2]Centre for Offshore Oil and Gas Environmental Research

Grain Size Normalisation of Sediment Heavy Metal Data
Yeats, P.A.—Marine Environmental Sciences Division, Bedford Institute of Oceanography, Fisheries and Oceans Canada, Dartmouth, NS, Canada

Using Ba/Al Ratios to Estimate Baseline Barium Sediment Concentrations and Monitor Changes in Barium Concentrations on a Regional Scale in Newfoundland's Offshore Oil Field
Veinott, G. —Northwest Atlantic Fisheries Centre, Fisheries and Oceans Canada, St. John's, NL, Canada

Biosea – A Research Project to Provide Knowledge and Data for Predicting and Monitoring Potential Biological Effects in the Arctic by Oil Industry
Sanni, S.[1] and M. Buffagni[2] — [1]Akvamiljø, Randaberg, Norway; [2]ENI, Milano, Italy

Environmental Monitoring of Oil and Gas Activities in Deep Waters West of Britain: Challenges and New Approaches
Hartley, J.P.—Hartley Anderson Ltd., Blackstone, Dudwick, Ellon, Aberdeenshire, UK

The Cuttings Pile Issue: How to Provide the Necessary Information to Assess the Impacts of Cuttings and Mud Accumulations around Offshore Installations?
Kjeilen, G. and S. Westerlund—RF-Rogaland Research, Norway

Transport Properties of Discharged Synthetic-based Drilling Wastes
Niu, H.—Faculty of Engineering and Applied Science, Memorial University of Newfoundland, St. John's, NL, Canada

Assessing Environmental Fate and Behaviour of Oil Discharges in Marine Ecosystem: Using Fugacity Model
Khan, M.I. and M.R. Islam—Faculty of Engineering, Dalhousie University, Halifax, NS, Canada

The Role of Mass Balance Models in Offshore Oil and Gas Environmental Effects Monitoring
Warren, C. and D. Mackay—Canadian Environmental Modelling Centre, Trent University, Peterborough, ON, Canada

Can In-situ Invertebrate Communities Directly or Indirectly Affect Toxicity Test Results?
Paine, M.D.[1], E.M. DeBlois[2], S. Whiteway[2], F. Power[3], and U. Williams[3] — [1]Paine, Ledge and Associates, North Vancouver, BC, Canada; [2]Jacques Whitford Environment Ltd., St. John's, NL, Canada; [3]Petro-Canada, St. John's, NL, Canada

Livebottom Impact Reduction, Mitigation, and Monitoring for the Construction of a Subsea Natural Gas Pipeline in the Gulf of Mexico
Ellsworth, S. W.[1], D.E. Martin[2], R. Schaul[3], and J. Schmidt[1] — [1]ENSR International; [2]Williams Companies; [3]Sea Byte, Inc.

Aspects of Polycyclic Aromatic Hydrocarbons in Offshore Sediments in the Azeri Sector of the Caspian Sea
Boehm, P.D.[1], C.L. Maxon[2], F.C. Newton[2], J.S. Brown[1], and Y. Galperin[2] — [1]Battelle Memorial Institute, Waltham, MA, USA; [2]Battelle Memorial Institute, Carlsbad, CA, USA

Monitoring for PAH Exposure Using Bile Metabolites and Concentrations in Tissues: Continuing to Raise Questions
Hellou, J.[1], T.K. Collier[2], F. Ariese[3], and J. Leonard[1] — [1]Marine Chemistry Section, Marine Environmental Sciences Division, Bedford Institute of Oceanography, Fisheries and Oceans Canada, Dartmouth, NS, Canada; [2]Northwest Fisheries Science Center, National Marine Fisheries Service, NOAA, Seattle, WA, USA; [3]Vrije Universiteit, Instituut voor Milieuvraagstukken, Amsterdam, Netherlands

Ongoing Monitoring Programs Relevant for the Offshore Oil and Gas Industry in Norway, Emphasising Methods and Trends Seen in the Perspective of the Development of Hydrocarbon and Produced Water Discharges
Melbye, A.—SINTEF Applied Chemistry, Marine Environmental Technology, Trondheim, Norway

Evaluation of Risks from Produced Water Discharges in Atlantic Canada
Lee, K.[1], K. Azetsu-Scott[2], P. Yeats[3], S. Niven[3], G. Wohlgeschaffen[1], J. Dalziel[3], T. Milligan[3], C. Hannah[2], and A. Vezina[2] — Bedford Institute of Oceanography, Fisheries and Oceans Canada, Dartmouth, NS, Canada: [1]Centre for Offshore Oil and Gas Environmental Research; [2]Ocean Sciences Division; [3]Marine Environmental Sciences Division

Environmental Modelling of Produced Water and Indicators for EEM
Berry, J.A.[1] and P.G. Wells[1,2] — [1]School for Resource and Environmental Studies, Dalhousie University, Halifax, NS, Canada; [2]Environmental Conservation Branch Environment Canada, Dartmouth, NS, Canada

Assessing the Impact of Produced Water Discharges in the North Sea – Comparison of Field Monitoring and Modelling
Durell, G.S.[1], T. Røe-Utvik[2], S. Johnsen[3], T. Frost[3], and J. Neff[1] — [1]Battelle, Duxbury, MA, USA; [2]Norsk Hydro, Bergen, Norway; [3]Statoil Research Centre, Trondheim, Norway

Biological Effects in Pelagic Ecosystems—An Overview of the ICES Workshop BECPELAG
Hylland, K., G. Becker, J. Klungsøyr, T. Lang, A. McIntosh, B. Serigstad, J.E. Thain, K.V. Thomas, T.I.R. Utvik, D. Vethaak, and W. Wosniok—NIVA, Oslo, Norway

The Experience from Using In-situ Deployment of Live Organisms and Passive Samplers during the ICES BECPELAG Workshop
Thain, J.E., K. Hylland, B. Serigstad, G. Becker, J. Klungsøyr, T. Lang, A. McIntosh, K.V. Thomas, T.I.R. Utvik, D. Vethaak, and W. Wosniok—CEFAS, Burnham on Crouch, Essex, UK

An Assessment of the In Vitro Oestrogen Receptor Agonist Potency and Alkylphenol Content of Produced Water Discharges into the North Sea
Thomas, K.V., J. Balaam, M.R. Hurst, and J.E. Thain—CEFAS, Burnham on Crouch, Essex, UK

Development and Field Validation of the EIF as a Produced Water Discharge Management Tool
Frost, T.K. and S. Johnsen—Statoil Research Centre, Norway

Bioassay-directed Fractionation and Chemical Identification of Complex Effluents from Oil Production Activities
Tollefsen, K.[1], M. Grung[1], C. Dye[2], M. Berntssen[3], T. Hartnick[4], L. Norrgren [5], J. Balaam [6], and K. Thomas [6] — [1]Norwegian Institute for Water Research (NIVA), Norway [2]Norwegian Institute for Air Research

(NILU), Norway; [3]Norwegian Institute for Fisheries and Seafood Research (NIFES), Norway; [4]Jordforsk, Norway; [5]Swedish Agricultural University Sweden (SLU); [6]Centre for Environment, Fisheries & Aquaculture Science (CEFAS), UK

Evaluation of bblt, a Drilling Mud Dispersal Model, at North Triumph (Scotian Shelf)
Hannah, C., A. Drozdowski, K. Muschenheim, and J. Loder— Bedford Institute of Oceanography, Fisheries and Oceans Canada, Dartmouth, NS, Canada

In-situ Deployment of Fish, Mussels, and Passive Samplers to Monitor Accumulation and Effects of Contaminants
Serigstad, B.—Institute of Marine Research, Bergen, Norway

Potential Effects of Produced Water Discharges from Offshore Oil and Gas Production Activities on the Early Life Stages of Haddock (*Melanogrammus aeglefinus*), Lobster (*Homarus americanus*), and Sea Scallop (*Placopecten magellanicus*)
Querbach, K.[1], G. Maillet[1], P. Cranford[1], C. Taggart[2], K. Lee[1], and J. Grant[2] — [1]Bedford Institute of Oceanography, Fisheries and Oceans Canada, Dartmouth, NS, Canada; [2]Department of Oceanography, Dalhousie University, Halifax, NS, Canada

Impacts of Nutrient Inputs from Produced Water on Marine Planktonic Production: An Ecosystem Modelling Approach
Khelifa, A.[1], M. Pahlow[2], A. Vézina[2], K. Lee[3], and C. Hannah[4] — [1]Department of Oceanography, Dalhousie University, Halifax, NS, Canada; [2]Biological Oceanography Section, Bedford Institute of Oceanography, Fisheries and Oceans Canada, Dartmouth, NS, Canada; [3]Centre for Offshore Oil and Gas Environmental Research, Bedford Institute of Oceanography, Fisheries and Oceans Canada, Dartmouth, NS, Canada; [4]Coastal Ocean Science, Bedford Institute of Oceanography, Fisheries and Oceans Canada, Dartmouth, NS, Canada

Seasonal and Spatial Patterns of Marine Bird and Mammal Occurrences Recorded on Offshore Support Vessel Transects on the Grand Banks
Montevecchi W.A., C.M. Burke, G.K. Davoren, and F.K. Wiese— Memorial University of Newfoundland, St. John's, NL, Canada

Interaction and Remediation of Oil Spills in Ice Infested Waters: An Experimental and Numerical Investigation
Bjorndalen, N.[1], S. Mustafiz[1], A. Basu[1], K. Lee[2], and M.R. Islam[1] — [1]Dalhousie University, Halifax, NS, Canada; [2]Centre for Offshore Oil and Gas Environmental Research, Bedford Institute of Oceanography, Fisheries and Oceans Canada Dartmouth, NS, Canada

Poster Presentations Abstracts

Screening Assessment of Environmental Effects of PEMEX Operations in Sonda de Campeche, Mexico
Brandt, C.A.[1], M.C. Gonzalez[2], L.A. Soto[3], M.E. Gallegos[4], J. Neff[5], K. Steinmaus[6], J.A. Ward[6] — [1]Battelle, Pacific Northwest Laboratories, Richland, WA, USA; [2]Instituto Mexicano del Petróleo, México; [3]Universidad Nacional Autónoma de México, Instituto de Ciencias del Mar y Limnologia, México; [4]Universidad Autónoma Metropolitana, Departamento de Hidrobiologia, México; [5]Battelle, Duxbury Operations, Duxbury, MA, USA; [6]Battelle, Marine Sciences Laboratory, Sequim, WA, USA

Using Genetic Diversity as an Indicator in Marine Monitoring Programs
Montagna, P.A.—University of Texas Marine Science Institute, Port Aransas, TX, USA

Short-term Effect of IPAR-3® and Diesel on Immune Responses in Mussels
Hamoutene D.[1], J. Payne[1], A. Rahimtula[2], and K. Lee[3] — [1]Science Branch, Fisheries and Oceans Canada, St John's, NL, Canada; [2]Biochemistry Department, Memorial University of Newfoundland, St John's, NL, Canada; [3]Centre for Offshore Oil and Gas Environmental Research, Bedford Institute of Oceanography, Fisheries and Oceans Canada, Dartmouth, NS, Canada

Marine Bird and Mammal Surveys from Offshore Support Vessels and Ships of Opportunity on the Grand Banks
Burke, C. M., G.K. Davoren, W.A. Montevecchi, and F.K. Wiese—Biopsychology Programme, Memorial University of Newfoundland, St. John's, NL, Canada

An Integrated Approach for Environmental Decision-making for Offshore Oil and Gas Operations
Worakanok, T., H. Niu, and V. Pennell—Faculty of Engineering and Applied Science, Memorial University of Newfoundland, St. John's, NL, Canada

The Results of Ten Years of Beached Bird Surveys on Sable Island
Lucas, Z.—Halifax CRO, Halifax, NS, Canada

The ICES Workshop on Biological Effects in Pelagic Ecosystems (BECPELAG): In-situ Techniques
Thain, J., G. Becker, K. Hylland, J. Klungsøyr, T. Lang, A. McIntosh, B. Serigstad, K.V. Thomas, T.I.R. Utvik, D. Vethaak, and W. Wosniok—CEFAS Burnham on Crouch, Essex, UK

ICES Biological Effects Monitoring In Pelagic Ecosystems (BECPELAG) Workshop: Field Sampling
Hylland, K., G. Becker, J. Klungsøyr, T. Lang, A. McIntosh, B. Serigstad, J. Thain, K. Thomas, T.I.R. Utvik, D. Vethaak, and W. Wosniok—CEFAS, Burnham on Crouch, Essex, UK

The ICES Workshop on Biological Effects of Contaminants in Pelagic Ecosystems (BECPELAG): In-vitro Bioassays
Thomas, K.V., G. Becker, K. Hylland, J. Klungsøyr, T. Lang, A. McIntosh, B. Serigstad, J. Thain, T.I.R. Utvik, D. Vethaak, and W. Wosniok — CEFAS, Burnham on Crouch, Essex, UK

Monitoring Birds and Marine Mammals on Offshore Vessels and Drilling Rigs
Farwell, S.—Oil and Gas Observer Program Canada, Dartmouth, NS, Canada

Cohassett EEM Program
Full, S.—EnCana Corporation, Halifax, NS, Canada

Partitioning, Exposure and Effects Associated with Environmental Mixtures of Contaminants, Including Petroleum and Combustion Hydrocarbons
Hellou, J.—Marine Chemistry Section, Marine Environmental Sciences Division, Bedford Institute of Oceanography, Fisheries and Oceans Canada Dartmouth, NS, B2Y 4A2, Canada

Long-term Response of Benthos to Production Platforms in the Gulf of Mexico
Montagna, P.A.[1] and D.E. Harper[2] — [1]University of Texas Marine Science Institute, Port Aransas, TX, USA; [2]Texas A&M University at Galveston, Galveston, TX, USA

Monitoring and Regulating Explosive Based Winter Seismic Exploration in Waterbodies Not Frozen to the Bottom, Mackenzie Delta-Beaufort Region, Northwest Territories 2000 - 2002
Cott, P.A.[1] and B.W. Hanna[2] — [1]Fisheries and Oceans Canada, Inuvik, NWT, Canada; [2]Fisheries and Oceans Canada, Yellowknife, NWT, Canada

Lipid Soluble Vitamins A and E as Integrative Measures of Fish Health for EEM
Palace, V.P., J. Werner, K. Wautier, C.L. Baron, R.E. Evans, and D.G. Cobb—Fisheries and Oceans Canada, Winnipeg, MB, Canada

The Environmental Sensitivity of Cold Water Corals
Gass, S.E. and J.M. Roberts—Scottish Association for Marine Science, Dunstaffnage Marine Laboratory, Dunbeg, Oban, Argyll, Scotland

Environmental Design: Reuse of Offshore Oil and Gas Production Platforms
Collins, D.—Department of Environmental Studies and Architecture, University of Waterloo, Waterloo, ON, Canada

WebTide and WebDrogue: Open Source Tidal Predictions and Trajectory Modelling
Hannah, C., D. Greenberg, S. Oakey, F. Dupont, and J. Chaffey—Bedford Institute of Oceanography, Fisheries and Oceans Canada, Dartmouth, NS, Canada

Conflict of Oil and Gas Activities in Cape Breton, Nova Scotia, Canada: An Approach of Resolution
Mehedi, M.Y. and M.R. Islam—Faculty of Engineering, Dalhousie University, Halifax, NS, Canada

Author Index

Index